Organic Synthesis via Examination of Selected Natural Products

Organic Synthesis via Examination of Selected Natural Products

Editor

David J Hart

The Ohio State University, USA

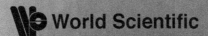

World Scientific

NEW JERSEY • LONDON • SINGAPORE • BEIJING • SHANGHAI • HONG KONG • TAIPEI • CHENNAI

Published by

World Scientific Publishing Co. Pte. Ltd.
5 Toh Tuck Link, Singapore 596224
USA office: 27 Warren Street, Suite 401-402, Hackensack, NJ 07601
UK office: 57 Shelton Street, Covent Garden, London WC2H 9HE

British Library Cataloguing-in-Publication Data
A catalogue record for this book is available from the British Library.

Cover image credits:

Sponge:

Reproduced from Office of Ocean Exploration and Research, National Oceanic and Atmospheric Administration, U.S. Department of Commerce
http://oceanexplorer.noaa.gov/explorations/03bio/logs/hirez/lasonolide1_hires.jpg

Rauwolfia serpentina:

Reproduced with permission from Dr. Aruna Radhakrishnan

ORGANIC SYNTHESIS VIA EXAMINATION OF SELECTED NATURAL PRODUCTS

ISBN-13 978-981-4313-70-4
ISBN-10 981-4313-70-X

Typeset by Stallion Press
Email: enquiries@stallionpress.com

Printed in Singapore.

This book is dedicated to

My Teachers and Mentors

Mark Green
Richard Lawton
William Dauben
David Evans
John Swenton;

My Parents

Harold and Geraldine Hart;

My Students and Colleagues at The Ohio State University;

My Family;

and Especially My Wife

Rose,

the Spice of My Life.

Contents

Preface

This book is based on a graduate level course I taught at The Ohio State University in the autumn of 2006. The course consisted of twenty-eight 50-minute classes over a period of ten weeks (Chemistry 941). Students enrolled in the course were largely in their second or third years of graduate school. All had taken a three-quarter "organic reactions" sequence and a two-quarter "physical organic chemistry" sequence. I expected students to bring a sizeable toolbox of reactions, and a sound understanding of mechanistic organic chemistry, with them to the classroom. My goal was to introduce students to the field of organic synthesis with a focus on my own interest in natural products synthesis.

This book follows the sequence of topics I discussed in Chemistry 941. I have done little to modify the slides that I used as the basis of lectures. I have merely added text to accompany each slide. Several homework assignments were presented during the quarter, and I have added many more problems that I hope readers will find interesting and instructive.[a] An index has also been appended.[b]

My view of organic synthesis has naturally been influenced by my own experiences. I have been influenced by my teachers and, if they read this book, they will see themselves reflected on many of these pages. I have also been influenced by other books on the topic of organic synthesis, a few of which appear below:

- Ireland, R. E. *Organic Synthesis*, Prentice-Hall, 1969 (147 pages)

[a] A partial answer key is available to course instructors. Please contact sales@wspc.com

[b] The index within the book is abbreviated and only provides information found in the text pages (odd-numbered pages). A thorough index of the slides (even-numbered pages) with sections organized by compound, reagent type, reaction type, and subject is available online at http://www.worldscibooks.com/chemistry/7815.html

- Fleming, I. *Selected Organic Syntheses: A Guidebook for Organic Chemists*, John Wiley and Sons, 1973 (227 pages)
- Warren, S. *Organic Synthesis: The Disconnection Approach*, John Wiley and Sons, 1982 (391 pages)
- Wyatt, P.; Warren, S. *Organic Synthesis — Strategy and Control*, John Wiley and Sons, 2007 (909 pages)
- Corey, E. J.; Cheng, X-M. *The Logic of Chemical Synthesis*, John Wiley and Sons, 1989 (436 pages)
- Nicolaou, K. C.; Sorensen, E. J. *Classics in Total Synthesis: Targets Strategies, Methods*, VCH, 1996 (798 pages)
- Nicolaou, K. C.; Snyder, S. A. *Classics in Total Synthesis II: Targets, Strategies, Methods*, John Wiley and Sons, 2003 (560 pages)

I was most influenced by the Ireland and Fleming books (now out of print), because they were published at the time I was developing an interest in organic chemistry and organic synthesis. To those who have read the Ireland and Fleming books, their influence will be apparent in my selection of topics. Also, my selection of topics is simply a reflection of my own interests. In no way do I mean to slight the many chemists not cited herein who have made landmark contributions to the field of natural products synthesis.[c]

I have provided references on the slides for the papers that form the basis of this book.[d] These references are not repeated in the text, but additional references have been provided at points where I think they could be useful. The reader can always refer to the papers that form the basis of this book for additional details and citations. I have also tried to provide reaction yields when they were easily gleaned from the papers. I did not make any attempt to extract yield information from experimental data and have taken yields reported by authors at face value.

Throughout the text I will refer to the "slides" by topic and number. For example there are six slides associated with the introduction. When I refer to Introduction-1, the reader should look at the first slide associated with the introduction. Slides appear on left-hand (even) pages with the accompanying text located on adjacent right-hand (odd-numbered) pages. A Table of Contents is provided that should help the reader move from one topic to another in a non-linear manner. Now let us begin.

---

[c] For some compilations of syntheses that you might find interesting see: Anand, N.; Bindra, J. S.; Ranganathan, S. *Art in Organic Synthesis*, John Wiley and Sons, 1970 (427 pages). Bindra, J. S.; Bindra, R. *Creativity in Organic Synthesis: Volume 1*, Academic Press, Inc., 1975 (322 pages).

[d] A PowerPoint presentation of each chapter, with selected structures, bonds and comments highlighted in color, is available online at http://www.worldscibooks.com/chemistry/7815.html

CHAPTER 1

Introduction

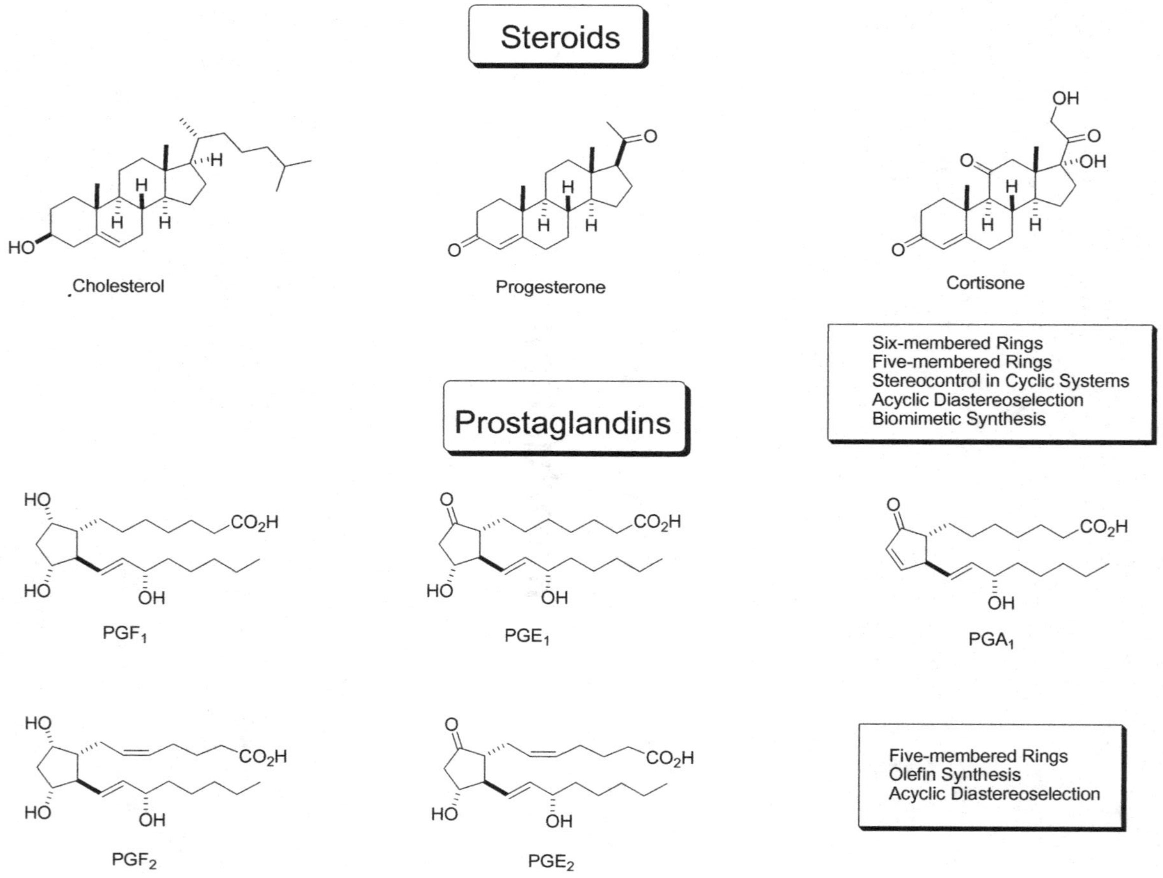

Introduction-1

Introduction-1

One of my objectives is to provide a sense of the history of organic synthesis.[1] I am not a historian, so the reader should understand that this is only my perspective of the field. It is my own sense of this history, however, that leads me to select steroids as the first family of natural products for discussion (Introduction-1). These compounds were clearly of interest to early practitioners of organic synthesis because of their biological activity, structural complexity, and central role they played in the development of the field of biosynthesis. On the other hand, one might argue that steroids were early targets for synthesis because some steroids, such as equilinen and estrone (Steroids-2), were not too complex, and thus were achievable synthetic goals given the tools available at the time. Our discussion will begin with the Woodward synthesis of cholesterol (circa 1950; the steroid we all love to eat) because it is one of the earliest examples of complex natural products synthesis. It also introduces a number of strategies that are still widely used in carbocycle synthesis (synthesis of 6-membered rings, synthesis of 5-membered rings, stereocontrol in cyclic systems). This synthesis will also touch upon other topics, such as acyclic diastereoselection, that are still of contemporary interest. We will next move to the topic of biomimetic synthesis to see how thinking about a biosynthetic pathway to a natural product can influence the design of synthetic pathways to that natural product. It will also illustrate how persistence and attention to detail can play an important role in realizing a given synthetic strategy in the laboratory (see Introduction-1 for targets).

The second broad topic for discussion will be prostaglandins (Introduction-1). Interest in this family of natural products was once again stimulated by their importance in biology. The focus here will be on methods for 5-membered ring synthesis, olefin synthesis, and once again we will encounter the topic of acyclic diastereoselection. Although steroids and prostaglandins have remarkably different structures, we will see that they have some structural features in common, and there are strategic parallels to be found in their synthesis when these structural features are addressed.

Return to Steroids: The Sidechain Stereochemistry Problem

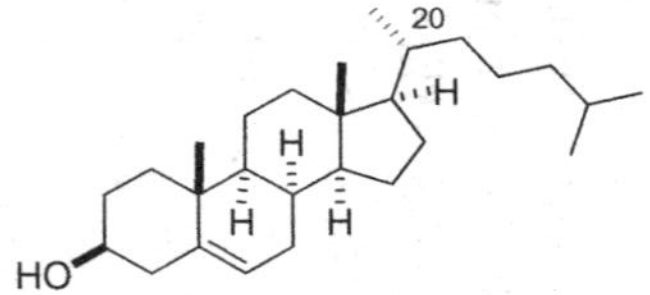

C_{20} of Cholesterol

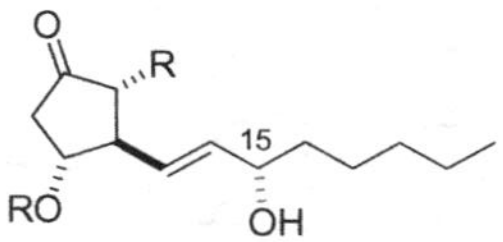

C_{15} of Prostaglandins

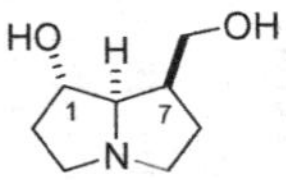

Pyrrolizidine Alkaloids

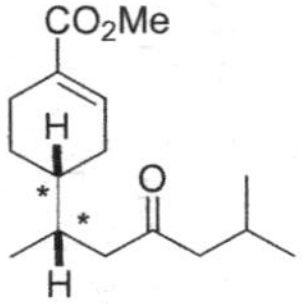

erythro

The Juvabiones

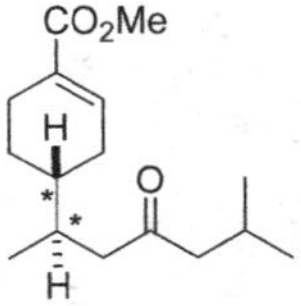

threo

Difunctional Relationships

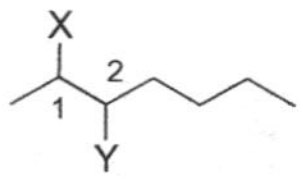

1,2-Difunctional Relationship

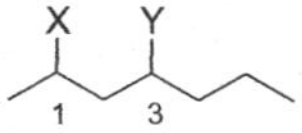

1,3-Difunctional Relationship

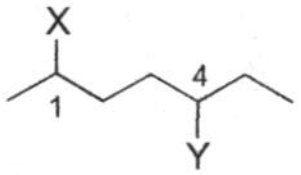

Even Difunctional Relationship

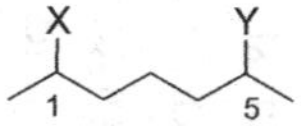

Odd Difunctional Relationship

Introduction-2

Introduction-2

I will next move from a family of compounds to a general problem in organic synthesis that I call "the terpenoid sidechain stereochemistry problem" (Introduction-2). This is really a relatively simple example of a problem in acyclic diastereoselection. The central question here is: When a stereogenic center does not reside in a ring, how can one control stereochemistry at this center relative to other stereogenic centers in the molecule? Acyclic diastereoselection is a general problem presented to the synthetic chemist by many families of natural products. For example, it arises in the steroids at C_{20}, in the prostaglandins at C_{15}, and at the exocyclic carbon of the sesquiterpene ester juvabione. It arises, in a less obvious manner, at C_1 and C_7 of the pyrrolizidine alkaloids. We will revisit this problem again and again throughout this book.

At this point we will examine a topic that can provide insight into strategies that have been adopted for the synthesis of a host of molecules. The topic here is that of "difunctional relationships". For example, the juvabiones have three functional groups: ketone, ester and alkene. The central question here is: When two functional groups are present in a target, does their spatial relationship provide a clue as to what chemistry can be used to construct that relationship and/or the ease with which the relationship can be constructed? The ideas set forth in this section are derived entirely from a concept I learned as a postdoc in the laboratories of David Evans (then at the California Institute of Technology), although any misrepresentation of these ideas are my responsibility alone.

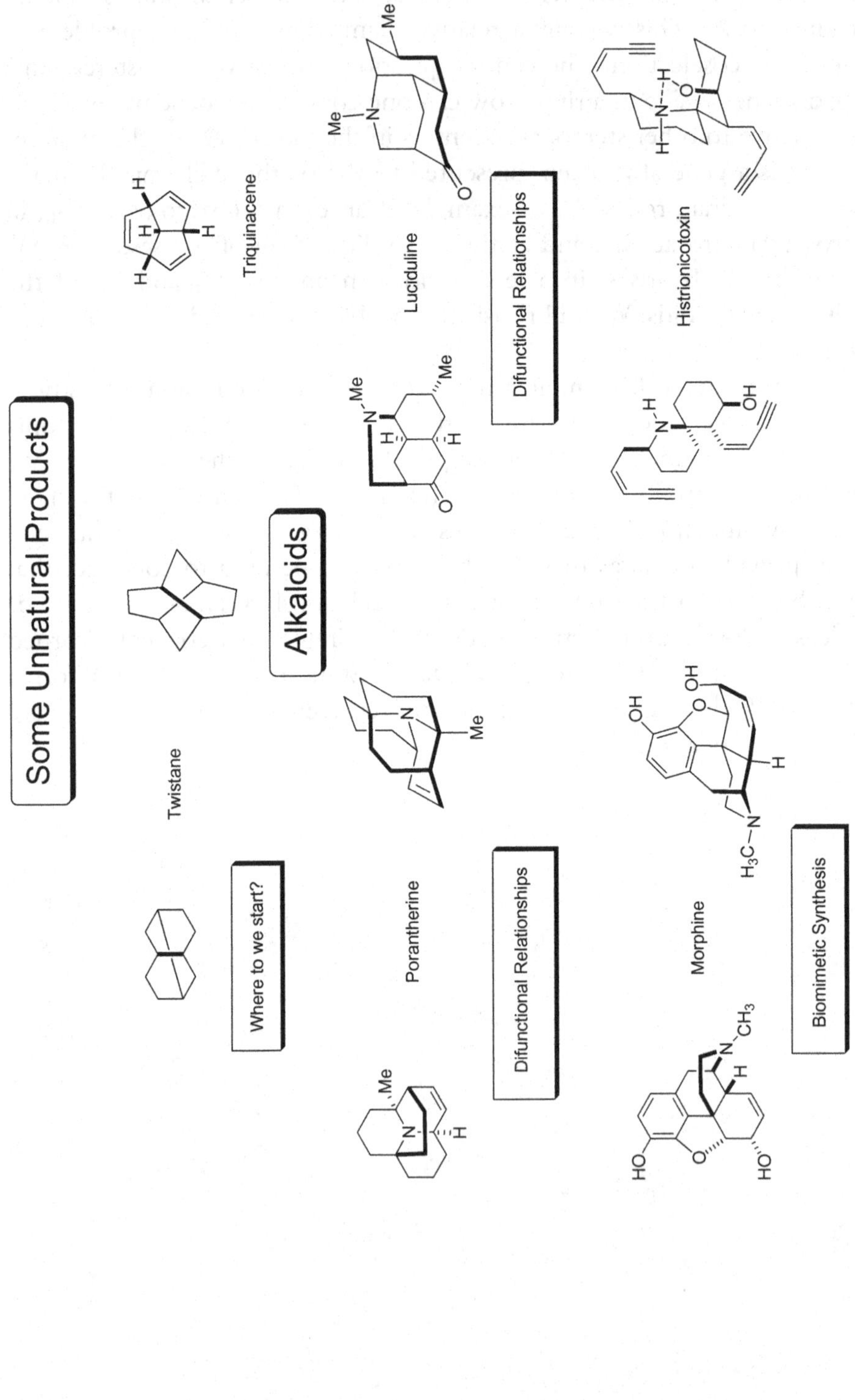

Introduction-3

Introduction-3

Targets to be addressed in the next section are twistane and triquinacene (Introduction-3). These are unnatural products. Twistane consists of 6-membered rings and triquinacene of 5-membered rings. Both have low levels of functionality. How can the concept of examining "difunctional relationships" be used to develop strategies (or analyze syntheses) of these compounds? We will also look at the alkaloids porantherine and luciduline with the same question in mind.

I think that it is important that aspiring practioners of synthesis lose any fear of heteroatoms they might have early in the game. Thus we will next spend some time talking about alkaloids, nitrogen-containing natural products. We will already have seen such compounds (the pyrrolizidine alkaloids) earlier in the book, so this section will serve in part to revisit strategies within the context of "new" targets. For example we will revisit the notion of "biomimetic synthesis" with morphine (arguably the most important pain-killer used in medicine) and the concept of difunctional relationships within the context of histrionicotoxin and pumiliotoxin-C, neurotoxins used as a defense secretion by a certain frog species.

More Alkaloids

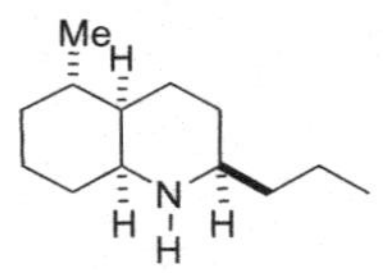

Pumiliotoxin-C

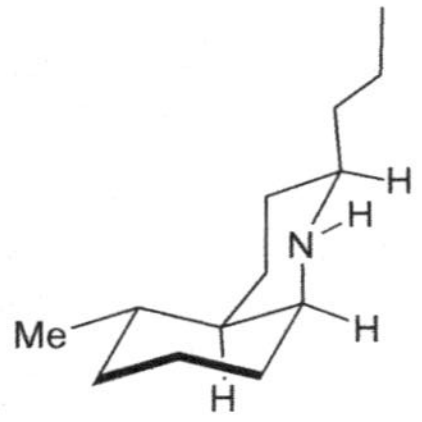

Kinetics vs. Thermodynamics for Stereocontrol

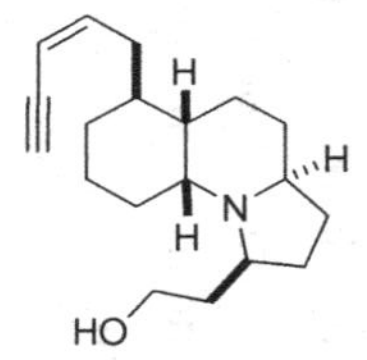

Gephyrotoxin

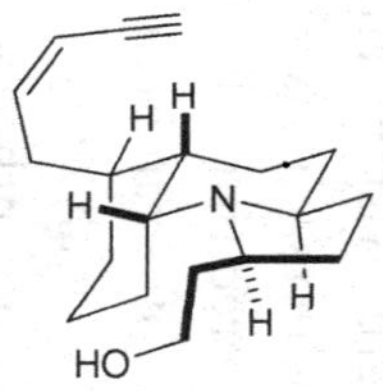

Stereoelectronic Effects

A Classic in Alkaloid Synthesis

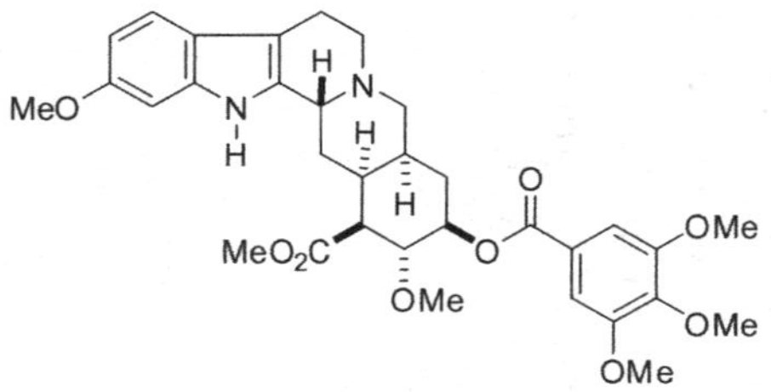

Reserpine

Introduction-4

Introduction-4

We will continue with alkaloids by discussing a personal favorite (gephyrotoxin) and a classical target (reserpine) (Introduction-4). As with all of the natural (and unnatural) product targets discussed in this book, numerous total syntheses of gephyrotoxin and reserpine have been reported. It is instructive to compare syntheses, and look for strategic differences and similarities, and this will be done here as well as throughout this book.

Cecropia Juvenile Hormone: Stereoselective Olefin Synthesis

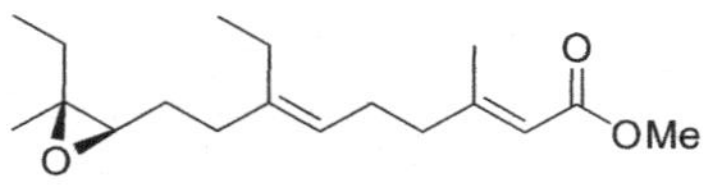

Cecropia JH

Synthesis and Structure Determination
Claisen Rearrangements
Fragmentations
Methodology Driven Synthesis
Sigmatropic Rearrangements

Macrocyclic Natural Products and Ionophores

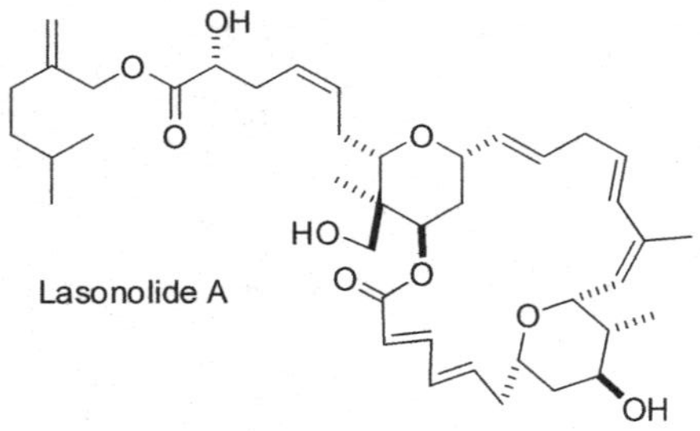

Lasonolide A

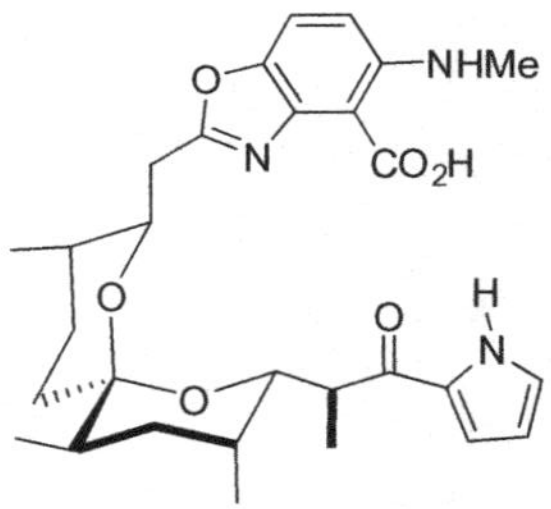

Calcimycin (A-23187)

Total Synthesis and Structure Determination
Olefin Synthesis
Macrolide Synthesis

Acyclic Diastereoselection
Thermodynamics and Stereocontrol

Introduction-5

Introduction-5

We will then leave alkaloids and move back in time to a target that became important in the 1960's, Cecropia juvenile hormone (Introduction-5). Approaches to this deceptively simple structure constitute a study in stereoselective tri-substituted olefin synthesis. Given the importance of olefins (both synthesis and chemistry of) to "modern" organic synthesis, I think that a visit to this "old" topic will be instructive, and will help set the stage for a discussion of targets of more contemporary interest. In addition, it will focus on the important role synthesis plays in structure determination, and the stimulus natural products can provide for the development of new synthetic methodology.

We will continue with olefin chemistry by considering lasonolide A (Introduction-5). Lasonolide A is a target that will reinforce how synthesis is still an important tool for determining structure. This target will allow us to briefly look at some modern organometallic chemistry as applied to the problem of stereoselective olefin synthesis. In addition it will be used to introduce the topic of macrolide synthesis (macrocyclic lactones).

We will move on to the ionophores, a large family of natural products that present many synthetic challenges. From the many targets one might discuss in this chapter, I have chosen the historically significant calcimycin (A23187). This target will provide us with a look at several strategies for synthesizing molecules with multiple stereogenic centers.

A Polypropionate Target

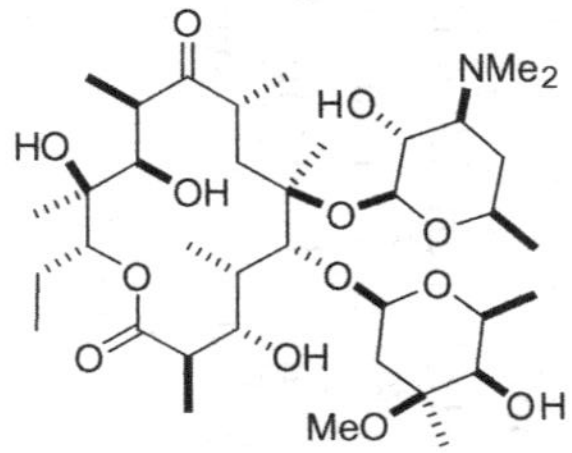

Erythromycin A

Macrolide Synthesis
Acyclic Diastereoselection

Strategy and Tactics

Strategy = Plans

- Assembling Carbon Skeleton
- Oxidation States at Carbon
- Stereochemical Issues
- Tactical Flexibility

Tactics = Execution of Strategy

- Reagent Selection
- Functional Group Compatibility
- Functional Group Interconversions
- Stereochemical Issues

Introduction-6

Introduction-6

Although macrolides are of contemporary interest, the first major steps toward successful macrolide syntheses were reported in the 1970's. Thus, for historical reasons, we will look at two approaches to the classical target erythromycin A and its aglycone, erythronolide-A.

The erythromycins are examples of "polypropionates", natural products biosynthetically derived largely from propionic acid units via a series of condensation reactions. Many natural products, broadly called polyketides, share this biosynthetic origin. These compounds are decorated with multiple stereogenic centers, and acyclic diastereoselection problems that are much more complex than "the terpenoid sidechain stereochemistry problem" will surface with erythromycin, including the problem of asymmetric synthesis.

Finally, throughout this book I will use the terms "strategy" and "tactics" when discussing syntheses. These terms are not new and, in fact, there is a series that bares the title *Strategies and Tactics in Organic Synthesis.*[2] In a broad sense what I will mean by strategy is "the plan" and by tactics I mean "execution of the plan". Some of the general features of strategy and tactics are outlined on the slide (Introduction-6). Whereas there are some distinct differences between the two terms, there is also some overlap (for example both deal with stereochemical issues), so I feel it is important not to get too rigid with definitions here. Nonetheless I hope this will help the reader keep these issues clear when I begin to use these terms. Let's have a look at the steroids as targets for synthesis.

References

1. For a retrospective view see: Nicolaou, K. C.; Vourloumis, D.; Winsdsinger, N.; Baran, P. S. "The art and science of total synthesis at the dawn of the twenty-first century" *Angew. Chem., Int. Ed.* **2000**, *39*, 44–122.
2. *Strategies and Tactics in Organic Synthesis*, Lindberg, T., Ed.; Academic Press; 1984, Vol. 1 (370 pages). *Strategies and Tactics in Organic Synthesis*, Lindberg, T., Ed.; Academic Press; 1989, Vol. 2 (469 pages). *Strategies and Tactics in Organic Synthesis*, Lindberg, T., Ed.; Academic Press; 1991, Vol. 3 (544 pages). *Strategies and Tactics in Organic Synthesis*, Harmata, M. Ed.; Elsevier; 2004, Vol. 4 (415 pages). *Strategies and Tactics in Organic Synthesis*, Harmata, M. Ed.; Elsevier; 2004, Vol. 5 (486 pages). *Strategies and Tactics in Organic Synthesis*, Harmata, M. Ed.; Academic Press; 2007, Vol. 7 (532 pages).

CHAPTER 2

Steroids

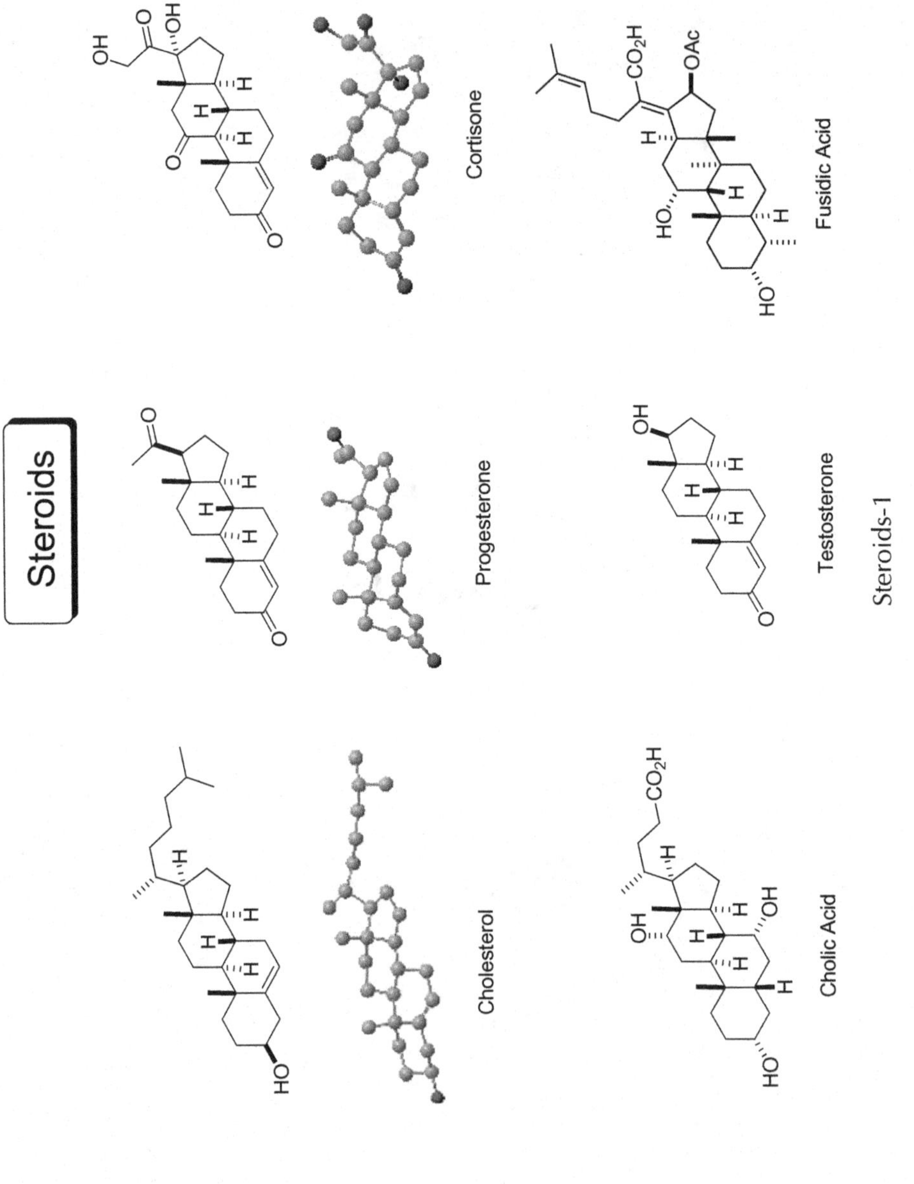
Steroids
Cholesterol
Progesterone
Cortisone
Cholic Acid
Testosterone
Fusidic Acid
Steroids-1

Steroids-1

The steroids we will consider as targets for synthesis are shown here (Steroids-1). Cholesterol (and derivatives) is an important component of cell membranes. Gallstones are mainly cholesterol. It is one of the steroids most familiar to the layperson. Other steroids of much interest include the sex hormones (structurally related to progesterone) and the corticosteroids, represented here by cortisone.

Most steroids share a common tetracyclic ring system, but are adorned differently in terms of oxidation state at various carbons. The three targets we will consider are only the tip of the iceberg. Fused ring systems with different ring juncture stereochemistry (cis vs trans), different sidechains, and different oxidation patterns are common. A few examples are shown here (see problems for some questions about fusidic and cholic acids).

Woodward, R. B.; Sondheimer, F.; Taub, D.; Heusler, K.; McLamore, W. M. "Total Synthesis of Steroids", *J. Am. Chem. Soc.* **1952**, *74*, 4223.

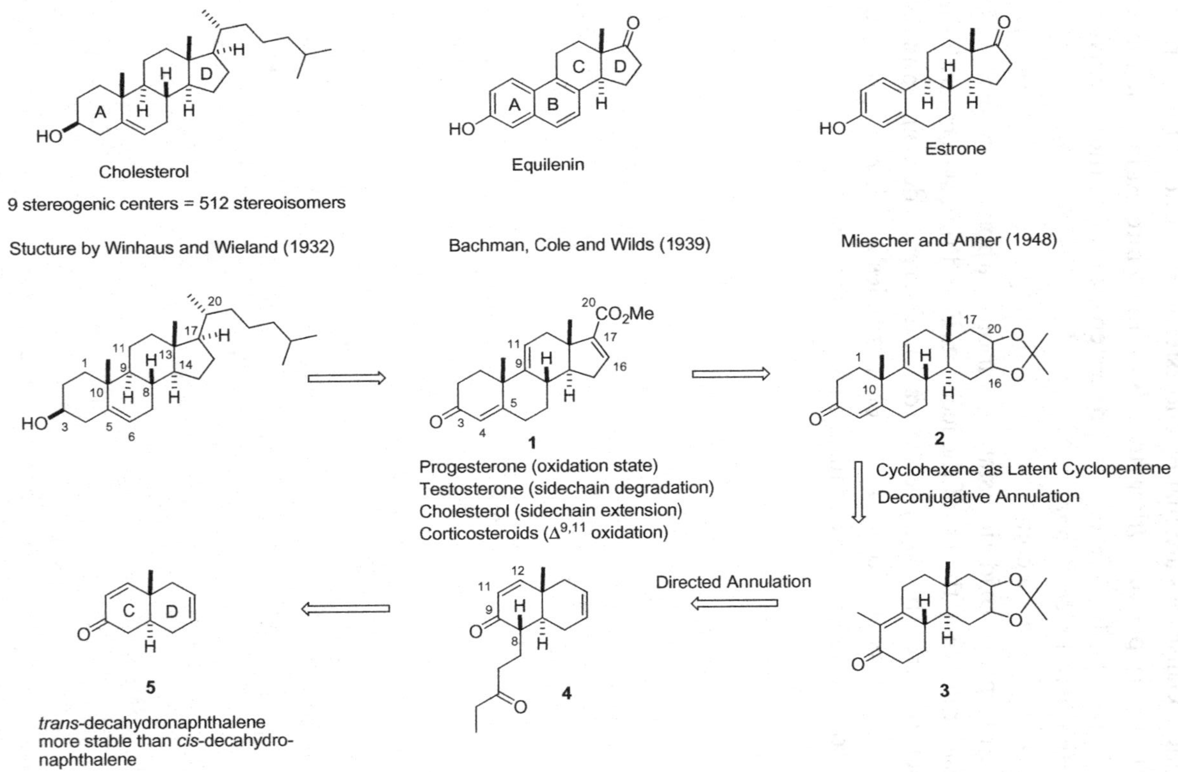

Steroids-2

Steroids-2

The first steroids to be prepared by total synthesis were equilenin and estrone (Steroids-2). These targets have progressively more complex stereochemistry. Equilenin has only two stereogenic centers and thus, four stereoisomers are possible. Estrone has four stereogenic centers (16 possible isomers).[1]

Cholesterol has 9 stereogenic centers and thus, in principle, $2^9 = 512$ stereoisomers (including enantiomers) are possible. Although some of these are geometrically impossible given the bonding requirements of carbon, it is easy to see that cholesterol presents stereochemical problems that are at least an order of magnitude more complex that equilenin and estrone. The Woodward group (Harvard) reported the first total synthesis of cholesterol in the early 1950s. Their objective was not simply to prepare cholesterol, but to pass through intermediates that could be used to access a host of other steroids including progesterone and the corticosteroids. They eventually settled on tetracyclic compound **1** as a key target.

At this point I will digress a bit. Throughout this book, I will undoubtedly interpret or present certain things in a manner that may seem inaccurate to the reader. Well, the reader may be correct. For example, I have no idea how Woodward and his students settled on **1** as a key intermediate. In hindsight, it is an excellent choice, but whether or not this was part of the initial design of the synthesis, or "evolved" as a key intermediate as research proceeded, I have no clue. The bottom line is that the reasons behind making a choice (for example of an intermediate or a reagent) are not always clear in print, and can even be inaccurate in print due to the tendency we all have as human beings to want to sound smarter than we are.

Certainly more important, choice of a key intermediate can be critical if the synthetic objective is a family of molecules. "Can I get to the target structure(s) from this intermediate?" is a question that one asks regularly when designing (or developing) a synthetic strategy. In this case, **1** is very close to progesterone and testosterone. Reaching both targets calls for reduction chemistry at the C_9-C_{11} and C_{16}-C_{17} double bonds, and some D-ring modifications, including an oxidative degradation at C_{17} to reach testosterone. To reach cholesterol from **1**, deconjugative reduction of the A-ring enone must be accomplished, as well as the aforementioned olefin reductions, and installation of the C_{17} side chain with control of stereochemistry at C_{20}. One can imagine using the C_9-C_{11} double bond as a functional handle for reaching cortisone from **1**. Finally, one can imagine a number of tactics for accomplishing all of the aforementioned transformations. It is important to avoid strategies that have to be scrapped if a single tactic fails. Having lots of options is desireable.

Working backwards to develop a synthetic strategy is also common. It is called retrosynthetic analysis.[2] As an example, one path to **1** might be through an intermediate of type **2**. In a forward direction, hydrolysis of the acetal, cleavage of the resulting diol, aldol dehydration of the resulting dialdehyde, and adjustment of oxidation state at C_{20} would accomplish the necessary transformation. Of course this would have to be done in a manner that was compatible other functional groups. The details of this will emerge when we examine the actual synthesis. Working backwards again, the D-homosteroid nucleus of **2** [the term "D-homosteroid" is loosely derived from "homologous series" which generally means one more CH_2] was to be prepared from tricyclic enone **3** via a deconjugative alkylation to establish the C_1-C_{10} bond, followed by an aldol dehydration to afford the A-ring enone. Enone **3** was to be derived from **4** via an aldol dehydration, and **4** was to come from dienone **5**, the first key intermediate in this synthetic strategy. There are several critical aspects to selection of **5** as an early intermediate. One is that cyclohexenes can always be converted to 1-acylcyclopentenes via oxidative cleavage of the double bond, followed by an aldol-dehydration reaction sequence. Since a cyclohexene has a lower level of functionality than an acylcyclopentene (one functional group vs two functional groups), it is often useful to package an acylcyclopentene as a cyclohexene (or cyclohexene derivative such as an acetonide). This is part of the generally useful advice to "keep functionality at as low a level as possible throughout a synthesis". Why? Problems of functional group compatibility associated with a given transformation are minimized. Another reason why **5** was chosen as a key intermediate is that, at the time, little was known about how to reliably prepare *trans*-fused hexahydroindans (the CD ring system of the steroids we have considered thus far). In fact, it was known that *cis*-hexahydroindan (the parent 6/5 fused ring hydrocarbon) was more stable than *trans*-hexahydroindan. On the other hand, it was known that *trans*-decalin (the parent 6/6 fused ring hydrocarbon) was more stable than *cis*-decalin. Thus, it was decided to prepare the 6/6 ring system, hoping that thermodynamics could be used to establish stereochemistry at the ring juncture, and knowing that the cyclohexene could serve as a latent 5-membered ring. Now let's examine the synthesis.

Steroids-3

The synthesis began with a Diels-Alder reaction between 1,3-butadiene and quinone **6** to give *cis*-hexahydronaphthalene 7 (Steroids-3).[3] The nature of the Diels-Alder reaction gave the *cis*-cycloadduct, but base-mediated epimerization via an intermediate enolate, provided *trans*-hexahydronaphthalene isomer **8**. In spite of the aforementioned plan, however, it turned out there was little difference in thermodynamic stability between 7 and **8**. Nonetheless, seeding a basic solution of 7 with **8**, accomplished selective crystallization of the desired trans isomer in near quantitative yield (an application of LeChatelier's Principle). Adjustment of oxidation states in the C-ring was accomplished by lithium aluminum hydride reduction to give an intermediate diol, followed by acid promoted enol ether hydrolysis and β-elimination of the C_{12} hydroxyl group to introduce the C_{11}-C_{12} double bond. The C_8 hydroxyl group was then removed via a reduction procedure that is now common, but in the 1950s represented new synthetic methodology.

Note that the conversion of **8** to **5** involves only reduction chemistry. Reduction-oxidation chemistry permeates syntheses and thus, a general consideration of the topic is worthwhile. Let's look in detail at the changes in oxidation state that occur during this transformation. One convenient way of comparing oxidation states at carbon is to count the number of bonds from carbon to oxygen. For example, methane (an alkane) has no bonds to oxygen, methanol (an alcohol) has one bond to oxygen, and formaldehyde (an aldehyde), formic acid (an acid) and carbon dioxide have 2–4 bonds to oxygen, respectively. Of course not all functional groups contain oxygen. But functional groups that can be converted to alcohols by substitution reactions (alkyl halides, sulfides, selenides, amines, azides, to mention a few) are really at the alcohol oxidation state. The same goes for functional groups at the ketone or aldehyde oxidation state (imines, oximes, hydrazones, thioacetals). Alkenes are a bit trickier. One way to look at alkenes is that hydration of an alkene (a reaction that does not involve redox chemistry) converts one carbon to the alkane oxidation state and the other to the alcohol oxidation state. Conversely, dehydration of an alcohol introduces an alkene (without redox chemistry). So in alkenes, the oxidation state at one carbon is "alkane" and at the other carbon "alcohol". It is not always possible to decide which carbon one should regard at which oxidation state. That depends on substituents. Now let's look at the conversion of **8** to **5**. In **8**, C_8 and C_{12} are clearly at the ketone oxidation state. Carbon-9 is best classified as being at the ketone oxidation state, because the "natural" course of hydration of an enol ether converts it to a ketone (or aldehyde). Conversely, enol ethers are derivatives

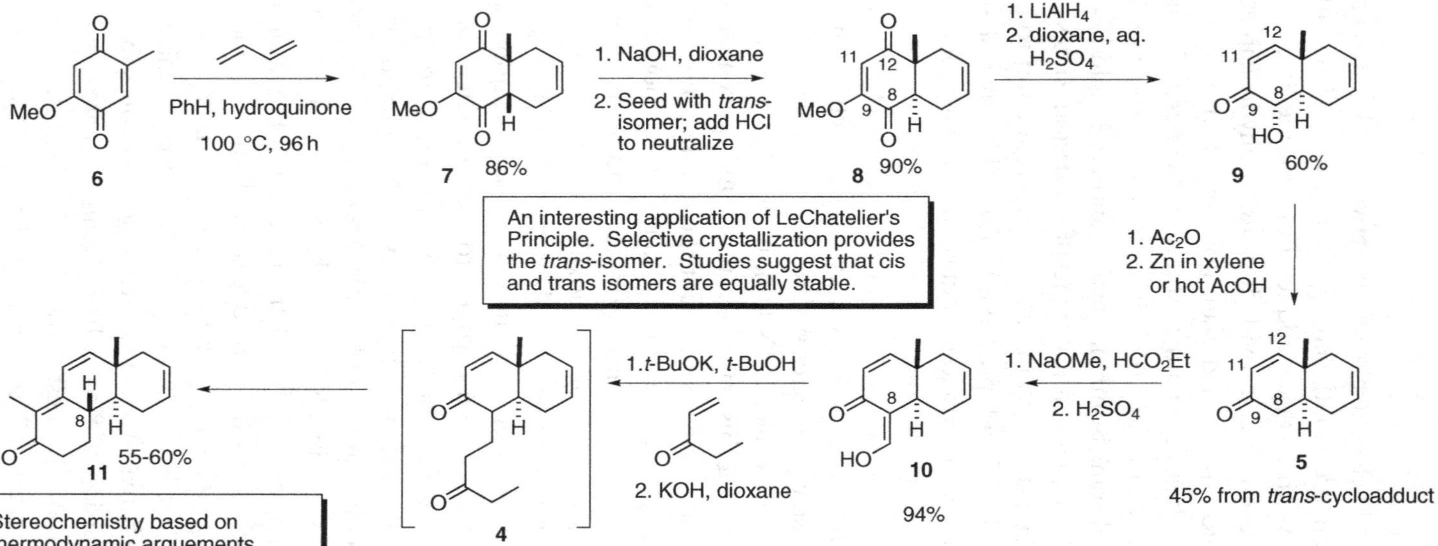

An interesting application of LeChatelier's Principle. Selective crystallization provides the *trans*-isomer. Studies suggest that cis and trans isomers are equally stable.

Stereochemistry based on thermodynamic arguements. The "guess" turned out to be correct. This was about the same time that Barton published his seminal papers on conformational analysis of cyclic systems.

Notice the use of Claisen condensation, conjugate addition, and aldol dehydration reactions to construct the six-membered ring (annulation chemistry). We will return to possible reasons for choice of this strategy for construction of the B-ring at a later date. Note that the same fundamental strategy is used for construction of the A-ring.

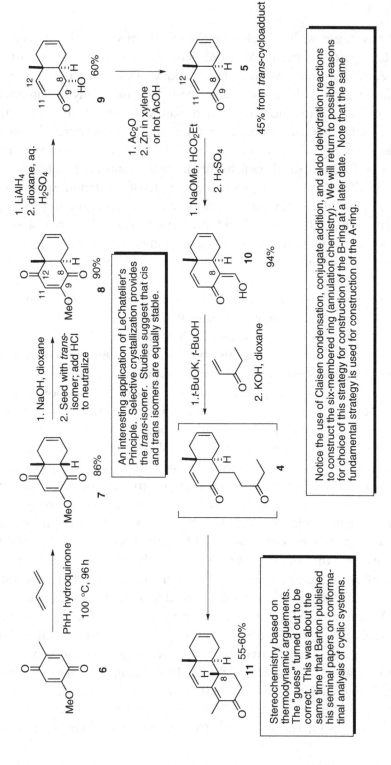

of ketones (or aldehydes). This leaves C_{11} at the alkane oxidation state. In **5**, the oxidation states at C_9 and C_8 are "ketone" and "alkane", respectively. The substituent at C_9 (a carbonyl group) renders C_{12} electron-deficient relative to C_{11}. Thus it is reasonable to regard C_{12} to be at the "alcohol" oxidation state and C_{11} at the "alkane" oxidation state. Note that hydration of conjugated enones normally occurs to provide β-hydroxy ketones. So, in the conversion of **8** to **5**, the oxidation state at C_{12} changes from "ketone" to "alcohol" and the oxidation state at C_8 changes from "ketone" to "alkane". This transformation requires "one reduction" at C_{12}, "two reductions" at C_8, and no redox chemistry at either C_9 or C_{11}. Whereas this analysis is "after the fact" in this case, a useful question to ask as one plans a functional group transformation (or series of transformations) is: What oxidation state changes occur? This can reveal whether redox chemistry is needed or not, and guide the practitioner in choice of reagents (reducing agents vs oxidizing agents). Whereas we have only considered oxidation states at carbon here, similar analyses can be applied to oxidation states at heteroatoms. Let's continue with the synthesis.

The conversion of enone **5** to dienone **11** is an example of an "annulation" or "annelation" reaction.[4] The words mean the same thing: to build one ring onto another. The only difference is that annulation is derived from the Greek "annulus", and annelation is derived from the French "anneler".[5] The type of reaction used here is commonly called a "Robinson Annelation". It involves conjugate addition of the enolate of α-formylketone **10** to ethyl vinyl ketone (a Michael Reaction),[6] followed by an aldol-dehydration sequence. This added what will become the B-ring of the steroids onto a pre-existing C-ring. Why was **5** converted to **10** prior to conducting the annulation? Both **5** and ethyl vinyl ketone (the annulation reagent) are ketones of comparable acidity. Conversion of **5** to **10** "activates" C_8 relative to the annulation reagent, defining **10** as the nucleophile and ethyl vinyl ketone the electrophile in the Michael reaction. Deformylation of the intermediate non-enolizable 1,3-dicarbonyl compound formed in the Michael addition, gave conjugate adduct **4**, which continued on to **11** under the basic reaction conditions. It is notable that the Woodward group had only limited analytical tools available at the time this synthesis was conducted. Nowadays one would use NMR methods, or perhaps crystallography, to assign stereochemistry. In this case, it was presumed that C_8 would be under thermodynamic control and this alone was the basis for assignment of stereochemistry at the newly formed stereogenic center. The "guess" turned out to be correct.

The next portion of the plan called for removal of the C_{11}-C_{12} double bond and deconjugative annulation of the A-ring to the pre-existing B-ring.

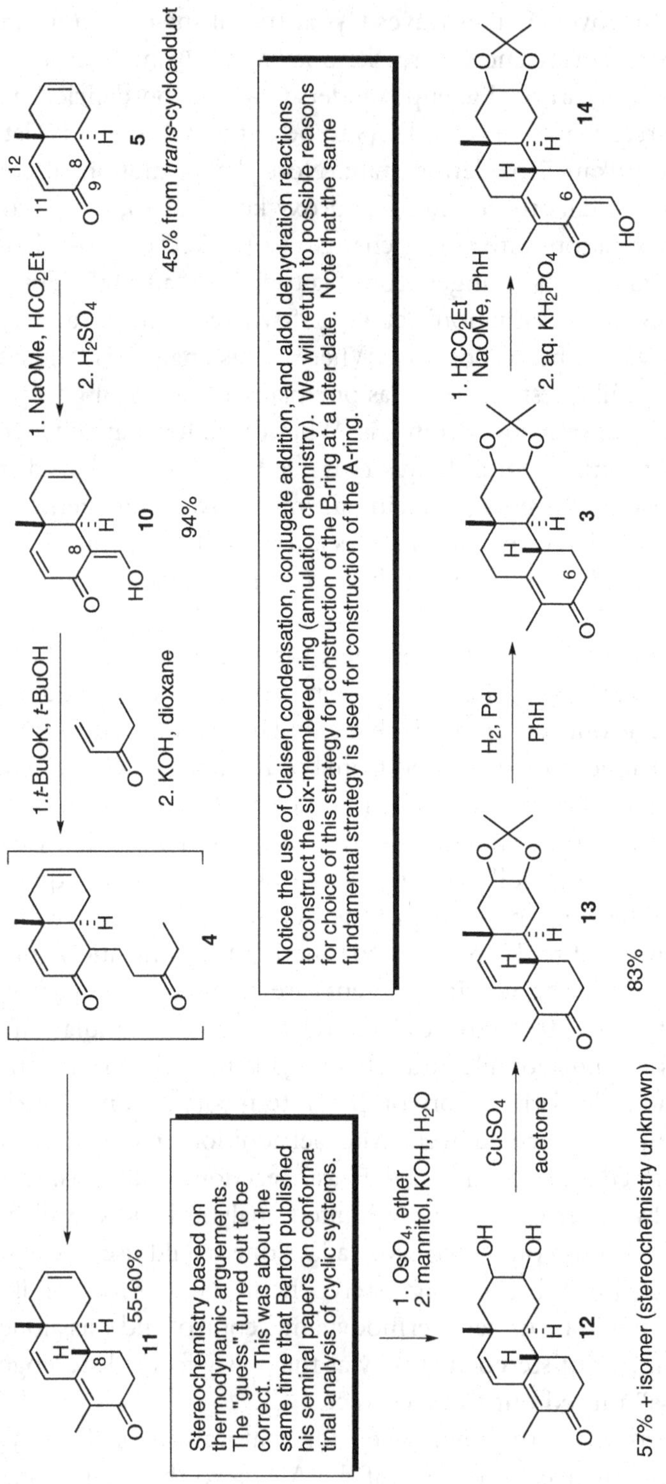

Steroids-3 (*Continued*)

First, however, the D- and C-ring double bonds had to be differentiated. This was accomplished by vicinal dihydroxylation of the D-ring olefin using osmium tetroxide (stoichiometric amounts) followed by protection of the resulting diol as acetonide **13** (an acetal).[7] The vicinal dihydroxylation proceeded with modest stereoselectivity and good regioselectivity, presumably due to steric effects. The stereochemistry of the diol was never established, but this did not thwart completion of the synthesis. A practical point that I want to make here is that although both diastereomers of diol **12** could have been carried through the synthesis in principle, this was not done in practice. Only the major isomer was used. Although one is sometimes tempted to say that "stereocontrol will not matter because the stereochemistry will be destroyed by the time I reach my target", this is generally not a good approach to take. In reality, characterization of mixtures is difficult and most successful syntheses do not carry mixtures through multiple steps. This does not mean that one should not attempt a reaction if there is some doubt about its stereochemical course. It is good, however, to have options available so the required control can be achieved. In reality, however, separations of stereoisomer mixtures are frequently needed to execute a synthesis of a complex molecule. Continuing with the synthesis, hydrogenation of the sterically less hindered of the remaining double bonds proceeded smoothly to provide **3**. Annulation of the A-ring onto the B-ring first required that C_6 be "blocked". This was accomplished by conversion of **3** to vinylogous amide **15** (Steroids-4) via intermediate C_6-hydroxymethylene ketone **14**.

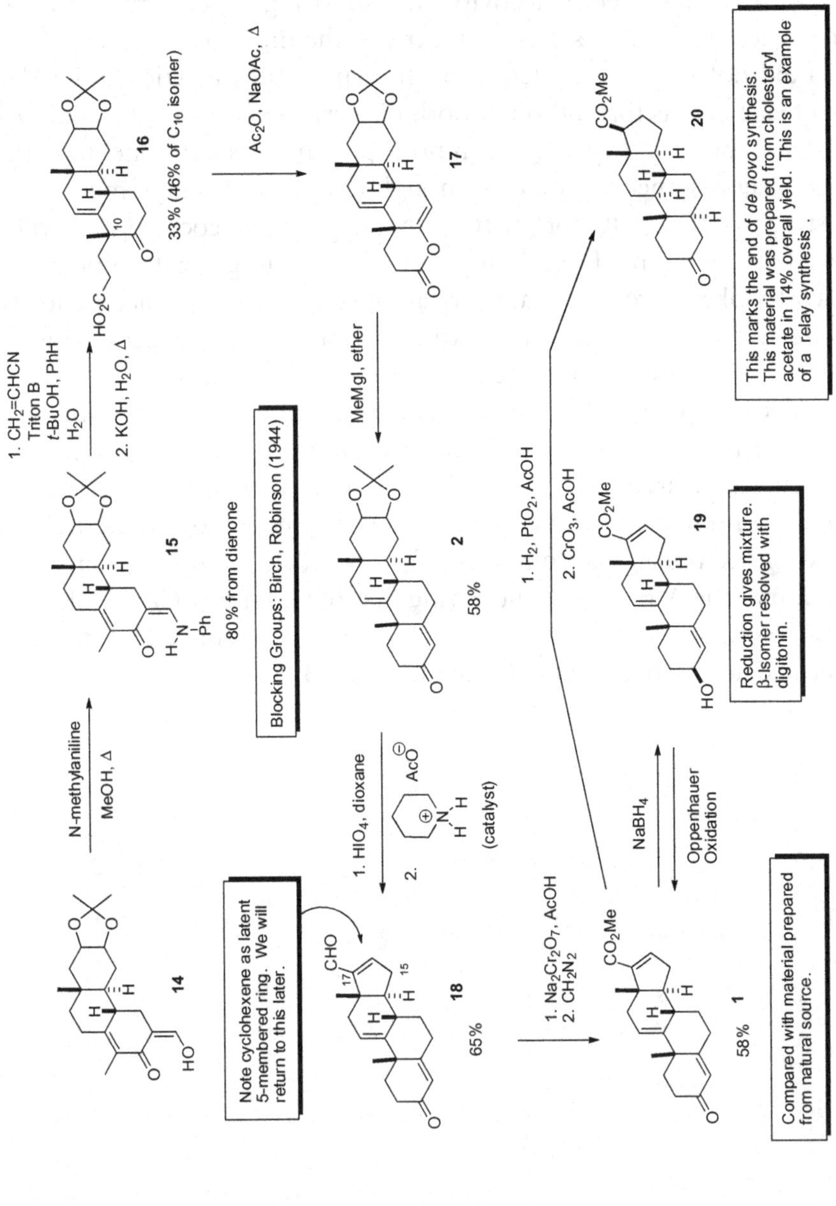
N-methylaniline
MeOH, Δ
1. CH2=CHCN
Triton B
t-BuOH, PhH
H2O
2. KOH, H2O, Δ
33% (46% of C10 isomer)
80% from dienone
Blocking Groups: Birch, Robinson (1944)
Ac2O, NaOAc, Δ
MeMgI, ether
58%
1. HIO4, dioxane
(catalyst)
Note cyclohexene as latent 5-membered ring. We will return to this later.
65%
1. Na2Cr2O7, AcOH
2. CH2N2
1. H2, PtO2, AcOH
2. CrO3, AcOH
NaBH4
Oppenhauer Oxidation
Reduction gives mixture. β-Isomer resolved with digitonin.
This marks the end of de novo synthesis. This material was prepared from cholesteryl acetate in 14% overall yield. This is an example of a relay synthesis .
Compared with material prepared from natural source.
58%
14
15
16
17
2
18
1
19
20

Steroids-4

Steroids-4

Treatment of **15** with base and acrylonitrile, followed by hydrolysis of the intermediate nitrile, gave conjugate adduct **16**. This reaction was complicated by formation of the C_{10} epimer of **16** as the major product. Woodward indicates that little selectivity was expected in this step. We will see (Steroids-9) that a modification of this approach that solved this problem was eventually developed by others. Annulation of the A-ring was completed by treating **16** with acetic anhydride, followed by opening of the resulting enol ester **17** with methyl magnesium iodide to give an intermediate 1,5-diketone. Aldol-dehydration occurred under the basic reaction conditions to provide tetracycle **2**.

Keeping in mind that **1** was projected as a key intermediate for the synthesis of a variety of steroids, the Woodward group next turned to contraction of the D-ring to the required acyl cyclopentene. The diol was liberated and cleaved to a dialdehyde using periodic acid. The dialdehyde was then subjected to piperidinium ion mediated intramolecular aldol-dehydration to provide **18**. The aldol-dehydration also provided the regioisomeric enal as a minor product. The regioselectivity of this reaction was attributed to the apparent steric accessibility of the C_{17} methylene in the intermediate dialdehyde relative to the C_{15} methylene. The synthesis of **1** was completed in a straightforward manner.

This synthesis provided racemic **1**. To conclusively establish its structure, **1** was reduced with sodium borohydride to provide **19** (this reaction gave a separable mixture of diastereomers), which was resolved and oxidized to return a single enantiomer of **1**. This material was identical in all respects to a sample of material prepared by degradation of a corticosteroid.

The Woodward group conducted most of their remaining work with material prepared by degradation of cholesteryl acetate to tetracycle **20**, which was also prepared from synthetic **1** as shown in (Steroids-4). A synthesis of a natural product from an intermediate prepared by degrading a natural product, prepared in turn by total synthesis, is known as a relay synthesis. This is a less common practice nowadays than in the early days of natural products synthesis, but it is not unheard of even today. We will see this practice again.

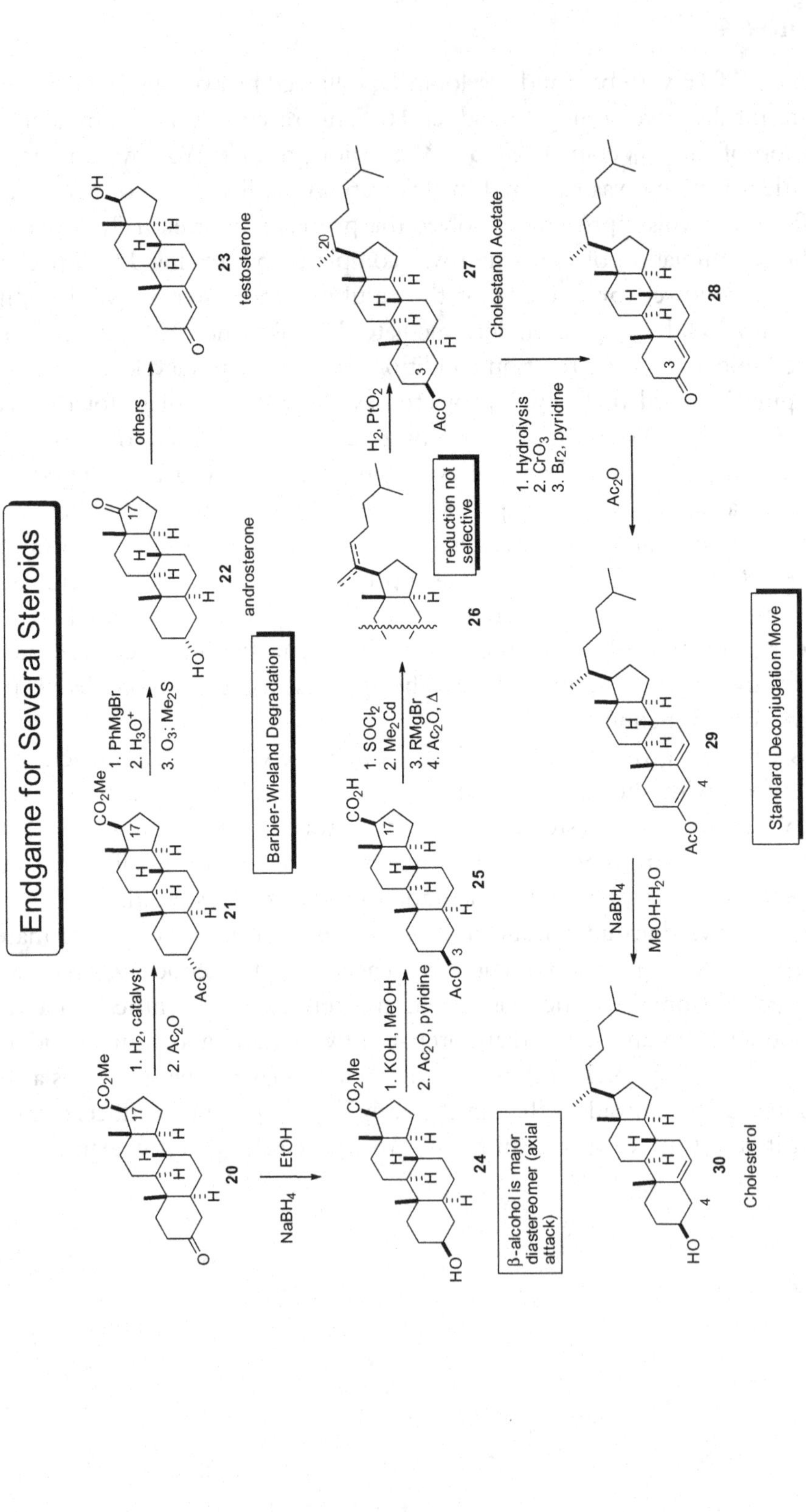
Endgame for Several Steroids
20
NaBH4 EtOH
1. H2, catalyst
2. Ac2O
21
1. PhMgBr
2. H3O+
3. O3; Me2S
22
androsterone
others
23
testosterone
24
β-alcohol is major diastereomer (axial attack)
1. KOH, MeOH
2. Ac2O, pyridine
25
Barbier-Wieland Degradation
1. SOCl2
2. Me2Cd
3. RMgBr
4. Ac2O, Δ
26
reduction not selective
H2, PtO2
27
Cholestanol Acetate
1. Hydrolysis
2. CrO3
3. Br2, pyridine
28
Ac2O
29
Standard Deconjugation Move
NaBH4
MeOH-H2O
30
Cholesterol

Steroids-5

Steroids-5

The Woodward group converted this key intermediate (**20**) into a variety of natural steroids including progesterone (Steroids-1). Key intermediate **20** was also converted to androsterone (**22**) by reduction of the C_3 ketone followed by Barbier-Wieland degradation of the D-ring ester to a C_{17} ketone.[8] Androsterone had previously been converted to testosterone (**23**) by others.

The cholesterol synthesis was accomplished by initial reduction of **20** with sodium borohydride to provide **24**. Ester hydrolysis and esterification of the resulting C_3 alcohol gave **25**. The C_{17} acid was converted to a mixture of olefins **26**, followed by unselective reduction of the olefins to provide cholestanyl acetate (**27**) after separation of the C_{20} diastereomers. The lack of diastereoselectivity in this transformation begs the question: How can stereochemistry at an acyclic stereogenic center be controlled relative to other stereogenic centers? We will return to this question later in detail when we consider syntheses of a natural product called juvabione, but for now let us continue. Hydrolysis of **27** provided cholestanol, which had previously been converted to cholesterol by other research groups. This conversion involved the preparation of **28** via bromination-dehydrobromination of a C_3 ketone. Enone **28** was converted to cholesterol by what is now a standard deconjugation move in terpenoid synthesis. Dienol acetate **29** was reduced with sodium borohydride, the intermediate enolate protonated at C_4, and the resulting β,γ-unsaturated ketone was then reduced by the $NaBH_4$ (pseudoaxial delivery of hydride) to provide the homoallylic alcohol substructure of cholesterol (**30**).

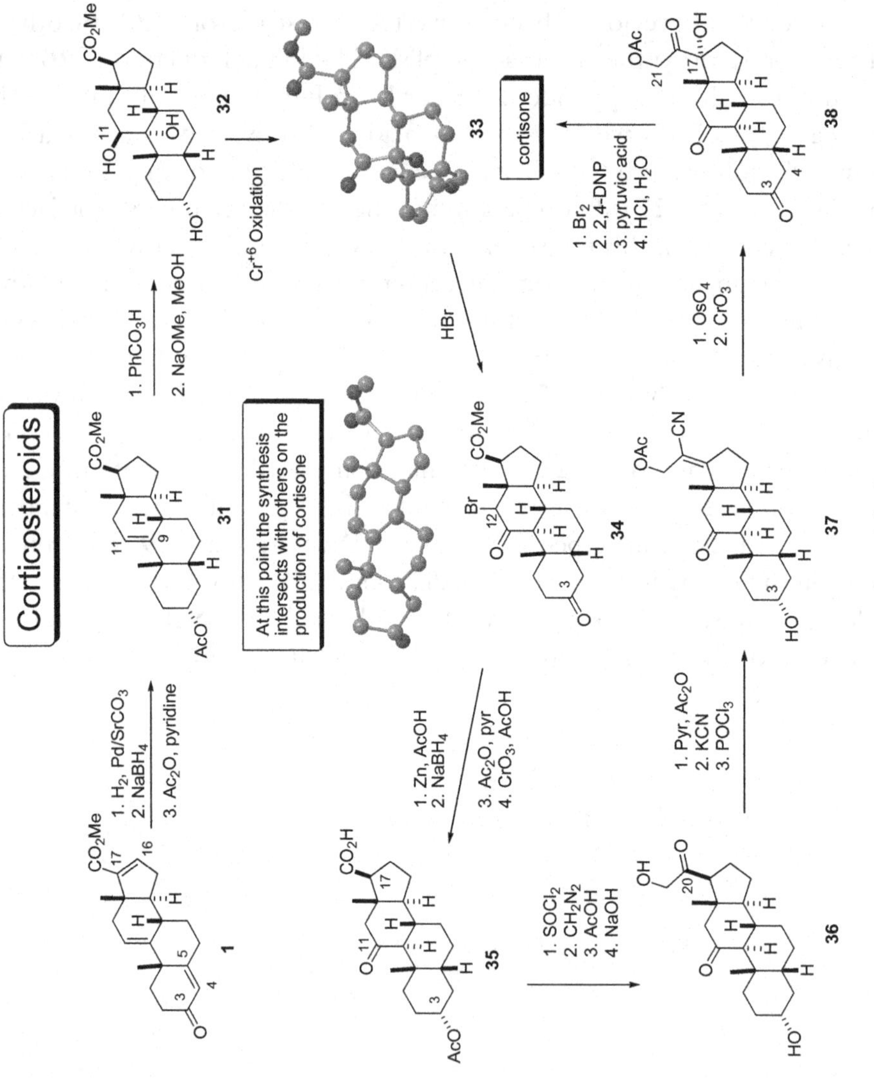
Corticosteroids
1. H_2, Pd/$SrCO_3$
2. $NaBH_4$
3. Ac_2O, pyridine
1. $PhCO_3H$
2. NaOMe, MeOH
Cr^{+6} Oxidation
At this point the synthesis intersects with others on the production of cortisone
HBr
1. Zn, AcOH
2. $NaBH_4$
3. Ac_2O, pyr
4. CrO_3, AcOH
1. $SOCl_2$
2. CH_2N_2
3. AcOH
4. NaOH
1. Pyr, Ac_2O
2. KCN
3. $POCl_3$
1. OsO_4
2. CrO_3
1. Br_2
2. 2,4-DNP
3. pyruvic acid
4. HCl, H_2O
cortisone
1
31
32
33
34
35
36
37
38

Steroids-6

Steroids-6

As a final target, we will examine the conversion of intermediate **1** to cortisone (Steroids-1) as outlined in Steroids-6. The critical transformation was conversion of **1** to **31**, which had previously been "connected" with cortisone by a series of transformations reported from other laboratories. Hydrogenation of **1** resulted in reduction of the C_4-C_5 and C_{16}-C_{17} olefins to provide material with the required C_{17} stereochemistry and a mixture of *cis*- and *trans*-fused AB-ring systems. Sodium borohydride reduction of the C_3 ketone and acetylation of the resulting alcohol provided **31**. The overall yield of this transformation was approximately 15% due to lack of stereoselectivity in the hydrogenation.

The rest of the cortisone synthesis involves a series of transformations reported by other groups over the period extending from approximately 1940–1950, but we will review these here for the purpose of comparison when we consider other approaches to cortisone. The C_{11} oxygen, that is so characteristic of the corticosteroids, was introduced by epoxidation of the sterically most accessible face of the olefin followed by ring opening to provide **32**. Oxidation of the secondary alcohols gave hemiacetal **33**. Treatment of **33** with HBr gave dione **34**, a transformation that is interesting from the standpoint of mechanism. Reduction of the C_{12} bromide (recall the conversion of **8** to **9**) was followed by reduction of the ketones, esterification of the sterically most accessible C_3 alcohol, and oxidation of the C_{11} alcohol to provide **35**. The C_{17} side chain was then introduced. Acid **35** was first converted to **36** using a four-reaction sequence. The primary alcohol was acylated, the C_{20} ketone was converted to a cyanohydrin, and dehydration of the C_{20} alcohol then provided **37**. Vicinal dihydroxylation of the unsaturated nitrile, followed by elimination of HCN to reveal the C_{20} ketone, and oxidation of the C_3 secondary alcohol, gave **38**. The A-ring stereochemistry and oxidation state was then adjusted by initial bromination at C_4, dehydrohalogenation to provide the 2,4-DNP of the A-ring enone, and subsequent hydrolysis of the 2,4-DNP and C_{21} acetate to provide cortisone (Steroids-1).

Wieland, P.; Ueberwasser, H.; Anner, G.; Miescher, K. "Uber die Herstellung des $\Delta^{8,14}$-1,7-Dioxo-8,11-dimethyldodecahydrophenanthrens. Totalsynthetische Versuche in der Steroidreihe I", *Helv. Chim. Acta* **1953**, *50*, 376.

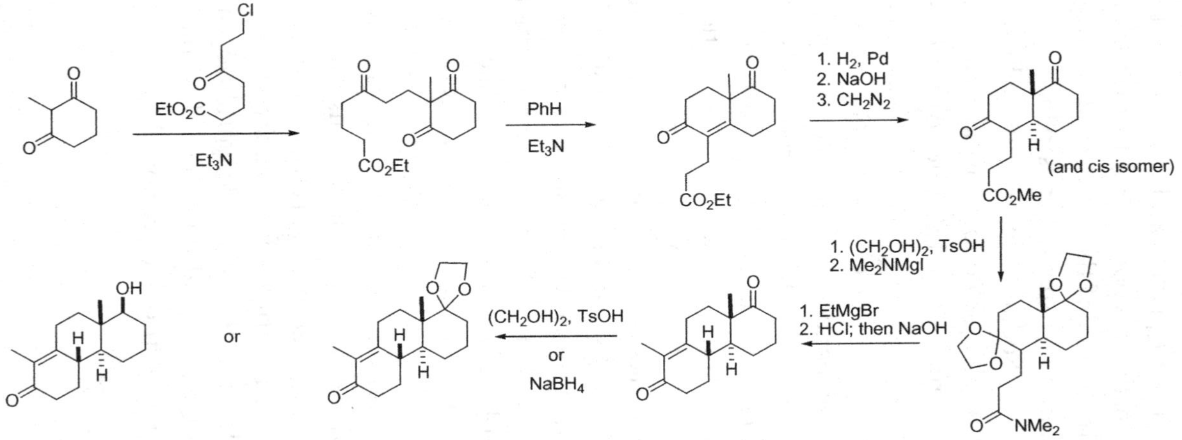

Wiland, P.; Anner, G.; Miescher, K. "Die sterische Verknupfung eines $\Delta^{8,14}$-1,7-Dioxo-8,11-dimethyldodecahydrophenanthrens mit den Steroiden. Totalsynthetische Versuche in der Steroidreihe II" *Helv. Chim. Acta* **1953**, *50*, 646.

Wieland, P. Ueberwasser, H.; Anner, G.; Miescher, K. "Totalsynthese von D-Homo-steroiden. Totalsynthetische Versuche in Steroidreihe III" *Helv. Chim. Acta* **1953**, 50, 1231.

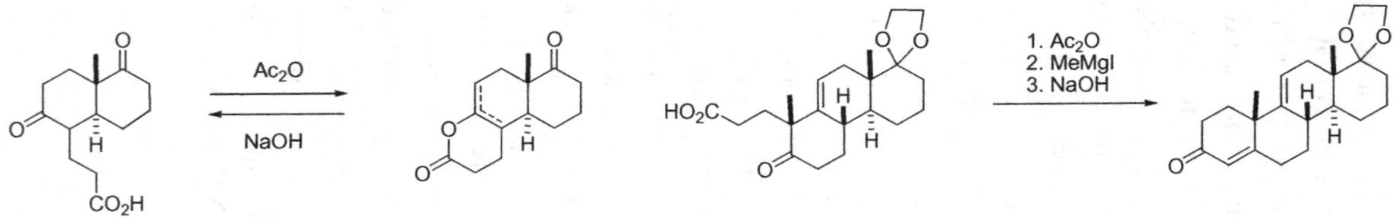

Steroids-7

Steroids-7

I picked the Woodward approach to steroids as the first synthesis to discuss because I want to revisit strategies that appear in that work within the context of other targets. But it is important to recognize that a number of other groups were working in the area, using conceptually similar approaches (in part). I will mention a few of these approaches here. The A- and B-rings in the synthesis we have just considered were introduced by annulation reactions. This type of chemistry was introduced by Robinson and ultimately used in his approach to steroids. It was also explored in detail by Wieland during the course of his studies directed toward the structure determination and synthesis of steroids. Some reactions reported by the Wieland group appear in Steroids-7 without much comment. These approaches resemble the Woodward approach in that the steroid ring system is assembled starting with the D-ring and working toward the A-ring. Other similarities involve use of a deconjugative Michael addition to introduce the incipient A-ring, application of the keto acid to cyclohexenone transformation to construct the A-ring, and synthesis of a D-homosteroid that could be ring contracted to provide the required 5-membered ring. Finally, if this work were being done today, additional tactics would be available that might provide better control of stereochemistry at the CD-ring juncture.

An Industrial Scale Synthesis of Steroids

For some related chemistry see: Velluz, L.; Nomine, G.; Mathieu, J.; Toromanoff, E.; Bertin, D.; Tessier, J.; Pierdet, A. "Sur l'acces stereospecifique, par synthese totale, a la serie 19-nor-steroide. La 19-nor-testosterone de synthesis" *Comptes Rend. Acad. Sci.* **1960**, *250*, 1084. The following material was abstracted from "Selected Organic Syntheses" by Ian Fleming with permission from Wiley-Blackwell. The synthesis relies in part on annulation chemistry developed by Wieland and Miescher.

Strategy: (1) Build from D-ring toward the A-ring using annulation chemistry. (2) Obtain desired enantiomer by enzymatic reduction of a pro-chiral diketone. (3) Use A-ring enone rather than AB-cis system throughout introduction of C_{11} ketone of corticosteroids. (4) Use C_{17} ketone as handle for introduction of D-ring sidechain.

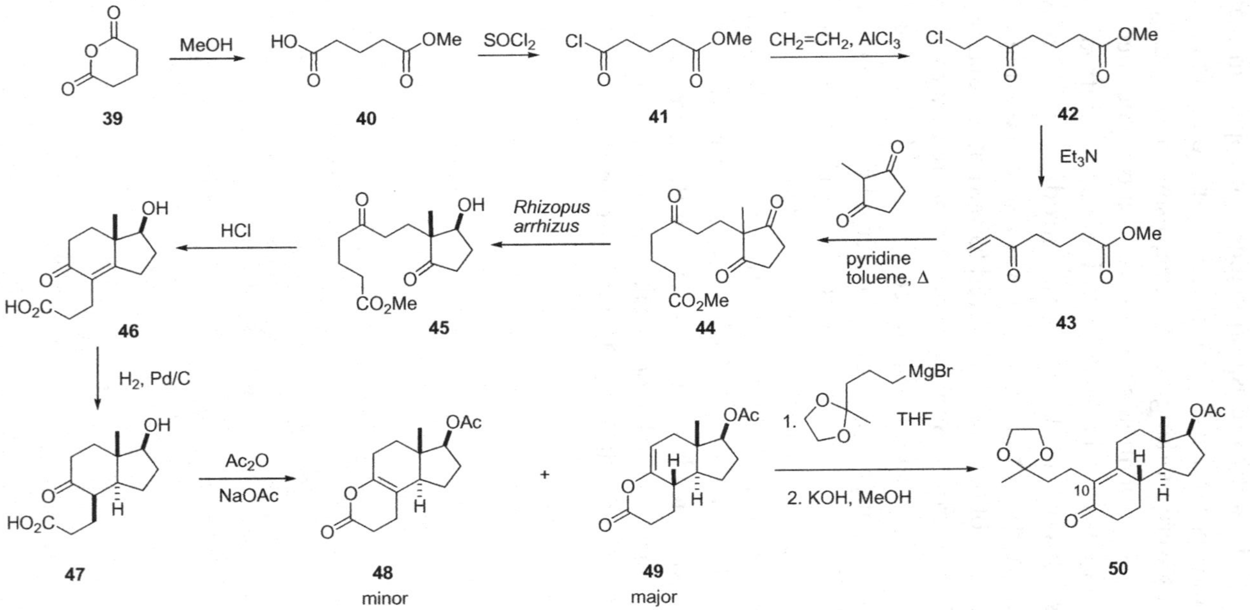

Steroids-8

An industrial scale approach to steroids was developed by both Merck (US)[9] and Ciba (Swiss) during the 1950s. The Ciba approach is summarized in Steroids 8–10. The critical aspects of the strategy were: (1) to build from the D-ring toward the A-ring using annulation chemistry, (2) to obtain the desired enantiomer by enzymatic reduction of a pro-chiral diketone, (3) to use an A-ring enone rather than an AB-*cis* system throughout introduction of the C_{11} ketone of the corticosteroids, and (4) to use a C_{17} ketone as a handle for introduction of the D-ring sidechain.

The synthesis began with the preparation of annulation reagent **43** from glutaric anhydride (**39**). Conjugate addition of 2-methylcyclopentan-1,3-dione to **43** provided trione **44**. This trione has a plane of symmetry bisecting the 1,3-dicarbonyl substructure. The two carbonyl groups are identical and reduction of one ketone with an achiral reducing agent provides a racemic mixture of **45** (an equal mixture of enantiomers). But use of an enantiopure reducing agent can lead to an unequal mixture of enantiomers. In this case, reductases (enzymes) produced by the bacterium *Rhizopus arrhizus* led to enantioselective reduction of **44** to provide **45**. Acid promoted aldol-dehydration of **45** was accompanied by hydrolysis of the methyl ester to provide hexahydroindan **46**. Catalytic hydrogenation gave **47**. The hydrogenation of enones of type **46** to give either a *cis*- or *trans*-fused ring system is known. It is a catalyst and substrate dependent process and, in practice, one must evaluate a number of tactics to determine how to best accomplish this transformation with the desired stereoselectivity.[10] Continuing with the synthesis, the B-ring was assembled using the enol lactone annulation procedure we saw in the Woodward synthesis. The difference between **50** and the BCD-ring system intermediate in the Woodward synthesis (see **3** in Steroids-3), is that the subsequent deconjugative alkylation reaction must install the C_{10} methyl group rather than the remaining carbons of the A-ring.

There is a small general lesson that can be discussed here. If you have to introduce two different groups to the same carbon during the course of a synthesis, you have to pick an order-of-introduction. If introducing the groups in one order shows a stereochemical preference, introducing the groups in the opposite order can often produce the other stereochemical result. In the Woodward synthesis, installation of the C_{10} methyl group preceeds introduction of the A-ring (via a deconjugative conjugate addition). In the Ciba approach, installation of the C_{10} methyl group follows introduction of the A-ring (precursor group). Whereas the "methyl first" approach gave a mixture of stereoisomers, the "methyl second" approach gave the desired stereochemical result.

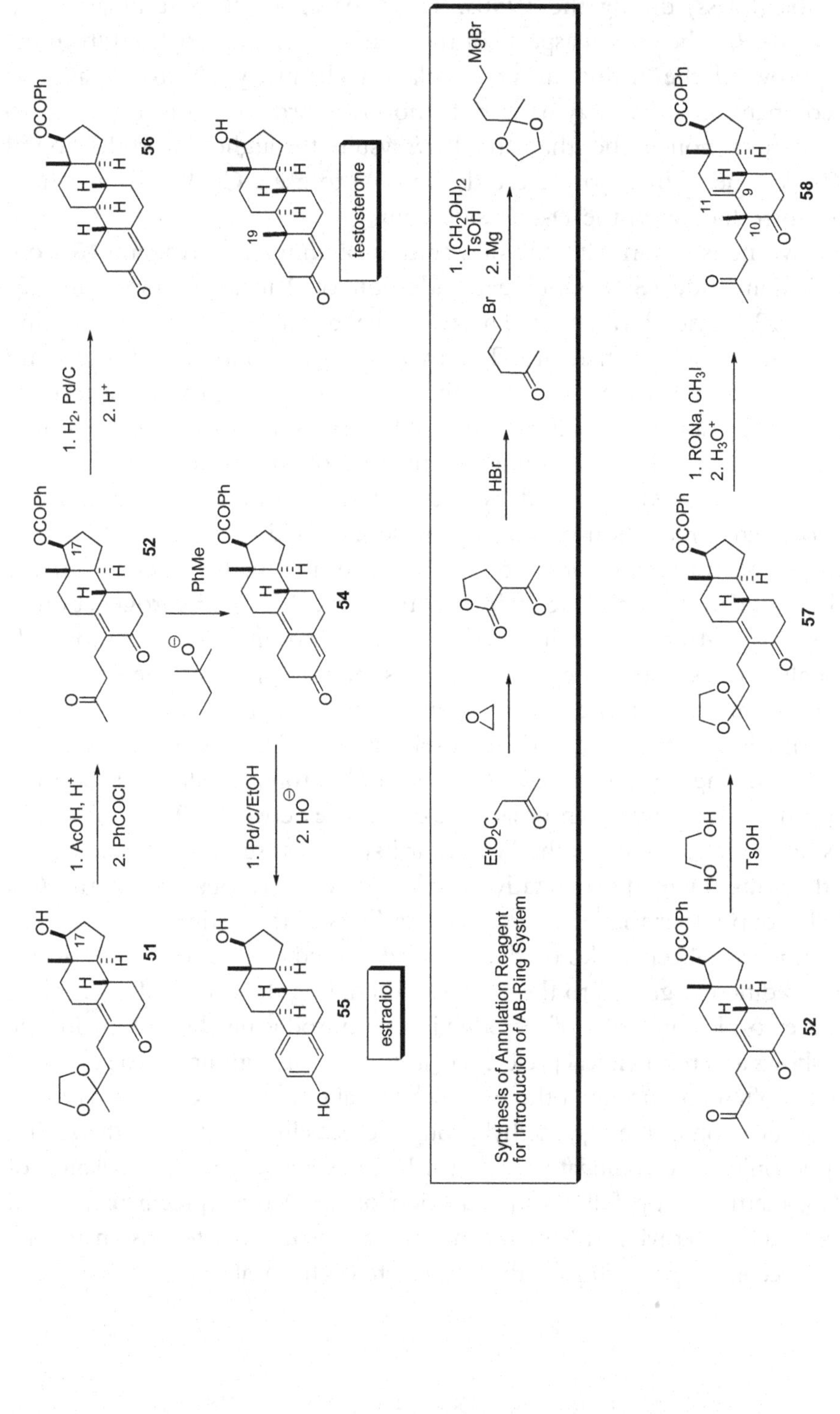
51
1. AcOH, H+
2. PhCOCl
52
1. H2, Pd/C
2. H+
56
PhMe
54
1. Pd/C/EtOH
2. HO
55
estradiol
testosterone
Synthesis of Annulation Reagent
for Introduction of AB-Ring System
HBr
1. (CH2OH)2
TsOH
2. Mg
MgBr
TsOH
57
1. RONa, CH3I
2. H3O+
58

Steroids-9

Steroids-9

The A-ring was introduced by hydrolysis of the acetate to provide D-ring alcohol **51**. Hydrolysis of the acetal and esterification of the C_{17} alcohol gave benzoate **52**. As we shall see, **52** proved to be a key intermediate in the synthesis of corticosteroids. But first let's take a little side trip to some simpler steroids. Treatment of **52** with *t*-amyloxide gave dienone **54** via an aldol-dehydration reaction. This steroid-like molecule is actually at the phenol oxidation state, and thus, Pd-mediated isomerization of the dienone followed by benzoate hydrolysis gave estradiol. On the other hand, catalytic hydrogenation of enone **52** followed by an acid-promoted aldol-dehydration gave enone **56**. This is the benzoate of 19-nortestosterone. Thus, this chemistry provided an efficient route to a number of steroid analogs.

Continuing with the path to cortisone, **52** was converted to acetal **57** (also available from **51**). Deconjugative alkylation of **57** gave **58** with the correct stereochemistry at C_{10} and the C_9-C_{11} double bond in place for use in corticosteroid synthesis. It appears that the stereochemical course of this alkylation involves "axial" attack of the electrophile (methyl iodide) on the intermediate dienolate. One might wonder why this stereochemistry was not also observed in Woodward's conversion of **15** to **16**. One possible explanation is that the conjugate addition might be reversible, and thus the reaction is controlled by thermodynamics, whereas the alkylation of **57** should clearly be contolled by kinetics. It is also possible that the blocking group in **15** exerts some influence that is not clearly understood. Finally, I note that it is possible that these reactions are not totally selective, and that information remains a trade secret.

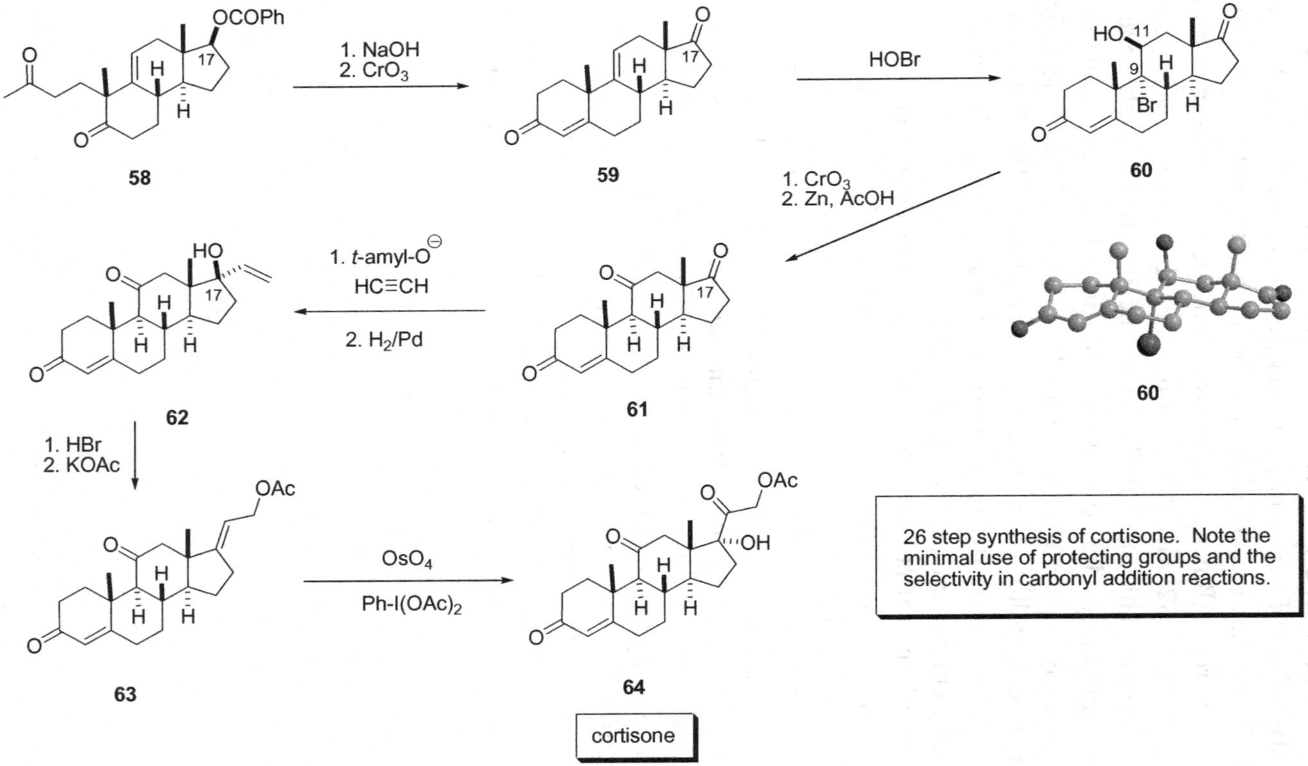

For a review with a good discussion of reactivity principles for 6-membered rings see: Velluz, L.; Walls, J.; Nomine, G. *Angew Chem. Int. Ed.* **1965**, *4*, 181–200

Steroids-10

The synthesis of cortisone continues with an aldol-dehydration reaction, ester hydrolysis, and oxidation of the resulting C_{17} alcohol to provide **59**. Treatment of **59** with HOBr results in addition to the most electrophilic of the two double bonds. Bromonium ion formation from the most accessible face of the olefin (recall the stereochemistry of epoxidation of **31**) followed by *anti*-periplanar opening of the intermediate with water afforded **60**. Oxidation of the C_{11} alcohol and reductive removal of the C_9 bromide gave **61**. Addition of acetylide to the C_{17} ketone followed by partial hydrogenation of the alkyne gave allylic alcohol **62**. The stereochemical course of the acetylide addition is typical for C_{17} steroidal ketones. Allylic rearrangement accompanied treatment of **63** with HBr and the resulting primary allylic bromide was converted to acetate **63**. Finally, treatment of **63** with osmium tetroxide and iodobenzene diacetate resulted in vicinal dihydroxylation of the most electrophilic olefin, and oxidation of the intermediate secondary alcohol, to provide cortisone acetate. Two remarkable features of this synthesis are the minimal use of protecting groups and the use of low-tech chemistry.

Biosynthesis of Steroids

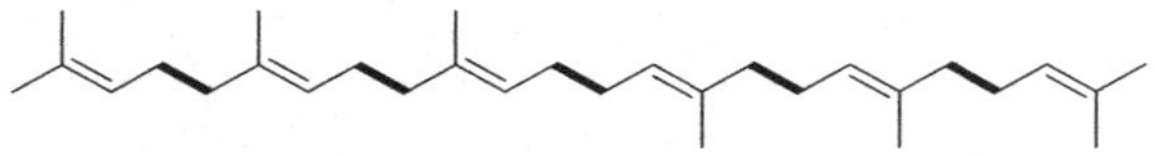

squalene

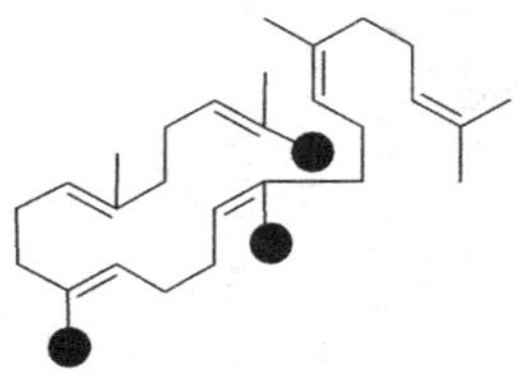

Sir Robert Robinson (1934)

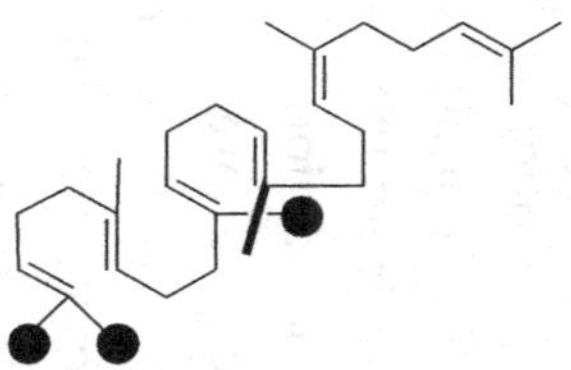

Woodward and Bloch (1953)

Labelling studies show that squalene is derived from acetic acid. 18 Carbons come from methyl groups and 12 carbonscome from carboxyl carbons. Whereas squalene has 30 carbons, cholesterol has only 27 carbons, and other steroids have even fewer carbons. Therefore “degradation” must accompany (or follow) cyclization. Robinson was the first to propose a manner in which squalene might cyclize to provide the tetracyclic ring system of the steroids. This was followed by an alternate hypothesis by Bloch and Woodward. The two proposals suggest different labelling patterns in tetracyclic products. For example, if one thinks about cholesterol as the product, the two proposals suggest that the “black” methyl groups would have to be lost. These are different for the two proposals. Experiments eventually disproved the Robinson hypothesis and supported the Bloch-Woodward hypothesis. For an early experiment see Woodward, R. B.; Bloch, K. “The Cyclization of Squalene in Cholesterol Synthesis” *J. Am. Chem. Soc.***1953**, *75*, 2023-2024. The following (abbreviated) pathway from squalene oxide to lanosterol is now standard in undergraduate textbooks.

squalene oxide (R = prenyl group)

Steroids-11

Natural products syntheses of numerous compounds have been inspired by nature. In other words, the pathway taken by nature to a given natural product often suggests a synthetic strategy that one might follow in the laboratory. Such syntheses are sometimes refered to as "biomimetic syntheses" and numerous examples permeate the literature. The notion of mimicing nature is quite old, but it is time-tested, often used, and still always stimulating to think about this approach to designing a laboratory synthesis of a new natural product.

There are at least two fundamental approaches that have been used to design biomimetic syntheses: (1) determine the biosynthetic pathway and mimic it as best possible and (2) imagine the biosynthetic pathway and develop the "biomimetic synthesis". The second approach is frequently followed because it is a lot of work to determine a biosynthetic pathway. There are a lot of well-established biosynthetic transformations (much like well-established laboratory reactions), however, and one can frequently propose a reasonable biosynthesis.

Steroids, and the terpenoids from which they are biosynthetically derived, were among the first natural products to be tackled in the laboratory via a biomimetic approach. Lanosterol, a triterpene (30 carbons), is the tetracyclic precursor of steroids such as cholesterol. Squalene is the biosynthetic precursor of lanosterol. Labelling studies show that squalene is derived from acetic acid, as are all terpenoids. Eighteen of the 30 carbons of squalene come from methyl groups (of acetic acid) and 12 carbons come from carboxyl carbons. Whereas squalene has 30 carbons, cholesterol has only 27 carbons, and other steroids have even fewer carbons. Therefore "degradation" must accompany (or follow) the cyclization of squalene to the tetracyclic triterpenes. Sir Robert Robinson was the first to propose a manner in which squalene might cyclize to provide the tetracyclic ring system of the steroids.[11] This was followed by an alternate hypothesis by Bloch and Woodward. The two proposals suggest different labelling patterns for tetracyclic products. For example, if one thinks about cholesterol as the product, the two proposals suggest that the "black" methyl groups would have to be lost (Steroids-11). These are different for the two proposals. Experiments eventually disproved the Robinson hypothesis and supported the Bloch-Woodward hypothesis (stereochemical details added by others). The abbreviated pathway from squalene to lanosterol is now standard in undergraduate textbooks (see Steroids-11 and Steroids-12).

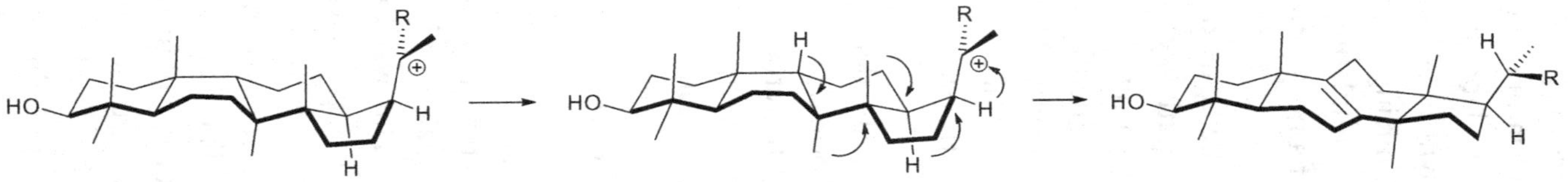

lanosterol

Stork-Eschenmoser Hypothesis

This biosynthetic pathway led a number of groups to suggest that polyolefin cyclizations might be used to prepare fused carbocycles including steroids. Of course the biosynthesis is mediated by enzymes that may predispose squalene oxide (or other isoprenoids) for cyclizations. Stork and Eschenmoser hypothesized that the stereoselective synthesis of terpenoids and steroids via polyolefin cyclizations should be possible if such cyclizations proceeded via synchronous processes. In a seminal paper Stork and Burgstahler said "It is the purpose of this paper to draw attention to the important conclusion that *concerted* cyclization to structurally identical hydronaphthalenes will result in a *trans* system if the bicyclic precursor is an open chain triene convertible to a cation of type **B** and a *cis* product if the precursor is monocyclic (type **A**). If the reactions are not concerted, mixtures of products will result".

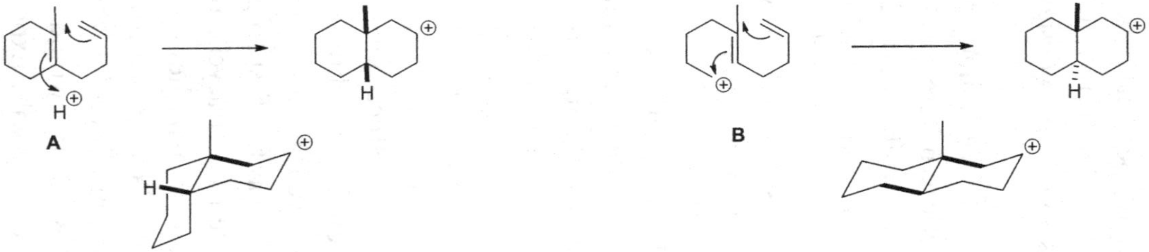

Stork, G.; Burgstahler, A. W. "The Stereochemistry of Polyene Cyclization" *J. Am. Chem. Soc.* **1955**, *77*, 5068. Eschenmoser, A.; Ruzicka, L.; Jeger, O.; Arigoni, D. "On Triterpenes: A Stereochemical Interpretation of the Biogenetic Isoprene Rule for the Triterpenes" *Helv. Chim. Acta* **1955**, *38*, 1890-1904 (original in German). For an English translation and an interested update see Eschenmoser, A.; Arigoni, D. "Revisited after 50 Years: The Stereochemical Interpretation of the Biogenetic Isoprene Rule for the Triterpenes" *Helv. Chim. Acta* **2005**, *88*, 3011–3050.

Steroids-12

This biosynthetic pathway to the steroids led a number of groups to suggest that polyolefin cyclizations might be used to prepare fused carbocycles including steroids. Of course the biosynthesis is mediated by enzymes that may predispose squalene oxide (or other isoprenoids) for cyclizations. In 1955, Stork and Eschenmoser independently hypothesized that the stereoselective synthesis of terpenoids and steroids via polyolefin cyclizations should be possible if such cyclizations proceeded via synchronous processes. There were many other players in the game and I will not attempt to tell the story in any detail. Suffice it to say, in a seminal paper, Stork and Burgstahler (then at Harvard University) said "It is the purpose of this paper to draw attention to the important conclusion that concerted cyclization to structurally identical hydronaphthalenes will result in a trans system if the bicyclic precursor is an open chain triene convertible to a cation of type **B**, and a cis product if the precursor is monocyclic (type **A**). If the reactions are not concerted, mixtures of products will result". At the same time, Eschenmoser, Ruzicka, Jeger and Arigoni (ETH-Switzerland) made comparable suggestions about the steric course of acid-catalyzed cyclizations in their landmark paper that discussed the biogenetic origins of triterpenes from polyolefins (squalene). This is a fascinating story that has been elegantly described by Eschenmoser and Arigoni on the 50th anniversary of these papers. What I will present next is an attempt by one group (that of William S. Johnson at Wisconsin and Stanford) to translate into practice the notion that one might make steroids in the lab by polyolefin cyclizations. During the course of the next few pages we will see how an idea can evolve from simple beginnings to a high level of complexity. Whereas the syntheses we will ultimately see do not follow the precise pathway nature takes to the steroids, they do revolve around stereoselective polyolefin cyclizations, and in that manner, can be called "biomimetic syntheses".[12]

Polyolefin Cyclization Route to Steroid Total Synthesis

Many groups have contributed to this research area, but the most extensive contributions come from the group of William S. Johnson (Wisconsin and Stanford) and we will focus on those contributions. His research had modest beginnings and it is worth a look at the "evolution" of his approach to steroids.

Johnson, W. S.; Bailey, D. M.; Owyang, R.; Bell. R. A.; Jaques, B.; Crandall, J. K. "Cationic Cyclizations Involving Olefinic Bonds. II. Solvolysis of 5-Hexenyl and *trans*-5,9-Decadienyl *p*-Nitrobenzenesulfonates" *J. Am. Chem. Soc.* **1964**, *86*, 1959.

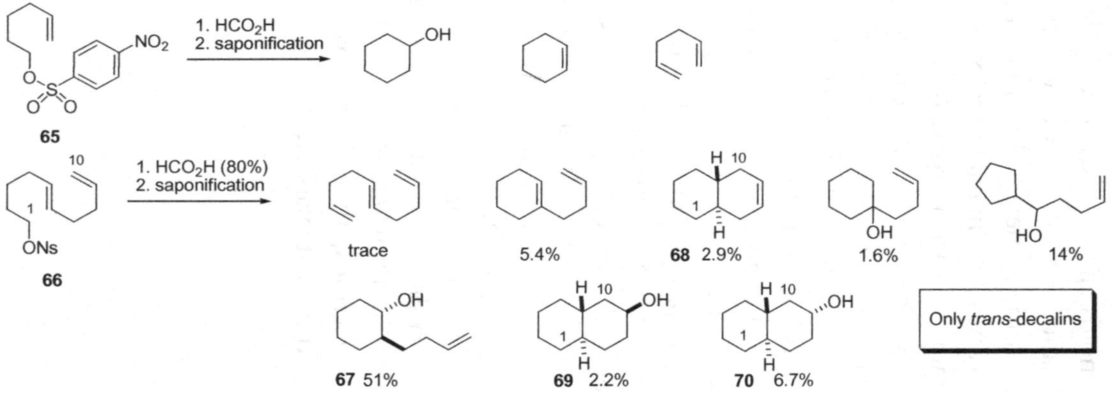

Only *trans*-decalins

Johnson, W. S.; Lunn, W. H.; Fitzi, K. "Cationic Cyclizations Involving Olefinic Bonds. IV. The Butenylcyclohexenol System" *J. Am. Chem. Soc.* **1964**, *86*, 1972.

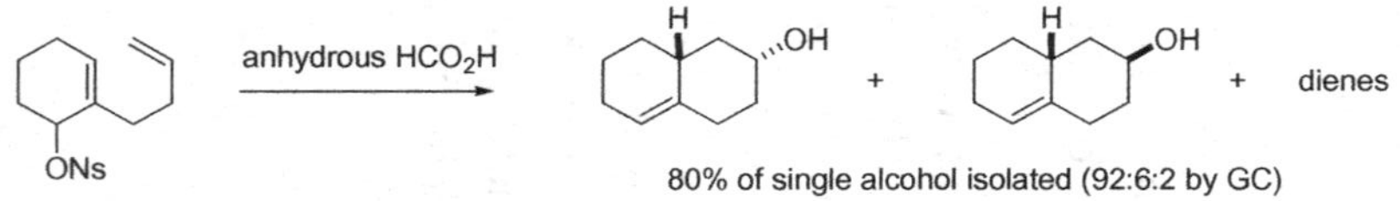

80% of single alcohol isolated (92:6:2 by GC)

Hint of a new cyclization initiation group

Steroids-13

The Johnson group began by examing the solvolysis of 5-hexenyl nosylate (**65**) in formic acid. This reaction gave only small amounts of cyclohexanol and cyclohexene, but illustrated that that cyclization was possible. Solvolysis of diene **66** gave cyclohexanol **67** as the major product. This product results from *anti*-addition of the electrophilic carbon (C_1) and oxygen nucleophile (formate) across the *trans*-olefin. Decalins **68–70** were produced in low yields, but it was notable that only *trans*-fused decalins were produced. This observation was consistent with an *anti*-addition of electrophilic carbon (C_1) and nucleophilic carbon (C_{10}) across the *cis*-olefin (see **B** in Steroids-12).

Johnson, W. S.; Crandall, J. K. "Cationic Cyclizations Involving Olefinic Bonds. V. Solvolysis of *cis*-5,9-Decadienyl *p*-Nitrobenzenesulfonate" *J. Am. Chem. Soc.* **1964**, *86*, 2085.

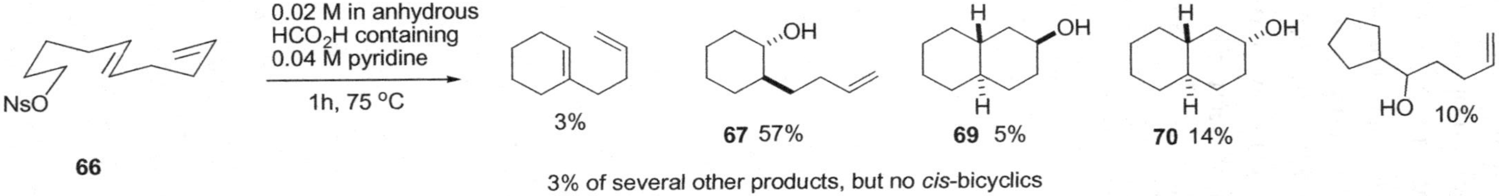

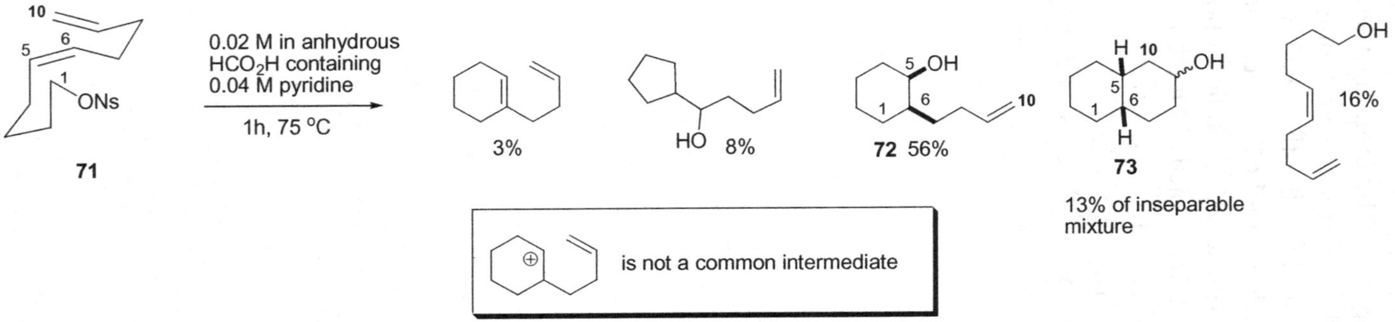

Johnson, W. S.; Owyang. R. "Olefinic Cyclizations. VI. Formolysis of Some Branched-Chain Alkenyl *p*-Nitrobenzenesulfonates" *J. Am. Chem. Soc.* **1964**, *86*, 5593. Johnson, W. S.; Crandall, J. K. "Olefinic Cyclizations. VII. Formolysis of *cis*- and *trans*-5,9-Decadienyl *p*-Nitrobenzenesulfonate and of Some Isomeric Monocyclic Esters" *J. Org. Chem.* **1965**, *30*, 1785–1790 (full paper of earlier work).

Steroids-14

Finally, solvolysis of diene **71** gave cyclohexanol **72** and *cis*-decalins **73** as the most notable products (Steroids-14). Once again, the stereochemistry of these products suggest that *anti*-addition of electrophile and nucleophile across the *cis*-olefin was dictating the stereochemical course of cyclization reactions. This experiment also showed that the solvolyses of **66** and **71** do not pass through a common monocyclic intermediate.

More Cation Cyclization Initiators

Johnson, W. S.; Neustaedter, P. J.; Schmiegel, K. K. "Olefinic Cyclizations. VIII. The Butenylmethylcyclohexanol System" *J. Am. Chem. Soc.* **1965**, *87*, 5148–5157.

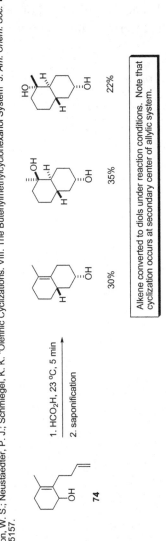

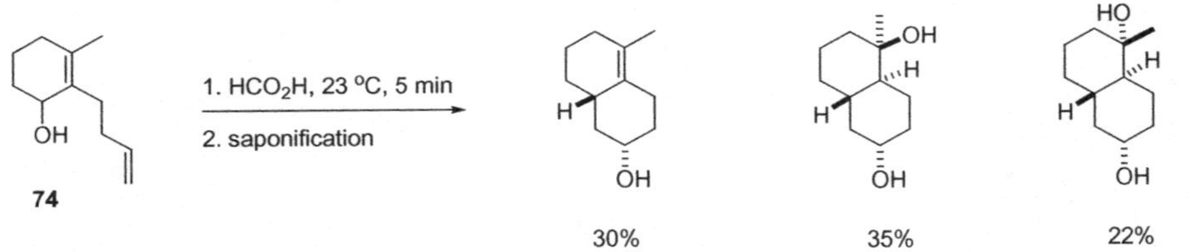

Alkene converted to diols under reaction conditions. Note that cyclization occurs at secondary center of allylic system.

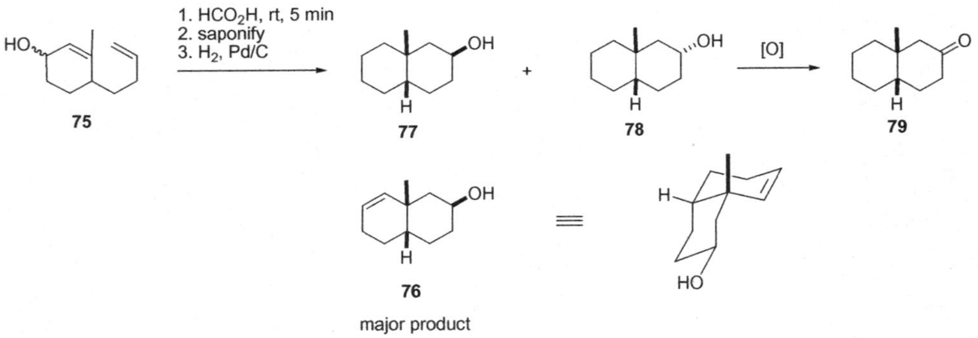

Steroids-15

The Johnson group also examined other cyclization initiators. For example the use of allylic alcohols gave high yields of monocyclization products (bottom of Steroids-13) via a presumably symmetrical allylic carbocation. Unsymmetrical carbocations derived from allylic alcohols **74** and **75** led to mixtures of products, but once again in good yield and with good stereocontrol. For example, solvolysis of **75** gave enol **76** as the major product after formate ester saponification. Catalytic hydrogenation of the olefin and oxidation of the secondary alcohol converted all cyclization products to *cis*-decalone **79**.

Advances Based on Improved Tactics

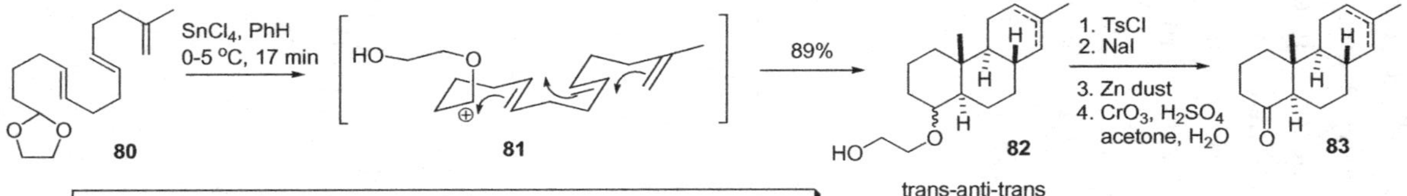

First example of tricycle formation with stereocontrol across two ring fusions

Johnson, W. S.; Kinnel, R. B. "Stereospecific Tricyclization of a Polyolefinic Acetal" *J. Am. Chem. Soc.* **1966**, *88*, 3861–3862

Synthesis of a Steroid

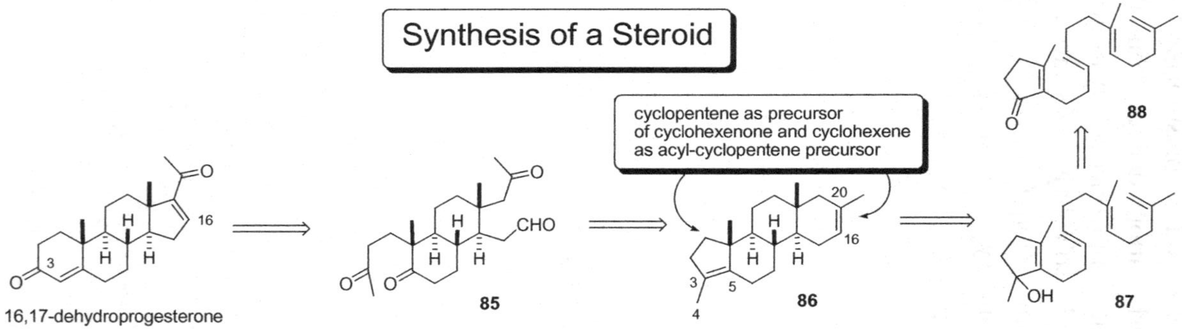

Notice elements of the annulation methodology finding its way into the strategy.

Johnson, W. S.; Semmelhack, M. F.; Sultanbawa, M. U. S.; Dolak L. A. "A New Approach to Steroid Total Synthesis. A Nonenzymatic Biogenetic-Like Olefinic Cyclization Involving the Stereospecific Formation of Five Asymmetric Centers" *J. Am. Chem. Soc.* **1968**, *90*, 2994–2996.

Steroids-16

A series of studies were conducted to develop new tactics for conducting polyolefin cyclizations. New initiating groups, solvents and acid promotors were examined. For example, acetal **80** was efficiently converted to **82**, whose tricyclic structure resembles the ABC-ring system of the steroids, presumably via an intermediate cation of type **81** (Steroids-16). Advances of this type set the stage for the total synthesis of steroids via polyolefin cyclizations. We will focus largely on progesterone (see Steroids-1 or Steroids-19 for structure).

Let's look at the Johnson group's plan in a retrosynthetic manner. The penultimate (next to last) intermediate in one approach to progesterone was to be 16,17-dehydroprogesterone (**84**). This compound was to be derived from tetracarbonyl compound **85** via a double aldol-dehydration. This highly functionalized compound was to be packaged, in a relatively benign, manner as diene **86**. This strategy contains some familiar elements: (1) use of a cyclohexene precursor to the D-ring (2) use of aldol-dehydration chemistry to prepare the A-ring enone. The cyclopentene to cyclohexenone transformation is also interesting. We have previously seen a cyclohexene used as a precursor of a 5-membered ring, and now we see (or will see) that a cyclopentene can be used to rapidly prepare a 6-membered ring. This is general transformation. At this point the retrosynthetic plan diverges from what we have previously seen, and converges with Johnson's polyolefin cyclization studies. The plan was to use a symmetrical allylic carbocation initiator (to be generated from **87**). Cyclization was expected to occur via a poly-chairlike conformation to add the BCD-rings to the cyclopentene (the A-ring precursor). Termination of the cyclization by loss of a proton from C_{16} was expected to provide **86**. It is notable that the regiochemistry of this final proton loss was not guaranteed. Tertiary alcohol **87** was expected to come from cyclopentenone **88**. We will not consider the strategy for construction of **88** in detail, but simply note that the key issues are construction of the cyclopentenone (which was to be accomplished by aldol-dehydration of a 1,4-diketone) and construction of the *E*-1,2-disubstituted and *E*-trisubstituted olefins. Let's go directly to the synthesis.

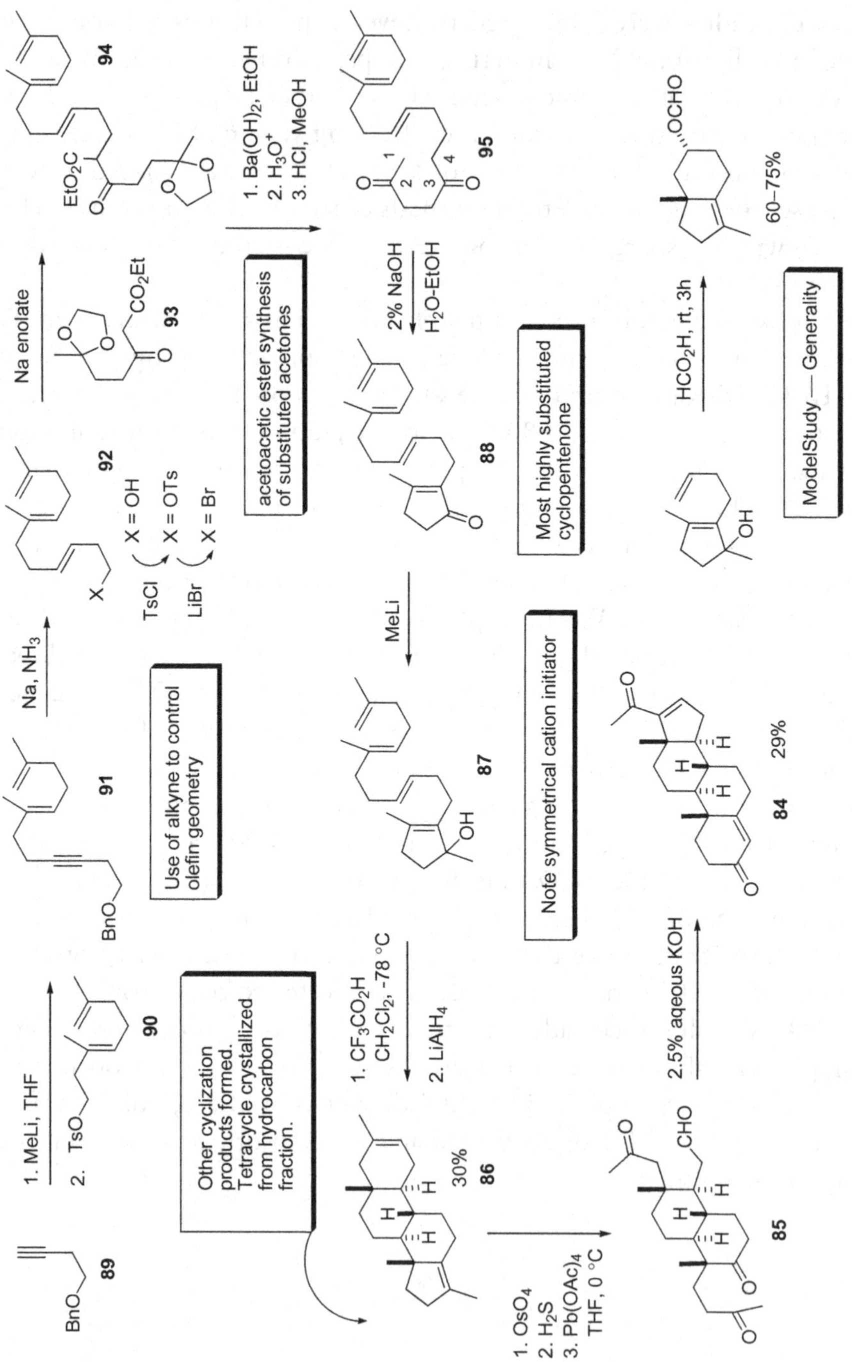

BnO
89
1. MeLi, THF
2. TsO
90
BnO
91
Na, NH3
Use of alkyne to control olefin geometry
X
X = OH
TsCl
X = OTs
LiBr
X = Br
92
Na enolate
93
CO2Et
EtO2C
94
1. Ba(OH)2, EtOH
2. H3O+
3. HCl, MeOH
acetoacetic ester synthesis of substituted acetones
95
2% NaOH
H2O-EtOH
88
Most highly substituted cyclopentenone
MeLi
87
OH
Other cyclization products formed. Tetracycle crystallized from hydrocarbon fraction.
1. CF3CO2H CH2Cl2, -78 °C
2. LiAlH4
86
30%
Note symmetrical cation initiator
1. OsO4
2. H2S
3. Pb(OAc)4 THF, 0 °C
CHO
85
2.5% aqeous KOH
84
29%
HCO2H, rt, 3h
OCHO
60–75%
ModelStudy — Generality

Steroids-17

Steroids-17

We have seen that in polyolefin cyclizations, olefin stereochemistry translates to fused-carbocycle stereochemistry (for example **80** to **82**). Thus, this approach to steroids (and fused carbocycles in general) requires methods for stereoselective olefin synthesis. We will see some examples of such methodology on the next few pages, and will revisit this topic throughout the book (in particular see the *Cecropia* Juvenile Hormone syntheses in Chapter 11). In the current case, the *E*-disubstituted olefin was prepared using the well-known dissolving metal reduction of alkynes. Terminal alkyne **89** was converted to the corresponding acetylide which reacted with homoallylic tosylate **90** to give **91**. Sodium in ammonia was then used to reduce the alkyne and remove the benzyl protecting group to provide **92** (X = OH). The alcohol was converted to the corresponding bromide **92** (X = Br) via an intermediate tosylate, and alkylation of **93** with this bromide then gave **94**. Ester hydrolysis, decarboxylation, and ketal hydrolysis provided 1,4-diketone **95**. The conversion of **92** to **95** is a variation of the "acetoacetic ester synthesis of substituted acetones". This tactic uses a *β*-dicarbonyl as the nucleophile, rather than an unactivated ketone, to control enolate regiochemistry and problems with proton transfers that often complicate alkylations of simple ketone enolates (also compare with the conversion of **5** to **4** on Steroids-3).[13] Aldol-dehydration of **95** provided **88** as the major product (C_4 as nucleophile and C_2 as electrophile) with possibly a trace of the product derived from the alternate aldol-dehydration. Addition of methyllithium to **88** gave **87**. Treatment of **87** with trifluoroacetic acid gave a variety of products from which **86** was crystallized in 30% yield (after conversion of trifluoroacetates to alcohols for separation purposes; additional **86** was available by dehydration of tetracyclic alcohols obtained from the cyclization of **87**; tricyclic products were also formed). The synthesis of **84** was completed by cleavage of the olefins and aldol-dehydration of the intermediate **85**.

This biomimetic route to progesterone was elegant, but there was room for improvement. For example, could a cyclization terminating group be developed that would lead directly to a 5-membered D-ring? One solution to this problem was to replace the terminating olefin with an alkyne. The idea was that if the alkyne cyclized to provide a 5-membered ring, capture of an intermediate vinyl cation would directly provide the C_{17}-acetyl group of progesterone (see cyclization substrate **99** in Steroids-18).

Total Synthesis of *dl*-Progesterone

This synthesis addresses problems associated with construction of the D-ring in the synthesis of *dl*-16,17-dehydroprogesterone. Can one terminate the polyolefin cyclization to directly afford a 5-membered ring?

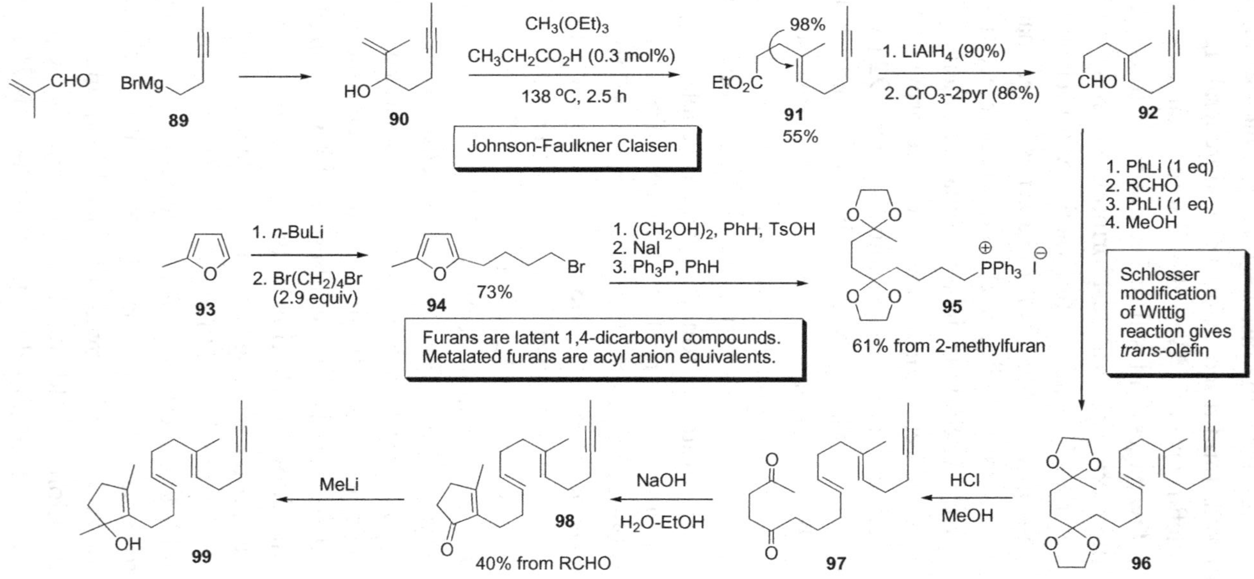

Johnson, W. S.; Gravestock, M. B.; McCarry, B. E. "Acetylenic Bond Participation in Biogenetic-Like Olefinic Cyclizations. II. Synthesis of *dl*-Progesterone" *J. Am. Chem. Soc.* **1971**, *93*, 4332–4334

Steroids-18

The synthesis of the cyclization substrate (**99**) began with addition of Grignard reagent **89** to methacrolein to provide allylic alcohol **90**. A variation of the Claisen rearrangment, developed by the Johnson group for the synthesis of squalene and other terpenoids, was used to establish stereochemistry in trisubstituted olefin **91**.

Once again I will digress. Pattern recognition plays a big role in synthesis design. Realization that **91** is a δ,γ-unsaturated carbonyl compound, and knowing that such substructures can always be prepared by a Claisen rearrangement (although this does not mean it is always the best method to use) probably played a role in the choice of this approach to **99**. There is actually another way to recognize that a Claisen rearrangement might be considered as a path from **90** to **91**. Do you want to do an S_N2' reaction on an allylic alcohol? Always consider a sigmatropic rearrangment as a possibility. This process "enforces" the regiochemistry of such a substitution reaction. The Claisen rearrangement is like an "enforced" S_N2' reaction of an enolate on an allylic alcohol derivative.

Returning to the synthesis, an oxidation state adjustment gave aldehyde **92**, which was then coupled with phorphorane **95** to provide *E*-olefin **96**. Wittig reactions between "unstabilized phorphoranes" such as **95** and aldehydes usually provide *Z*-alkenes. The Schlosser modification, however, is a nice variation that provides *E*-olefins with good stereoselectivity, and that is what was used in this case.[14] The origin of **95** is also interesting. The masked 1,4-dicarbonyl group in **95** has its origin in the choice of 2-methylfuran (**93**) as the point of departure. Hydrolysis of the two "enol ethers" present in furan liberates the 1,4-dicarbonyl. Finally, the organolithium intermediate derived from **93** is a nice example of how 2-metalated furans can function as acyl anion equivalents.[15] Conversion of **96** to **99** followed chemistry we have already seen.

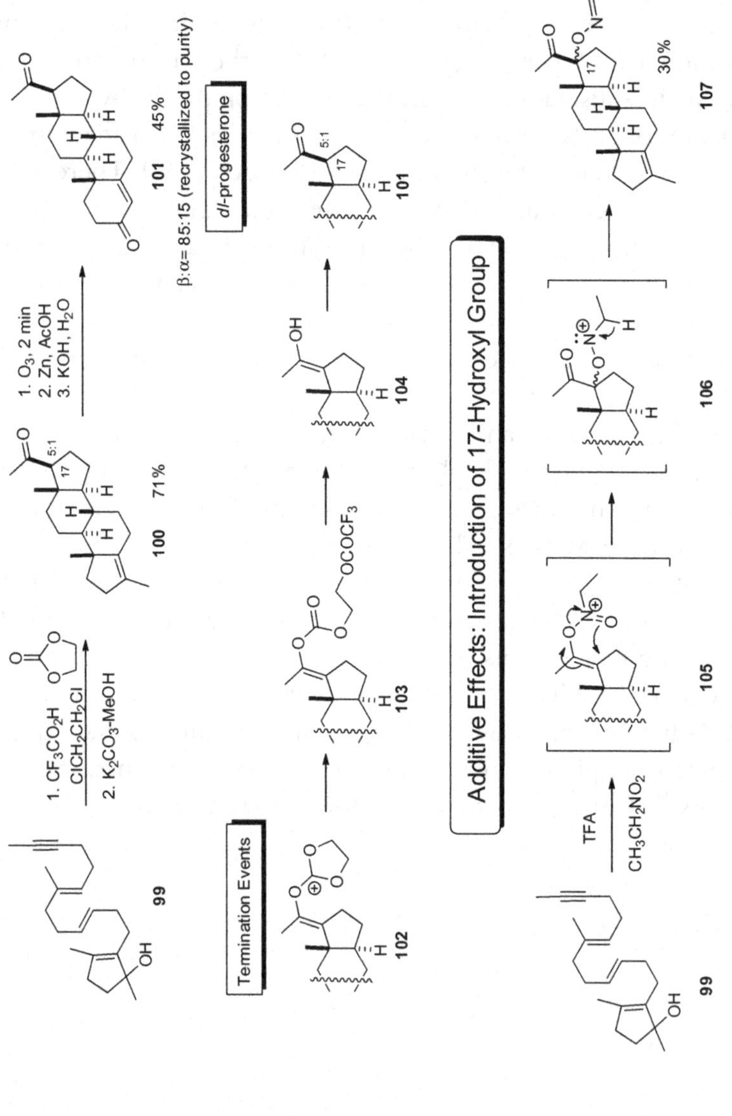

Morton, D. R.; Gravestock, M. B.; Parry, R. J.; Johnson, W. S. "Acetylenic Bond Participation in Biogenetic-Like Olefinic Cyclizationos in Nitroalkane Solvents. Synthesis of the 17-Hydroxy-5β-pregnan-2-one System" *J. Am. Chem. Soc.* **1973**, *95*, 4418–4419.

Steroids-19

Steroids-19

The cyclization of **99**, initiated using trifluoroacetic acid and terminated using an excess of ethylene carbonate, gave **100** in 71% as a 5:1 mixture at C_{17}. The termination events presumably involve capture of an intermediate vinylic carbocation to **103** via highly stabilized carbocation **102**. Hydrolysis of the trifluoroacetate, an intramolecular acyl transfer, and tautomerization of the resulting enol **104** provided the products. The synthesis of progesterone (**101**) was then completed using the now familiar transformation of the cyclopentene to the A-ring enone. Use of nitroethane as a solvent also led to an interesting termination event with introduction of a C_{17} hydroxyl group (as an oxime ether and mixture of stereoisomers), presumably via the sequence shown at the bottom of Steroids-19.

Corticosteroids

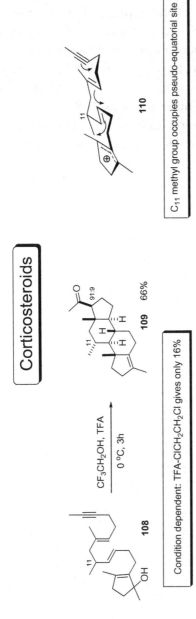

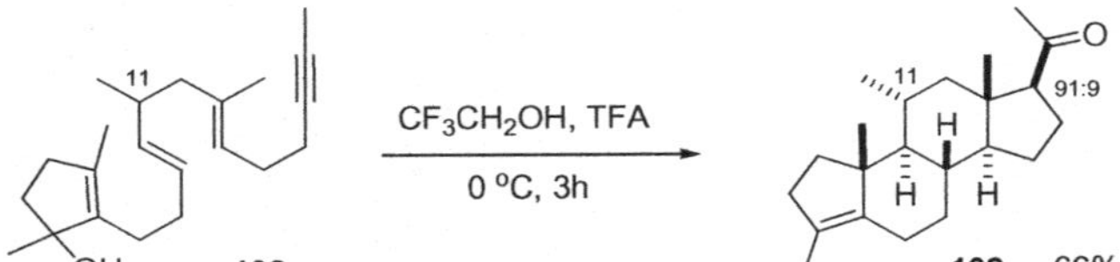

108 **109** 66%

110

Condition dependent: TFA-ClCH$_2$CH$_2$Cl gives only 16%

C_{11} methyl group occupies pseudo-equatorial site

Johnson, W. S.; DuBois, G. E. "Biomimetic Polyene Cyclizations. Asymmetric Induction by a Chiral Center Remote from the Initiating Cationic Center. 11α-Methylprogesterone" *J. Am. Chem. Soc.* **1976**, *98*, 1038–1039.

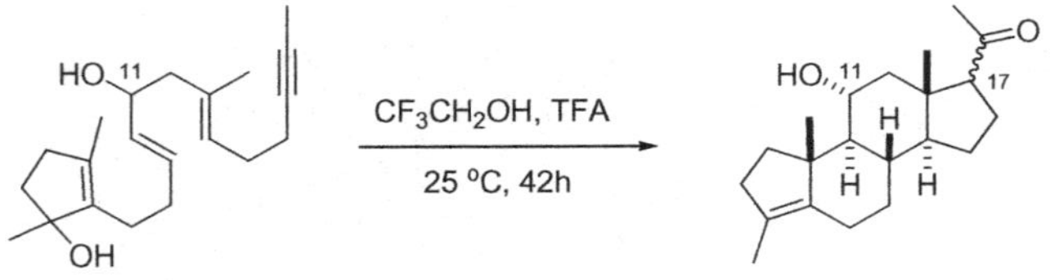

111 **112** 29-35%

Johnson, W. S.; Escher, S.; Metcalf, B. W. "A Stereospecific Total Synthesis of Racemic 11α-Hydroxyprogesterone via a Biomimetic Polyene Cyclization" *J. Am. Chem. Soc.* **1976**, *98*, 1039–1041.

Steroids-20

The Johnson group next adapted their polyolefin cyclizations to accommodate preparation of corticosteroids. Initial work directed at this goal involved the preparation and cyclization of *rac*-**108**. This material contains one stereogenic center (C_{11} methyl group). The idea was to see if this single stereogenic center would induce relative stereochemistry at the six stereogenic centers formed in the cyclization. Indeed this was the case and, in accord with prediction, the C_{11} methyl group was equatorially disposed in the product (**109**) (and presumably the cyclization transition state **110**). When the C_{11} methyl group was replaced with a hydroxyl group (*rac*-**111**), the cyclization gave *rac*-**112** with an α-hydroxyl group at C_{11}. The cyclization of **111** was much slower than the cyclization of **108**, suggesting a build-up of positive charge near the electron-withdrawing hydroxyl group during the rate-determining events of the cyclization process.

Asymmetric Synthesis of C_{11}-Oxygenated System

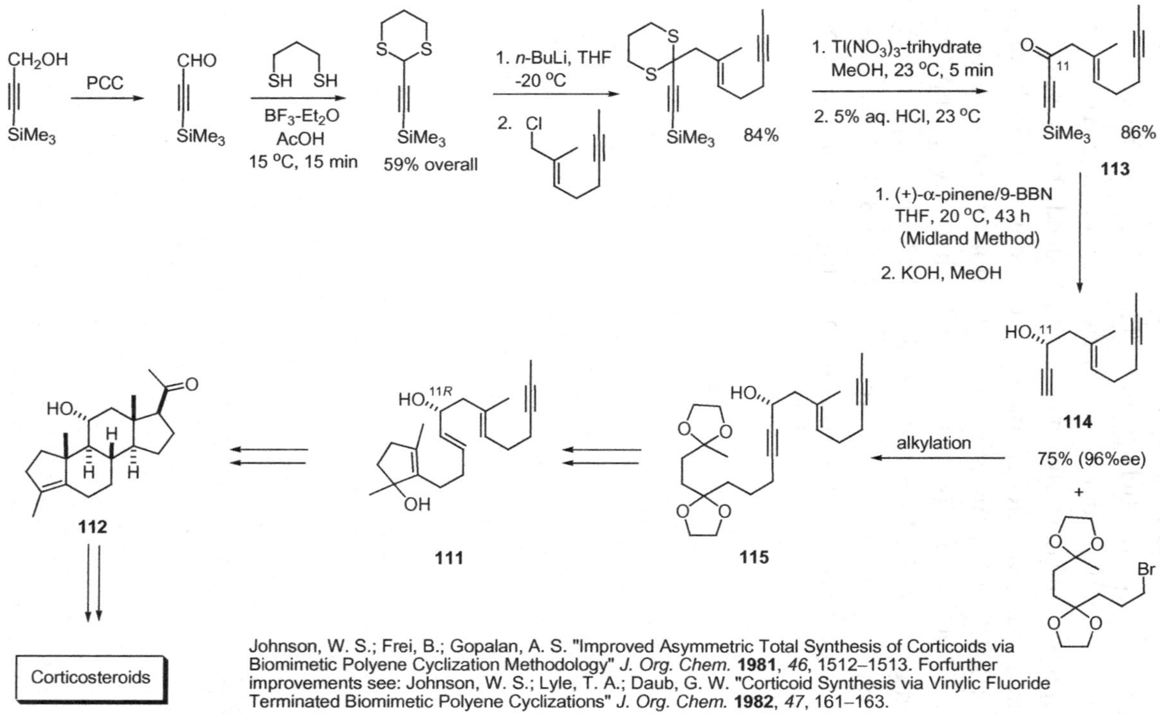

Johnson, W. S.; Frei, B.; Gopalan, A. S. "Improved Asymmetric Total Synthesis of Corticoids via Biomimetic Polyene Cyclization Methodology" *J. Org. Chem.* **1981**, *46*, 1512–1513. Forfurther improvements see: Johnson, W. S.; Lyle, T. A.; Daub, G. W. "Corticoid Synthesis via Vinylic Fluoride Terminated Biomimetic Polyene Cyclizations" *J. Org. Chem.* **1982**, *47*, 161–163.

Steroids-21

The synthesis of **112** was then modified to provide a single enantiomer. This called for an asymmetric synthesis of cyclization substrate **111**. This was accomplished by Midland reduction of ketone **113** to provide **114** with excellent enantioselectivity (Steroids-21). Alkylation of **114** with the appropriate bromide (prepared from 2-methylfuran according to the procedures described on Steroids-18), followed by a few well-precedented reactions, gave **115**, and thence **111** and **112**. Application of the Midland reduction is notable. This is a relatively early application of a reagent-controlled asymmetric synthesis. It is also notable that the Midland method works extremely well on alkyl alkynyl ketones (because they look like aldehydes to the reagent) and thus, is well-suited to this application.[16]

Effect of Cation Stabilizing Groups

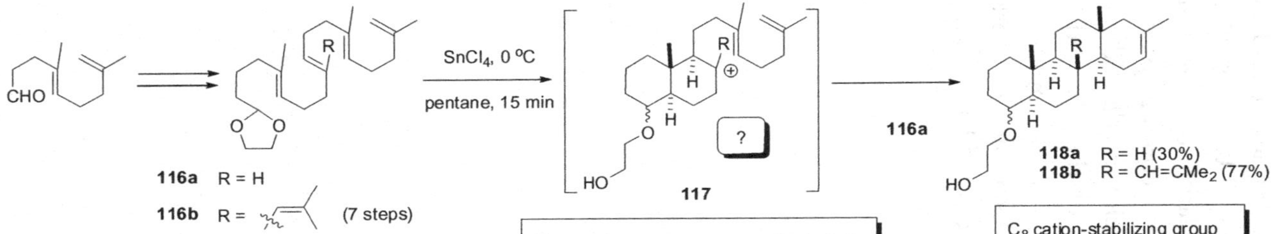

Suggests enzymes may operate similarly

C_8 cation-stabilizing group increases yield in cyclization

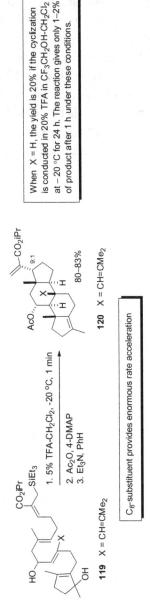

Johnson, W. S. Telfer, S. J.; Cheng, S.; Schubert, U. "Cation-Stabilizing Auxilliaries: A New Concept in Biomimetic Polyene Cyclization" *J. Am. Chem. Soc.* **1987**, *109*, 2517–2518. An extension to another cyclization system and the corticosteroids is shown below: Johnson, W. S.; Lindell, S. D.; Steele, J. *J. Am. Chem. Soc.* **1987**, *109*, 5852.

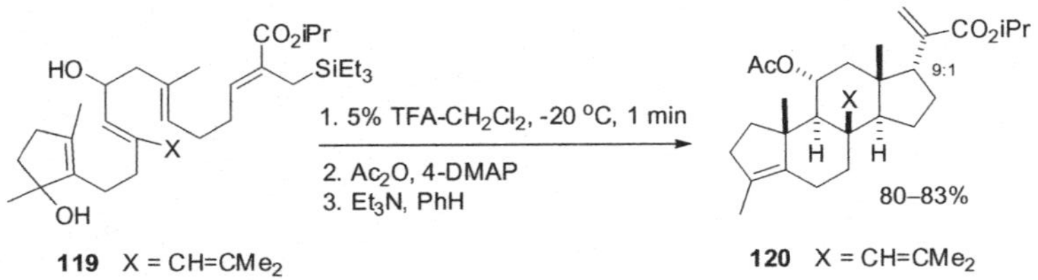

When X = H, the yield is 20% if the cyclization is conducted in 20% TFA in CF_3CH_2OH-CH_2Cl_2 at – 20 °C for 24 h. The reaction gives only 1–2% of product after 1 h under these conditions.

C_8-substituent provides enormous rate acceleration

Steroids-22

We will end this section with the series of studies shown on Steroids-22. These studies were designed to determine the effect of a carbocation-stabilizing group at C_8 on the rate of steroid-targeted polyolefin cyclizations. The effects were remarkable. The influence on yield is seen in the cyclizations of **116a** and **116b**. The substrate with carbocation-stabilizing 2-methylpropenyl group at C_8 (**116b**) gave a much higher yield of tetracycle (**118b**) than did the "parent" substrate **116a**. This observation was translated to a system relevant to the corticosteroids. Thus **119** was converted to **120** in excellent yield. This is notable because of the low yield (slow cyclization rates) encountered with 11-hydroxy substrate **111**, and the poor performance of the "parent" substrate related to **119** (see Steroids-22). The conversion of **119** to **120** also illustrates yet another termination tactic, the use of an allylsilane to direct the regiochemical course of D-ring formation.

References and Notes

1. Anner, G.; Miescher, K. "Steroids. LXXIII. Total Synthesis of Natural Estrone" *Experientia* **1948**, *4*, 25–26. Bachmann, W. E.; Cole, W.; Wilds, A. L. "Total Synthesis of the Sex Hormone Equilenin" *J. Am. Chem. Soc.* **1939**, *61*, 974–975. Johnson, W. S.; Petersen, J. W.; Gutsche, C. D. "A New Method of Producing Fused Ring Structures Related to the Steroids. Synthesis of Equilenin" *J. Am. Chem. Soc.* **1945**, *67*, 2274–2275.
2. For an introduction to "retrosynthetic analysis" see Corey, E. J.; Cheng, X-M. "The Logic of Chemical Synthesis", John Wiley and Sons, 1989 (pages 5–16)
3. For some early reviews that the author (DJH) has found useful when teaching the fundamentals of the Diels-Alder reaction see: Sauer, J. "Diels-Alder Reactions. II. Reaction Mechanism" *Angewandte Chem. Int. Ed. Eng.* **1967**, *6*, 16–33. Sauer, J.; Lang, D.; Wiest, H. "Diels-Alder Reaction. II. The Addition Capacity of Cis-Trans Isomeric Dienophiles in Diene Additions" *Chem. Ber.* **1964**, *97*, 3208–3218. Sauer, J.; Wiest, H.; Mielert, A. "Diels-Alder Reaction. I. Reactivity of Dienophiles Towards Cyclopentadiene and 9,10-Dimethylanthracene" *Chem. Ber.* **1964**, *97*, 3183–3207.
4. Jung, M. E. "A Review of Annulation" *Tetrahedron* **1976**, *32*, 3–31.
5. du Feu, E. C.; McQuillin, F. J.; Robinson, R. "Synthesis of Substances Related to the Sterols. XIV. A Simple Synthesis of Certain Octalones and Ketotetrahydrohydrindenes which may be of Angle-Methyl-Substituted Type. A Theory of the Biogenesis of the Sterols" *J. Chem. Soc.* **1937**, 53–60.
6. Bergmann, E. D.; Ginsburg, D.; Pappo, R. "The Michael Reaction" *Organic Reactions* **1959**, *10*, 179–555.
7. Van Rheenan, V.; Kelly, R. C.; Cha, D. Y. "An Improved Catalytic Osmium Tetroxide Oxidation of Olefins to *cis*-1,2-Glycols using Tertiary Amine Oxides as the Oxidant" *Tetrahedron Lett.* **1976**, *17*, 1973–1976.
8. Wieland, H. "Hydrogenation and Dehydrogenation" *Ber.* **1912**, *45*, 484–493. Barbier, P.; Locquin, R. "Method of Decomposing Various Saturated Mono- and Dibasic Acids" *Comptes Rend.* **1913**, *156*, 1443–1446.
9. Hirschmann, R. "The Cortisone Era: Aspects of its Impact. Some Contributions of the Merck Laboratories" *Steroids* **1992**, *57*, 579–592.
10. Hajos, Z. G.; Parrish, D. R. "Stereocontrolled Synthesis of *trans*-Hydrindan Steroidal Intermediates" *J. Org. Chem.* **1973**, *38*, 3239–3243.
11. Robinson, R. "Structure of Cholesterol" *Chem. Ind. (London)* **1934**, 1062–1063.
12. Johnson, W. S. "A Fifty-Year Love Affair with Organic Chemistry" in Profiles, Pathways, and Dreams, American Chemical Society (Washington DC), **1998** (229 pages)

13. House, H. O.; Trost, B. M. "The Chemistry of Carbanions. IX. The Potassium and Lithium Enolates Derived from Cyclic Ketones" *J. Org. Chem.* **1965**, *30*, 1341–1348. House, H. O.; Trost, B. M. "The Chemistry of Carbanions. X. The Selective Alkylation of Unsymmetrical Ketones" *J. Org. Chem.* **1965**, *30*, 2502–2512.
14. Vedejs, E.; Peterson, M. J. "The Wittig Reaction: Stereoselectivity and a History of Mechanistic Ideas (1953–1995)" *Advances in Carbanion Chemistry*, **1996**, *2*, 1–85. Vedejs, E.; Peterson, M. J. "Stereochemistry and Mechanism in the Wittig Reaction" *Topics in Stereochemistry* **1994**, *21*, 1–157. Maryanoff, B. E.; Reitz, A. B. "The Wittig Olefination Reaction and Modifications Involving Phosphoryl-Stabilized Carbanions. Stereochemistry, Mechanism, and Selected Synthetic Aspects" *Chemical Reviews* **1989**, *89*, 863–927.
15. Meyers, A. I. "Heterocycles in Organic Synthesis" Wiley-Interscience, **1974** (332 pages). Degl'Innocenti, A.; Pollicino, S.; Capperucci, A. "Synthesis and Stereoselective Functionalization of Silylated Heterocycles as a New Class of Formyl Anion Equivalents" *Chemical Communications* **2006**, 4881–4893. Yus, M.; Najera, C.; Foubelo, F. "The Role of 1,3-Dithianes in Natural Product Synthesis" *Tetrahedron* **2003**, *59*, 6147–6212. Albright, J. D. "Reactions of Acyl Anion Equivalents Derived from Cyanohydrins, Protected Cyanohydrins and α-Dialkylaminonitriles" *Tetrahedron* **1983**, *39*, 3207–3233. Seebach, D.; Corey, E. J. "Generation and Synthetic Applications of 2-Lithio-1,3-dithianes" *J. Org. Chem.* **1975**, *40*, 231–237.
16. Midland, M. M.; Greer, S.; Tramontano, A.; Zderic, S. A. "Chiral Trialkylborane Reducing Agents. Preparation of 1-Deuterio Primary Alcohols of High Enantiomeric Purity." *J. Am. Chem. Soc.* **1979**, *101*, 2352–2355. Midland, M. M.; McDowell, D. C.; Hatch, R. L.; Tramontano, A. "Reduction of α,β-Acetylenic Ketones with B-3-Pinanyl-9-borabicyclo[3.3.1]nonane. High Asymmetric Induction in Aliphatic Systems" *J. Am. Chem. Soc.* **1980**, *102*, 867–9.

Problems

1. Draw the most stable conformations of cholic acid and fusidic acid (Steroids-1). Describe the stereochemistry at each ring junction and the conformation of each six-membered ring. (Steroids-1)
2. The energy difference between the chair conformations of methylcyclohexane is 1.8 kcal mol^{-1}. Use this information to estimate the energy difference between (a) *cis*-decalin and *trans*-decalin and (b) the two chair-chair conformations of each of the *cis*-decalins shown below. (Steroids-2)

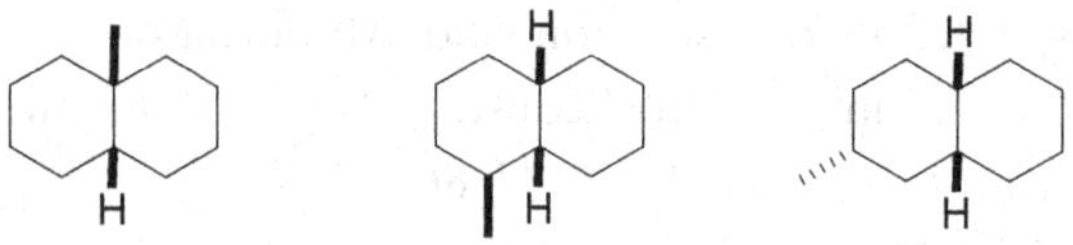

3. Provide a mechanism for the reduction of the acetate of ketol **9** to ketone **5** with Zn/AcOH. (Steroids-3)
4. The following transformation is called a Pummerer rearrangement. Identify the oxidation state changes that occur at specific atoms during this reaction. Explain why no oxidizing agents are needed to accomplish this transformation. (Steroids-3)

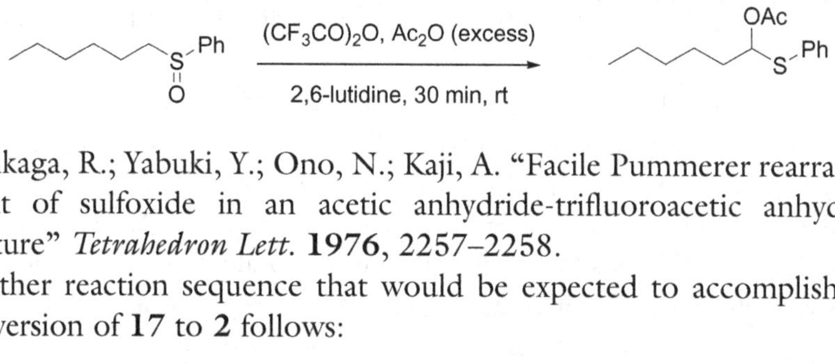

Tanikaga, R.; Yabuki, Y.; Ono, N.; Kaji, A. "Facile Pummerer rearrangement of sulfoxide in an acetic anhydride-trifluoroacetic anhydride mixture" *Tetrahedron Lett.* **1976**, 2257–2258.

5. Another reaction sequence that would be expected to accomplish the conversion of **17** to **2** follows:

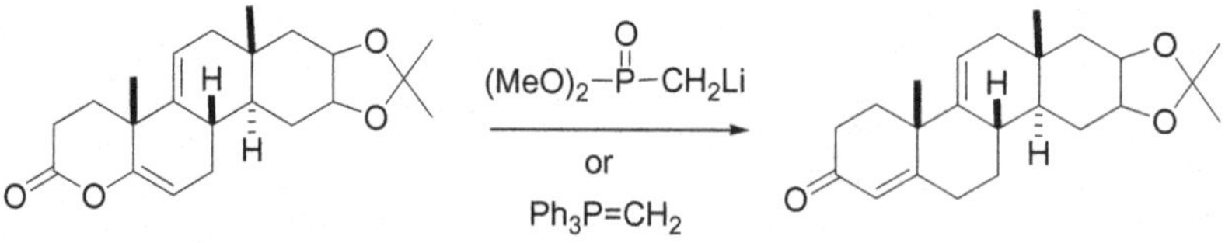

Provide a mechanistic interpretation of this transformation. See Henrick, C. A.; Boehme, E.; Edwards, J. A.; Fried, J. H. "Reaction of phosphoranes and phosphonate anions with enol lactones. A new method for the preparation of cyclic α,β-unsaturated ketones" *J. Am. Chem. Soc.* **1968**, *90*, 5926–5927 for more information. (Steroids-4)

6. Suggest a mechanism for the conversion of **33** to **34**. (Steroids-6)
7. Provide structures (most stable conformation) for the intermediates *en route* from **34** to **35**. (Steroids-6)
8. Provide structures for the intermediates *en route* from **38** to cortisone. Provide a mechanism for the step that introduces the C_4-C_5 double bond. (Steroids-6)
9. Certain cationic transition metal complexes, for example [Ir(cod)(PCy_3)py]$^+$ PF_6^-, can be used to "direct" hydrogenations. Two examples are shown below.

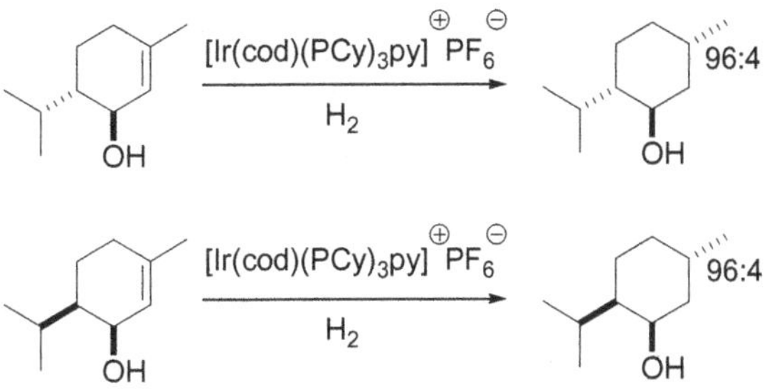

Show how this observation, in conjunction with other chemistry, can be used to accomplish the following "*trans*-perhydroindan-producing" transformation. (Steroids-8)

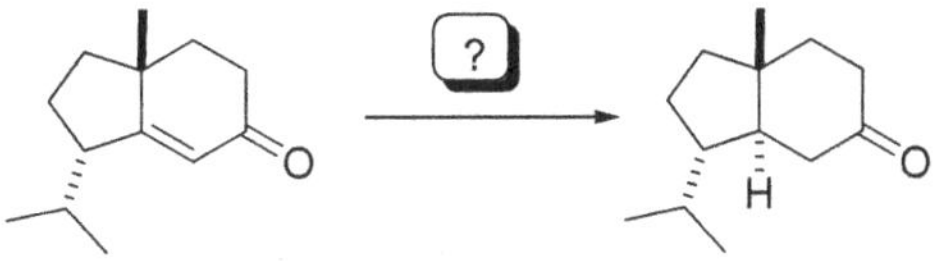

Crabtree, R. H.; Davis, M. W. "Directing effects in homogeneous hydrogenation with [Ir(cod)(PCy_3)(py)]PF_6" *J. Org. Chem.* **1986**, *51*, 2655. See also Evans, D. A.; Morrissey, M. M. "Rhodium(I)-catalyzed hydrogenation of olefins. The documentation of hydroxyl-directed stereochemical control in cyclic and acyclic systems" *J. Am. Chem. Soc.* **1984**, *106*, 3866.

10. Alkylation of enolates derived from cyclohexanone enolates occurs largely with "axial" entry of the electrophile. Propose stereoselective syntheses of **A** and **B**. (Steroids-8)

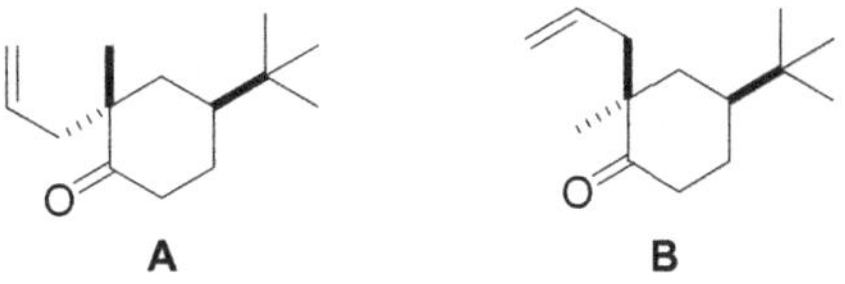

11. Provide the structures isolated after each of the four steps used to convert **82** → **83**. (Steroids-16)
12. Show the conformation of the allylic carbocation (derived from **87**) that would lead to diene **86**. What other tetracyclic dienes might be anticipated? (Steroids-16)
13. The Claisen rearrangement is known to proceed largely through a chair-like transition state. Explain why the rearrangement of **C** is more selective (in terms of olefin geometry) than **D**. Predict the product expected from "Ireland-Claisen" rearrangment of **E**. (Steroids-16)

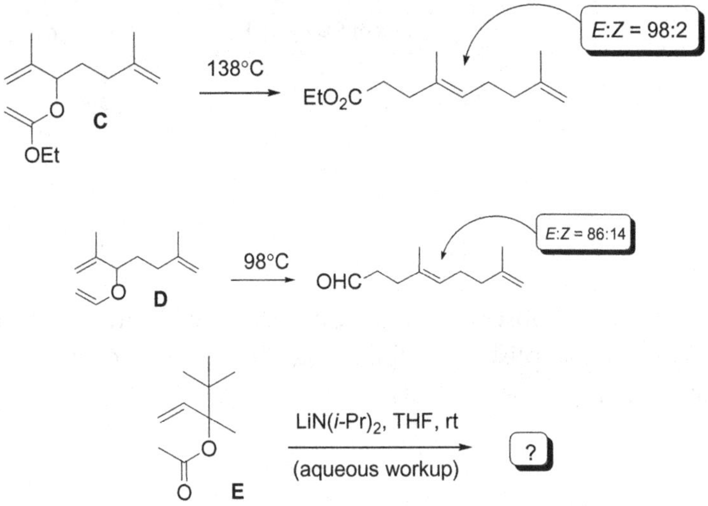

Perrin, C. L.; Faulkner, D. J. "Cis-Trans ratios in Claisen and Cope rearrangements" *Tetrahedron Lett.* **1969**, 2783. Ireland, R. E.; Willard, A. K. "Stereoselective Generation of Ester Enolates" *Tetrahedron Lett.* **1975**, *16*, 3975–3978. Ireland, R. E.; Mueller, R. H.; Willard, A. K. "Ester Enolate Claisen Rearrangement. Construction of the Prostanoid Skeleton" *J. Org. Chem.* **1976**, *41*, 986–996. Chillous, S.; Hart, D. J.; Hutchinson, D. K. "A Non-resolutive Approach to the Preparation of Configurationally Pure Difunctional Molecules" *J. Org. Chem.* **1982**, *47*, 5418–5420.

14. Propose a mechanism for the following transformation. (Steroids-22)

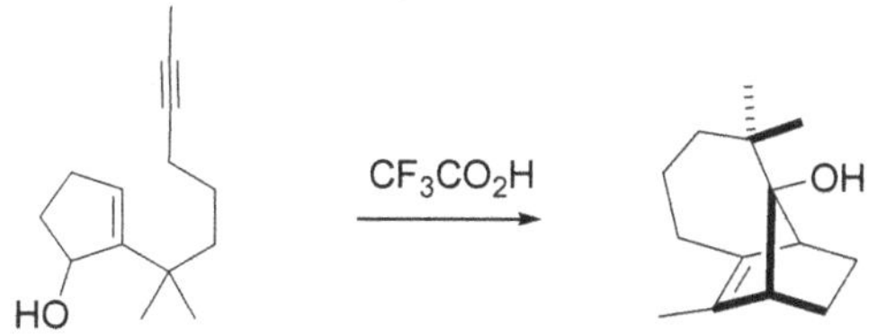

Volkmann, R. A.; Andrews, G. C.; Johnson, W. A. "Novel Synthesis of Longifolene" *J. Am. Chem. Soc.* **1975**, *97*, 4777.

15. Propose a synthesis of the allylic alcohol used in Problem 14. (Steroids-22)
16. Propose a mechanism for the following transformation. (Steroids-22)

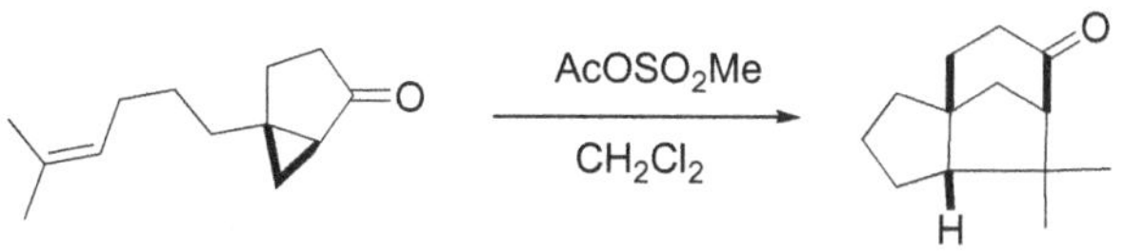

Corey, E. J.; Balanson, R. D. "Simple synthesis of (±)-cedrene and (±)-cedrol using a synchronous double annulation process" *Tetrahedron Lett.* **1973**, *14*, 3153–3156.

17. Propose syntheses of the following compounds.

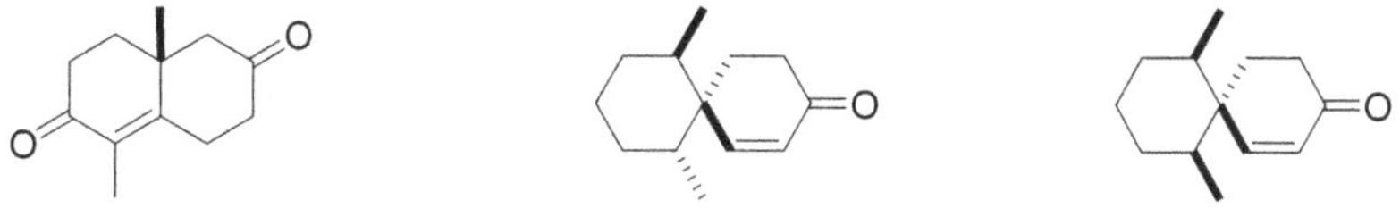

CHAPTER 3

Prostaglandins

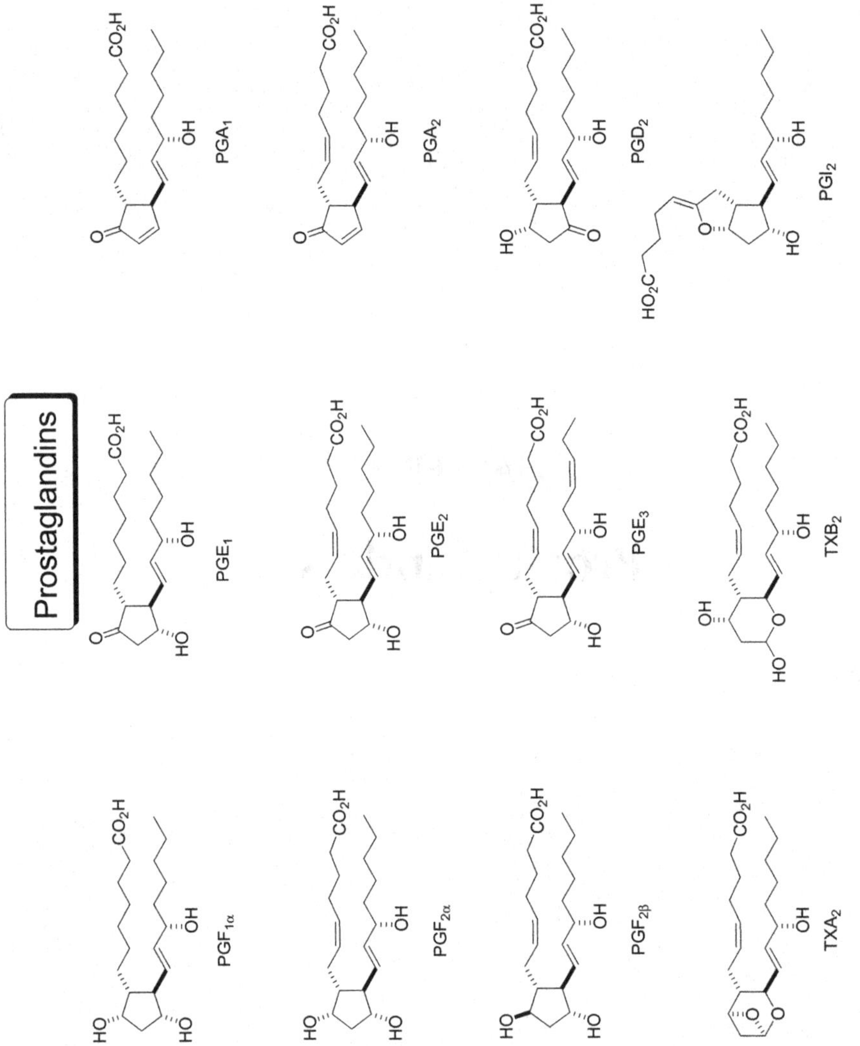

Prostaglandins-1

Prostaglandins-1

Prostaglandins are hormones derived from the C_{20} fatty acid arachidonic acid. The compounds were first isolated from the prostate gland and hence their name.[1] A key structural feature of the prostaglandins is the presence of a five-membered ring with the carbons in a variety of oxidation states, adorned with two adjacent (carbon) sidechains. The thromboxanes are a closely related family of fatty acid derived hormones, two examples of which are shown here. We will focus on the prostaglandins.

Prostaglandins: Some History and a Brief Perspective

Prostaglandins are hormones derived from the C_{20} fatty acid arachidonic acid. The compounds were first isolated from the prostate gland and hence their name (Samuelsson and Bergstrom). Prostaglandins are made in cells by a series of enzyme-catalyzed reactions. A critical step is the reaction of arachidonic acid with oxygen to produce PGG_2, a highly reactive peroxide-containing prostaglandin. In this reaction, promoted by an enzyme called cyclooxygenase (COX), C_9 and C_{11} of arachidonic acid bond to oxygen, and C_8 and C_{12} bond to one another, to form the cyclopentane ring that is characteristic of all prostaglandins. Once PGG_2 is formed, it is reduced to PGH_2, which is subsequently converted to PGE_2 and a host of other prostaglandins.

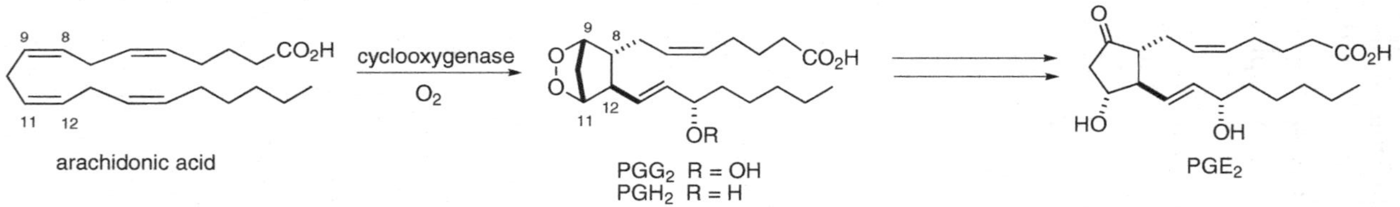

Prostaglandins are widespread in tissue, and elicit a wide variety of physiological responses. They play important roles in many life processes such as digestion, blood circulation, and reproduction. It has also been established that injured cells in the body produce and release prostaglandins that, in this situation, give rise to the phenomena of inflammation and pain. In 1969, Dr. John Vane and his collaborators at the Royal College of Surgeons in London discovered that aspirin inhibited production of prostaglandins by injured tissue. Eventually it was shown that this happens because aspirin binds to cyclooxygenase, inhibiting the conversion of arachidonic acid to PGG_2. This stops the production of prostaglandins, resulting in a reduction of inflammation and pain. One side effect of taking aspirin is stomach irritation and this is also a result of inhibition of prostaglandin synthesis. In turns out that PGE_2 protects the cells of the stomach wall by stimulating formation of a protective layer of mucous. PGE_2 also helps regulate acid levels in the stomach and, in its absence, hydrochloric acid production rises. Thus, it is understandable how inhibition of PGE_2 release in the stomach by orally ingested aspirin could lead to an upset stomach. Of course, this does not keep aspirin from being useful. The discovery of how aspirin works has inspired medicinal chemists to search from new COX inhibitors that might behave as painkillers. Celebrex and (the now infamous) Vioxx are COX inhibitors that were developed for this purpose.

Because prostaglandins were only isolable in small amounts from natural sources, synthesis played a key role in the determination of their structures (including absolute stereochemical assignments) and in furnishing material for biological studies. In response to this need (in part), many methods for the preparation of highly decorated cyclopentanes were developed beginning in the 1960's. Of course developments in the area of terpenoid chemistry also contriubuted to this burst of activity (do a search for cyclopentane-containing terpenoids and see what you find). This section will be devoted to selected methods that were developed for the preparation of prostaglandins with a focus on DJH's favorites.

Prostaglandins-2

A critical step in the biosynthesis of prostaglandins is the reaction of arachidonic acid with oxygen to produce PGG_2, a highly reactive peroxide-containing prostaglandin. In this reaction, promoted by an enzyme called cyclooxygenase (COX), C_9 and C_{11} of arachidonic acid bond to oxygen, and C_8 and C_{12} bond to one another, to form the cyclopentane ring that is characteristic of all prostaglandins.[2] Once PGG_2 is formed, it is reduced to PGH_2, which is subsequently converted to PGE_2 and a host of other prostaglandins.

Prostaglandins are widespread in tissue, and elicit a wide variety of physiological responses. They play important roles in many life processes such as digestion, blood circulation, and reproduction. It has also been established that injured cells in the body produce and release prostaglandins that, in this situation, give rise to the phenomena of inflammation and pain. In 1969, Dr. John Vane and his collaborators at the Royal College of Surgeons in London discovered that aspirin inhibited production of prostaglandins by injured tissue.[3] Eventually it was shown that this happens because aspirin binds to cyclooxygenase, inhibiting the conversion of arachidonic acid to PGG_2. This stops the production of prostaglandins, resulting in a reduction of inflammation and pain. One side effect of taking aspirin is stomach irritation, and this is also a result of inhibition of prostaglandin synthesis. In turns out that PGE_2 protects the cells of the stomach wall by stimulating formation of a protective layer of mucous. PGE_2 also helps regulate acid levels in the stomach and, in its absence, hydrochloric acid production rises. Thus, it is understandable how inhibition of PGE_2 release in the stomach by orally ingested aspirin could lead to an upset stomach. Of course, this does not keep aspirin from being useful. The discovery of how aspirin works has inspired medicinal chemists to search from new COX inhibitors that might behave as painkillers. Celebrex and (the now infamous) Vioxx are COX inhibitors that were developed for this purpose.[4] Because prostaglandins were only isolable in small amounts from natural sources, synthesis played a key role in the determination of their structures (including absolute stereochemical assignments) and in furnishing material for biological studies. In response to this need (in part), many methods for the preparation of highly decorated cyclopentanes were developed beginning in the 1960's. Of course, developments in the area of terpenoid chemistry also contributed to this burst of activity (do a search for cyclopentane-containing terpenoids and see what you find). This section will be devoted to selected methods that were developed for the preparation of prostaglandins with a focus on some of my favorites.

Prostaglandin Synthesis: Some Contributions from the Corey Group

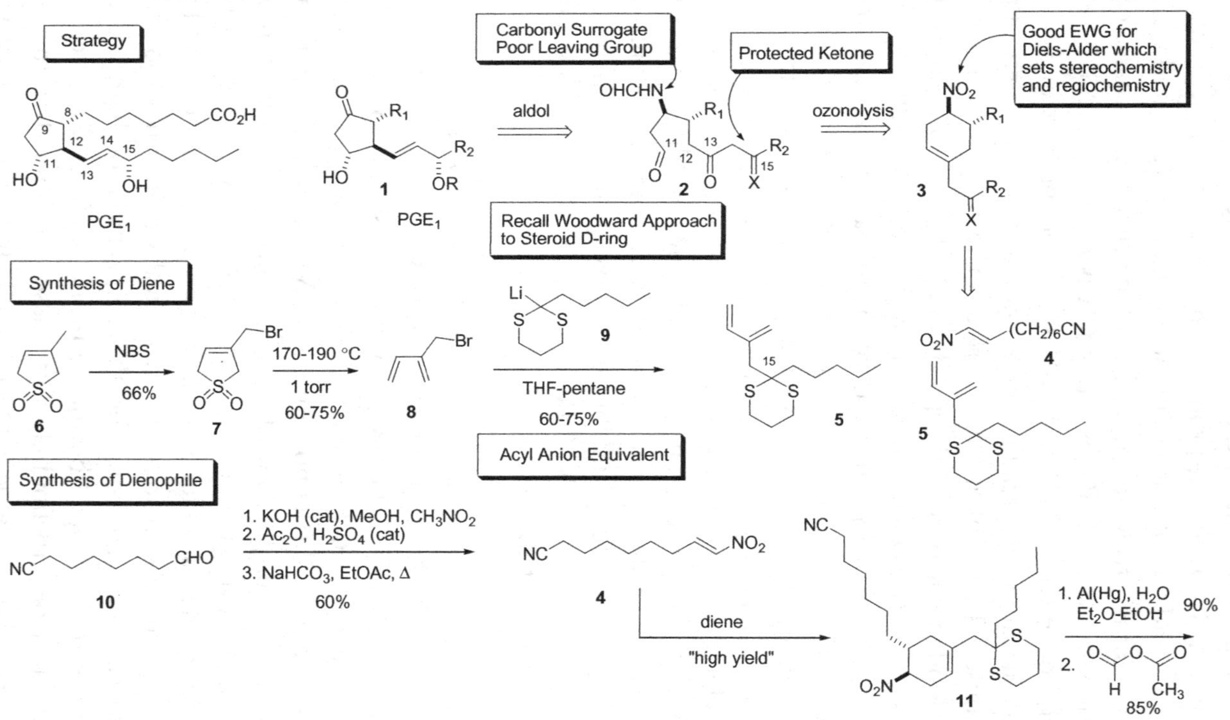

Corey, E. J.; Andersen, N. H.; Carlson, R. M.; Paust, J.; Vedejs, E.; Vlattas, I.; Winter, R. E. K. "Total Synthesis of Prostaglandins. Synthesis of the Pure *dl*-E_1, -F_1 , F_1 , A_1, and B_1 Hormones" *J. Am. Chem. Soc.* **1968**, 90, 3245-3247.

Prostaglandins-3

Prostaglandins-3

We will start with some contributions to prostaglandin synthesis from the Corey group, perhaps the major contributors to this area of natural products synthesis. The structures of many prostaglandins were initially not on firm ground. Early synthetic objectives in this area were tied to a desire to establish structure, including stereochemical details. Thus, early approaches to prostaglandins did not address all stereochemical issues in a selective manner. Rather they were designed to provide a number of compounds (whose stereochemistry could be assigned using chemical and spectroscopic methods) that could be compared with natural materials to establish both structure and possibly structure-activity relationships from the standpoint of biology.

The first synthesis reported by the Corey group afforded members of the PGE, PGF and PGA families of prostaglandins (see Prostaglandins-1). The strategy is outlined here within the context of PGE_1 (short-handed as structure **1**). The plan was to prepare this compound from 1,6-dicarbonyl compound **2** via an intramolecular aldol condensation. The details behind the choice of **2** as an intermediate are interesting. We can speculate that the choice of a formamide as a precursor to the C_9 ketone, and the inclusion of a protected ketone at C_{15}, were designed to minimize potential problems with β-elimination reactions during the aldol condensation. Protection of the C_{15} ketone would also differentiate it from the C_{13} ketone and facilitate reduction chemistry that would be needed at C_{13} subsequent to formation of the 5-membered ring. From the standpoint of tactics, mild conditions would be required to avoid dehydration (to an acylcyclopentene) during intramolecular aldol condensation. Intermediate **2** was to be prepared from cyclohexene **3**, which was to come from a Diels-Alder reaction between nitroalkene **4** and 2-substituted-1,3-butadiene **5**. This diene-dienophile pair is well-matched to provide the required regiochemistry. It is notable that this strategy is simply a variation of the cyclohexene-to-cyclopentane strategy we saw several times during our examination of steroid syntheses (see the Woodward D-ring synthesis on Steroids-4, and the Johnson D-ring synthesis on Steroids-17).

In the forward direction, diene **5** was prepared by alkylation of metallated 1,3-dithiane **9** with allylic bromide **8**. In this reaction, **9** plays the role of an "acyl anion equivalent".[5] We will talk about equivalencies in more detail in Chapter 6, but at this point it is worth noticing that the dithiane will eventually emerge as the C_{15} protected ketone. Dienophile **4** was prepared by an "aldol-dehydration" reaction between nitromethane and aldehyde **10**, a reaction known as the Henry reaction.[6] The Diels-Alder reaction between **4** and

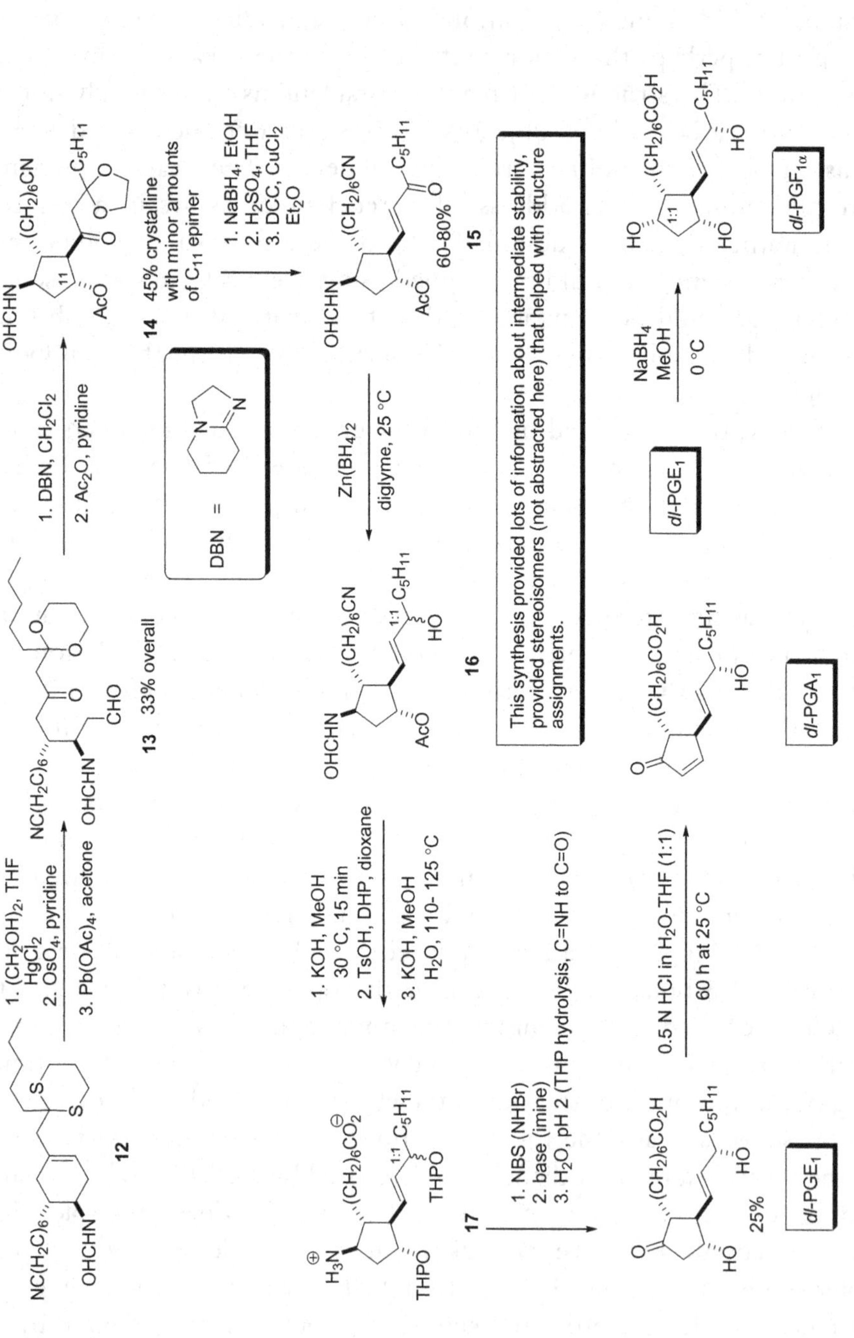

Prostaglandins-4

5 gave **11** in good yield. Reduction of the nitro group to an amino group was accomplished using aluminum amalgam. The resulting amine was protected as the formamide (**12**) using formic acetic anhydride. The selective reduction of the nitro group in the presence of a nitrile, an olefin, and a dithiane is notable. One reason why it is useful to have an arsenal of reactions for accomplishing any single functional group transformation, is to be able to have at least one transformation that can be used in the presence of any other functional group.

Prostaglandins-4

The synthesis continued with conversion of the thioacetal to the corresponding acetal, vicinal dihydroxylation of the olefin, and oxidative cleavage of the resulting diol with lead tetraacetate to provide keto-aldehyde **13**, a specific example of generic intermediate **2** (see Prostaglandins-3). The intramolecular aldol condensation was accomplished using mild basic conditions (amidine base DBN) to provide β-acetoxy ketone **14** as the major stereoisomer after acylation of the intermediate alcohol. One might imagine that thermodynamics are responsible for the stereoselectivity observed in the ring-forming reaction.

Protected 1,3-diketone **14** was converted to enone **15** using a standard reduction-hydrolysis-dehydration reaction sequence. Reduction of the C_{15} ketone provided a mixture of diastereomeric alcohols (**16**). The acetate was hydrolyzed under mild conditions to provide the C_{11} alcohol. The two hydroxyl groups were then protected as THP ethers. The nitrile was then converted to the carboxylate salt with concomitant hydrolysis of the formamide under vigorous basic conditions to provide **17**. Oxidation of the amino group and hydrolysis of the resulting imine gave *dl*-PGE_1 along with equal amounts of *dl*-15-*epi*-PGE_1. Dehydration of *dl*-PGE_1 (a β-elimination) gave *dl*-PGA_1 and reduction of *dl*-PGE_1 with sodium borohydride provided *dl*-$PGF_{1\alpha}$ and *dl*-$PGF_{1\beta}$ (the C_9 diastereomer). This synthesis gave racemic material, but it also revealed a lot about intermediate stability, afforded stereoisomers that helped with structure assignments of the natural products, and provided materials for biological evaluation. Needless to say, numerous other syntheses shortly followed this report. We will next examine a so-called "second generation aldol approach" reported by the Corey group.

Second Generation Aldol Approach

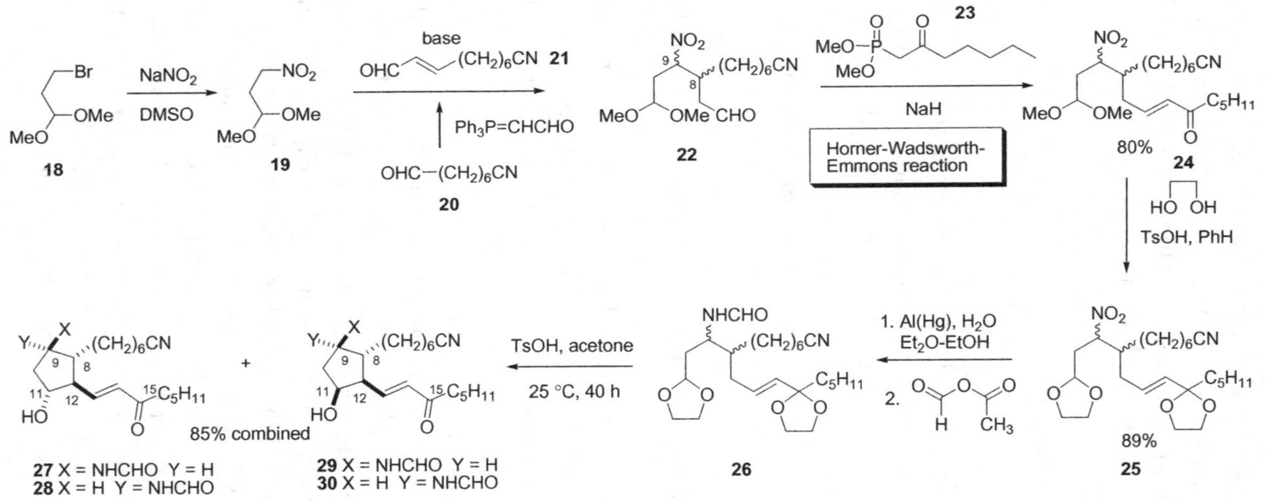

Advanced to PGE_1, 15-*epi*-PGE_1, 11-*epi*-PGE_1, 11,15-*epi*-PGE_1, PGA_1 and 15-*epi*-PGA_1

Separation schemes for **27-30** were developed. Materials were advanced using methodology related to that described in the Diels-Alder approach. Whereas this synthesis appears to be shorter, it is truly stereorandom. The upside is that it provided stereoisomers for biological evaluation.

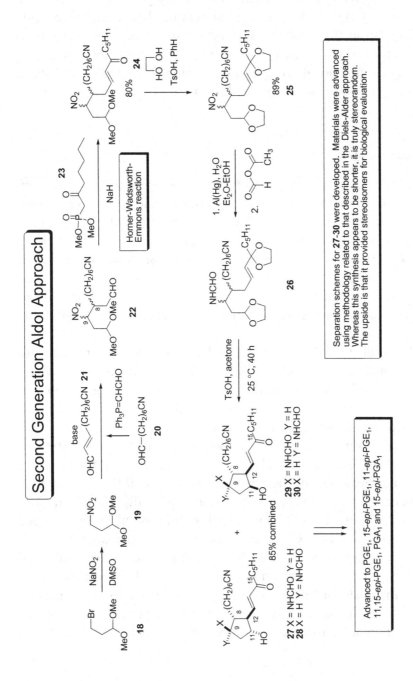

Corey, E. J.; Vlattas, I.; Andersen, N. H.; Harding, K. "A New Total Synthesis of Prostaglandins of the E_1 and F_1 Series Including 11-Epiprostaglandins" *J. Am. Chem. Soc.* **1968**, 90, 3247-3248.

Prostaglandins-5

The key feature of this synthesis is the use of an acid-promoted vinylogous aldol condensation to provide compounds **27–30** from key intermediate **26**. The synthesis of **26** begins with a conjugate addition between nitro-acetal **19** and α,β-unsaturated aldehyde **21** to provide a mixture of diastereomeric nitro-aldehydes **22**. A Horner-Wadsworth-Emmons (HWE) reaction installed what will become the C_{12} sidechain. Protection of the C_{15} ketone was followed by conversion of the nitro group to a C_9 formamide (**26**), as a racemic mixture of diastereomers at C_8 and C_9. Treatment of **26** with *p*-toluenesulfonic acid in acetone (presumably with some water present) gave a mixture of vinylogous aldol products **27–30** in excellent yield. Separation schemes were developed and the individual isomers were advanced to a host of prostaglandins (and prostaglandin stereoisomers) using methodology related to Diels-Alder approach we examined above. The downside of this synthesis is that it was truly stereorandom. The upside is that it was truly stereorandom; it provided stereoisomers for biological evaluation. Finally, this approach does a better job at introducing C_{13}-C_{15} at the desired oxidation states. We will see that the HWE reaction became a popular method for introduction of the C_{12} sidechain.

A Practical Synthesis

1. All primary PGs from single precursor
2. Control stereochemistry
3. Resolution at early stage

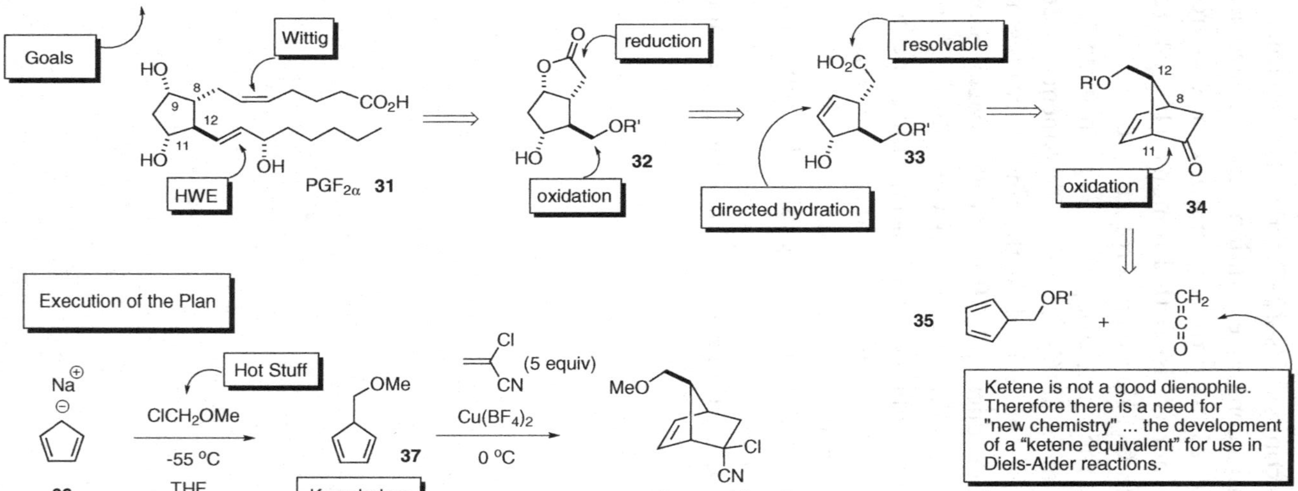

Execution of the Plan

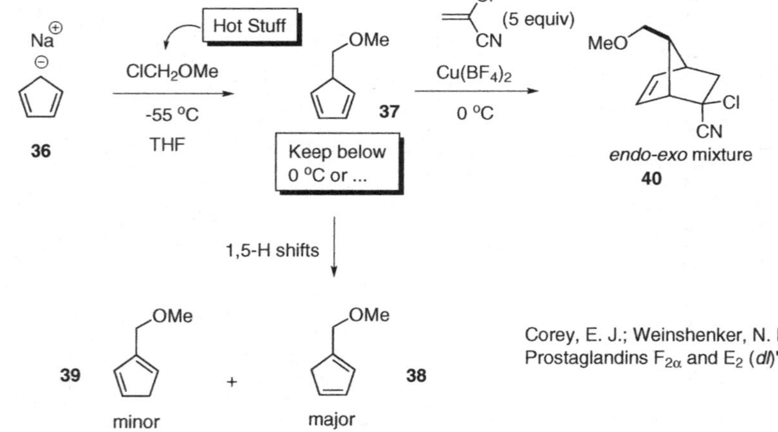

Corey, E. J.; Weinshenker, N. M.; Schaaf, T. K.; Huber, W. "Stereo-Controlled Synthesis of Prostaglandins $F_{2\alpha}$ and E_2 (*dl*)" *J. Am. Chem. Soc.* **1969**, *91*, 5675.

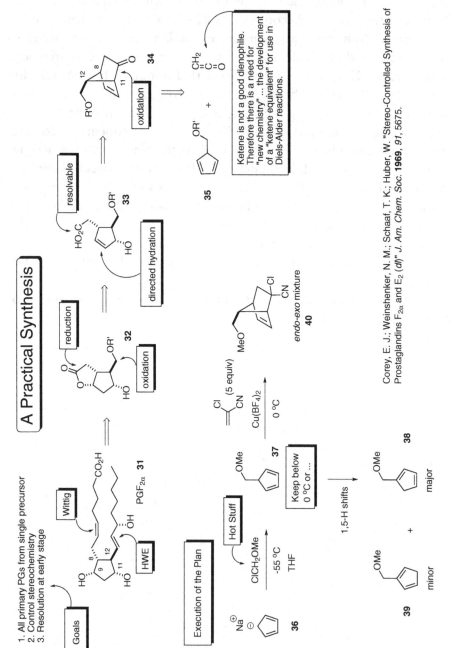

Prostaglandins-6

Prostaglandins-6

As the structures of prostaglandins were established, a need for practical syntheses developed. The Corey group set as a goal the synthesis of all primary prostaglandins from a single precursor, with control over stereochemistry, and providing the required enantiomer at an early stage of the synthesis. The strategy that was followed is illustrated here within the context of an approach to $PGF_{2\alpha}$ (**31**). It was felt that lactone **32** might provide access to **31** (and many of the prostaglandins shown at the beginning of this chapter). Attachment of the "upper" C_8 sidechain might be accomplished by reduction of the lactone to a lactol followed by an olefination reaction. Attachment of the "lower" C_{12} sidechain might be accomplished by oxidation of C_{13} to an aldehyde followed once again by olefination methodology. Lactone **32** was to be derived from acid **33**. The carboxyl group was to provide a handle for resolution and also help in the "directed hydration" of the olefin needed to convert **33** to **32**. Hydroxy acid **33** was to be derived from **34** by a Baeyer-Villiger oxidation, and **34** was to come from a Diels-Alder reaction between a cyclopentadiene of type **35** and an appropriate dienophile.

This was a very daring strategy, particularly in the early stages of the proposed synthesis. It was known that ketene, the "dienophile" that would directly lead from **35** to **34**, did not partake in Diels-Alder reactions. It was also known that 5-substituted cyclopentadienes of type **35** easily isomerize to the more stable 1- and 2-substituted isomers via 1,5-hydride shifts (sigmatropic rearrangements). Thus, the proposed Diels-Alder strategy could not be accomplished directly and required use of some "indirect" tactics.

In the first Corey prostaglandin synthesis we saw the use of an "acyl anion equivalent". Now we will see another application of the "equivalency concept", use of a "real" reactant that, after some functional group manipulations, can be used to accomplish a transformation that could not be accomplished otherwise. Let's look at the tactics that were ultimately used by the Corey group. Sodium cyclopentadienide was alkylated with chloromethyl methyl ether to provide **37**. Chloromethyl methyl ether is extremely reactive and thus, this reaction could be accomplished at a low temperature, below the temperature at which **37** isomerizes to cyclopentadienes **38** and **39**. Cyclopentadiene **37** was then reacted with an excess of 2-chloroacrylonitrile, an extremely reactive dienophile. Under the influence of Lewis acid promotion (copper tetrafluoroborate) this diene-dienophile pair underwent cycloaddition, to give **40** at temperatures where diene isomerization did not occur. Although **40** was obtained as a mixture of *endo* and *exo* diasteromers, it is notable that the dienophile approached **37** from the sterically least

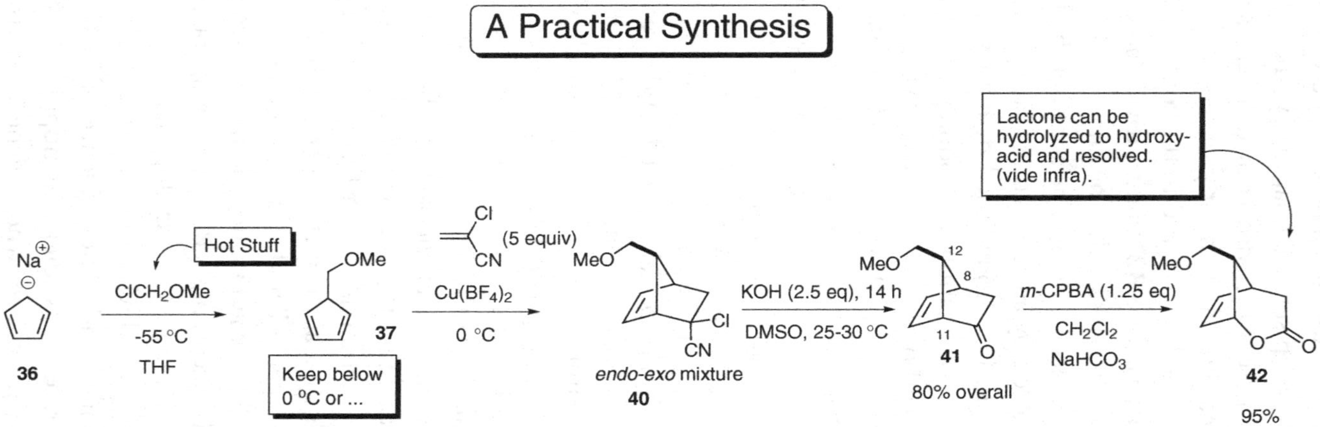

Corey, E. J.; Weinshenker, N. M.; Schaaf, T. K.; Huber, W. "Stereo-Controlled Synthesis of Prostaglandins $F_{2\alpha}$ and E_2 (*dl*)" *J. Am. Chem. Soc.* **1969**, *91*, 5675.

Prostaglandins-6 (*Continued*)

hindered face (opposite the methoxymethyl group) to provide the required relative stereochemical relationship between what was to become C_8, C_{11} and C_{12} of $PGF_{2\alpha}$. Cycloadduct **40** was almost there. All that had to be done was to convert the α-chloronitrile to the corresponding cyanohydrin. The cyanohydrin would then presumably afford the desired ketone **41**. This transformation was accomplished in good yield by treating **40** with KOH in DMSO. Whereas it seemed likely that a cyanohydrin should be an intermediate in the conversion of **40** to **41**, it did seem unlikely that the transformation of **40** to the corresponding cyanohydrin would take place via an intermolecular S_N2 type of process. The mechanistic details of this reaction were eventually revealed 14 years after the Corey group reported their synthesis![7]

Before proceeding, it is important to note that it is the conversion of the α-chloronitrile to a ketone (a functional group transformation) that establishes the α-chloroacrylonitrile as a ketene equivalent in Diels-Alder chemistry. α-Chloroacrylonitrile is not the only ketene equivalent that has been used in Diels-Alder chemistry, nor in this strategy for prostaglandin synthesis. Reviews have been written on this topic.[8] What is perhaps most important is the development of the concept of "equivalencies" that was occuring at the time this synthesis was undertaken.[9] We will see this concept again.

The synthesis continued with Baeyer-Villiger oxidation of **41** to provide **42**. Notice the fortunate selectivity here between the Baeyer-Villiger oxidation and olefin epoxidation.

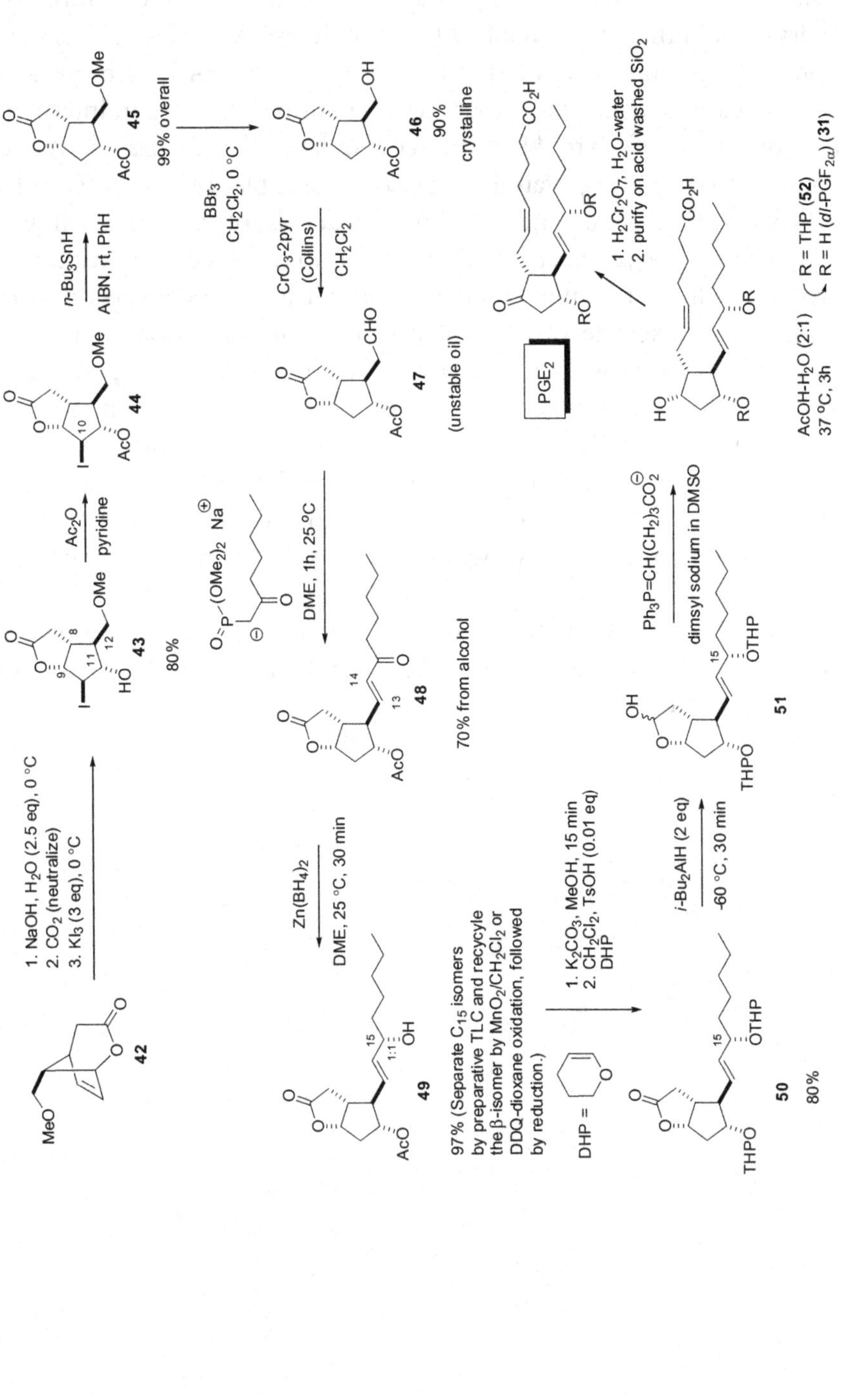

Prostaglandins-7

Prostaglandins-7

Conversion of **42** to intermediate **45** (a specific example of strategic structure **32**) required regio- and stereoselective hydration of the olefin, and a transesterification. This was accomplished by lactone hydrolysis, careful neutralization using carbon dioxide to sequester excess sodium hydroxide, and an electrophilic addition (using an I^+ source) to provide iodolactone **43**. The alcohol was then protected and the iodide was reduced to complete the synthesis of **45**. To extend the "equivalency concept" we have been developing, we could say that the iodine serves as an "equivalent" to a proton during the conversion of **42** to **45**. Whereas a proton would certainly have been more direct, the use of iodine (which presumably forms an iodonium ion intermediate rather than the carbocation expected from a protonation) may help control regiochemistry and stereochemistry.[10] The tactic used to reduce the iodide is also notable. This process occurs via a radical intermediate at C_{10}, thus avoiding β-elimination reactions that would most likely have plagued metal-mediated reductions (for example Zn/AcOH), or troubles associated with the sterics (2° iodide) and selectivity (lactone and ester incompatibility) that would have accompanied hydride reductions (for example $LiAlH_4$ or $LiEt_3BH$).

The synthesis continued with ether cleavage, oxidation of the resulting primary alcohol **46**, and olefination of the intermediate aldehyde **47**. Reduction of the ketone **48** gave **49** as a mixture of diastereomers at C_{15}. These were separated and the "wrong" C_{15} diastereomer could be recycled via an oxidation-reduction sequence. Adjustment of protecting groups (in preparation for a carbonyl reduction) gave **50**. The lactone was reduced to the corresponding lactol **51** using reliable methodology and a Wittig olefination installed the C_8 sidechain with control of olefin geometry. Removal of the alcohol protecting groups completed the synthesis of *dl*-$PGF_{2\alpha}$ (**31**).

Tactical Improvements

The synthesis of prostaglandins via the "Corey Lactone" was adopted by chemical industry and used to prepare kilogram quantities of materials. The synthesis met most of the stated goals, but there was room for tactical improvements. Some problems to be addressed were: (1) How was the key hydroxy acid to be resolved? (2) Could asymmetry be introduced into the Diels-Alder reaction such that resolution would be unnecessary? (3) Could the need for stoichiometric use of n-Bu_3SnH be avoided? (4) Could the C_{15} stereochemistry problem be solved? Research directed toward solving these tactical problems led the the development of generally useful chemistry.

The Resolution

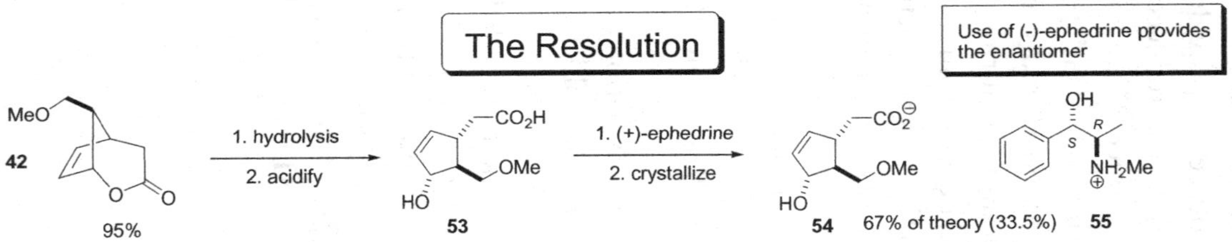

Corey, E. J.; Schaaf, T. K.; Huber, W.; Koelliker, U.; Weinshenker, N. M. "Total Synthesis of Prostaglandins $F_{2\alpha}$ and E_2 as the Naturally Occuring Forms" *J. Am. Chem. Soc.* **1970**, *92*, 397-398.

Diels-Alder Problem

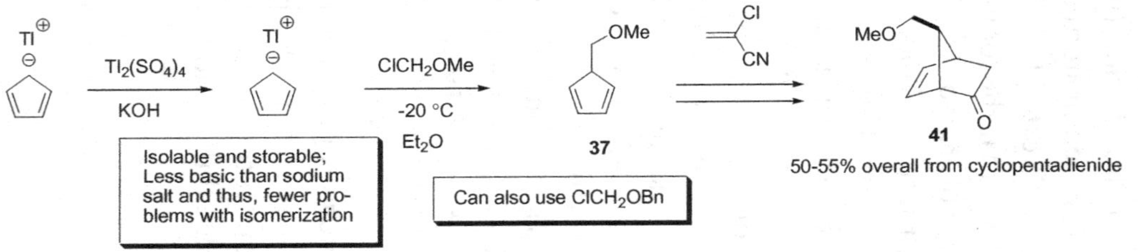

Corey, E. J.; Koellinker, U.; Neuffer, J.; "Methoxymethylation of Thallous Cyclopentadienide. A Simplified Preparation of a Key Intermediate for the Synthesis of Prostaglandins" *J. Am. Chem. Soc.* **1971**, *93*, 1489.

Prostaglandins-8

Synthesis of prostaglandins via the "Corey Lactone" (**45**) was adopted by chemical industry and used to prepare kilogram quantities of materials. The synthesis met most of the stated goals, but there was room for tactical improvements. Some problems to be addressed were: (1) How was the key hydroxy acid to be resolved? (2) Could asymmetry be introduced into the Diels-Alder reaction such that resolution would be unnecessary? (3) Could the need for stoichiometric use of *n*-Bu_3SnH be avoided? (4) Could the C_{15} stereochemistry problem be solved? Research directed toward solving these tactical problems led to the development of generally useful chemistry.

Resolution of **42** was accomplished by traditional methods. Conversion of **42** to hydroxy acid **53** was followed by resolution as its ephedrine salt. Variants of the Diels-Alder reaction were developed. For example, use of thallium cyclopentadienide (rather than sodium cyclopentadienide) provided a more stable, less basic nucleophile for use in generation of **37**.

Another Ketene Equivalent

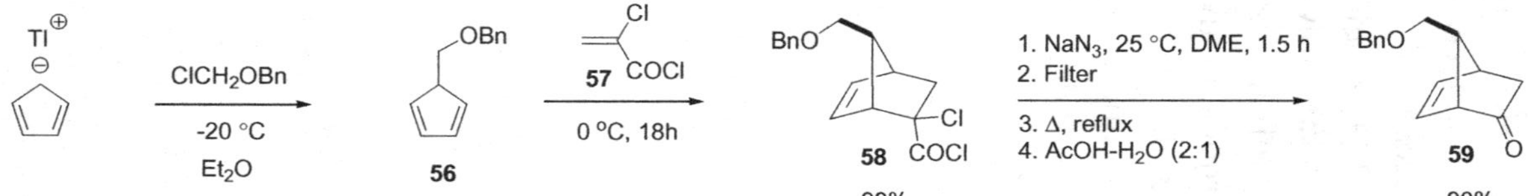

Acid chlorides are extremely hot dienophiles.

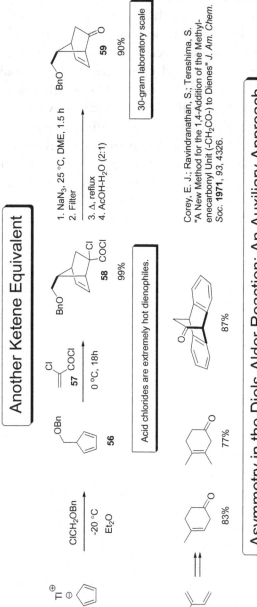

30-gram laboratory scale

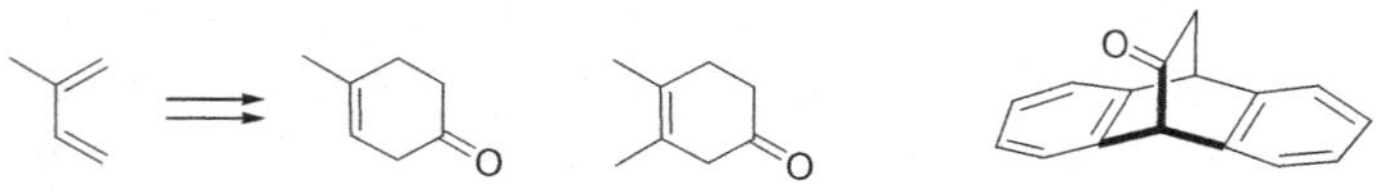

Corey, E. J.; Ravindranathan, S.; Terashima, S. "A New Method for the 1,4-Addition of the Methylenecarbonyl Unit ($-CH_2CO-$) to Dienes" *J. Am. Chem. Soc.* **1971**, *93*, 4326.

Asymmetry in the Diels-Alder Reaction: An Auxiliary Approach

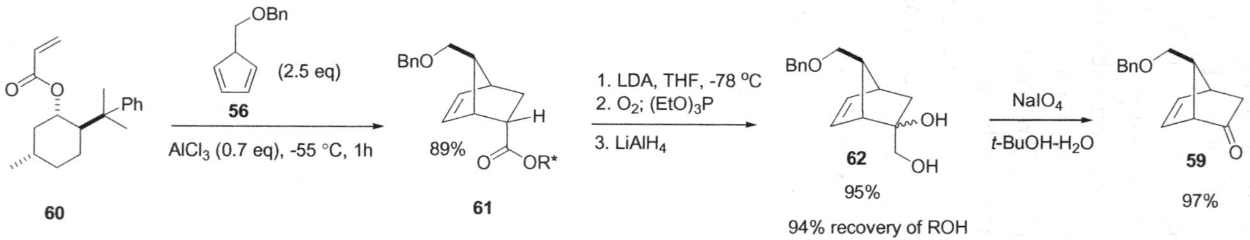

Corey, E. J.; Ensley, H. E. "Preparation of an Optically Active Prostaglandin Intermediate via Asymmetric Induction" *J. Am. Chem. Soc.* **1975**, *97*, 6908

Prostaglandins-9

Prostaglandins-9

The temperature at which the Diels-Alder could be conducted was lowered by developing α-chloroacryloyl chloride (**57**) as the ketene equivalent. Acid chlorides are ***much*** more reactive than their nitrile or ester counterparts in Diels-Alder reactions, and this is an excellent example of putting this kinetic fact to use.[11] A Curtius rearrangment was used as the key functional group transformation in establishing the ketene equivalency.

Ester **60** was then developed as a ***chiral auxiliary*** for use with diene **56**. This diene-dienophile pair reacted at a very low temperature, under the influence of a Lewis acid promoter, to provide **61** in excellent yield and with high diastereoselectivity. The equivalency of **60** with ketene was established by sequential α-hydroxylation of the enolate of **61**, reduction of the ester to give diol **62**, and periodiate cleavage to afford **59** as a single enantiomer.

Synthesis of Auxiliary

Model for Asymmetric Induction (for enantiomer)

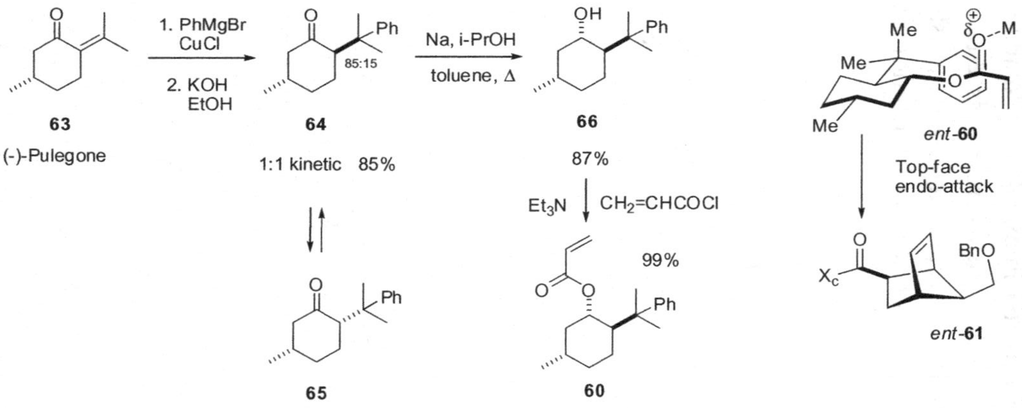

1. Potential for π-stacking
2. Minimize dipole-dipole repulsion in conformation of O-acyl bond
3. Minimze steric effects in complexation of Lewis acid
4. The non-catalyzed process shows little asymmetric induction

The Tin Hydride Problem

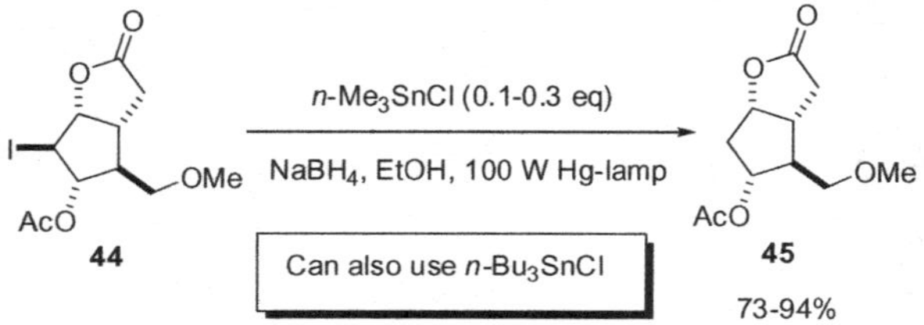

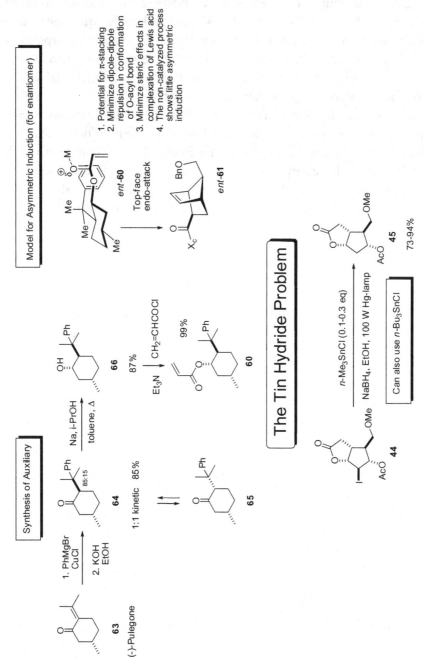

Corey, E. J.; Suggs, J. W. "A Method for Catalytic Dehalogenations via Trialkyltin Hydrides" *J. Org. Chem.* **1975**, *40*, 2554-2555

Prostaglandins-10

Prostaglandins-10

Chiral ketene equivalent **60** was prepared from pulegone (**63**), a common monoterpene. Both enantiomers of **63** are known and thus, both enantiomers of **60** are available. A model that rationalizes the observed diastereoselectivity follows. The model emphasizes three points: (1) The ester reacts from a conformation that minimizes dipole-dipole repulsion in terms of conformation around the *O*-acyl bond. This is normally the lowest energy conformation for any ester. (2) Steric effects are minimized in the presumed reactive complex between **60** and the Lewis acid. The metal complexes opposite the large ester alkyl group, and the vinyl dienophile reacts from an *s*-trans conformation to minimize metal-vinyl group interactions. (3) π-Stacking contributes to shielding of one face of the olefin from the diene. It is notable that the non-catalyzed process shows little asymmetric induction.[12]

The use of stoichiometric amounts of tin hydride in the conversion of **44** to **45** could be perceived as a problem due to the toxicity of organotin compounds. The problem was addressed by developing a method that was catalytic in tin. Basically the trimethyltin iodide produced during this free radical reduction was recycled through trimethyltin hydride via reduction with sodium borohydride.

The C_{15} Stereochemistry Problem

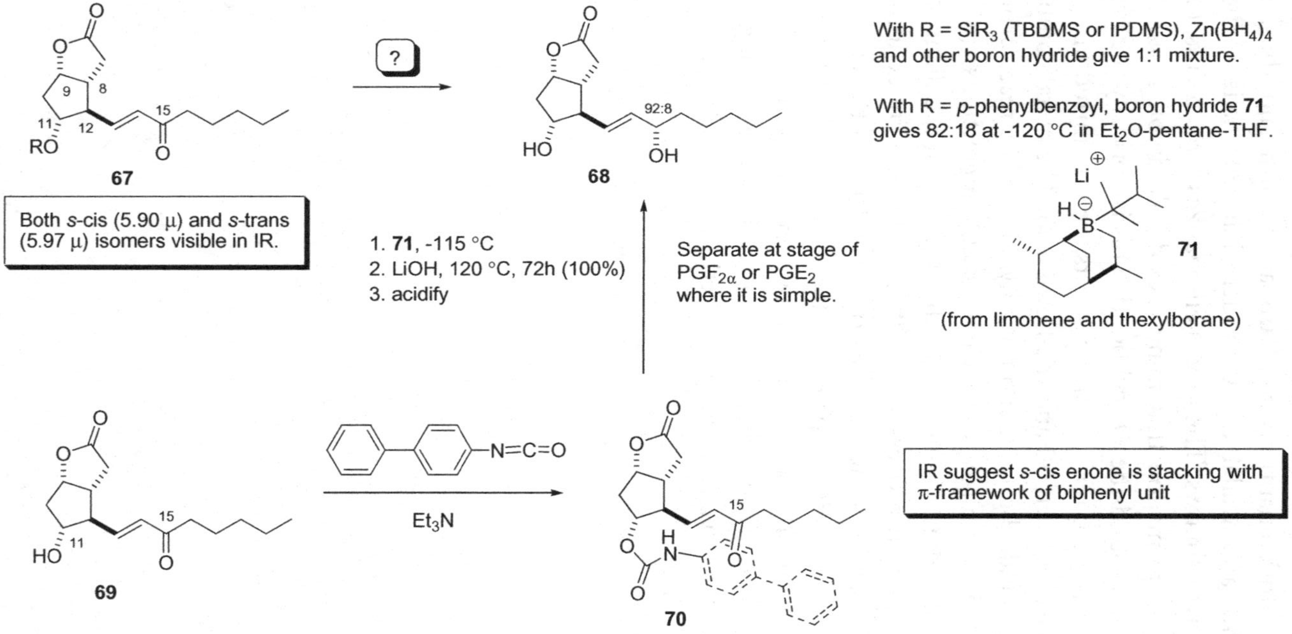

Corey, E. J.; Becker, K. B.; Varma, R. V. "Efficient Generation of the 15*S* Configuration in Prostaglandin Synthesis. Attractive Interactions in Stereochemical Control of Carbonyl Reduction" *J. Am. Chem. Soc.* **1972**, *94*, 8616.

Prostaglandins-11

The C_{15} stereochemistry problem represents a "general" stereochemical problem in synthesis. Consider the conversion of enone **67** to allylic alcohol **68**. Compound **67** has four stereogenic centers (C_8, C_9, C_{11} and C_{12}). Reduction of C_{15} to an alcohol could generate two diastereomers. In principle, the aforementioned stereogenic centers could influence the reduction, resulting in formation of one diastereomer (for example **68**) in preference to the other (for example 15-*epi*-**68**). This type of diastereoselection does not have to rely on the use of chiral reducing agents and is called "relative asymmetric induction". It is notable that this approach can be applied to one enantiomer of **67**, or racemic **67**. If complete stereoselectivity is observed with one enantiomer of **67**, a single enantiomer of **68** would be obtained. The same reaction starting with racemic **67** would afford a racemic mixture of **68**.

A second approach to this problem involves use of a single enantiomer of a chiral reducing agent. This approach involves "reagent control" of stereochemistry. If we use a reagent that will reduce an enone of type **67** to the *S*-alcohol, regardless of stereogenic centers elsewhere in the molecule, then reduction of **67** will give **68**. Of course if we start with the enantiomer of **67**, we will not obtain the enantiomer of **68**. We will obtain the enantiomer of 15-*epi*-**68**. This type of diastereoselection must use a single enantiomer of a chiral reducing agent. This approach is effective if one is working with single enantiomers of reaction substrates. It is not effective if one is working with racemic material.

How do these approaches work with prostaglandin substrates of type **67**? The bottom line is that the first approach is only marginally successful. The four ring stereogenic centers in **67** are remote from C_{15}. The two faces of the ketone are only marginally different (review stereochemical definitions such as *re* and *si* from your earlier studies of stereochemistry). The only functional handle for differentiating the diastereotopic faces of the carbonyl group is the C_{11} protecting group. A number of protecting groups were surveyed in conjunction with a variety of reducing agents. The idea was clearly to use a "big" protecting group that would selectively block one diastereotopic face of the C_{15} carbonyl group. Ultimately it was found that the combination of urethane **70** and reducing agent **71** (single enantiomer derived from (+)-limonene) gave the desired isomer **68** with 92:8 diastereoselectivity. The explanation invoked for success was π-π stacking of the biphenyl unit with the enone as shown here.[13] IR spectroscopy suggested that not only did π-complexation occur, but it occurred selectively with the enone in the *s*-cis conformation. It

Reagent Control of Stereochemistry

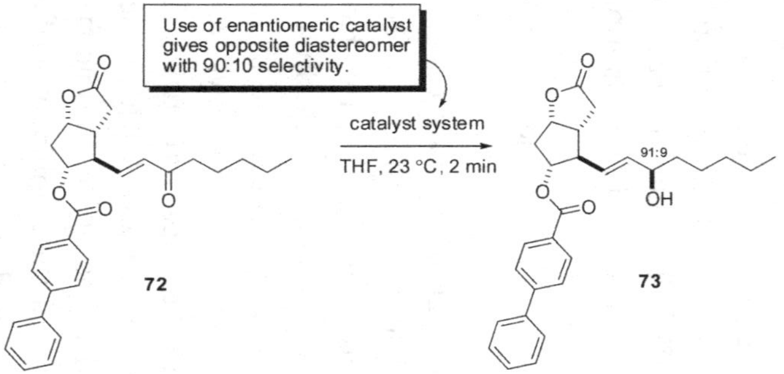

Catalyst System

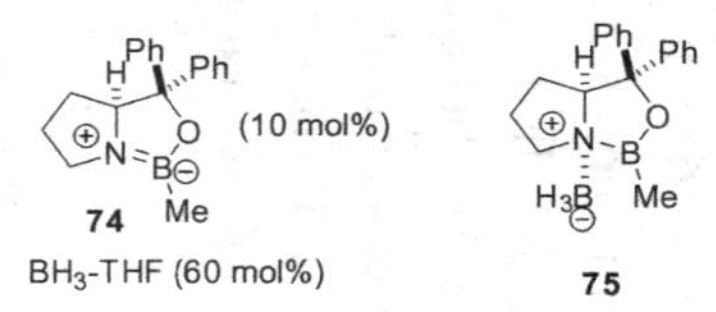

Complex

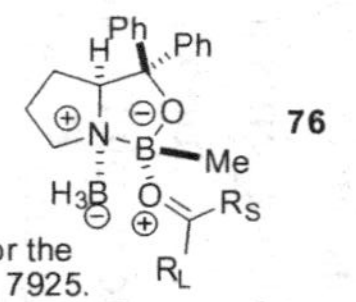

Corey, E. J.; Bakshi, R. R.; Shibata, S.; Chen, C.-P; Singh, V. K. " A Stable and Easily Prepared Catalyst for the Enantioselective Reduction of Ketones. Applications to Multistep Syntheses" *J. Am. Chem. Soc.* **1987**, *109*, 7925.
See also Itsuno, S.; Ito, K.; Hirao, A.; Nakahama, S. "Asymmetric Reduction of Aliphatic Ketones with the Reagent Prepared from (*S*)-(-)-2-Amino-3-methyl-1,1-diphenylbutan-1-ol and Borane" *J. Org. Chem.* **1984**, *49*, 555-557. For a review see Corey, E. J.; Helal, C. J. *Angew. Chem. Int. Ed.* **1998**, *37*, 1986-2012.

Inversion of C_{15} Stereochemistry

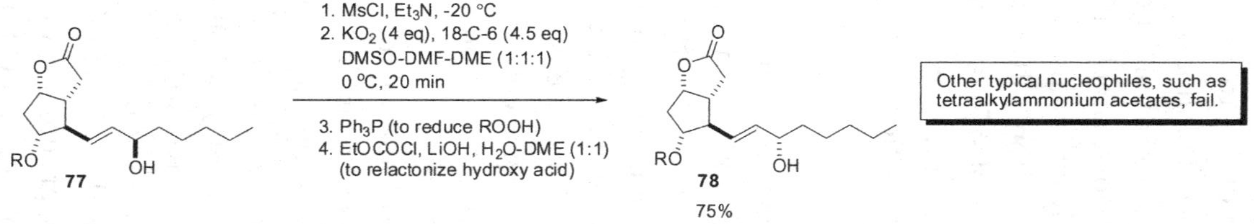

Corey, E. J.; Nicolaou, K. C.; Shibasaki, M.; Machida, Y.; Shiner, C. S. "Superoxide ion as a Synthetically Useful Oxygen Nucleophile" *Tetrahedron Lett.* **1975**, 3183-3186.

would be interesting to see if computational chemistry (not really developed in 1972) could have guided the synthetic effort associated with this somewhat Edisonian approach to the problem. Finally, note that the reducing agent (**71**) is chiral (and a single enantiomer) and thus, some "reagent control" may also be operating in this example.

Prostaglandins-12

Now let's consider an example of reagent control of stereochemistry. Treatment of **72** with a catalytic amount of **74** (derived from proline) and "stoichiometric" amounts of borane-THF complex gives **73**. Use of *ent*-**74** gives 15-*epi*-**73** with the same magnitude (but opposite sense) of asymmetric induction. It is clear that the reduction reagent ignores everything but the carbonyl group (and its *environs*) and controls the sense of the reduction. In this reaction, **74** really plays the role of a *pre*-catalyst with **75** being the actual catalyst. An explanation for the stereochemical sense of the reduction involves: (1) initial complexation the Lewis acidic boron with the Lewis basic carbonyl group. Complexation occurs on the convex face of the azaboraoxabicyclo[3.3.0]octane and opposite R_L of the carbonyl compound (the vinyl group in this case); (2) intramolecular transfer of hydride to the carbonyl group, and (3) "ligand exchange" at nitrogen and boron to regenerate the catalyst.

Note that if a reagent selects for the undesired stereoisomer, as in the conversion of **72** → **73**, this can sometimes be corrected by inverting configuration. In the present case the Corey group develop potassium superoxide as a reagent for accomplishing this task (inversion of 2° alcohol **77** to **78**).

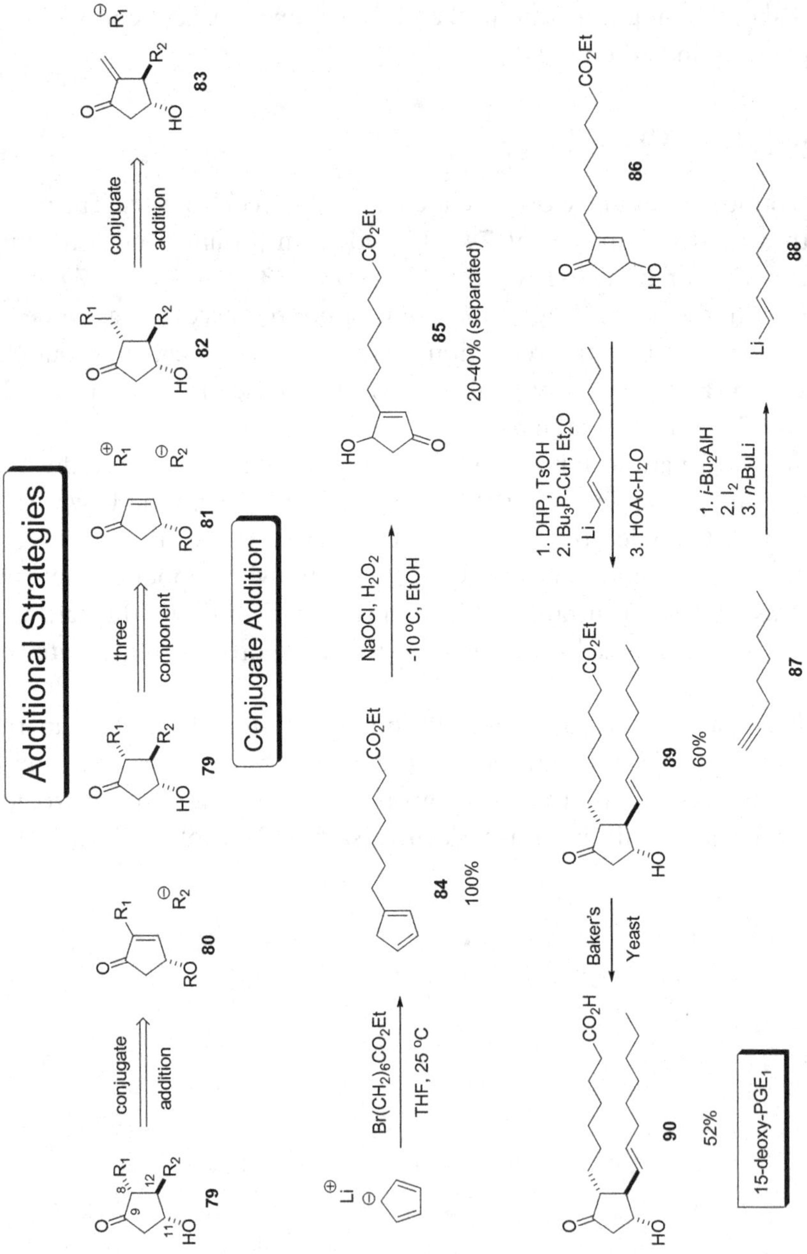

Sih, C. J.; Salomon, R. G.; Price, P.; Peruzotti, G. Sood, R. "Total Synthesis of *dl*-15-Deoxyprostaglandin E_1" *J. Chem. Soc., Chem. Commun.* **1972**, 240

Prostaglandins-13

Prostaglandins-13

Many other strategies have been adopted for the synthesis of prostaglandins. Three are shown here. One approach involves conjugate addition of the C_{12} sidechain (R_2) to a cyclopentenone of type **80** (**80** → **79**). Presumably the C_{11} substitutent would control the stereochemistry at C_{12} (addition of nucleophile from opposite side of ring) and thermodynamics could then be used to control stereochemistry at the epimerizable C_8 center. If the absolute stereochemistry at C_{11} is fixed, and the absolute stereochemistry at C_{15} in the nucleophile (R_2) is also set, control of stereochemistry between the ring stereogenic centers in the PGE series of prostaglandins can be controlled based on choice of the coupling partners. A related "three-component coupling" approach involves conjugate addition of the C_{12} sidechain to enone **81** followed by trapping of the enolate with the C_8 sidechain as an electrophile to provide **79**. Stereochemistry would presumably be controlled by kinetics, that is addition of nucleophile and electrophile from the face of the five-membered ring opposite vicinal substitutents. A third approach involves preparation of an α-methylene cyclopentanone (**83**) and introduction of the remainder of the C_8 sidechain via a conjugate addition. We will look at the first two strategies in some detail, followed by a glance at the α-methylene cyclopentanone strategy.

All of these syntheses have convergency in common. For that matter, the "Corey Lactone" approach to prostaglandins is also convergent. In a convergent synthesis, several fragments are prepared and then assembled using key bond constructions. Convergency miminizes the longest linear sequence of reactions and can increase overall yield. For example, a 10-step sequence with a 90% yield at each step would give a 35% yield of final product. If the same target could be prepared by conducting a 5-step sequence, 4-step sequence, and a 1-step coupling reaction, each proceeding in 90% yield, the overall yield would be 53%. Some targets are amenable to convergent approaches and others are not. But if possible, convergency is desireable.

Let's start with a synthesis of racemic 15-*deoxy*-PGE_1 (**90**) developed by the Sih group at the University of Wisconsin. The key reaction was the addition of vinyllithium **88** to the THP ether derived from enone **86**. The vinyllithium reagent was prepared by hydroalumination-iodination of 1-octyne (**87**) followed by a transmetallation. Enone **86** was prepared in two steps from cyclopentadiene and ethyl 7-bromoheptanoate. Notice that the alkylation of lithium cyclopentadienide with this bromoester provided 1-substituted cyclopentadiene **84** in quantitative yield. The isomerization that had to be avoided in the Corey lactone approach was a necessary part of these tactics.

Adaptation for Asymmetric Synthesis

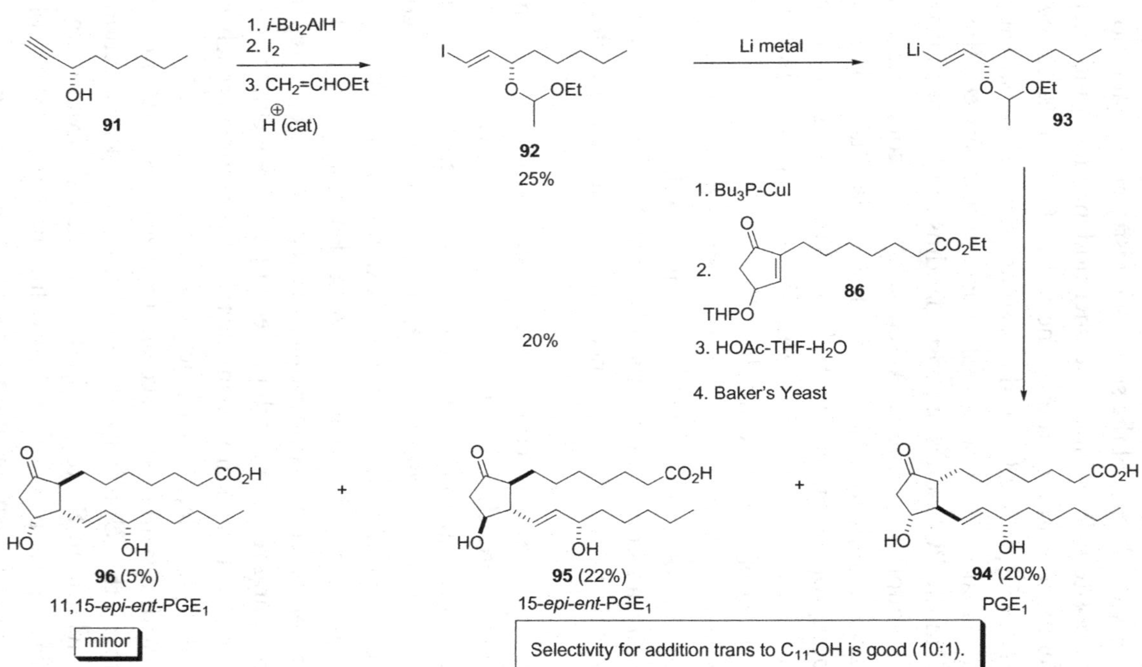

Selectivity for addition trans to C_{11}-OH is good (10:1).

Sih, C. J.; Price, P.; Sood, R.; Salomon, R. G.; Peruzzotti, G.; Casey, M. "Total Synthesis of Prostaglandins. II. Prostaglandin E_1" *J. Am. Chem. Soc.* **1972**, 94, 3643

Prostaglandins-14

Treatment of **84** with singlet oxygen, generated from bleach and hydrogen peroxide, gave **85** and **86** as a separable mixture. This transformation actually takes place by a cycloaddition between $^1O_2$ and the 2-substituted cyclopentadiene, presumably in equilibrium with **84**. The intermediate cyclic peroxide presumably undergoes fragmentation to provide the isomeric cyclopentenones. Protection of **86** as its THP ether was followed by addition of **88** as the corresponding cuprate, and protonation of the enolate to provide **89**. The synthesis of **90** was completed by enzymatic hydrolysis of the ethyl ester.

Prostaglandins-14

The Sih approach was readily adapted to an asymmetric synthesis of PGE_1. Alcohol **91** was prepared as a single enantiomer and converted to **92** using standard methodology. Note that installation of the ethoxyethyl ether (EE) generates a new stereogenic center. Vinyl iodide **92** was actually a mixture of diastereomers. This is a practical problem associated with EE and THP protecting groups that can complicate characterization of intermediates. Nontheless, vinyllithium **93** was prepared (once again as a presumed mixture of diastereomers), converted to the corresponding cuprate, and reacted with racemic **86** (2 equivalents of cuprate). Hydrolysis of the THP protecting group and enzymatic hydrolysis of the ester provided PGE_1 (**94**) and 15-*epi-ent*-PGE_1 (**95**) as the major products and 11,15-*epi-ent*-PGE_1 (**96**) as a minor product. The origin of PGE_1 (**94**) and **96** was (*R*)-**86**, while the origin of **95** was (*S*)-**86**. The cuprate addition occurred trans to the C_{11} hydroxyl group with good selectivity. This synthesis is short, but is complicated by formation of mixtures because of the use of (*rac*)-**86**. So let's look at a strategically related synthesis that addresses this problem.

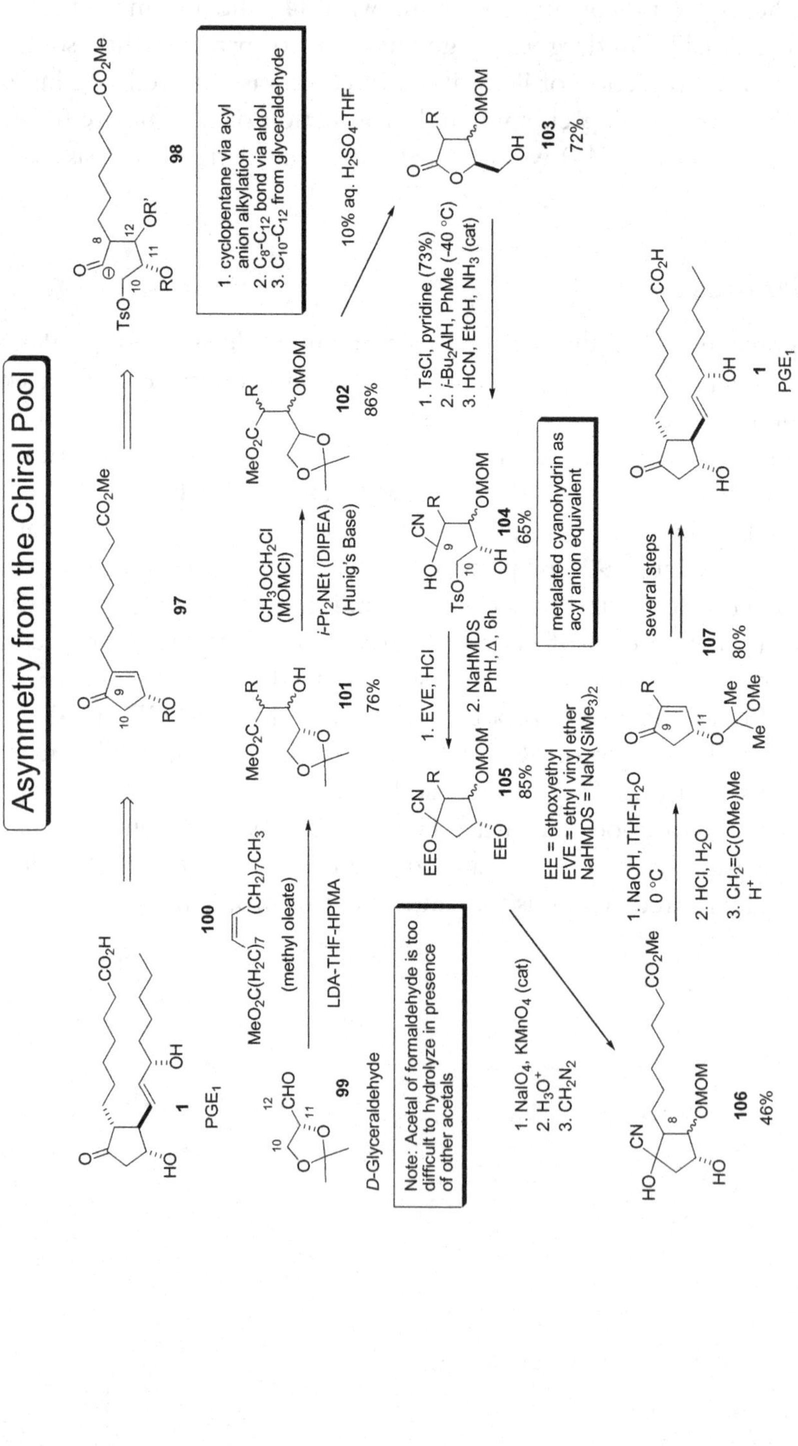

Stork, G.; Takahashi, T. "Chiral Synthesis of Prostaglandins (PGE$_1$) from *D*-Glyceraldehyde" *J. Am. Chem. Soc.* **1977**, *99*, 1275-1276

Prostaglandins-15

Prostaglandins-15

The Stork group developed several prostaglandin syntheses, one of which is outlined here. The convergent endgame (**97** → **1**) is related to the Sih synthesis we just reviewed. The synthesis of the cyclopentenone (**97**), however, is quite different. The plan was to prepare an acyl anion equivalent of type **98**.[14] An intramolecular alkylation would construct the 5-membered ring. This was to be followed by a β-elimination of the C_{12} substituent to introduce the double bond. There is a general strategic message here. When making an unsaturated ring (particularly one that will not accommodate a *trans*-double bond), make the ring first (make the σ-bond) and then introduce the double bond (make the π-bond). A double bond places geometric constraints on a ring-forming reaction (the substituents on the double bond involved in ring formation must have a *Z*-relationship). This can be problematic. We have seen this strategy before. For example, the aldol-dehydration route to cycloalkenes makes the ring first (σ-bond formation in an aldol reaction) and introduces the double bond second (π-bond formation via dehydration).

"Chiral pool" is a term used loosely to describe all enantiomerically pure materials that are available from natural sources in such abundance that they can be used as starting materials for synthesis.[15] Glucose (a carbohydrate), citronellol (a terpene), and proline (an amino acid) are all members of the "chiral pool". The Stork synthesis begins with D-glyceraldehyde acetonide (**99**), a common triose derivative. The chiral center in **99** is ultimately to become C_{11} of the prostaglandins. An aldol condensation between **99** and the enolate derived from methyl oleate (**100**) provided **101** as a mixture of diastereomers. Protection of the hydroxyl group and acetonide hydrolysis was accompanied by lactonization to give **103**. To convert **103** to an acyl anion equivalent of type **98**, the primary hydroxyl group had to be activated for displacement, the lactone had to be reduced to an aldehyde, and the "polarity" of the aldehyde had to be inverted because deprotonation of the aldehyde to directly provide **98** is not possible. These tasks were accomplished in sequence by formation of the primary tosylate, reduction of the lactone to the lactol (a cyclic hemiacetal form of the aldehyde) and cyanohydrin formation. We saw one application of the relationship between carbonyl compounds and their cyanohydrin derivatives in an earlier prostaglandin synthesis (Prostaglandins-6). Now we will see another. Cyanohydrin **104** was converted to the corresponding ethoxyethyl (EE) ether. Deprotonation of the nitrile was followed by an intramolecular alkylation to give **105**. Hydrolysis of the EE ether followed by reverse cyanohydrin formation would reveal the C_9 ketone and establish the metalated cyanohydrin derivative as an acyl anion

Three-Component Coupling Approach

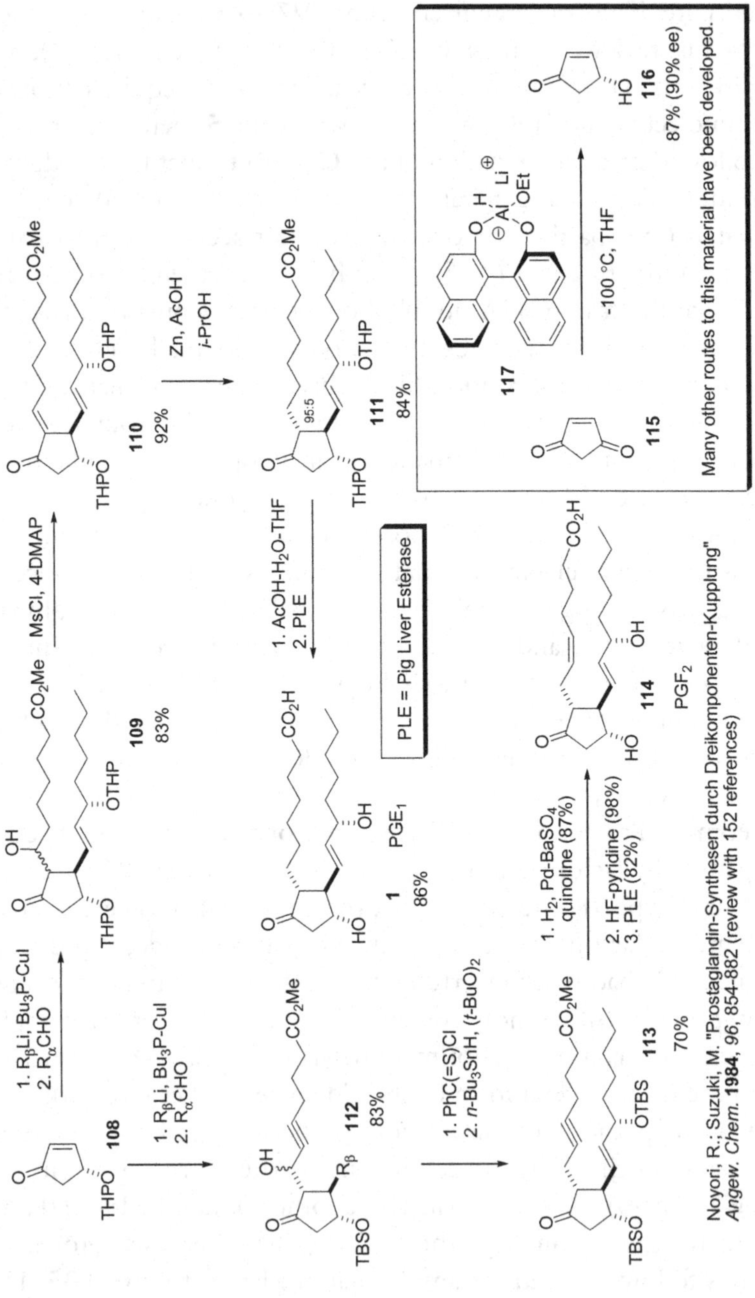

Noyori, R.; Suzuki, M. "Prostaglandin-Synthesen durch Dreikomponenten-Kupplung" *Angew. Chem.* **1984**, *96*, 854-882 (review with 152 references)

Prostaglandins-16

equivalent, but this was not done immediately. First the C_8 sidechain olefin was oxidatively cleaved to afford an acid, the EE protecting groups were removed, and the acid was converted to the corresponding methyl ester using diazomethane. We can now see the reason for selecting methyl oleate as the nucleophile in the starting aldol condensation. The olefin was serving as an equivalent of the sidechain carbonyl group (recall olefins serving as carbonyl precursors in several of the steroid synthesis seen in Chapter **2**). Notice that the olefin was benign throughout the transformation of **99** → **105**. Consider the difficulties that might have been encountered if an attempt had been made to carry C_1 through as an acid derivative. The synthesis of **107** (a specific example of **97**) was accomplished by transforming the cyanohydrin into a ketone, β-elimination of the C_{12} substitutent, and protection of the C_{11} hydroxyl group. Note that the acetal protecting group in **107** does not introduce a new stereogenic center (unlike EE and THP protecting groups). The synthesis of PGE_1 (**1**) was completed in several steps by addition of the lower sidechain in a manner similar to what we have already seen.

Prostaglandins-16

The Noyori group championed the three-component coupling approach to prostaglandins. This is illustrated here with syntheses of PGE_1 (**1**) and PGF_2 (**114**). Cuprate additions were used to introduce the C_{12} sidechain and aldol condensations of the resulting enolate were used to attach the C_8 sidechain. Use of an aldol condensation leaves the C_8 sidechain in too high an oxidation state for PGE_1. This was corrected by dehydration of **109** to **110**, followed by reduction of the electron-deficient alkene (without disturbing the relatively electron-rich C_{13}-C_{14} olefin). The PGF_2 synthesis also required some adjustments in oxidation state after the three-component coupling step. The aldol product (**112**) hydroxyl group was reduced using the Barton-McCombie reduction (radical intermediates).[16] Lack of complications from potential propargyl-allenyl radical isomerization is notable. Semi-hydrogenation of the alkyne was used to establish the C_5-C_6 *Z*-olefin. One of several enantioselective routes to **108** (via **116**) is shown without comment.

A Variation on the Three-Component Theme

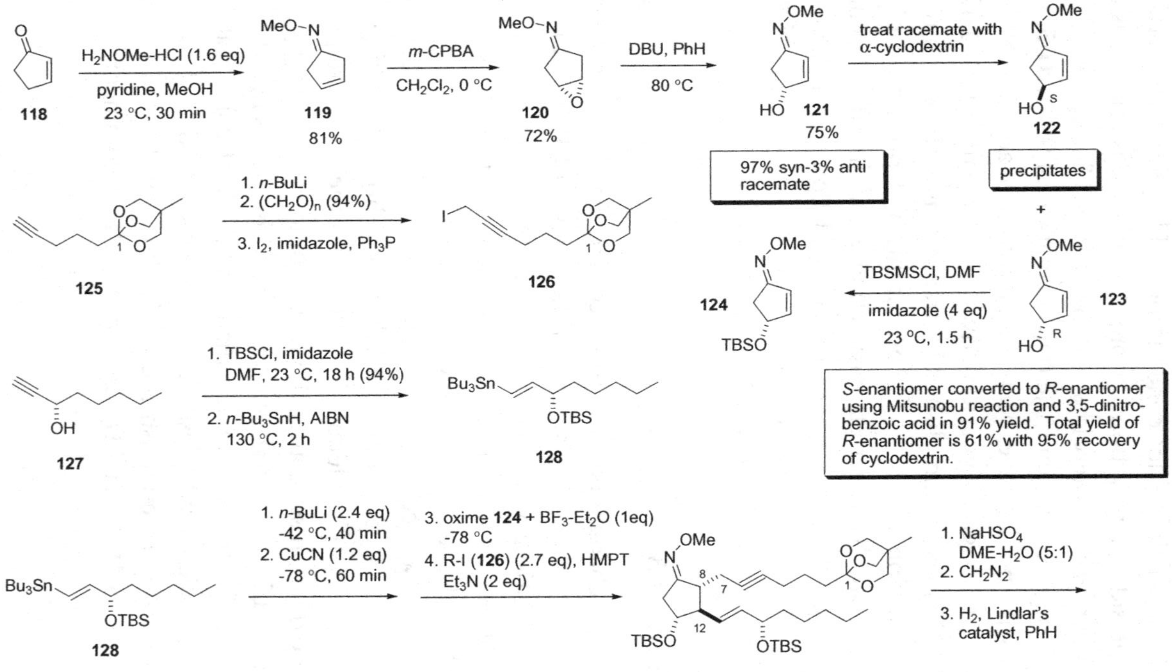

129 (60%) + 16% of recovered oxime ether

Corey, E. J.; Niimura, K.; Konishi, Y.; Hashimoto, S.; Hamada, Y. "A New Synthetic Route to Prostaglandins" *Tetrahedron Lett.* **1986**, *27*, 2199.

Prostaglandins-17

Prostaglandins-17

Another example of the three-component coupling approach to PGs is shown here. This variation uses oxime ether **124** as the cyclopentenone component, a mixed cuprate derived from vinylstannane **128** as the C_{12} sidechain, and introduces the C_8 sidechain via an alkylation reaction (rather that the aldol condensation used by Noyori) with iodide **126** serving as the electrophile. The preparation of **124** began with conversion of cyclopentenone (**118**) to oxime ether **119**. Deconjugation of the double bond was observed in this reaction, presumably because the carbonyl group of the β,γ-unsaturated cyclopentonone is more electrophilic (and thus undergoes oxime formation at a faster rate) than the carbonyl group of **118** itself. Deconjugation is a fairly common observation when derivatizing a conjugated enone. Epoxidation of **119** was followed by a β-elimination reaction to give **121** (and its enantiomer **122**) which were resolved by selective precipitation of **122** using α-cyclodextrin. The *R*-enantiomer (**123** = **121**) was converted to **124**. The *S*-enantiomer (**122**) was also kept in play using a Mitsunobu alcohol inversion-protection sequence. Iodide **126** was prepared from terminal alkyne **125** in a straightforward manner. Use of an orthoester protecting group for the incipient C_1 carboxyl group was unusual. This so-called OBO (triOxaBicycloOctane) protecting group was developed for this purpose as shown on Prostaglandins-18 without comment. Alcohol **127** was protected and converted to vinylstannane **128** using a free radical hydrostannylation. Stannane **128** was converted to the corresponding organolithium reagent by a transmetalation reaction, and subsequently to a mixed cuprate. Reaction of the cuprate with oxime ether **124**, followed by trapping of the resulting anion with iodide **126**, completed the three-component coupling to provide **129**.

This version of the three-component coupling provides the "carbon skeleton" of the prostaglandins with the required oxidation state at C_7. This is a consequence of introducing the C_8 sidechain via an *alkylation* reaction rather than an *aldol* reaction. One can speculate that the "enolate" derived from oxime ether **124** is better behaved in alkylation reactions than the enolate derived from ketone **108** (Prostaglandins-16), permitting this change in tactics. This may be because it is a better nucleophile or undergoes proton-transfer reactions (after alkylation) more slowly due to reduced acidity of the cyclopentanone oxime ether (from **124**) relative to the cyclopentanone (from **108**). There are trade-offs here with the differing tactics. The Corey approach eliminates the need for an oxidation state adjustment at C_7 (needed in the Noyori approach), but requires the use of more protecting group chemistry.

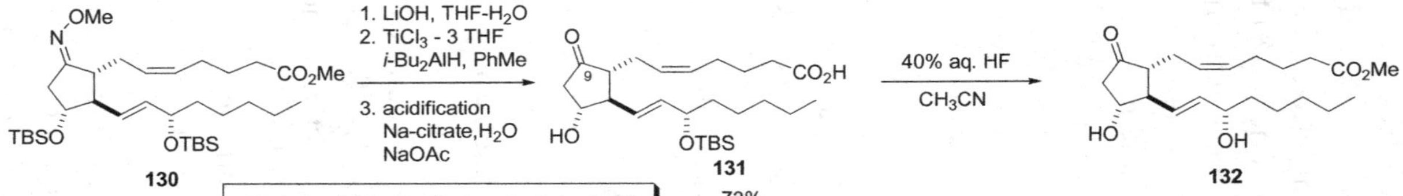

Ti reagent (black) effects reduction of N-O bond: C=NOMe to C=NH

OBO-Ester Protecting Groups

Corey, E. J.; Raju, N. "A New General Synthetic Route to Bridged Carboxylic Ortho Esters" *Tetrahedron Lett.* **1983**, *24*, 5571.

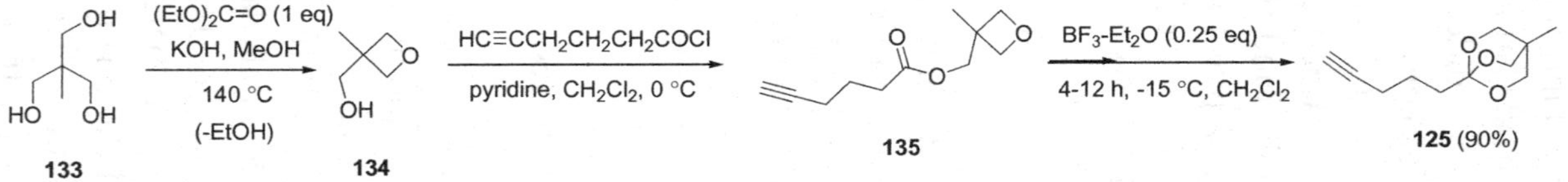

sidechain = $CH_2CH_2CH_2Br$ (91%), n-C_5H_{11} (90%), CH_2CH_2Ph (75%)

α-Methylenecyclopentane Strategy

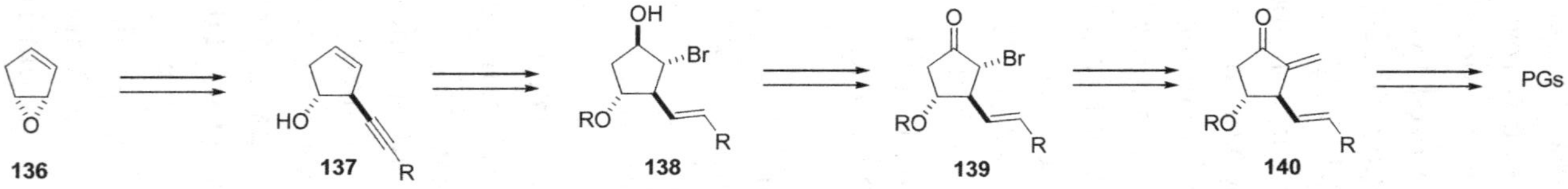

Stork, G.; Isobe, M. "A General Approach to Prostaglandins via Methylenecyclopentanones. Total Synthesis of *dl*-Prostaglandin $F_{2\alpha}$" *J. Am. Chem. Soc.* **1975**, *97*, 4745-4746.

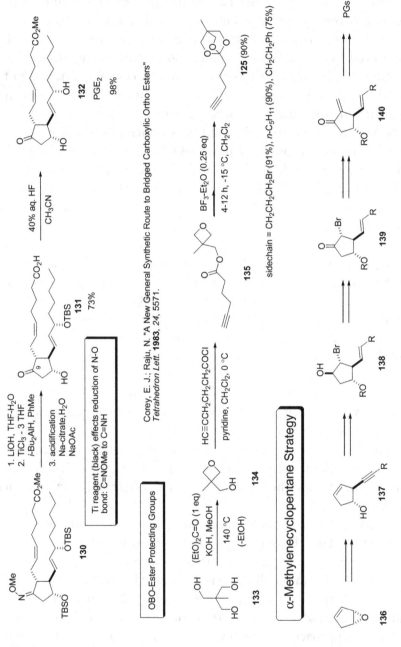

Prostaglandins-18

Prostaglandins-18

The Corey group version of the three-component coupling approach was completed by hydrolysis of orthoester **129**, conversion of the C_1 acid to the corresponding methyl ester, and semi-hydrogenation of the alkyne to give **130**. The ester was hydrolyzed and the oxime ether was then converted to the C_9 carbonyl group (**131**). The protecting groups were then removed to give PGE_2 (**132**).

The third strategy mentioned on Prostaglandins-13 was the α-methylenecyclopentane strategy. This route was introduced by the Stork group and followed the reaction sequence outlined at the bottom of this page. This is clearly a variation of the three-component coupling approach, but it is less efficient because two carbon-carbon bond-forming reactions are used to transform **139** to the PGs.

Organopalladium Chemistry and Prostaglandins

Holton, R. A. "Prostaglandin Synthesis via Carbopalladation" *J. Am. Chem. Soc.* **1977**, *99*, 8083-8085

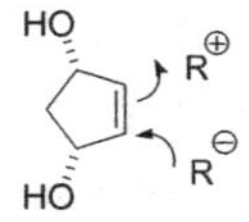

141

Unactivated olefin requires: (1) activation and (2) direction to control stereochemistry

Model Studies

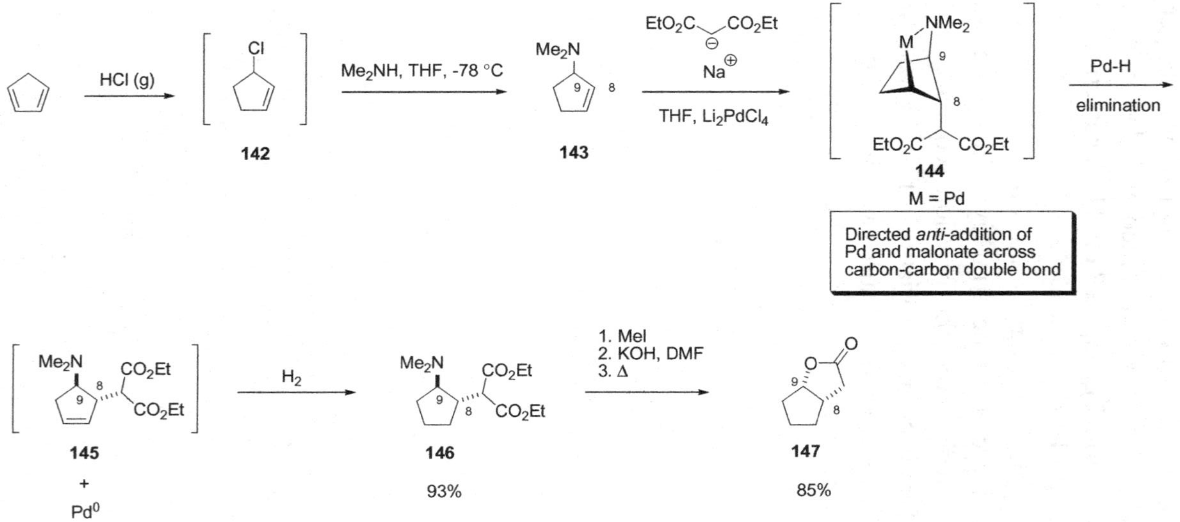

Prostaglandins-19

We will next focus on the ambitious strategy described by structure **141**. The idea is to develop an equivalent to **141** that would allow addition of one sidechain to an unactiviated olefin, and capture the reactive intermediate with the other sidechain. For the sake of simplicity, I have presented this as though one sidechain is introduced as a nucleophile and the other as an electrophile, but as we will see the strategy is not restricted to polar intermediates. The critical obstacles in this strategy are (1) identifying suitable reactions for use with an "unactivated" olefin and (2) using cyclopentane substitutents (for example the two hydroxyl groups) to control addition stereochemistry and regiochemistry. We will look at two approaches, one that involves organometallic intermediates, and the other involving free radical intermediates.

The Holten group (then at Purdue University) reported carbopalladation chemistry that was adaptable to this approach to prostaglandins. In model studies, it was found that allylic amine **143** reacted with diethyl sodiomalonate, in the presence of stoichiometric amounts of lithium tetrachloropalladate, to give **146** after exposure to hydrogen. This reaction apparently involves an "amine-directed" *anti*-electrophilic addition of electrophile (Pd^{II}) and nucleophile (malonate) across the olefin (**143** → **144**), followed by a *syn*-elimination of a palladium hydride (Pd-H) to afford **145**, concluding with hydrogenation of the double bond to provide **146**. Quaternization of the amine, conversion of the malonate to a carboxylate, and an intramolecular displacement reaction, gave lactone **147**. The similarity between **147** and the Corey lactone is clear.

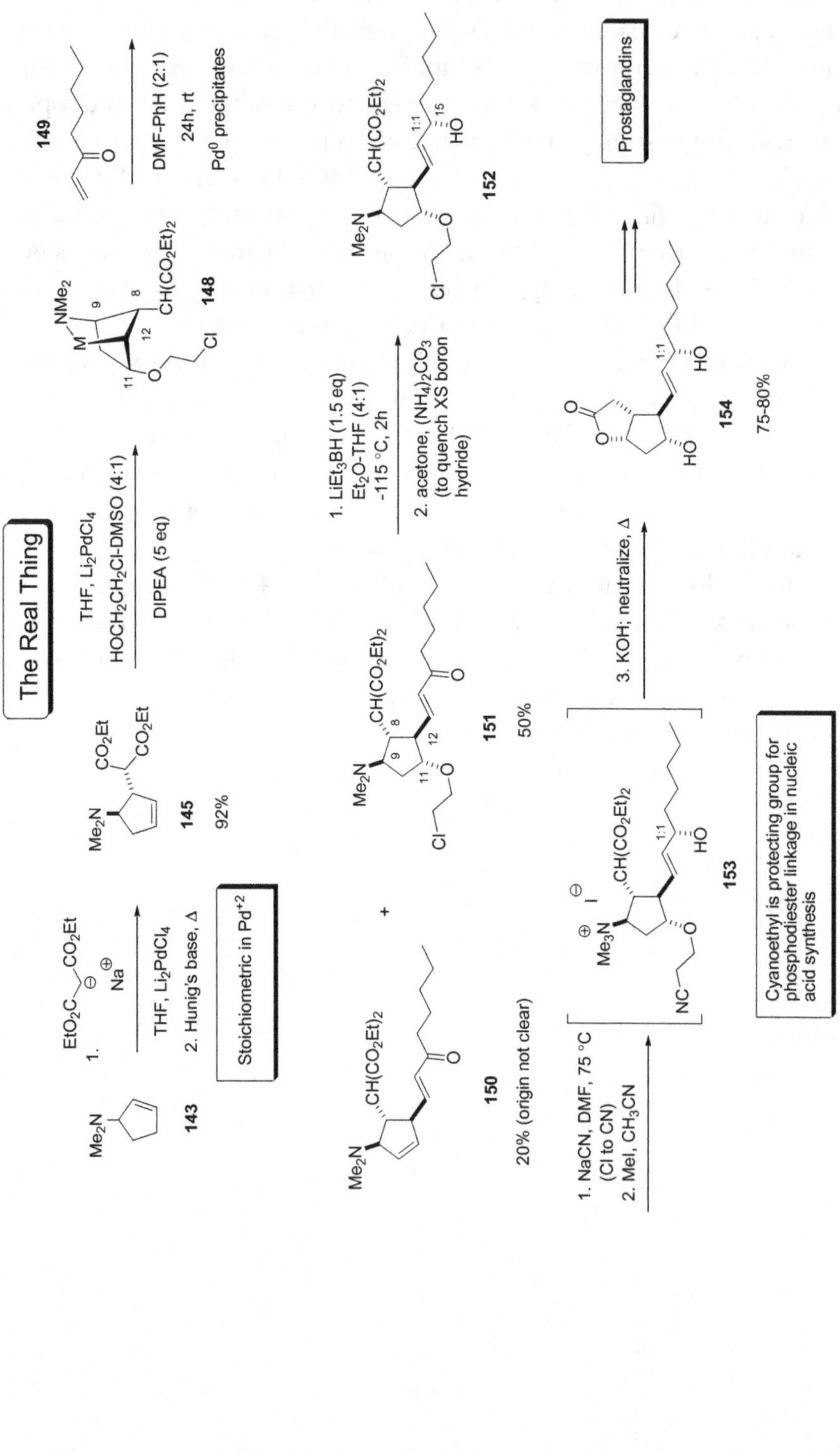

Prostaglandins-20

Prostaglandins-20

The key to adapting this chemistry to the "real thing" was harnessing electrophilic addition chemistry of aminoalkene **145**. Thus, omission of hydrogen allowed isolation of **145** in excellent yield. Repeating the "amine-directed" addition chemistry, with 2-chloroethanol as the nucleophile, presumably gave *anti*-oxypalladation product **148**. This presumed intermediate was unstable, but reacted with enone **149** to provide a 50% overall yield of enone **151** along with some **150** as a side product. The desired product (**151**) was presumably formed by a carbopalladation-dehydropalladation reaction that occurred with retention of configuration at C_{12} of intermediate **148**. This is a variation of the now well-known Heck reaction.[17] It is possible that **150** is derived from **145** via a π-allyl palladium intermediate, followed by the Heck-type reaction. The synthesis of a PG intermediate was accomplished from **151** by reduction of the C_{15} carbonyl group to give a mixture of diastereomeric alcohols **152**. This was followed by conversion of the chloride to a nitrile, to facilitate removal of this C_{11} hydroxyl protecting group. Quaternization of the amine, conversion of the malonate to a carboxylate, and lactone formation as in the model study, completed the synthesis of **154**. It is notable that deprotection of the C_{11} hydroxyl group via a retro-1,4-addition accompanied the malonate hydrolysis and decarboxylation. The Holten synthesis of prostaglandins was truly ground-breaking research as it established the potential of organopalladium chemistry as a tool for the synthesis of structurally complex natural products.

Stork's Free Radical Approach

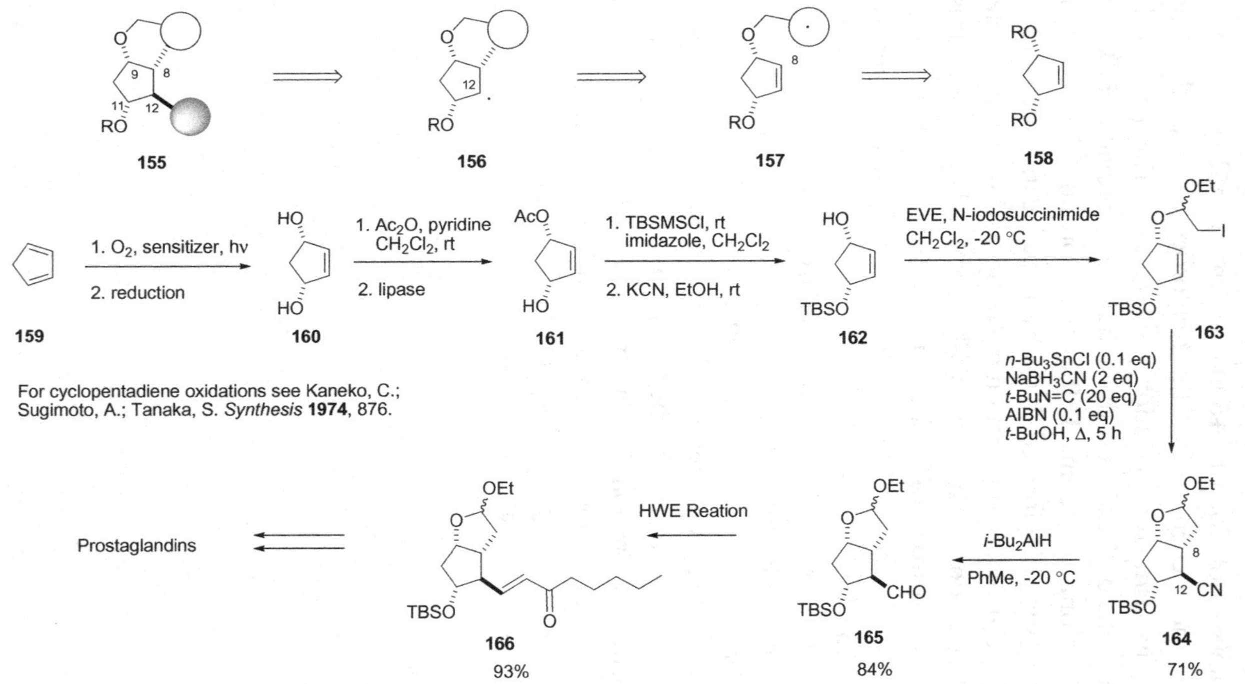

For cyclopentadiene oxidations see Kaneko, C.; Sugimoto, A.; Tanaka, S. *Synthesis* **1974**, 876.

Stork, G.; Sher, P. M.; Chen, H-L. "Radical Cyclization-Trapping in the Synthesis of Natural Products. A Simple, Stereocontrolled Route to Prostaglandin $F_{2\alpha}$" *J. Am. Chem. Soc.* **1986**, *108*, 6384.

Prostaglandins-21

A strategically related, but tactically different, synthesis of prostaglandins was developed by the Stork group.[18] This synthesis features the development of stereo- and regioselective free radical addition reactions for the functionalization of "unactivated olefins". The idea was to prepare a generic PG precursor of type **155** from a diol derivative of type **158**. It was imagined that one of the alkoxy groups could "direct" a free radical addition to C_8 of **157**, and that capture of the resulting C_{12} radical (**156**) from the convex face would give **155**. The initial free radical was to be derived from iodide **163**.

Iodide **163** was prepared from cyclopentadiene (**159**). A singlet oxygen cycloaddition was followed by reduction of the peroxide to provide diol **160**. Esterification and enzyme-mediated hydrolysis of the diester gave **161** as a single enantiomer. Protection of the C_{11} hydroxyl group, hydrolysis of the C_9 acetate, and functionalization of the resulting alcohol gave **163**. Treatment of **163** with tri-*n*-butyltin radicals (generated using a variation of the Corey-Suggs method we saw earlier), in the presence of a large excess of *tert*-butylisonitrile, gave nitrile **164**. This reaction presumably involves an initial 5-hexenyl radical cyclization, followed by addition of the resulting C_{12} radical to the isonitrile carbon. The resulting iminoyl radical expels a *tert*-butyl radical to generate **164**. The fate of the *tert*-butyl radical is most likely to continue chains (abstract H-atom from Sn) and to isomerize the *tert*-butyl isonitrile to *tert*-butyl nitrile (hence the use of excess isonitrile). Conversion of nitrile **164** to aldehyde **165** and HWE olefination provided intermediate **166**, whose potential as an intermediate in PG synthesis should now be obvious.

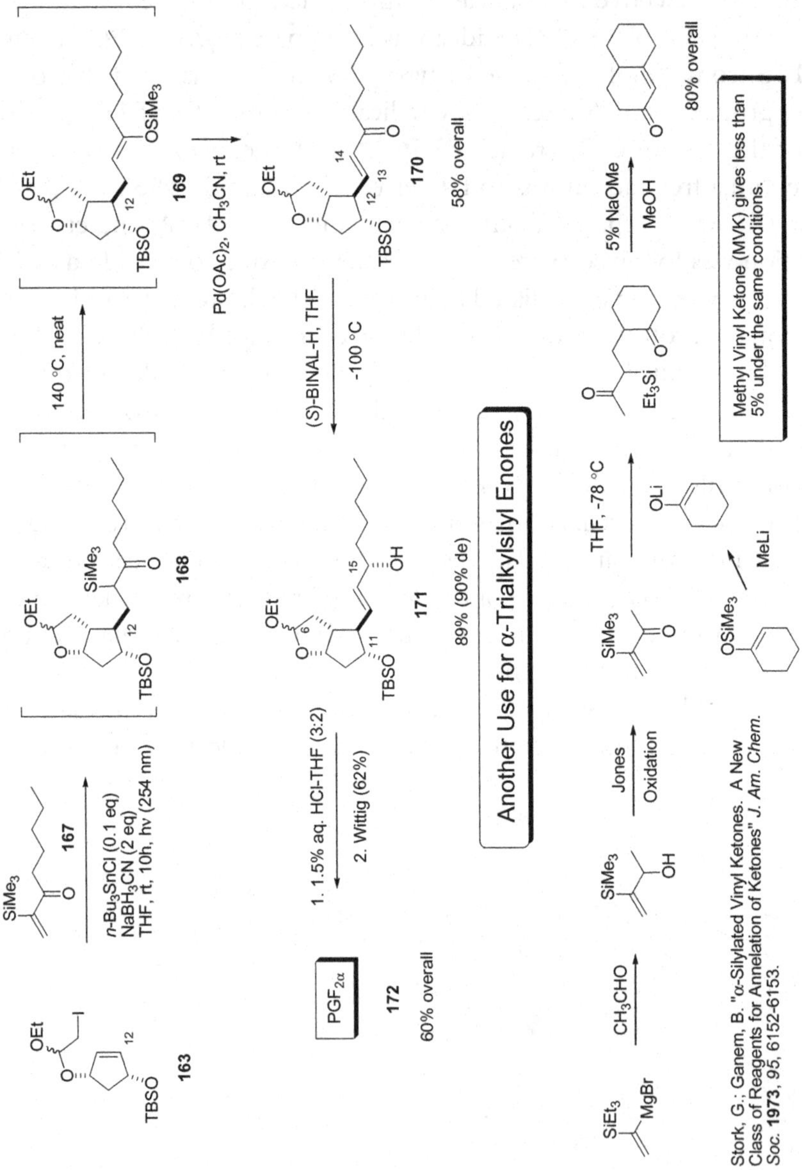

Stork, G.; Ganem, B. "α-Silylated Vinyl Ketones. A New Class of Reagents for Annelation of Ketones" *J. Am. Chem. Soc.* **1973**, *95*, 6152-6153.

Prostaglandins-22

Prostaglandins-22

The "isonitrile trap" approach requires two carbon-carbon bond-forming steps to introduce the C_{12} sidechain. A more convergent approach (one step for sidechain introduction) is shown here. Free radical generation from **163** in the presence of α-trimethylsilyl enone **167** provided (presumably) cyclization-trapping product **168**. Thermolysis of **168** resulted in "tautomerization" to α-trimethylsilyl enol ether **169**. Exposure of **169** to stoichiometric amounts of Pd^{II} introduced the C_{13}-C_{14} olefin in a "Saegusa oxidation" (α-palladation of the enol ether followed by a β-hydride elimination, much as in the Holton carbopalladation seen in Prostaglandins-20).[19] Application of a reagent-controlled reduction of the C_{15} ketone, hydrolysis of the C_6 acetal, deprotection at C_{11}, and Wittig olefination completed a synthesis of $PGF_{2\alpha}$ (**172**).

The Acyclic Diastereoselection Problem

Control of Vicinal Stereochemistry Through Control of Olefin Geometry

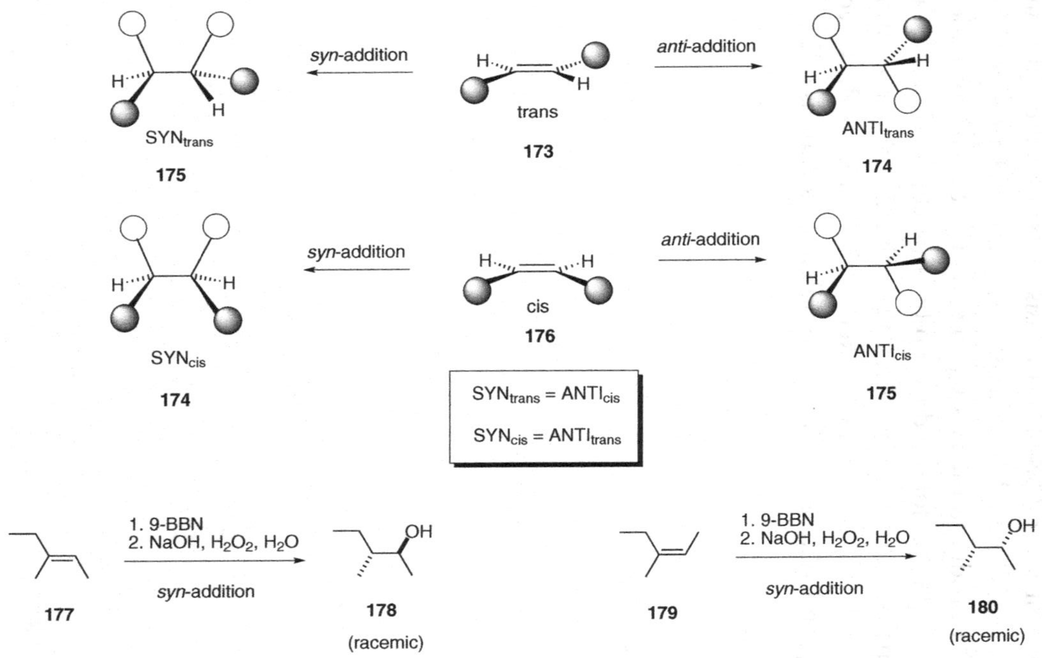

Prostaglandins-23

Prostaglandins-23

Let's look again at the problem of establishing the relative stereochemistry between the C_{15} and ring stereogenic centers in the prostaglandins. In the syntheses examined thus far, this problem has been handled in two ways: (1) use of ring stereogenic centers to control diastereoselectivity in reduction of a C_{15} ketone (Prostaglandins-11) and (2) independent control of absolute stereochemistry at C_{15} either before (Prostaglandins-12, 16, 17) or after (Prostaglandins-22) introduction of the C_{12} sidechain. Let's look at another approach to this acyclic diastereoselection problem that has proved useful well beyond the context of prostaglandin synthesis (as have the two strategies we have already studied). We will start by examining two general topics in stereocontrol: (1) control of vicinal stereochemistry through control of olefin geometry and (2) transfer of chirality.

There are a number of approaches to controlling vicinal stereochemistry. One useful method involves olefin addition chemistry. For example, *anti*-addition of two "groups" (white balls) to *trans*-olefin **173** affords **174**, whereas *syn*-addition of the same two groups to **173** gives **175**. Adducts **174** and **175** are diastereomers. Therefore the stereochemistry of the addition, "syn" or "anti", dictates vicinal stereochemistry in the product. If we start with the olefin geometrical isomer (**176**), the *anti*-addition provides **175** and the *syn*-addition gives **174**. Thus, starting olefin geometry also dictates vicinal stereochemistry in the product. The mechanism by which the addition takes place is of no real consequence, as long as it is a stereospecific process (syn or anti). Electrophilic addition reactions and cycloaddition reactions, for example, are two very different families of reactions that can be used in this approach to control of vicinal stereogenic centers.

The hydroboration-oxidation of olefins is one example of this approach. The process involves a stereospecific *syn*-addition of "H" and "OH" across the olefin π-bond. This reaction sequence is regioselective with unsymmetrical olefins. The diastereoselective conversion of *E*-olefin **177** to **178**, and the isomeric *Z*-olefin **179** to **180**, illustrates this process. Note that this discussion applies to relative control of stereochemistry. Whereas use of an enantioselective hydroborating agent might afford **178** and **180** as single enantiomers (or enriched in one enantiomer), **177** would still provide **178** and the diastereomeric olefin (**179**) would provide the diastereomeric alcohol (**180**).[20]

Transfer of Chirality

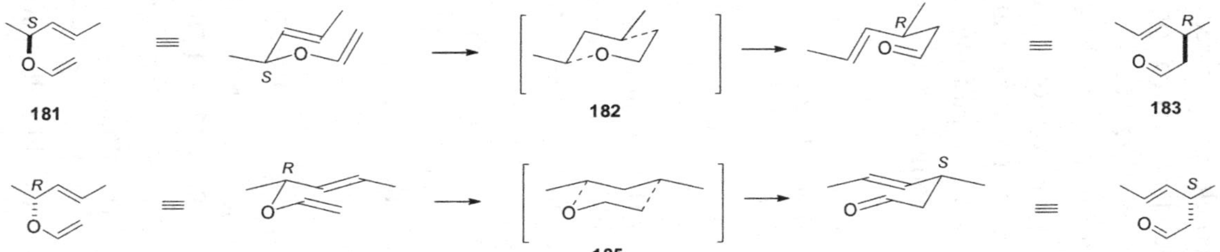

181 182 183

184 185 186

Stereoisomeric starting materials lead to stereoisomeric products. Understand that the absolute configuration of starting materials and products are not important. For example it is possible to have an and *R*/*S* pair of starting materials that rearrange to an *R*/*S* pair of products with the *R* starting materials providing the *R* product and the *S* starting material provides the *S* product. Many sigmatropic rearrangements take place with excellent transfer of chirality (not always 100% due to stereochemical leakage due to competing rearrangment transition states). The Claisen rearrangement (above) is one example. The Mislow-Evans rearrangement (sulfoxide to sulfenate ester as shown below) is another.

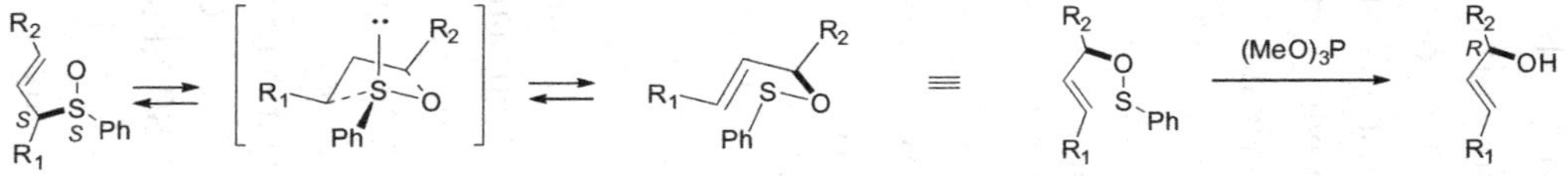

187 189 191 192

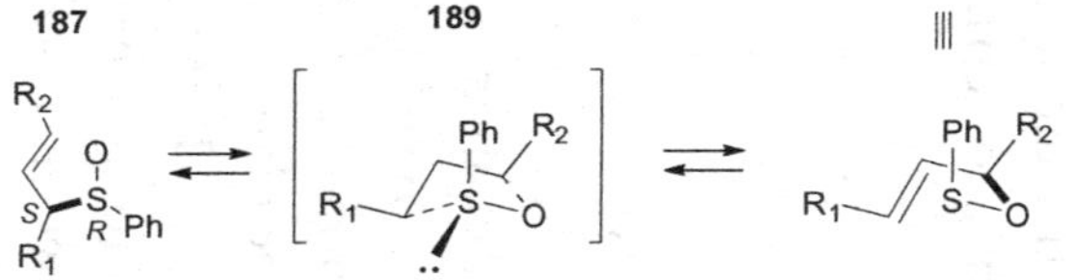

188 190 191

In principle, diastereomeric sulfoxides **187** and **188** can interconvert via **191** and also both afford a single allylic alcohol (**192**).

For a review on transfer of chirality see: Hill, R. K. "Chirality Transfer via Sigmatropic Rearrangements" in "*Asymmetric Synthesis*", Morrison, J. D., Ed.; Academic Press, **1984**, *3*, 503-572 (266 references).

Prostaglandins-24

What do we mean by transfer of chirality? Reactions in which the stereoselective formation of one stereogenic center is connected to the destruction of another stereogenic center involve transfer of chirality.[21] Sigmatropic rearrangements are a large family of intramolecular reactions that often can be used to transfer chirality. We will look at two examples: (1) the Claisen rearrangement (of allyl vinyl ethers to β,γ-unsaturated carbonyl compounds) and (2) the Mislow-Evans rearrangement (of allylic sulfoxides to allylic alcohols via sulfenate esters).

Thermal rearrangement of **181** provides **183**, presumably via chairlike transition state **182**. This reaction involves transfer of chirality from a carbon-oxygen bond to chirality in a carbon-carbon bond. The transfer of chirality is high. Stereoisomeric starting materials lead to stereoisomeric products. For example **184** (the enantiomer of **181**) affords **186** (the enantiomer of **183**). The designation of absolute configuration in starting materials and products are not important. For example it is possible to have an and *R*/*S* pair of starting materials that rearrange to an *R*/*S* pair of products where the *R* starting material provides the *R* product and the *S* starting material provides the *S* product.

Many sigmatropic rearrangements take place with excellent transfer of chirality (not always 100% due to stereochemical leakage due to competing rearrangment transition states). Claisen rearrangements (including many variations of the reaction shown here) are one example. The Mislow-Evans rearrangement (sulfoxide to sulfenate ester) is another. For example, in the presence of thiophilic reagents such as trimethyl phosphite, allylic sulfoxides of type **187** or **188** are converted to allylic alcohols **192** with excellent transfer of chirality. This reaction involves transfer of chirality from a carbon-sulfur bond to a carbon-oxygen bond. As with the Claisen rearrangement, the transfer of chirality is high because the transition state geometry in the critical [2,3]-sigmatropic rearrangement is well defined. The Evans-Mislow rearrangment is, in principle, a bit more complex than the Claisen from the standpoint of analysis. The sulfoxide sulfur represents a stereogenic center. Thus, **187** and **188** are diastereomers. Nonetheless, it is the stereogenicity at carbon that controls the rearrangment and thus, both diastereomers lead to **192** (largely) as single enantiomer. Note that the enantiomers of **187** and **188** should both provide the enantiomer of **192**. So what does this have to do with the acyclic diastereoselection problem in posed by the prostaglandins?

Application to Prostaglandin Problem

Taber, D. F. "Cyclopentanone Ring Formation with Control of Side Chain Stereochemistry. A Simple Stereoselective Route to the Prostaglandins" *J. Am. Chem. Soc.* **1977**, *99*, 3513-3514. For another application of the Mislow-Evans rearrangement to this problem see: Stork, G.; Raucher, S. *J. Am. Chem. Soc.* **1976**, *98*, 1583 and Kondo, K.; Umemoto, T.; Takahatake, Y.; Tunemoto, D. *Tetrahedron Lett.* **1977**, 113.

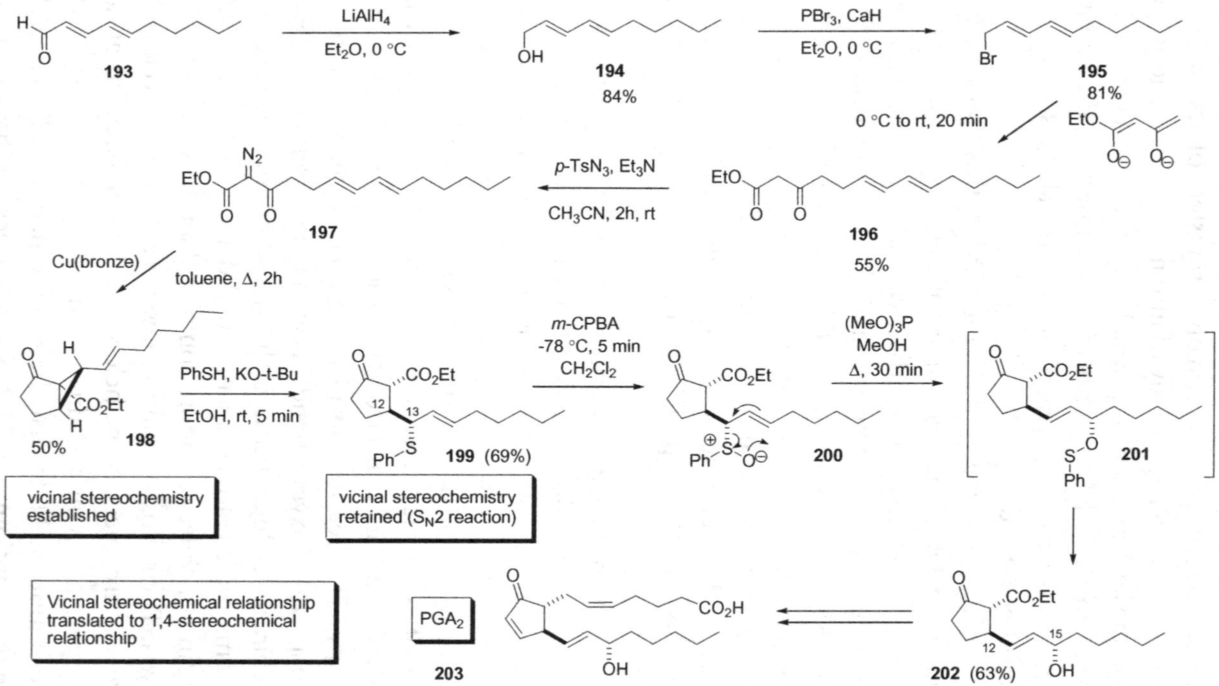

Prostaglandins-25

Prostaglandins-25

Within the context of a synthesis of PGA_2 (**203**), Taber united the two stereochemical topics we have just reviewed to establish the stereochemical relationship between ring stereogenic centers and the acyclic stereogenic center (C_{15}). There are three key steps in his plan: (1) Starting with diazo compound **197**, a *syn*-addition of a carbene to the C_{12}-C_{13} trans double bond established a well-defined vicinal stereochemical relationship between the incipient C_{12} and C_{13} carbons of PGA_2. (2) An S_N2 opening of **198**, using potassium thiophenoxide as a nucleophile, modified the vicinal stereochemical relationship (**198** → **199**). The overall conversion of **197** to **199** accomplished the *anti*-addition of a carbon electrophile and a sulfur nucleophile across a *trans*-olefin. (3) Oxidation of sulfide **199** provided a mixture of sulfoxides which underwent Mislow-Evans rearrangement to provide allylic alcohol **202** with transfer of chirality. The conversion of **199** to **202** involved translation of a 1,2-stereochemical relationship into a 1,4-stereochemical relationship with control of diastereoselectivity. The sequence of establishing 1,2-stereochemistry, followed by transfer of chirality to establish 1,4- (or other) stereochemical relationships is one we will see again.

Application to the Terpene Sidechain Stereochemical Problem

Trost, B. M.; Taber, D. F.; Alper, J. B. "An Approach to the Stereocontrolled Creation of an Acyclic Side Chain of Some Natural Products" *Tetrahedron Lett.* **1976**, 3857-3860.

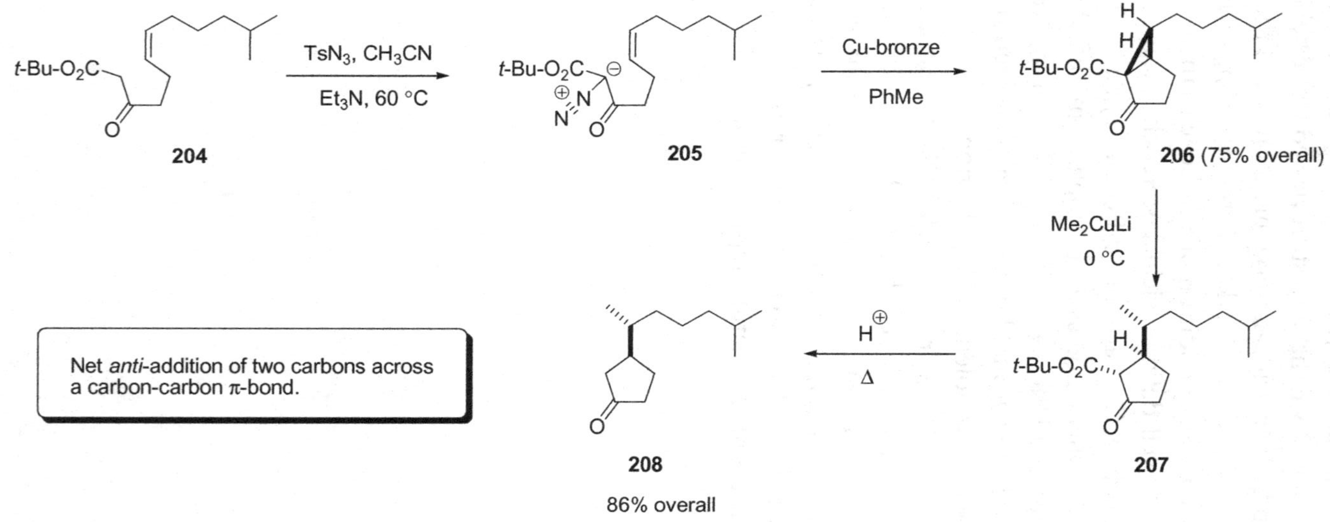

Net *anti*-addition of two carbons across a carbon-carbon π-bond.

Prostaglandins-26

Prostaglandins-26 illustrates how the vicinal stereocontrol exercise just examined within the context of prostaglandins has been applied to the steroid (or terpenoid) sidechain problem we encountered in Chapter 2. Diazoketone **205** (derived from **204**) was converted to **206** using an intramolecular carbenoid cyclopropanation of the *cis*-olefin. Opening of the cyclopropane with lithium dimethyl cuprate gave **207** which was then converted to **208**. The relationship between **208** and the D-ring of steroids (with appended C_{17} sidechain) is clear.

Geminally Activated Cyclopropanes as Carbon Electrophiles

Abraham, N. A. "Prostaglandin IX. A Simple Synthesis of Optically Active 11-Deoxyprostaglandins" *Tetrahedron Lett.* **1974**, 1393-1394.

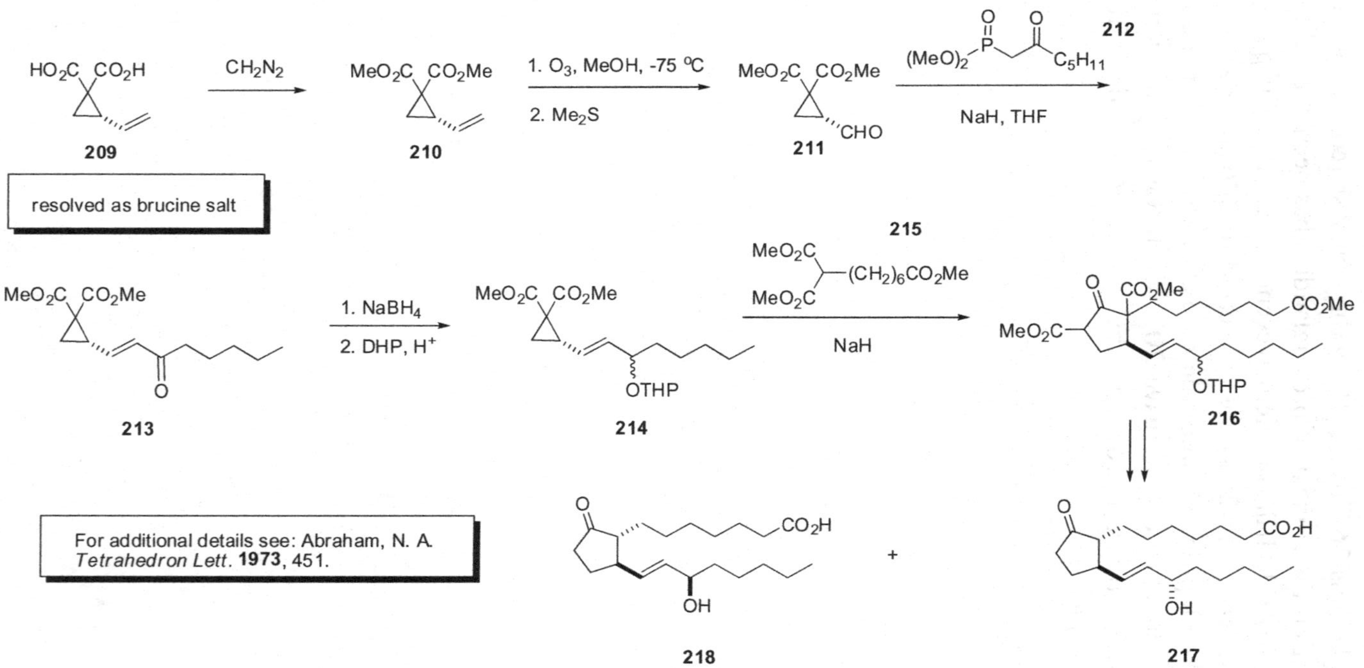

resolved as brucine salt

For additional details see: Abraham, N. A. *Tetrahedron Lett.* **1973**, 451.

Prostaglandins-27

Prostaglandins-27

Compounds of type **198** (Prostaglandins-25) and **206** (Prostaglandins-26) are known as geminally activated cyclopropanes. They are cyclopropanes with two electron-withdrawing groups positioned on one of the cyclopropane carbons. Geminally activated cyclopropanes are carbon electrophiles that react with nucleophiles in S_N2-like processes in which a carbon (positioned between the electron-withdrawing groups) behaves as a leaving group. It is this reactivity pattern that makes geminally activated cyclopropanes useful synthetic intermediates.

In our last look at prostaglandins, we can see this put to use in an approach to some prostaglandin derivatives. The plan was to construct PG derivatives from an intermediate of type **216**. This β-ketoester was to be derived from a Dieckmann condensation (construction of the C_9-C_{10} bond). The Dieckmann precursor was to be derived from the reaction of malonate **215** with geminally activated cyclopropane **214**. Note that control of absolute stereochemistry in this route leads back to **209**, which was resolved as its brucine salt. It is also notable that **214** reacts selectively with malonate **215** at the more hindered secondary (but allylic) carbon of the cyclopropane with clean inversion of configuration.

This brings to a close to our examination of routes to prostaglandins. I have chosen the geminally activated cyclopropane route as the last route to provide an entrée to our next topic, synthesis of pyrrolizidine alkaloids discussed in Chapter 4. This is not the only connection between these two topics. Prostaglandins and pyrrolizidines both contain 5-membered rings as important substructures, and the acyclic diastereoselection problem rears its head in both families of natural products. Thus, there will be some overlap in chemistry as well as some new strategies to study as we move forward.

References and Notes

1. Bergström, S.; Samuelsson, B. "Prostaglandins and Related Factors. XI. Isolation of Prostaglandin E1 from Human Seminal Plasma" *J. Biol. Chem.* **1962**, *237*, 3005–3006. Bergström, S.; Ryhage, R.; Samuelsson, B.; Sjovall, J. "Prostaglandins and Related Factors. XV. The structures of Prostaglandin E1, F1α, and F1β" *J. Biol Chem.* **1963**, *238*, 3555–3564. Samuelsson B.; Stallbert, G. "Prostaglandins and Related Factors. XVI. Structure and Synthesis of a Derivative of Prostaglandin E1" *Acta Chem. Scand.* **1963**, *17*, 810–816. Bergström, S. "Prostaglandins: Members of a New Hormonal System" *Science* **1967**, *157*, 382–391.
2. Samuelsson, B. "Biosynthesis of Prostaglandins" *Progr. Biochem. Pharmacol.* **1969**, *5*, 109–128. Bakhle, Y. S. "Structure of COX-1 and COX-2 Enzymes and Their Interaction with Inhibitors" *Drugs of Today* **1999**, *35*, 237–250.
3. Vane, J. R.; Botting, R. M. "The Mechanism of Action of Aspirin" *Thrombosis Research* **2003**, *110*, 255–258. Moncada, S.; Vane, J. R. "Mode of Action of Aspirin-like Drugs" *Advances in Internal Medicine* **1979**, *24*, 1–22.
4. Pennisi, E. "Building a Better Aspirin" *Science* **1998**, *280*, 1191–1192.
5. Yus, M.; Najera, C.; Foubelo, F. "The Role of 1,3-Dithianes in Natural Product Synthesis" *Tetrahedron* **2003**, *59*, 6147–6212. Groebel, B. T.; Seebach, D. "Umpolung of the Reactivity of Carbonyl Compounds Through Sulfur-Containing Reagents" *Synthesis* **1977**, 357–402. Seebach, D.; Corey, E. J. "Generation and Synthetic Applications of 2-Lithio-1,3-dithianes" *J. Org. Chem.* **1975**, *40*, 231–237.
6. Luzzio, F. A. "The Henry Reaction: Recent Examples" *Tetrahedron* **2001**, *57*, 915–945.
7. Shiner, C. S.; Fisher, A. M.; Yacoby, F. "Intermediacy of α-Chloro Amides in the Basic Hydrolysis of α-Chloro nitriles to Ketones" *Tetrahedron Lett.* **1983**, *24*, 5687–5690.
8. Aggarwal, V. K.; Ali, A.; Coogan, M. P. "The Development and Use of Ketene Equivalents in [4+2] Cycloadditions for Organic Synthesis" *Tetrahedron* **1999**, *55*, 293–312. Ranganathan, S.; Ranganathan, D.; Mehrotra, A. K. "Ketene Equivalents" *Synthesis* **1977**, 289–296.
9. Seebach, D. "Methoden der Reaktivitätsumpolung" *Angewandte Chem.* **1979**, *91*, 259–278. Evans, D. A.; Andrews, G. C. "Allylic Sulfoxides. Useful Intermediates in Organic Synthesis" *Acct. Chem. Res.* **1974**, *7*, 147–155.
10. For an excellent description of how to determine whether an electophile adds to an alkene to give a bridged intermediate or a carbocation see: Alder, R. W.; Baker, R.; Brown, J. M. "Mechanism in Organic Chemistry" Wiley-Interscience, **1971**, 297–302
11. For some very useful rate data see: Sauer, J. "Diels-Alder Reactions. II. Reaction Mechanism" *Angewandte Chem. Int. Ed. Eng.* **1967**, *6*, 16–33. Sauer, J.; Lang, D.;

Wiest, H. "Diels-Alder Reaction. II. The Addition Capacity of Cis-Trans Isomeric Dienophiles in Diene Additions" *Chem. Ber.* **1964**, *97*, 3208–3218. Sauer, J.; Wiest, H.; Mielert, A. "Diels-Alder Reaction. I. Reactivity of Dienophiles Towards Cyclopentadiene and 9,10-Dimethylanthracene" *Chem. Ber.* **1964**, *97*, 3183–3207.

12. Evans, D. A.; Helmchen, G.; Ruping, M.; Wolfgang, J. "Chiral Auxiliaries in Organic Synthesis" *Asymmetric Synthesis*, Christmann, M.; Braese, S., Eds.; Wiley-VCH, **2007**, 3–9.
13. Jones, G. B. "π-Shielding in Organic Synthesis" *Tetrahedron* **2001**, *57*, 7999–8016. Jones, G. B.; Chapman, B. J. "π-Stacking Effects in Asymmetric Synthesis" *Synthesis* **1995**, 475–497.
14. Albright, J. D. "Reactions of Acyl Anion Equivalents Derived from Cyanohydrins, Protected Cyanohydrins and α-Dialkylaminonitriles" *Tetrahedron* **1983**, *39*, 3207–3233.
15. Scott, J. W. "Readily Available Chiral Carbon Fragments and Their Use in Synthesis" in *Asymmetric Synthesis*, Morrison, J. D., Scott, J. W., Eds.; Academic Press, **1984**, *4*, 1–226.
16. Barton, D. H. R.; McCombie, S. W. "New Method for the Deoxygenation of Secondary Alcohols" *J. Chem. Soc., Perkin 1* **1975**, 1574–1585.
17. Link, J. T. "The Intramolecular Heck Reaction" *Organic Reactions* **2002**, *60*, 156–534.
18. Keck, G. E.; Burnett, D. A. "β-Stannyl Enones as Radical Traps: A Very Direct Route to $PGF_{2\alpha}$" *J. Org. Chem.* **1987**, *52*, 2958–2960.
19. Ito, Y.; Hirao, T.; Saegusa, T. "Synthesis of α,β-Unsaturated Carbonyl Compounds by Palladium(II)-Catalyzed Dehydrosilylation of Silyl Enol Ethers" *J. Org. Chem.* **1978**, *43*, 1011–1013.
20. Matteson, D. S. "The Use of Chiral Organoboranes in Organic Synthesis" *Synthesis* **1986**, 973–985. Masamune, S.; Kim, B. M.; Petersen, J. S.; Sato, T.; Veenstra, S. J.; Imai, T. "Organoboron Compounds in Organic Synthesis. 1. Asymmetric Hydroboration" *J. Am. Chem. Soc.* **1985**, *107*, 4549–4591. Egger, M.; Keese, R. "MNDO Analysis of Regio- and Stereoselectivity in Hydroboration" *Helv. Chim. Acta* **1987**, *70*, 1843–1854. For another interesting application of the Masamune reagent see: Imai, T; Tamura, T.; Yamamuro, A.; Sato, T.; Wollmann, T. A.; Kennedy, R. M.; Masamune, S. "Organoboron Compounds in Organic Synthesis. 2. Asymmetric Reduction of Dialkyl Ketones with (*R,R*)- or (*S,S*)-2,5-Dimethylborolane" *J. Am. Chem. Soc.* **1986**, *108*, 7402–7204 (and the following paper which discusses mechanistic details of these reactions).
21. Nubbemeyer, U. "Recent Advances in Asymmetric [3.3]-Sigmatropic Rearrangements" *Synthesis* **2003**, 961–1008. Hill, R. K. "Chirality Transfer via Sigmatropic Rearrangements" *Asymmetric Synthesis* **1984**, *3*, 503–572.

Problems

1. Propose a reaction sequence that will convert PhCHO to PhCDO via (a) a sequence in which the deuterium label is introduced as an electrophile and (b) a sequence in which the deuterium label is introduced as a nucleophile. (Prostaglandins-3)
2. Suggest alternative tactics for the conversion of **11** → **12** (Prostaglandins-3). What problems might be anticipated if this conversion was attempted using (a) H_2, Pd/C or (b) $LiAlH_4$. (Prostaglandins-3)
3. Provide the structures of intermediates *en route* from **14** → **15**. Suggest a mechanism for the dehydration step (DCC, $CuCl_2$, Et_2O). (Prostaglandins-4)
4. Suggest a mechanism for the conversion of **26** → **27–30**. (Prostaglandins-5)
5. Propose a mechanism for the conversion of **40** → **41** that does not rely on an intermolecular S_N2 reaction at C_6. (Prostaglandins-6)
6. Illustrate how the following reagents could serve as ketene equivalents for use in Diels-Alder reactions with 1,3-cyclopentadiene. (Prostaglandins-6)

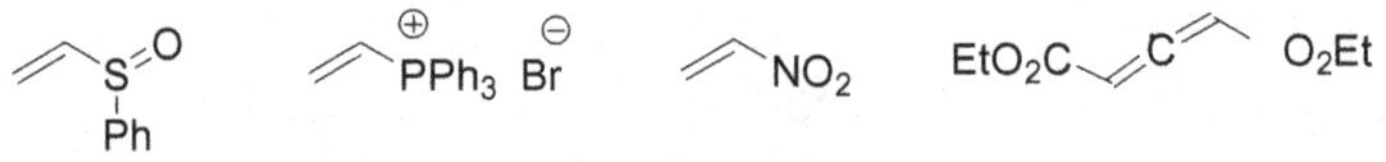

For references see: Williams, R. V.; Lin, X. "New ketene equivalents for the Diels-Alder reaction. Vinyl sulfoxide cycloaddition." *J. Chem. Soc., Chem. Commun.* **1989**, 1872–1873. Ruden, R.; Bonjouklian, R. "Cycloaddition of vinyl triphenyl-phosphonium bromide. New synthesis of cyclic phosphonium salts." *Tetrahedron Lett.* **1974**, *15*, 2095–2098. Ranganathan, S.; Ranganathan, D.; Mehrotra, A. K. "Nitroethylene as a versatile ketene equivalent. Novel one-step preparation of prostaglandin intermediates by reduction and abnormal Nef reaction." *J. Am. Chem. Soc.* **1974**, *96*, 5261–5262. Kozikowski, A. P.; Floyd, W. S.; Kuniak, M. P. "1,3-Diethoxycarbonylallene: an active dienophile and ethoxycarbonylketene equivalent in the synthesis of antibiotic C-nucleosides." *J. Chem. Soc., Chem. Commun.* **1977**, 582–583.

7. Illustrate how the following reagents could serve as "CH_2=CHOH" and "CH_2=$CHNH_2$" equivalents in Diels-Alder reactions with 1,3-cyclopentadiene. (Prostaglandins-6)

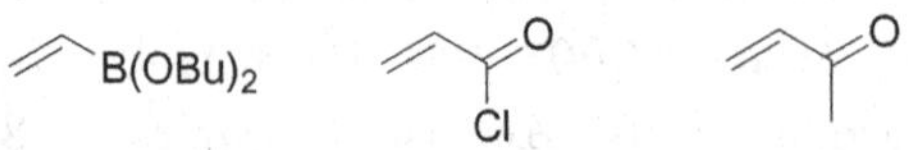

For a relevant review see: Vaultier, M.; Lorvelec, G.; Plunian, B.; Paulus, O.; Bouju, P.; Mortier, J. "Recent Developments in the Use of α,β-Unsaturated Boronates as Partners in Diels-Alder Cycloadditions" Royal Society of Chemistry Special Publication 253 (Contemporary Boron Chemistry, **2000**, 464–471. For another relevant article see: LeBel, N. A.; Cherluck, R. M.; Curtix, E. A. "Improved synthesis of amides from the Curtius reaction. Reaction of isocyanates and organolighium compounds" *Synthesis* **1973**, 678–679.

9. Provide the structures of intermediates in the conversion of **58** → **59**. (Prostaglandins-9)
10. Ethyl acetate is not miscible with water whereas γ-butyrolactone is freely soluble in water. Explain this observation and describe its relevance to the asymmetric induction model developed in Prostaglandins-10.
11. Suggest a mechanism for the conversion of **44** → **45**. (Prostaglandins-10).
12. Provide the structures of intermediates in the conversion of **77** → **78**. (Prostaglandins-12)
13. The Mitsunobu reaction is another method frequently used to invert alcohol stereochemistry [for a review see Mitsunobu, O. "The use of diethyl azodicarboxylate and triphenylphosphine in synthesis and transformation of natural products" *Synthesis* **1981**, 1–28]. An example is shown below.

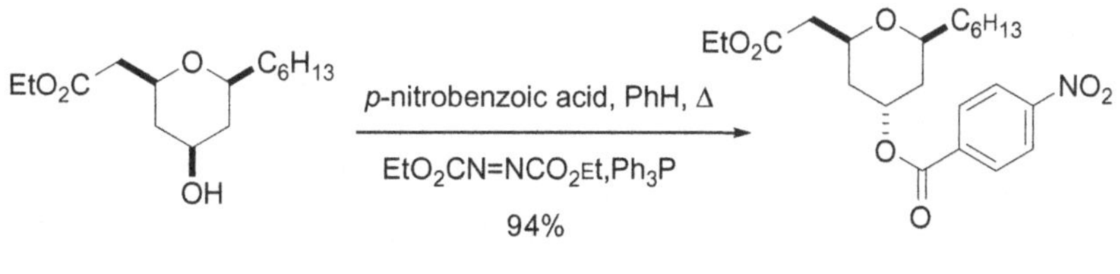

Provide a mechanism for this reaction. (Prostaglandins-12). Also see: Bose, A. K.; Lal, B.; Hoffman, W. A, III; Manhas, M. S. "Steroids IX. Facile inversion of unhindered sterol configuration" *Tetrahedron Lett.* **1973**, *14*, 1619–1622.

14. Explain how the reaction of aqueous NaOCl with hydrogen peroxide generates singlet oxygen. (Prostaglandins-13)
15. Provide mechanistic details for the conversion of **84** → **85** + **86**. (Prostaglandins-13)
16. Propose at least three approaches to the asymmetric (enantioselective) synthesis of propargylic alcohol **91**. (Prostaglandins-14)

17. Propose syntheses of the following butenolides. Keep in mind the benefits of late introduction (vs early introduction) of the carbon-carbon π-bond. (Prostaglandins-15)

For several relevant papers see: Bartlett, P. A. "Synthesis of β-acylacrylic esters and α,β-butenolides via β-keto sulfoxide alkylation" *J. Am. Chem. Soc.* **1976**, *98*, 3305–3312. Kende, A. S.; Toder, B. H. "Stereochemistry of deconjugative alkylation of ester dienolates. Stereospecific total synthesis of the litsenolides" *J. Org. Chem.* **1982**, *47*, 163–167. Yao, Z.-J.; Wu, Y.-L. "Total Synthesis of (10ξ,15R,16S,19S,20S,34R)-corossoline" *Tetrahedron Lett.* **1994**, *35*, 157–160.

18. Provide a mechanistic explanation for the following observation. (Prostaglandins-17)

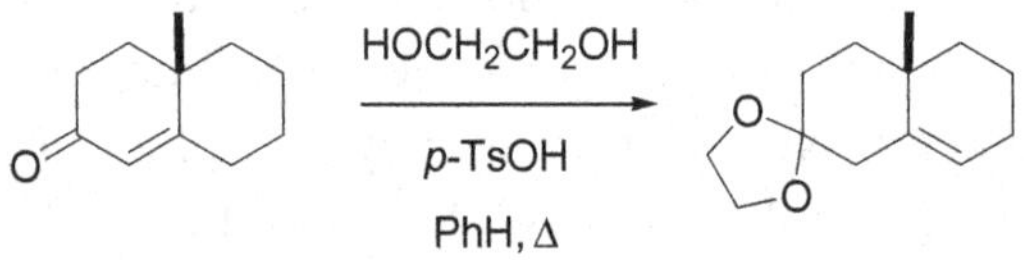

For references see: Babler, J. H.; Malek, N. C.; Coghlan, M. J. "Selective hydrolysis of α,β- and β,γ-unsaturated ketals: method for deconjugation of β,β-disubstituted α,β-unsaturated ketones" *J. Org. Chem.* **1978**, *43*, 1821–1823.

19. Provide a mechanistic explanation for the following observation [House, H. O.; Trost, B. M. "The chemistry of carbanions. X. The selective alkylation of unsymmetrical ketones." *J. Org. Chem.* **1965**, *30*, 2502–2512. House, H. O.; Trost, B. M. "The chemistry of carbanions. IX. The potassium and lithium enolates derived from cyclic ketones" *J. Org. Chem.* **1965**, *30*, 1341–1348.]:

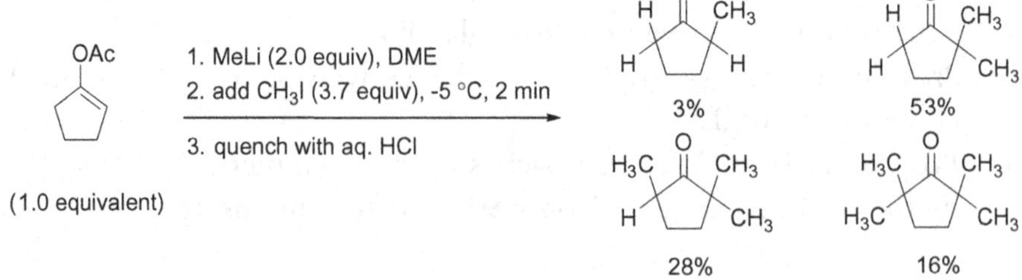

Describe the relevance of this observation to the choice of electrophile for introducing the C_8 sidechain in the three-component coupling approach to the prostaglandins. (Prostaglandins-17)

20. Provide a mechanism for the conversion of **153** → **154**. (Prostaglandins-20)
21. Provide a mechanism for the isomerization of *t*-BuN=C to *t*-BuC≡N during the conversion of **163** → **164**. (Prostaglandins-21)
22. Predict the products expected from the following hydroboration-oxidation reactions. (Prostaglandins-23) (Masamune, S.; Kim, B.; Peterson, J. S.; Sato, T.; Veenstra, S. J. "Organoboron compounds in organic synthesis. 1. Asymmetric hydroboration" *J. Am. Chem. Soc.* **1985**, *107*, 4549–4951).

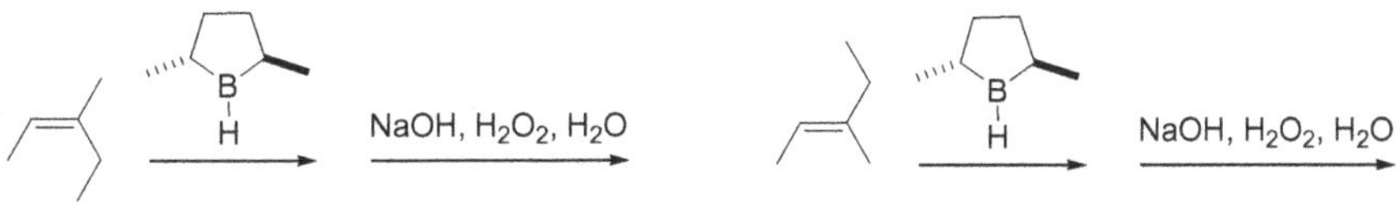

23. Each if the following reactions involves transfer of chirality via a sigmatropic rearrangement. Predict the stereochemistry of each product, or explain the stereochemical course of each reaction sequence, using some mechanistic rationale. (Prostaglandins-24)

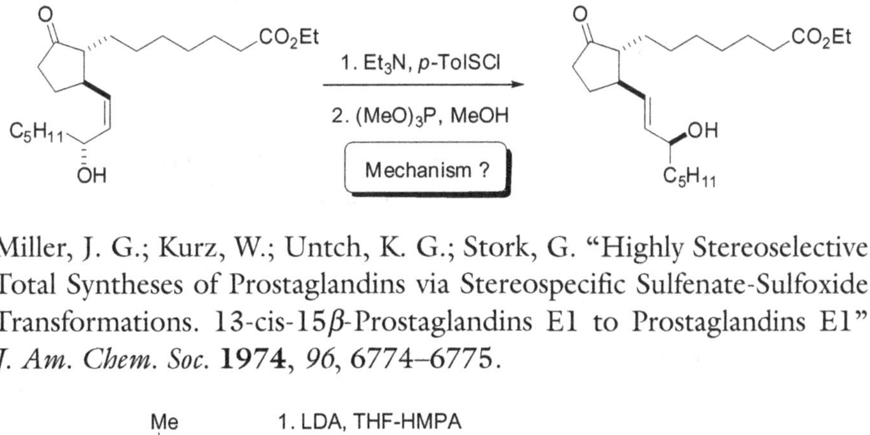

Miller, J. G.; Kurz, W.; Untch, K. G.; Stork, G. "Highly Stereoselective Total Syntheses of Prostaglandins via Stereospecific Sulfenate-Sulfoxide Transformations. 13-cis-15β-Prostaglandins E1 to Prostaglandins E1" *J. Am. Chem. Soc.* **1974**, *96*, 6774–6775.

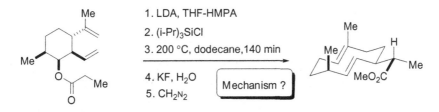

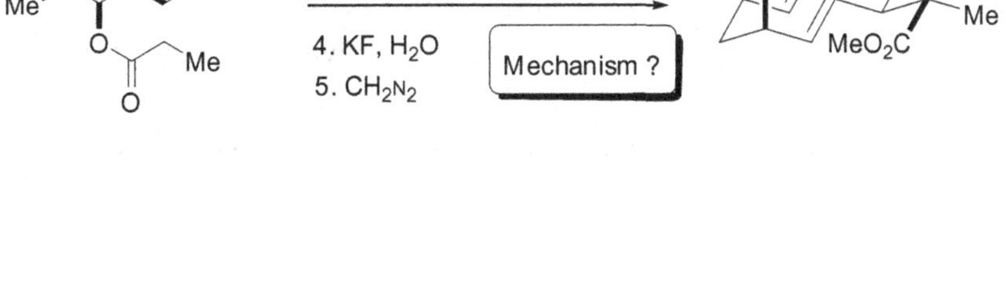

Raucher, S.; Chi, K. W.; Hwang, K. J.; Burks, J. E. "Synthesis via Sigmatropic Rearrangement. 12. Total Synthesis of (±)-Dihydrocostunolide via Tandem Cope-Claisen Rearrangement" *J. Org. Chem.* **1986**, *51*, 5503–5505.

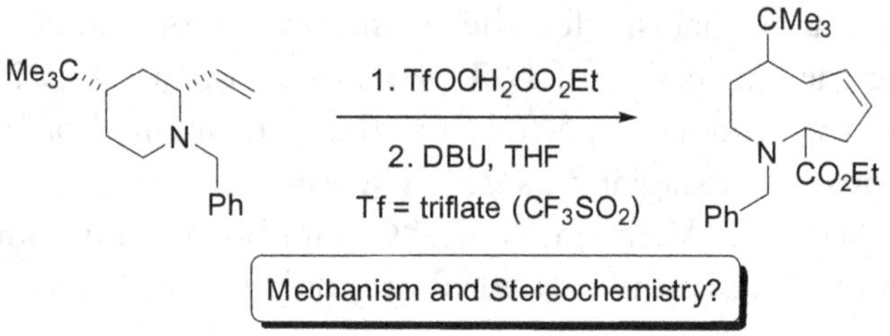

Vedejs, E.; Arco, M. J.; Renga, J. M. "Conformational Control of Olefin Geometry in 2,3-Sigmatropic Ring Expansion" *Tetrahedron Lett.* **1978**, *19*, 523–526.

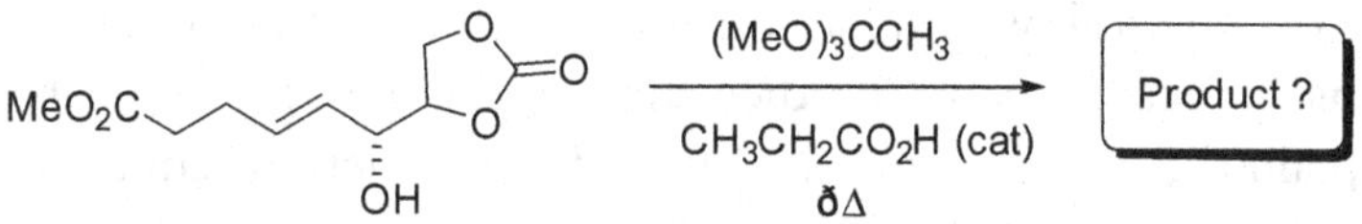

Stork, G.; Raucher, S. "Chiral Synthesis of Prostaglandins from Carbohydrates. Synthesis of (+)-15-(*S*)-prostaglandin A2" *J. Am. Chem. Soc.* **1976**, *98*, 1583–1584.

24. Outline a mechanism for the following transformation (note that the sequence must involve 1,4-asymmetric induction following by transfer of chirality to afford a 1,2-stereochemical relationship. (Prostaglandins-24)

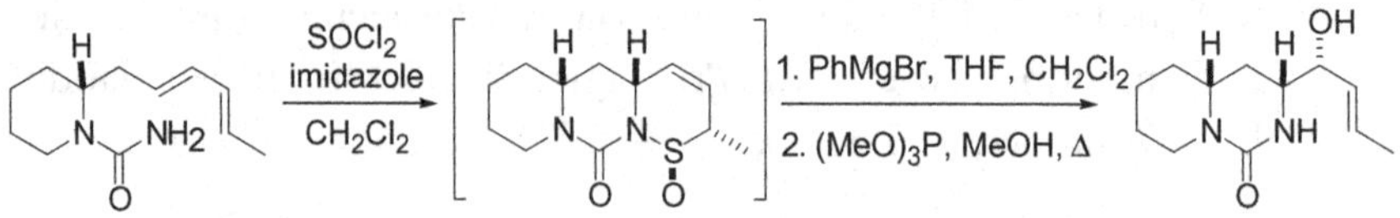

Heintzelman, G. R.; Fang, W. K.; Keen. S. P.; Wallace, G. A.; Weinreb, S. M. "Stereoselective total synthesis of the cyanobacterial hepatotoxin 7-epicylindrospermopsin: revision of the stereochemistry of cylindrospermopsin" *J. Am. Chem. Soc.* **2001**, *123*, 8851–8853. Heintzelman, G. R.; Parvez, M.; Weinreb, S. M. "A model study on the synthesis of the marine hepatotoxin cylindrospermposin" *Synlett* **1993**, 551–552.

25. Outline two syntheses of the diastereomer of **208** that use the strategy outlined in Prostaglandins-26.
26. Suggest a mechanism for the conversion of **214** → **216**. (Prostaglandins-27)

CHAPTER 4

Pyrrolizidine Alkaloids

Pyrrolizidine Alkaloids

Pyrrolizidine alkaloids are a large family of compounds that contain the pyrrolizidine substructure, usually decorated with a variety of hydroxyl groups and frequently with a macrocyclic structure bridging the C_1 and C_7 positions. Derivatives have been studied for use in cancer chemotherapy, but these studies have been discouraging due to toxicity problems. Four isomeric alkaloids that are representative of this family of compounds are shown below.

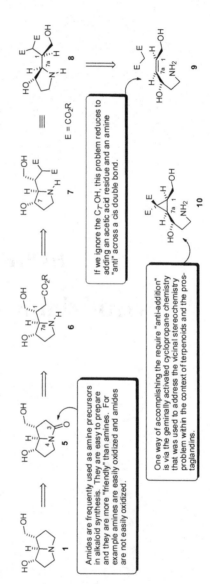

1 hastanecine; 2 dihydroxyheliotridane; 3 turneforcidine; 4 platynecine

The pyrrolizidine alkaloids pose a number of synthetic challenges and have served as a testing ground for synthetic methodology. One problem that appears within the targets shown above is the problem of vicinal stereocontrol. Since all of the stereogenic centers are located on ring carbons, the relationship between the pyrrolizidine vicinal stereochemistry and the steroid side chain problem might not be obvious. The following "plan" for the synthesis of pyrrolizidine alkaloids, however, reveals this relationship.

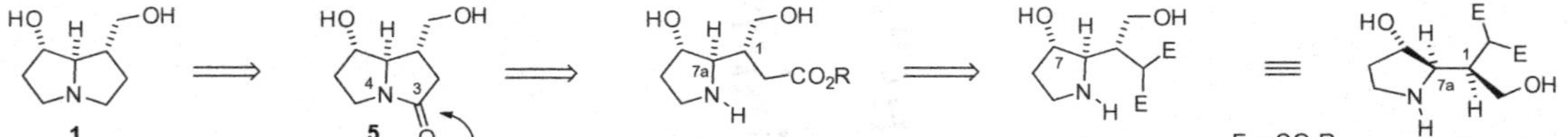

1 ⟹ 5 ⟹ 6 ⟹ 7 ≡ 8 (E = CO_2R)

Amides are frequently used as amine precursors in alkaloid synthesis. They are easy to prepare and they are more "friendly" than amines. For example amines are easily oxidized and amides are not easily oxidized.

If we ignore the C_7-OH, this problem reduces to adding an acetic acid residue and an amine "anti" across a cis double bond.

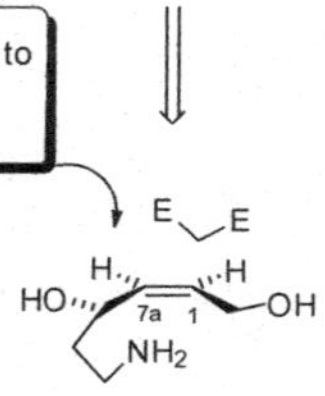

9

One way of accomplishing the require "anti-addition" is via the geminally activated cyclopropane chemistry that was used to address the vicinal stereochemistry problem within the context of terpenoids and the prostaglandins.

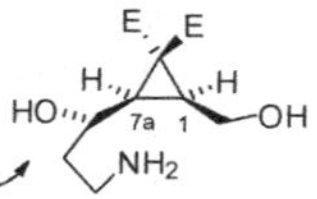

10

Pyrrolizidines-1

Pyrrolizidine alkaloids are a large family of compounds that contain the pyrrolizidine substructure, usually decorated with a variety of hydroxyl groups, and frequently with a macrocycle bridging the C_1 and C_7 positions.[1] Derivatives have been studied for use in cancer chemotherapy, but these studies have been discouraging due to toxicity problems. Four isomeric alkaloids that are representative of this family of compounds (**1–4**) are shown in Pyrrolizidines-1.

The pyrrolizidine alkaloids pose a number of synthetic challenges and have served as a testing ground for synthetic methodology. One problem that appears within these targets is that of vicinal stereocontrol. Since all of the stereogenic centers are located on ring carbons, the relationship between the pyrrolizidine vicinal stereochemistry problem and the steroid side chain problem might not be obvious. The following "plan" for the synthesis of pyrrolizidine alkaloids, however, reveals this relationship. Consider hastanecine (**1**). A "last step" in a possible synthesis of this alkaloid would be reduction of lactam **5**. It is notable that amides are frequently used as amine precursors in alkaloid synthesis. They are easy to prepare and are more "user friendly" than amines. For example, amines are easily oxidized and amides are not. The "friendliness" is directly related to the nitrogen lone pair being involved in bonding to the adjacent carbonyl carbon (in amides) and thus, not being as nucleophilic as the nitrogen lone pair of the corresponding amine. Continuing with the plan, one precursor of lactam **5** would be amino ester **6**. The relationship between the C_1-C_{7a} vicinal stereochemistry problem and the terpenoid sidechain problem emerges when we consider this structure. The stereogenic center at C_1 is exocyclic to the ring stereogenic center at C_{7a}. If we ignore the C_7-OH stereogenic center, and we recall the discussion presented in Prostaglandins-25, one solution to this vicinal stereochemistry problem is addition of an acetic acid residue and an amine "anti" across a cis double bond. One way of accomplishing this required "anti-addition" would be via the geminally activated cyclopropane chemistry used to address the vicinal stereochemistry problem within the context of terpenoids and the prostaglandins. Thus, cyclopropanation of **9** (or derivative thereof) followed by amine-opening of geminally activated cyclopropane **10** would give **8** (**7**) which could be then moved on toward hastanecine. Of course, relative asymmetric induction would be needed in the conversion of **9** to **10**. Let's see how this problem was handled in practice.

Danishefsky Synthesis of Hastanecine

Danishefsky, S.; McKee, R.; Singh, R. K. "Stereospecific Total Synthesis of *dl*-hastanecine and *dl*-dihydroxyheliotridane" *J. Am. Chem. Soc.* **1977**, *99*, 7711.
Also see: Danishefsky, S.; McKee, R.; Singh, R. K. "Kinetically Controlled Total Syntheses of *dl*-Trachelanthamidine and *dl*-Isoretronecanol" *J. Am. Chem. Soc.* **1976**, *98*, 4783-4788.

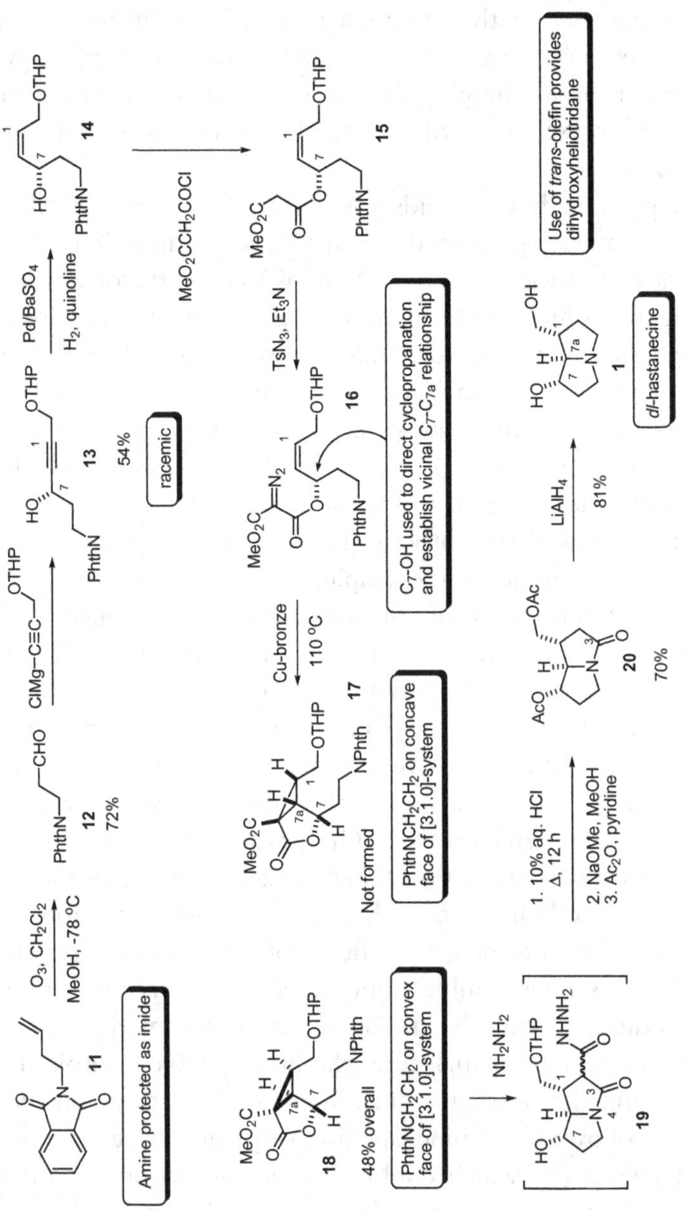

Pyrrolizidines-2

Relative stereochemistry between C_7 and C_{7a}-C_1 was to be established using an intramolecular cyclopropanation. The hope was that the carbene derived from diazoester **16** would undergo cyclopropanation to provide **18** rather than **17**. Cyclopropanes **17** and **18** are diastereomers and **18** was expected to be the more stable isomer. Why? These isomers differ in the relationship of the C_7 β-imidoethyl chain to the "fold" of the oxabicyclo[3.1.0]hexane ring system. In **18**, the sidechain is on the convex face of this ring system. In **17**, the sidechain is on the sterically more conjested concave face of the [3.1.0] ring system. For this reason it was assumed that **18** would be more stable than **17**, and it was hoped that some of the product energy difference would be felt in the transition states leading to these products, favoring formation of **18**.

Carbene precursor **16** was prepared from phthalimide **11**. Ozonolysis of the olefin gave aldehyde **12**. An acetylide addition gave **13** as a racemic mixture. Semihydrogenation of the alkyne, esterfication of **14** with methyl malonyl chloride, and diazotization of mixed malonate **15** gave **16**. The carbenoid insertion reaction proceeded according to plan to give **18**.

The phthalimido group was removed using hydrazine, liberating N_4 as a nucleophilic amine. This was accompanied by hydrazinolysis of the lactone to liberate the C_7 hydroxyl group. Opening of the geminally activated cyclopropane and cyclization of the intermediate aminoester provided **19**. This material was treated with aqueous acid to remove the THP protecting group, subjected to methanolysis and decarboxylation (retro crossed-Claisen condensation) to transform the malonic acid substructure to an acetic acid unit, and esterification to provide lactam **20**. The synthesis of hastanecine (**1**) was completed by LAH reduction of the lactam.

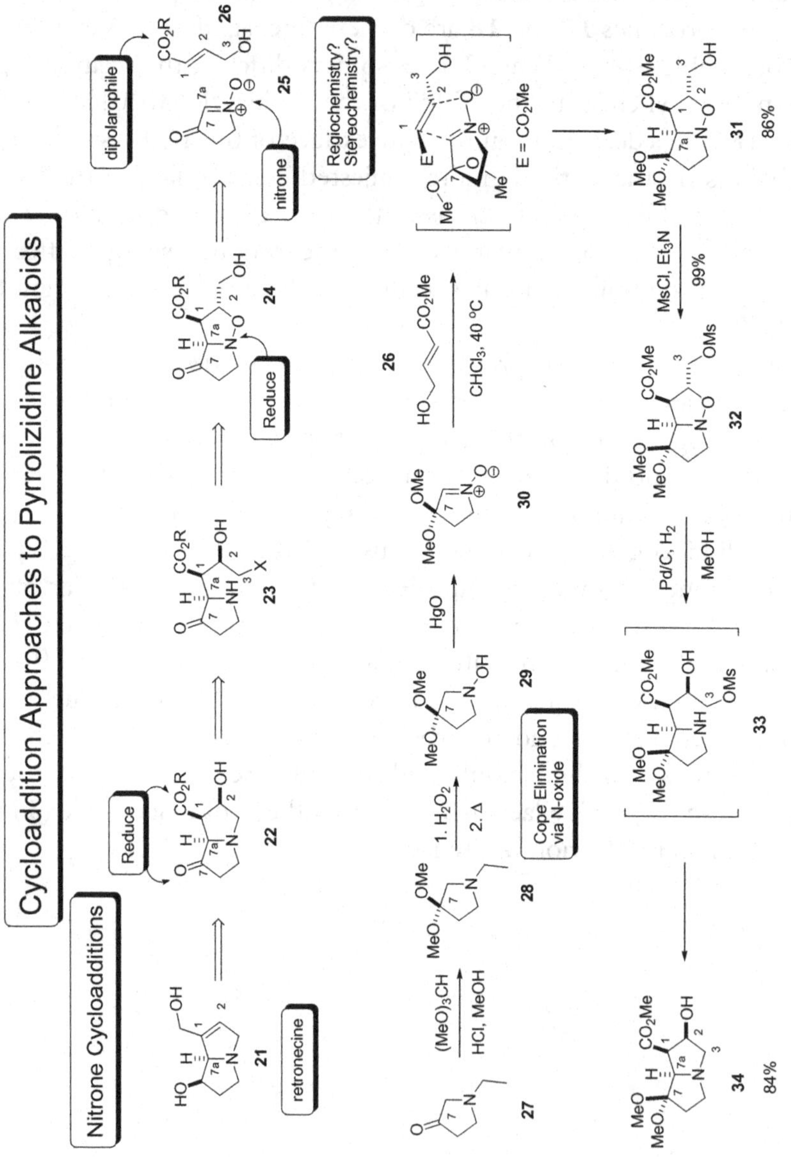

Pyrrolizidines-3

Pyrrolizidines-3

Now let's take a little side journey into the field of alkaloid synthesis. Alkaloids are loosely defined as natural products that contain a basic nitrogen. Some peptides that contain basic nitrogen (lysine and imidazole residues for example) are not classifed as alkaloids, and some amides that do not contain basic nitrogen (colchicine for example) have been called alkaloids, and that is why I use the phrase "loosely defined".[2,3] I have already mentioned that the pyrrolizidine alkaloids are popular synthetic targets and thus, at this point we will examine a few other approaches to these natural products.

Retronecine (**21**) is a pyrrolizidine alkaloid that has only two stereogenic centers, but does contain unsaturation in one of the 5-membered rings. It has been a popular synthetic target, in part because of its biological activity, as has been its C_7 epimer heliotridine (see **38** on Pyrrolizidines-4). We will look at three approaches, all of which revolve around cycloaddition chemistry. The first of these revolves around nitrone 1,3-dipolar cycloaddition chemistry.[4] The plan was to construct **21** from key intermediate **22**. This transformation requires reduction of a ketone at C_7 and an ester appended to C_1, and a dehydration reaction to introduce the C_1-C_2 double bond. Intermediate **22** was to be prepared from **23** via an intramolecular *N*-alkylation. Notice that the synthesis of the unsaturated ring of retronecine was to follow the aforementioned strategy: make the σ-bond first and then introduce the π-bond (Prostaglandins-15). Intermediate **23** contains a 1,3-NO relationship: C_{7a} is bonded to a nitrogen and C_2 is bonded to an oxygen. This "difunctional relationship" is a structural feature common to many alkaloids (we will see why this is so in Chapter 8). Thus, synthetic methods that establish a 1,3-NO relationship are of "general" importance to the field of alkaloid synthesis. One such method is the dipolar cycloaddition of a nitrone (an azomethine oxide) to an olefin. This is the reaction around which the Tufariello group designed their approach to the pyrrolizidine alkaloids.[5] The specific plan was to examine the reaction of nitrone **25** with dipolarophile **26** with the hope of obtaining cycloadduct **24** as the major product. Reduction of the weak nitrogen-oxygen bond was then to reveal an intermediate of type **23**. The regiochemical and stereochemical course of the cycloaddition was critical to the success of the plan.

It turned out that **25** was an impractical 1,3-dipole for implementation of the plan, but ketal **30** served admirably. This 1,3-dipole was prepared by conversion of **27** to the corresponding ketal **28**. Oxidation of **28** to the amine oxide, followed by a Cope elimination, gave hydroxylamine derivative **29**.

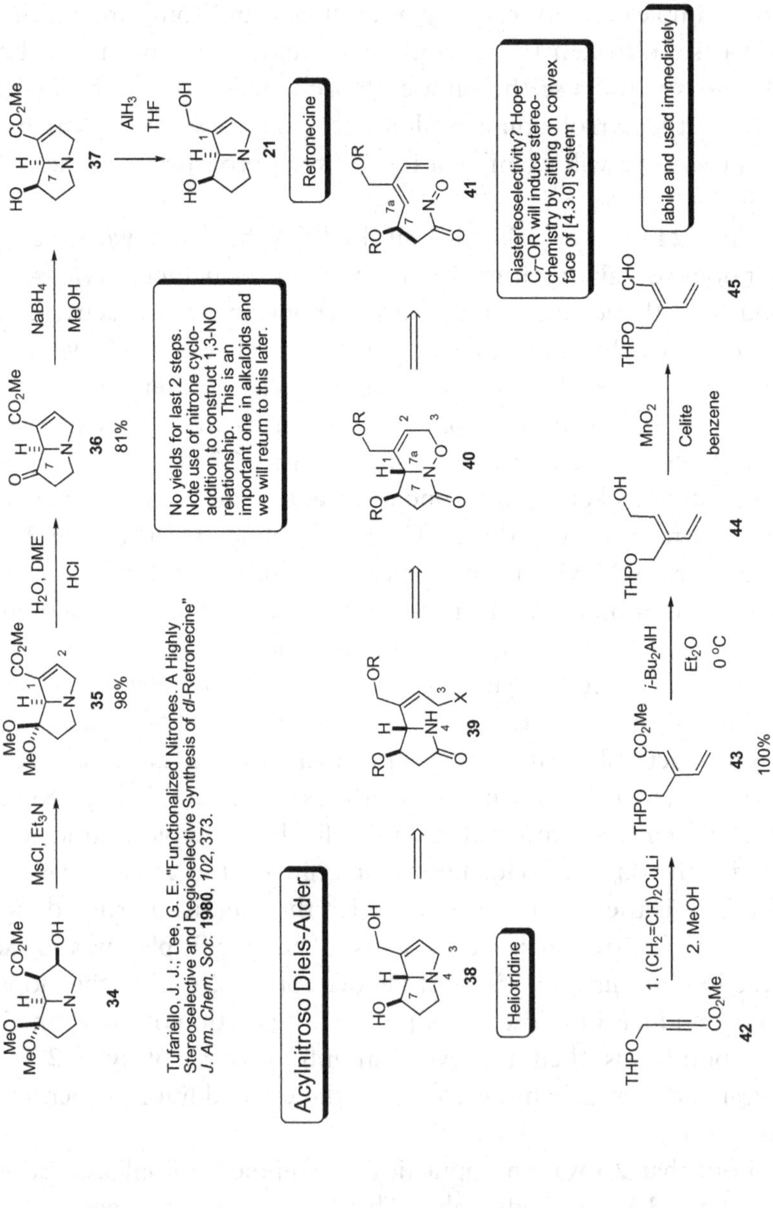

Pyrrolizidines-4

Oxidation of **29** provided nitrone **30** which, in the presence of methyl γ-hydroxycrotonate underwent cycloaddition with the desired regiochemistry and stereochemistry to provide **31**. The C_3 hydroxyl group was then activated for the subsequent intramolecular *N*-alkylation by formation of mesylate **32**. Hydrogenolysis of the N–O bond was followed by *in situ* *N*-alkylation to give **34** (a structural equivalent of **22**).

Pyrrolizidines-4

The synthesis of retronecine was completed using a 4-reaction sequence. The C_3 hydroxyl group of **34** was converted to a mesylate which underwent β-elimination to introduce the C_1-C_2 olefin. The acetal of **35** was hydrolyzed, and sodium borohydride reduction of the C_7 ketone from the convex face provided **37**. 1,2-Reduction of the unsaturated ester using alane finished the synthesis of retronecine (**21**).

The Keck group described an approach to the pyrrolizidine alkaloids that revolved around acylnitroso Diels-Alder chemistry. The plan, outlined within the context of an approach to heliotridine (**38**), was to once again use an intramolecular *N*-alkylation to construct the N_4–C_3 bond. *N*-Alkylation substrate **39** was to be prepared by reduction of the N–O bond of **40**, which was to result from an intramolecular cycloaddition of **41**. It was hoped that the C_7 substituent might induce relative stereochemistry at C_{7a} by occupying a site on the convex face of the incipient azaoxabicyclo[4.3.0]nonane ring system in the cycloaddition transition state. Note that the cycloaddition of **41** establishes the olefin geometry needed to support the intramolecular *N*-alkylation.

One of the key features of this synthesis was the development of tactics for the generation of the highly functionalized *N*-acylnitroso compound **41**. It was well known that such compounds behaved as dienophiles in 4+2 cycloaddition reactions, but it was also known that they were extremely reactive and had to be generated in the presence of the diene partner, ready for immediate use. The synthesis began with alkynoate **42**, which was converted to **43** using a cuprate conjugate addition. Reduction of the ester gave **44** and oxidation of the allylic alcohol using manganese dioxide provided aldehyde **45**, which was used immediately.

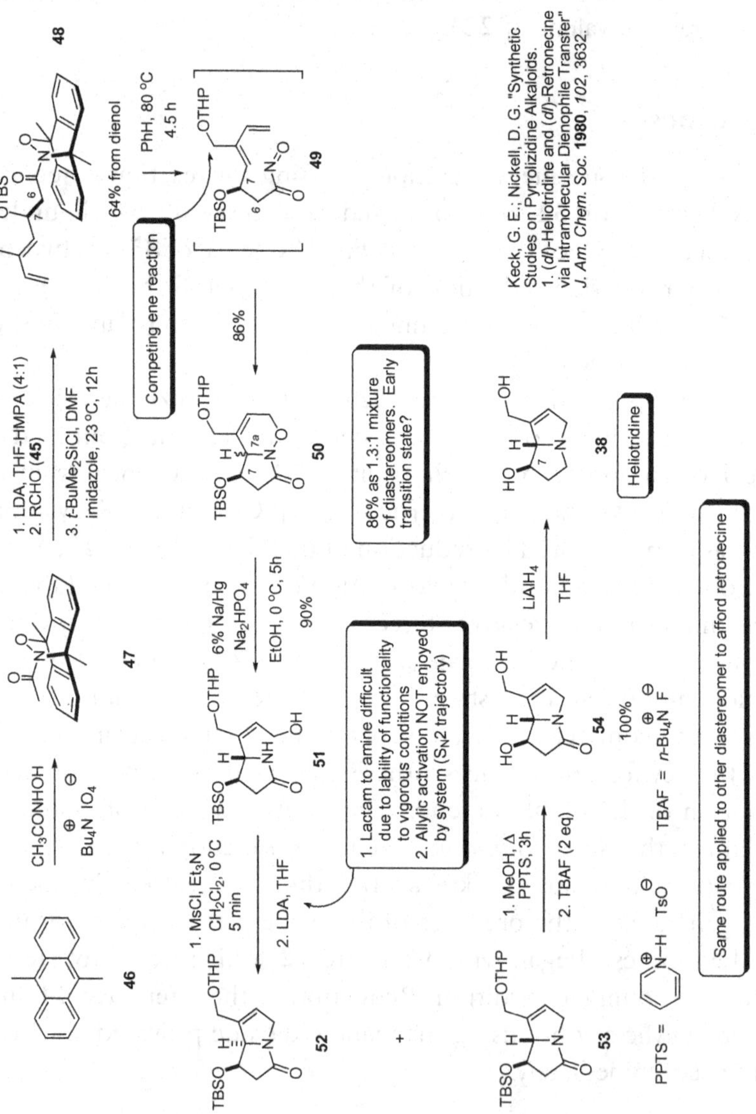

Pyrrolizidines-5

Pyrrolizidines-5

The reactivity of the acylnitroso group was "packaged" in the form of 9,10-dimethylanthracene cycloadduct **47**.[6] Conversion of this hydroxamic acid derivative to the corresponding enolate, followed by reaction with aldehyde **45** and protection of the resulting alcohol, gave **48** as a stable and easily characterized intermediate. Warming a benzene solution of **48** gave 9,10-dimethylanthracene and the desired intramolecular acylnitroso Diels-Alder cycloadduct **50** as a mixture of diastereomers at C_{7a}. The key intermediate **49** was presumably generated by a *retro*-Diels-Alder reaction. The level of diastereoselection in the cycloaddition was disappointing, and perhaps indicative of an early cyclization transition state in which the aforementioned steric effects were not well developed. Reduction of the N–O bond in **50** was followed by mesylate formation and *N*-alkylation to afford **52** and **53** as a separable pair of diastereomers. The synthesis of heliotridine (**38**) was completed by hydrolysis of the THP group, removal of the silicon protecting group from the C_7 alcohol, and reduction of lactam **54**. Subjecting **52** to the identical reaction sequence provided retronecine (**21**). Notice that as in the Danishefsky synthesis (and unlike the Tufariello synthesis) the basic pyrrolizidine nitrogen was packaged as a lactam until the last step of the synthesis.

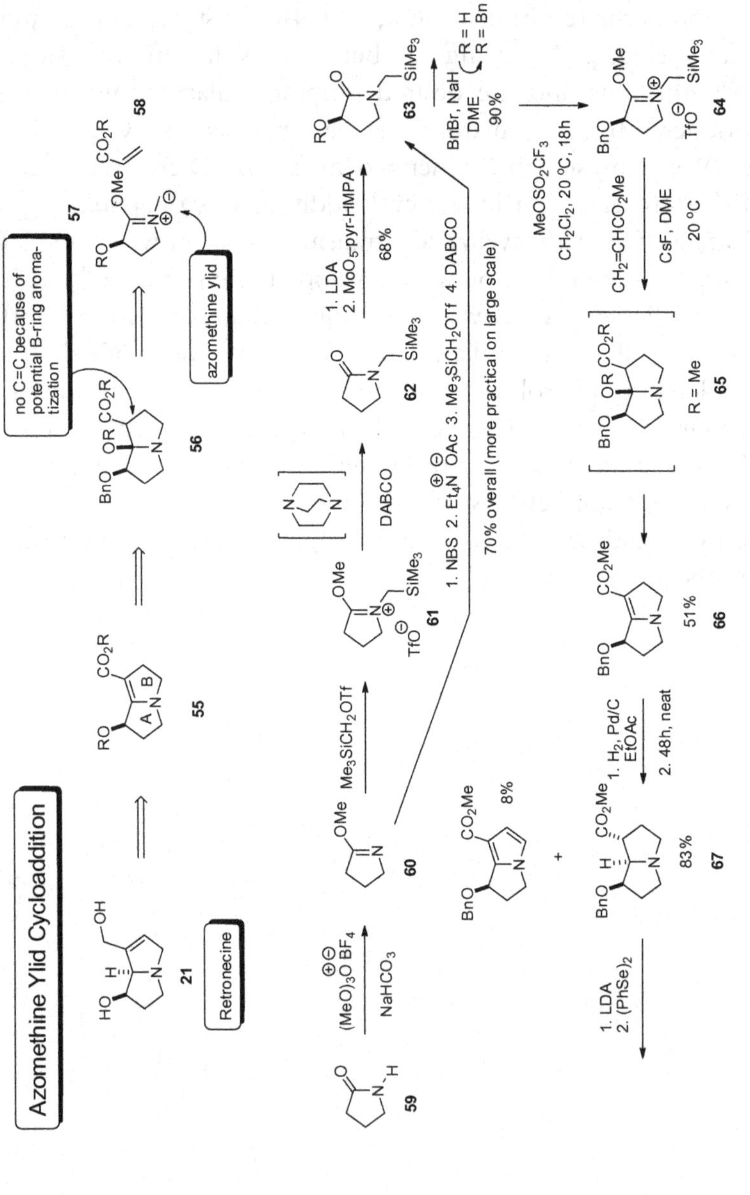

Vedejs, E.; Martinez, G. R. "Stereospecific Synthesis of Retronecine by Imidate Methylide Cycloaddition" *J. Am. Chem. Soc.* **1980**, *102*, 7993.

Pyrrolizidines-6

Pyrrolizidines-6

The final pyrrolizidine synthesis we will consider is Vedejs' approach to retronecine. This synthesis relies on a dipolar cycloaddition reaction for *direct* construction of the pyrrolizidine nucleus. The key reaction was to be a 1,3-dipolar cycloaddition between an azomethine ylid of type **57** and an acrylate of type **58**. It was projected that this would provide **56** (or **57** after loss of methanol), which would then be converted to retronecine (**21**) via a sequence involving late-stage introduction of the C_1-C_2 double bond.

The synthesis began with the conversion of lactam **59** to imidate **60**. *N*-Alkylation using trimethylsilylmethyl triflate, followed by demethylation of **61** with DABCO, gave **62**. Hydroxylation of the enolate derived from **62**, followed by Williamson etherification, gave **63** (R=Bn). Alternative tactics for moving forward from **60**, that were more amenable to scale-up, were also developed and are shown without comment. *O*-Alkylation of **63** provided **64**, the immediate precursor of azomethine ylid **57** (where R=Bn). Treatment of **64** with CsF in the presence of excess methyl acrylate gave **66**, presumably via intermediate cycloadduct **65**. The regiochemical course of the cycloaddition was precedented and stereochemistry was not an issue, given that the C_1 and C_{7a} stereogenic centers in **65** are destroyed in the subsequent elimination of methanol. Catalytic hydrogenation of **66**, followed by epimerization at C_1, gave **67** with good control of stereochemistry. The C_1-C_2 olefin was then introduced. Enolate formation and selenenylation gave **68**.

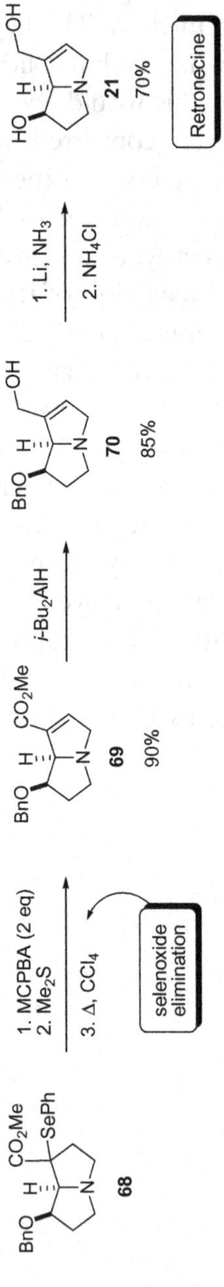

Pyrrolizidines-7

Pyrrolizidines-7

Oxidation of **68**, followed by elimination of the intermediate selenoxide, provided **69**. Reduction of the ester and removal of the benzyl protecting group from the C_7 alcohol completed the synthesis of retronecine (**21**). I note that my own personal interest in the Vedejs synthesis arises in part because the student who did the laboratory work (Greg Martinez) was an undergraduate student at UC Berkeley that I worked with closely for about a year when I was a graduate student.

The aforementioned syntheses are examples of research that was designed to develop and test synthetic methodology. My opinion is that the lasting value of these syntheses is the chemistry and stategies that were developed within the context of a target-oriented excercise, not the supply of material for biological evaluation or the determination of a structure. These strategies have all seen use in other settings. Now let's return to the "terpenoid sidechain problem".

References and Notes

1. Yoda, H. "Recent Advances in the Synthesis of Naturally Occuring Polyhydroxylated Alkaloids" *Current Organic Chemistry* **2002**, *6*, 223–243. Casiraghi, G.; Zanardi, F.; Rassu, G.; Pinna, L. "Recent Advances in the Stereoselective Synthesis of Hydroxylated Pyrrolizidines. A Review" *Org. Prep. Procedures Int.* **1996**, *28*, 641–682. Stevens, R. V. "General Methods of Alkaloids Synthesis" *Acc. Chem. Res.* **1977**, *10*, 193–198. Stevens, R. V. "Alkaloid Synthesis" *Total Synth. Nat. Prod.* **1977**, *3*, 439–554.
2. For one view of what constitutes an alkaloid see: Pelletier, S. W. "The Nature and Definition of an Alkaloid" Alkaloids: *Chemical and Biological Perspectives*, Pelletier, S. W., Ed.; John Wiley and Sons, **1982**, Vol. 1, pp 1–31.
3. There are many excellent monographs on alkaloid chemistry. For a thorough series see "Alkaloids" Academic Press; 1950–2007 (68 volumes).
4. Huisgen, R.; "1,3-Dipolar Cycloadditions" *Angew. Chem.* **1963**, *75*, 604–637.
5. Banerji, A.; Bandyopadhyay, D. "Recent Advances in the 1,3-Dipolar Cycloaddition Reactions of Nitrones" *J. Ind. Chem. Soc.* **2004**, *81*, 817–832. Black, D. St. Clair.; Crozier, R. F.; Davis, V. C. "1,3-Dipolar Cycloaddition Reactions of Nitrones" *Synthesis* **1975**, 205–221.
6. Kibayashi, C.; Aoyagi, S. "Nitrogenous Natural Products via *N*-Acylnitroso Diels-Alder Methodology" *Synlett* **1995**, 873–879. Corrie, J. E. T.; Kirby, G. W.; Sharma, R. P. "Formation of Aryl Isocyanates by Deoxygenation of Nitrosocarbonylarenes" *J. Chem. Soc., Chem. Commun.* **1975**, 915–916. Kirby, G. W.; Sweeny, J. G. "Nitrosocarbonyl Compounds as Intermediates in the Oxidative Cleavage of Hydroxamic Acids" *J. Chem. Soc., Chem. Commun.* **1973**, 704–705.

Problems

1. Use of the *trans*-isomer of **14** in this reaction sequence gave *dl*-dihydroxyheliotridane (**2**). Outline a synthesis of *trans*-**14** and provide the structure of intermediates *en route* from this olefin to **2**. This synthesis of **2** also gave a small amount of *dl*-platynecine (**4**). What is the origin of this minor product? (Pyrrolizidines-2)
2. Propose mechanisms for the conversions of (1) **28** → **29** and (2) **29** → **30**. (Pyrrolizidines-3)
3. Provide the structures of all dipolar cycloaddition products that would have been produced had the reaction of **26** with **30** shown no regioselectivity or stereoselectivity. (Pyrrolizidines-3)
4. Explain the stereochemical course of the conversion of (1) **36** → **37** and (2) **42** → **43**. (Pyrrolizidines-4)
5. The conversion of **51** → **52** did not enjoy the "allylic activation" (*N*-alkylation rate enhancement) that might have been anticipated. Provide an explanation for this observation. (Pyrrolizidines-5)
6. Aside from being an appropriate nucleophile for the demethylation of **61**, DABCO is nice for following the course of amine-promoted reactions in an NMR tube [compared with triethylamine, DBU, DBN or Hunig's base (i-Pr_2EtN)]. Why? (Pyrrolizidines-6)

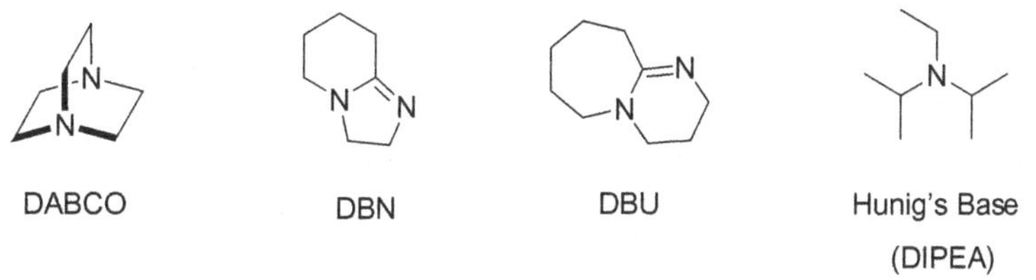

7. Provide the structure of intermediates *en route* from **60** → **63** via the "alternative tactics" that begin with NBS. (Pyrrolizidines-6)
8. Provide a rationale for the stereochemical course of the conversion of **66** → **67**. (Pyrrolizidines-6)
9. What is a problem that one might have anticipated in the reduction of **69** → **70**? (Pyrrolizidines-7)
10. Suppose you made the following observation [Heiba, E-A. I.; Dessau, R. M. "Free-radical isomerization. I. Novel rearrangement of vinyl radicals" *J. Am. Chem. Soc.* **1967**, *89*, 3772–3777]. How might you

develop this into an approach to the pyrrolizidine alkaloids (see text and papers cited therein for some targets)?

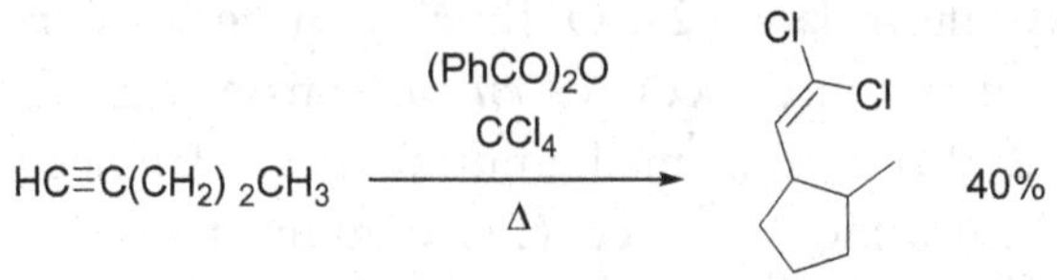

11. Suppose you made the following observation [Schoemaker, H. E.; Dijkink, J.; Speckamp, W. N. "Biomimetic *α*-acylimmonium cyclizations of unactivated olefins" *Tetrahedron* **1978**, *34*, 163–172. How might you develop this into an approach to the pyrrolizidine alkaloids?

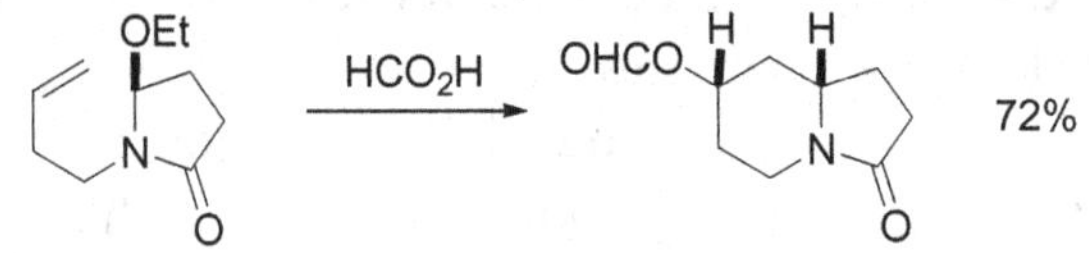

12. Predict the product of the following reaction [Smith, R.; Livinghouse, T. "An Expedient Synthetic Approach to the Physostigmine Alkaloids via Intramolecular Formamidine Ylide Cycloadditions" *J. Org. Chem.* **1983**, *48*, 1554–1555].

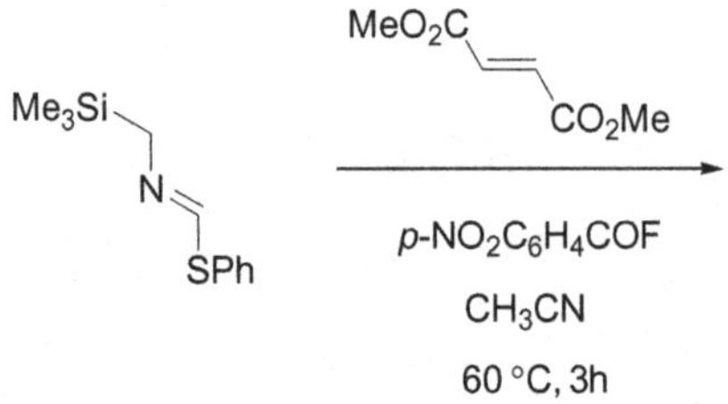

CHAPTER 5

Juvabione and the Vicinal Stereochemistry Problem

A Solution to the Cholesterol Side Chain Problem: Ester Alkylation

Wicha, J. "Synthesis of 25-Hydroxycholesterol from 3β-Hydroxyandrost-5-en-17-one. A Method for Stereospecific Construction of a Sterol Side-chain" *J. Chem. Soc., Chem. Commun.* **1975**, 968-969

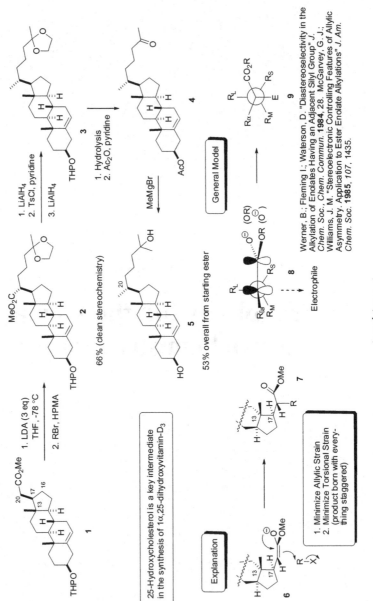

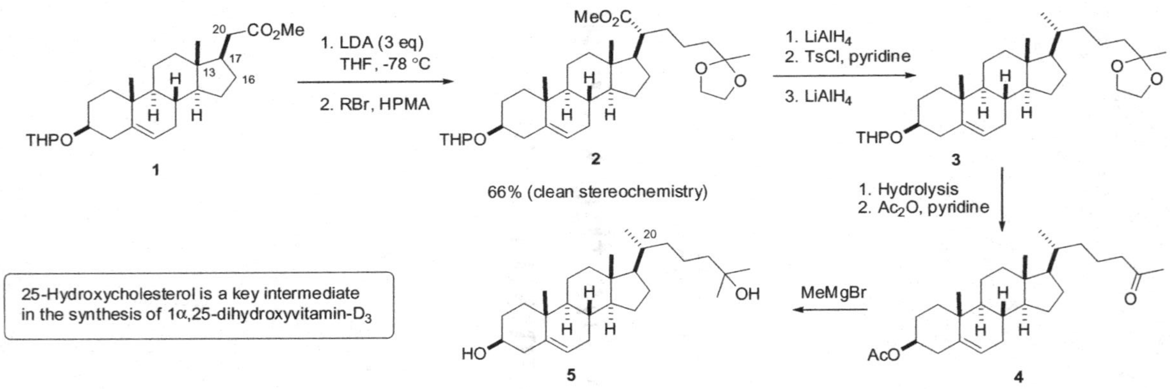

66% (clean stereochemistry)

53% overall from starting ester

25-Hydroxycholesterol is a key intermediate in the synthesis of 1α,25-dihydroxyvitamin-D_3

Explanation

1. Minimize Allylic Strain
2. Minimize Torsional Strain (product born with every-thing staggered)

General Model

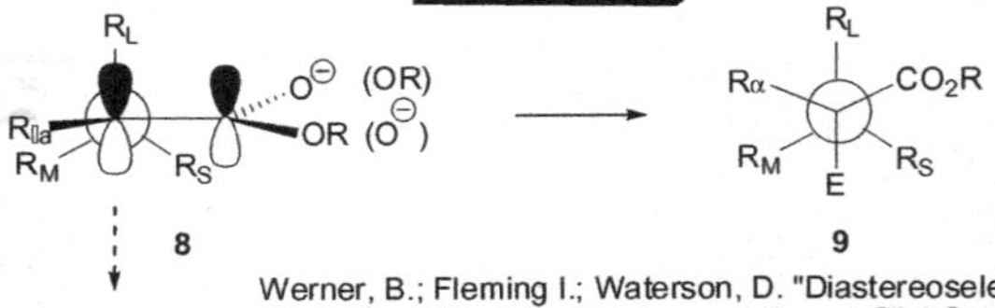

Werner, B.; Fleming I.; Waterson, D. "Diastereoselectivity in the Alkylation of Enolates Having an Adjacent Silyl Group" *J. Chem. Soc., Chem. Commun.* **1984**, 28. McGarvey, G. J.; Williams, J. M. "Stereoelectronic Controlling Features of Allylic Asymmetry. Application to Ester Enolate Alkylations" *J. Am. Chem. Soc.* **1985**, *107*, 1435.

Juvabione-1

In this chapter we will take a look at several syntheses of a sesquiterpene ester called juvabione. We will see that it became a target for synthesis because of structural and biological issues, and became a "test molecule" for the evaluating methodology for controlling vicinal stereochemistry in acyclic systems. But first let's once again examine the steroid sidechain problem.

One of the simplest solutions was developed by Jerzy Wicha (Poland). He found that alkylation of the enolate derived from **1** gave **2** with a high level of stereocontrol at C_{20}. The explanation for the high level of stereoselectivity was not obvious at the time, but work from the groups of Fleming and McGarvey provided an explanation several years later. The explanation is based on the following notions: (1) There is a size difference between the three substituents on the carbon β to the enolate carbonyl group. In the case of **1** this is C_{17}. One group is small (R_S=H), another medium (R_M=CH_2 at C_{16}) and the third group is large (R_L=C_{13} mimics a *tert*-butyl group in size). (2) The rate of alkylation is fastest from conformation **6**. This conformation is arguably the most stable conformation of the enolate. Allylic strain is minimized. The large group (C_{13}) is orthogonal to the plane defined by the enolate π-system. The medium-sized group (C_{16}) is positioned adjacent to the vicinal hydrogen. The small group (H) is positioned nearly in the same plane as the enolate oxygen. (3) Alkylation takes place such that as the new carbon-carbon bond is formed, everything is staggered along the C_{17}-C_{20} bond (torsional strain is minimized in the alkylation transition state). Thus **6** leads directly to **7**. In a more general sense, systems of type **8** will provide products of type **9**.

Additional Examples of Side Chain via Alkylation

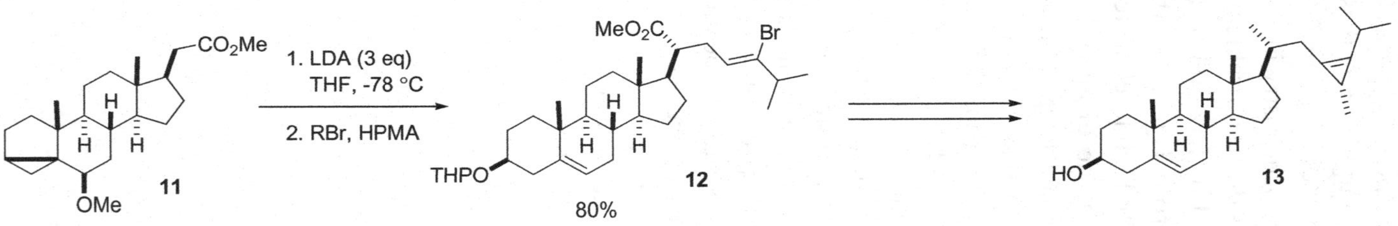

Kurek-Tyrlik, A.; Minksztym, K.; Wicha, J. "Synthesis of (23*R*)- and (23*S*)-23H-Isocalysterols. The First Synthesis of a Representative of Marine Sterols with a Cyclopropene Moiety in the Side Chain" *J. Am. Chem. Soc.* **1995**, *117*,1849.

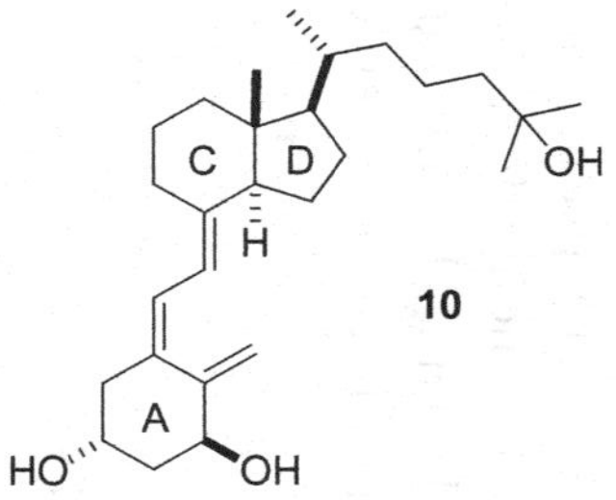

10

1α,25-dihydroxyvitamin-D_3

Juvabione-2

Juvabione-2

The original work of Wicha was used to prepare a precursor of 1α,25-dihydroxyvitamin-D_3 (**10**), but it has been applied to other steroids (for example **13**) as well. This is a simple and effective strategy for controlling C_{17}-C_{20} relative stereochemistry in steroids. The solution relies on what is called 1,2-asymmetric induction. One stereogenic center induces stereochemistry in the formation of an adjacent stereogenic center. Whereas this "solution" works in a well-defined set of situations, it is not a universal solution. Let's see why.

Juvabione: Vicinal Stereocontrol Problem

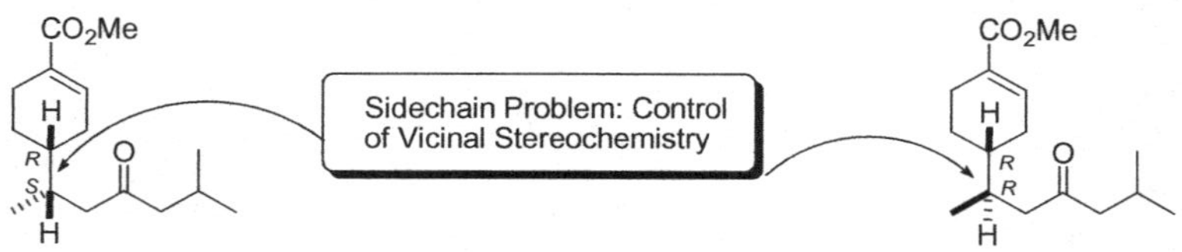

14 *erythro*-Juvabione

15 *threo*-Juvabione

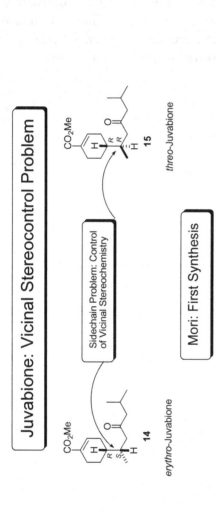

Mori: First Synthesis

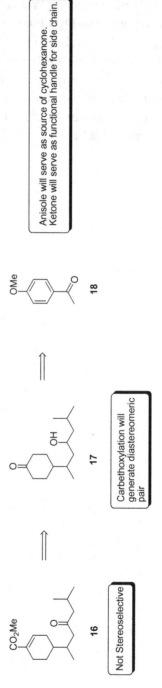

Mori, K.; Matsui, M.; "Synthesis of *dl*-Juvabione (Methyl *dl*-Todomatuate), A Sesquiterpene Ester with Juvenile Hormone Activity" *Tetrahedron Lett.* **1967**, 2515-2518.

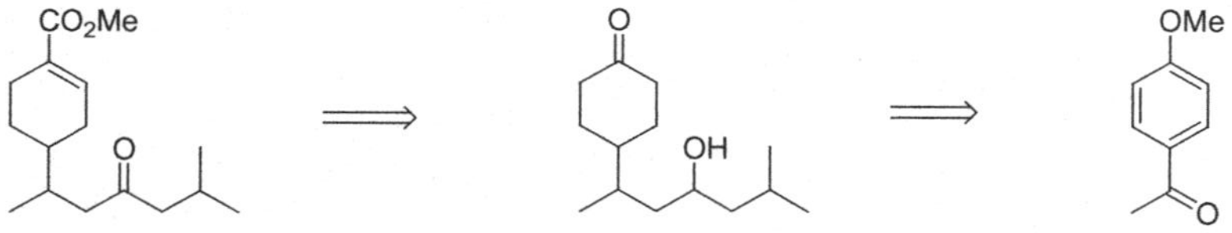

16 Not Stereoselective

17 Carbethoxylation will generate diastereomeric pair

18

Anisole will serve as source of cyclohexanone.
Ketone will serve as functional handle for side chain.

Juvabione-3

Juvabione-3

Synthetic organic chemistry became an important tool to the field of entomology in the 1960s when scientists began to learn about small molecules that played an important role in the life cycle of insects.[1] Of course, the "practical goal" was to use these chemicals to control insect populations. One family of molecules that became of interest are the "juvenile hormones". These are compounds that regulate larval development. The "juvabiones" (**14** and **15**) are diastereomeric sesquiterpene methyl esters that were first isolated from the balsam fir. They contain vicinal cyclic and exocyclic stereogenic centers. Establishing this stereochemical relationship is an issue that must be addressed in any stereoselective synthesis. Whereas initial efforts were developed to supply material for biology, and to help resolve stereochemical issues, these targets became a popular testing-ground for the development and application of reactions that generate vicinal stereochemical relationships. We will spend some time looking at a selection (not all inclusive) of syntheses that I have chosen to make points of some general interest.

Let's start with a question. Can the Wicha strategy be successfully applied to the juvabiones? I am not aware that anyone has tried, but I suspect the answer is "no". I will leave it for you to explain "why not" (or "why yes" if you disagree with me). The point I want to make here is that no one solution to a given problem will work for all variations of that problem. Thus there is usually a need for multiple solutions to a given problem (in this case a synthesis problem). I hope this "study" of a set of approaches (in some cases solutions) to the same problem is educational.

Let's start with the first total synthesis of the juvabiones reported by the Mori group in the 1960s. The goal here was to produce material. Thus, the synthesis was not stereoselective. The plan was to prepare a mixture of diastereomers and separate at some point. Thus **16** was to be prepared from **17** via dehydration of an intermediate cyanohydrin, tactics that were destined to lack selectivity.

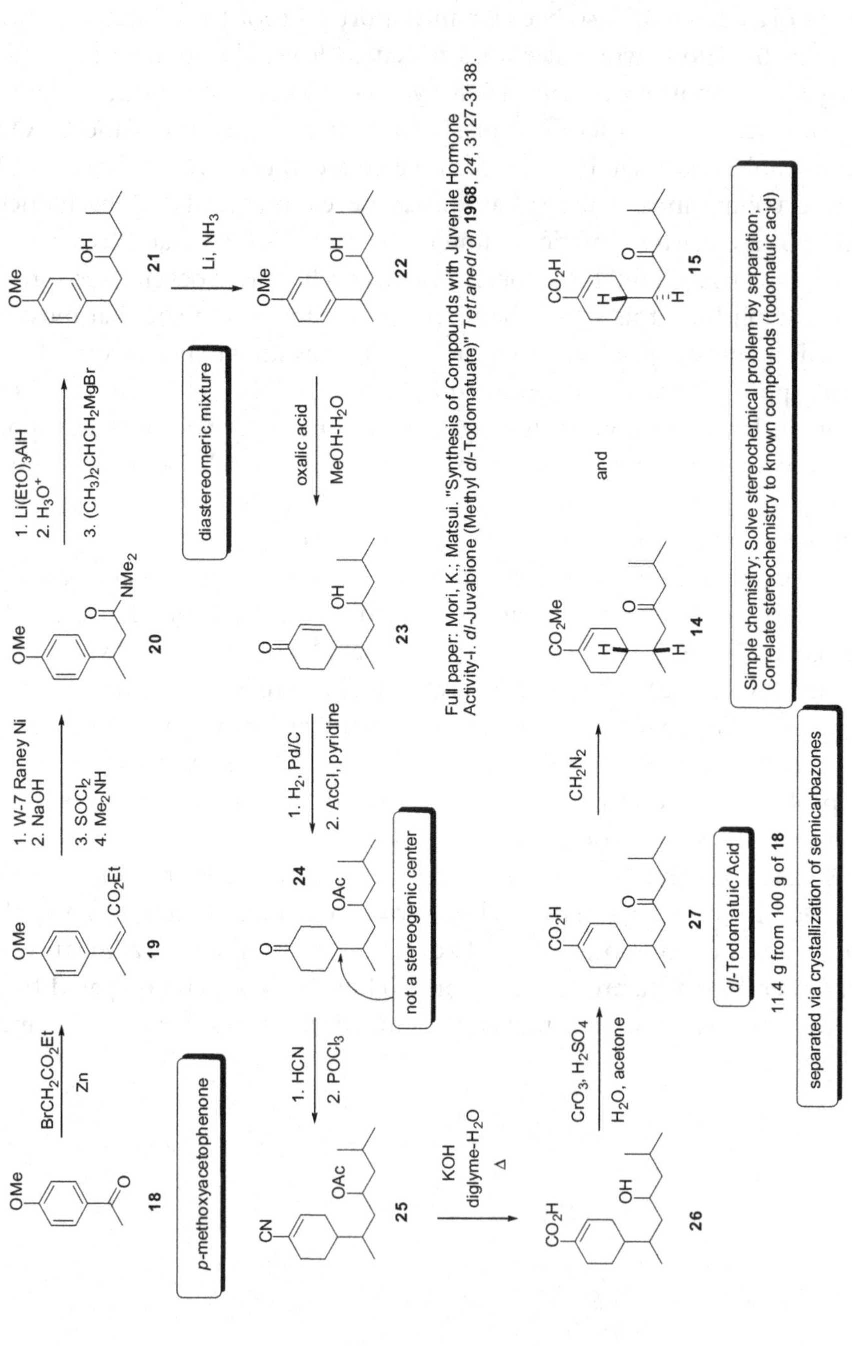

Full paper: Mori, K.; Matsui. "Synthesis of Compounds with Juvenile Hormone Activity-I. *dl*-Juvabione (Methyl *dl*-Todomatuate)" *Tetrahedron* **1968**, *24*, 3127-3138.

Juvabione-4

Juvabione-4

The Mori synthesis began with *p*-methoxyacetophenone (**18** = PMA). The acetyl group was converted into the "terpenoid sidechain" using a reliable reaction sequence. Birch reduction of **21**, followed by enol ether hydrolysis, reduction of the olefin, and protection of the sidechain alcohol as an acetate, gave **24** as a mixture of diastereomers. Note that the critical vicinal stereochemical relationship was established in the sequence leading from **21** to enone **23**, but without any diastereoselectivity. The conversion of **23** to **24** destroyed the vicinal stereochemical relationship. Conversion of **24** to the corresponding cyanohydrin, followed by dehydration to unsaturated nitrile **25**, once again introduced the vicinal stereochemistry, but again without selectivity. Hydrolysis of the mixture of nitriles to acids **26**, and oxidation of the sidechain alcohol, provided a racemic mixture of the diastereomeric keto acids **27**. Ketones **27** were converted to a separable mixture of semicarbazones (upon reaction with $NH_2NHCONH_2$). The separated semicarbazones were hydrolyzed and converted to the corresponding methyl esters **14** and **15**. The stereoisomers were correlated with known compounds (todomatuic acid).

This synthesis did not address the vicinal stereochemistry problem in a selective manner. The synthesis did provide an impressive amount of **27**, largely because it involved simple and reliable chemistry. The stereochemistry problem was solved using separation science.

Related Approaches to the Juvabiones

Ayyar, K. S.; Rao, G. S. K. "Studies in Terpenoids. IV. Synthetic Studies in Juvabiones and Analouges. Conversion of *ar*-(+)-turmerone to *ar*-(+)-juvabione" *Can. J. Chem.* **1968**, *46*, 1467.

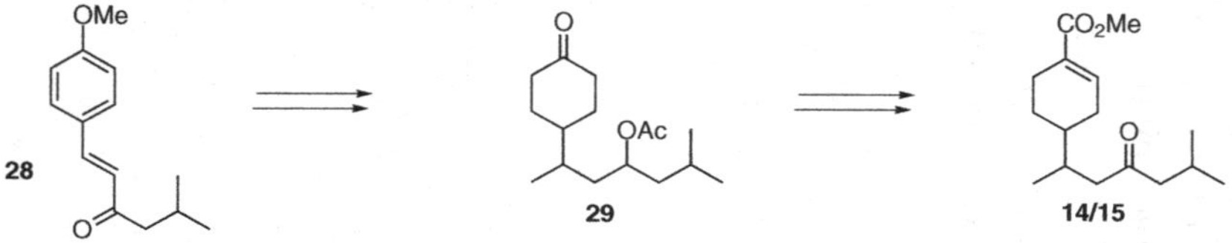

mixture of diastereomers

Ferrino, S. A.; Maldonado, L. A. "Further Extensions of the Kinetic Enolate Method for Terpenoid Synthesis" *Syn. Commun.* **1984**, *14*, 925-931.

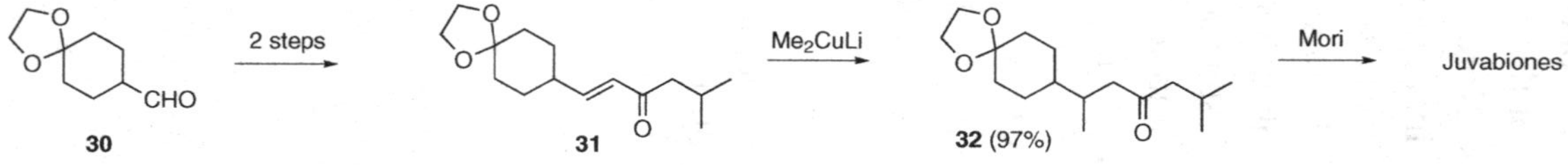

Feature is use of aldol condensation to introduce side chain as title implies. Synthesis does not address stereochemical issues.

Birch Approach: First Attempt to Seriously Address Diastereomer Problem

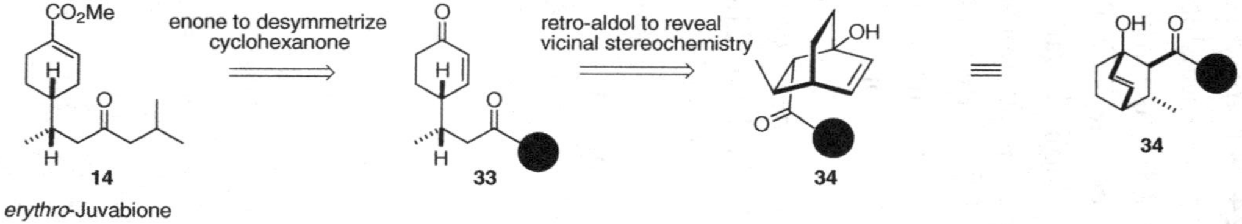

Juvabione-5

Two additional syntheses that did not address the vicinal stereochemistry problem are outlined here. They differ from the Mori synthesis in regard to the chemistry used to introduce the sidechain, but are similar in that they rely on reactions of a 4-substituted cyclohexanone to introduce the vicinal stereochemical relationship.

The first clear attempt to address the stereochemical problem presented by the juvabiones was reported by Birch. The plan was to use enone **33** to desymmetrize the 4-substituted cyclohexanone intermediates encountered thus far. The vicinal stereocenters in enone **33** were to be derived from **34** via a retro-aldol condensation. As we will see, **34** was to be prepared via a Diels-Alder reaction, a reaction that frequently has been used to establish vicinal stereochemistry with a high degree of selectivity.

Birch, A. J.; Macdonald, P. L.; Powell, V. H. "Reactions of Cyclohexadienes. Part VIII. Stereoselective and Non-Stereoselective Syntheses of *dl*-Juvabione" *J. Chem. Soc. (C)* **1970**, 1469.

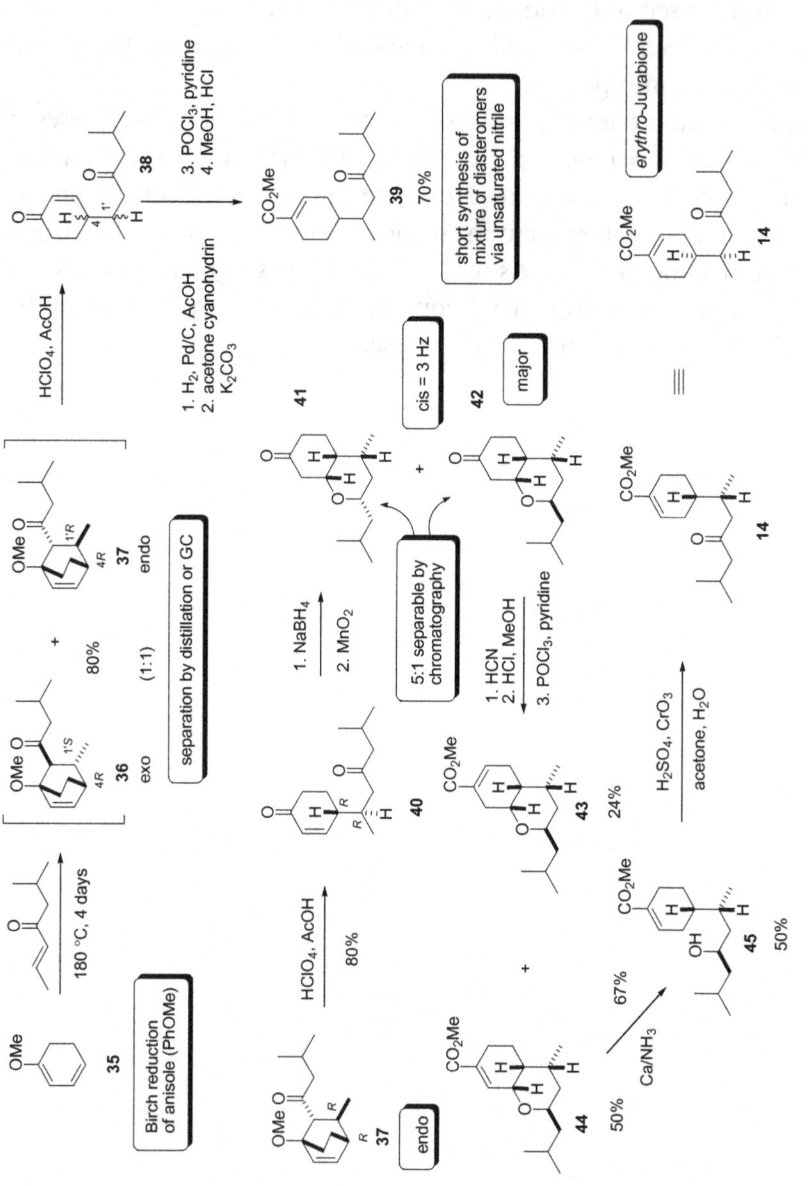

Juvabione-6

Juvabione-6

The Birch group synthesis began with a cycloaddition between diene **35** and *trans*-6-methyl-2-hepten-4-one to give a mixture of *endo* and *exo* cycloadducts. This reaction established the critical vicinal stereochemical relationship between C_4 and C_1'. In the *endo* cycloadduct, this relationship was *rel*-1′*R*,4*R*, and in the *exo* cycloadduct it was *rel*-1′*S*,4*R*.* I imagine the hope was that the *endo* cycloadduct would predominate (the norm for Diels-Alder reactions) leading to selective formation of **37**. One can imagine that given the technological advances that have taken place since 1970 (high pressure, Lewis acid catalysis),[2,3] it would now be possible to accomplish this reaction (or a variation using another dienophile) with high levels of diastereoselection. Nonetheless, in this instance, the diastereomers were either carried forward as a mixture in a non-selective synthesis, or separated before moving forward in a stereocontrolled manner.

The non-selective synthesis involved treatment of **36/37** with acid to effect a retro-aldol condensation to provide **38**. Reduction of the double bond, followed by a Mori-like endgame, gave **39** (**14/15**) as a mixture of diastereomers.

When the synthesis was continued with *endo*-cycloadduct **37**, the retroaldol condensation gave **40** as a single diastereomer. Reduction of the saturated ketone was accompanied by conjugate addition of the resulting alcohol to the enone to give **41** and **42** as a separable 5:1 mixture of diastereomers ("not necessarily respectively" in the words of the authors). We will just continue as though **42** was the major isomer as it is of little consequence to the outcome of the synthesis. Cyanohydrin formation and dehydration provided a separable mixture of unsaturated esters **43** and **44**. Treatment of **44** with calcium in ammonia gave **45** and oxidation of the alcohol completed a synthesis of *erythro*-juvabione (**14**). In principle, **41** could have been put through the same paces to provide additional material. Formation of **43** represents a more serious loss of material, as bringing it through to either **14** or **15** would require considerably more effort than conversion of **44** to **14**.

Overall the Birch strategy is creative and directly addresses the vicinal stereochemistry problem that is the focus of this chapter. The execution of the plan, however, is problematic. I imagine that with more time, and tools

---

*Cycloadducts **36** and **37** are produced as racemic mixtures. Thus the *RS* descriptors only indicate *relative* stereochemistry. Furthermore, I have assigned *RS* stereochemistry not according to the Cahn-Ingold-Prelog (CIP) convention, but to indicate the diastereomeric juvabione that would be produced from the indicated cycloadduct.

Ficini, J.; Touzin, A. M.; "Cycloaddition of an ynamine to cyclohexenones. Synthesis of aminobicyclo[4.2.0]octenones" *Tetrahedron Lett.* **1972**, 2093. Ficini, J.; Touzin, A. M.; "Stereochemical control of an asymmetric center formed a to the carboxyl function by hydrolysis of bicyclic enamines. Stereoselective synthesis of diastereoisomeric γ-keto acids" *Tetrahedron Lett.* **1972**, 2097. Ficini, J.; d'Angelo, J.; Noire, J. "Stereospecific Synthesis of *dl*-Juvabione" *J. Am. Chem. Soc.* **1974**, *96*, 1213.

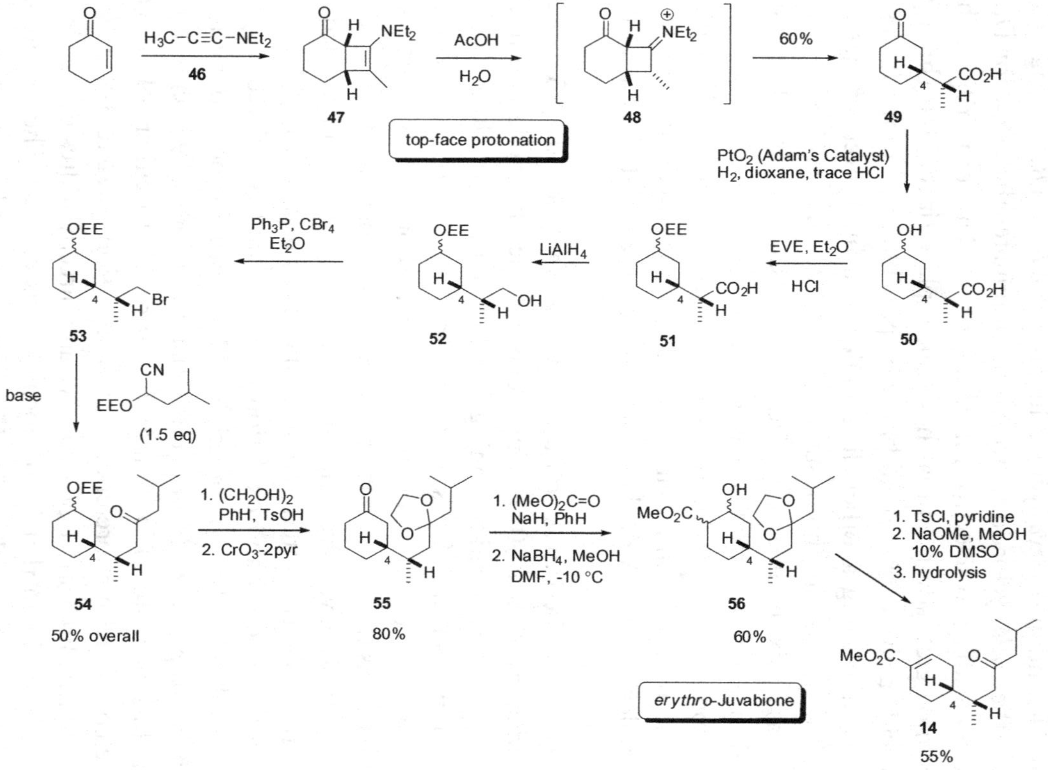

Juvabione-7

available today (40 years later), tactics could be identified that would realize the potential of this strategy.

Finally, I note the Birch synthesis is one example of a general strategy for controlling vicinal stereochemistry in which one (or both) of the stereogenic centers is not in a ring: (1) set the stereogenic centers in a ring and (2) open the ring. Let's look at another synthesis of *erythro*-juvabione that follows this general strategy.

Juvabione-7

Jacqueline Ficini's group (France) was very interested in ynamine chemistry and applied some of their discoveries to a synthesis of *erythro*-juvabione (**14**) as shown here. A formal [2+2]-cycloaddition of cyclohexenone with ynamine **46** gave **47**. This bicyclo[4.2.0]octene derivative has very distinct convex and concave faces. Thus, a kinetically controlled protonation of the enamine double bond occured from the less hindered (convex) face to presumably give iminium ion **48**, which underwent subsequent hydrolysis. A *retro*-Claisen condensation of the resulting strained-1,3-carbonyl compound provided **49**. The strategy for controlling vicinal stereochemistry is truly (1) set it in a ring and (2) open the ring. In this case the ring opening relied on the strain inherent in an intermediate 2-acylcyclobutanone.

The synthesis continued with reduction of the cyclohexanone to the alcohol oxidation state, taking it out of play for a series of reactions that constructed the sidechain (**49** → **54**). The sidechain ketone was then protected as an acetal, and the cyclohexanone was reinstalled by deprotection and oxidation of the cyclohexanol. Regioselective acylation of **55** under conditions of thermodynamic control, followed by reduction of the intermediate β-ketoester, gave **56** (for comparison see **3** → **14** on Steroids-3). Formation of the tosylate, a β-elimination, and ketal hydrolysis completed the synthesis of **14**.

One of the distinguishing features of this synthesis (relative to those we have examined thus far) is the use of a 3-substituted cyclohexanone as an intermediate rather than a 4-substituted cyclohexanone. This choice eliminates problems associated with passing through a symmetrical (**24**, **29**, **32**) or pseudosymmetrical (**42**) intermediates that either destroy stereochemistry at C_4 (for example see **23** → **24** or hydrogenation of **42**) or make maintaining control of this center difficult (see conversion of **42** → **45**).

Evans, D. A.; Nelson, J. V. "Stereochemical Study of the [3,3] Sigmatropic Rearrangement of 1,5-Diene-3-Alkoxides. Application to the Stereoselective Synthesis of *dl*-Juvabione" *J. Am. Chem. Soc.* **1980**, *102*, 774.

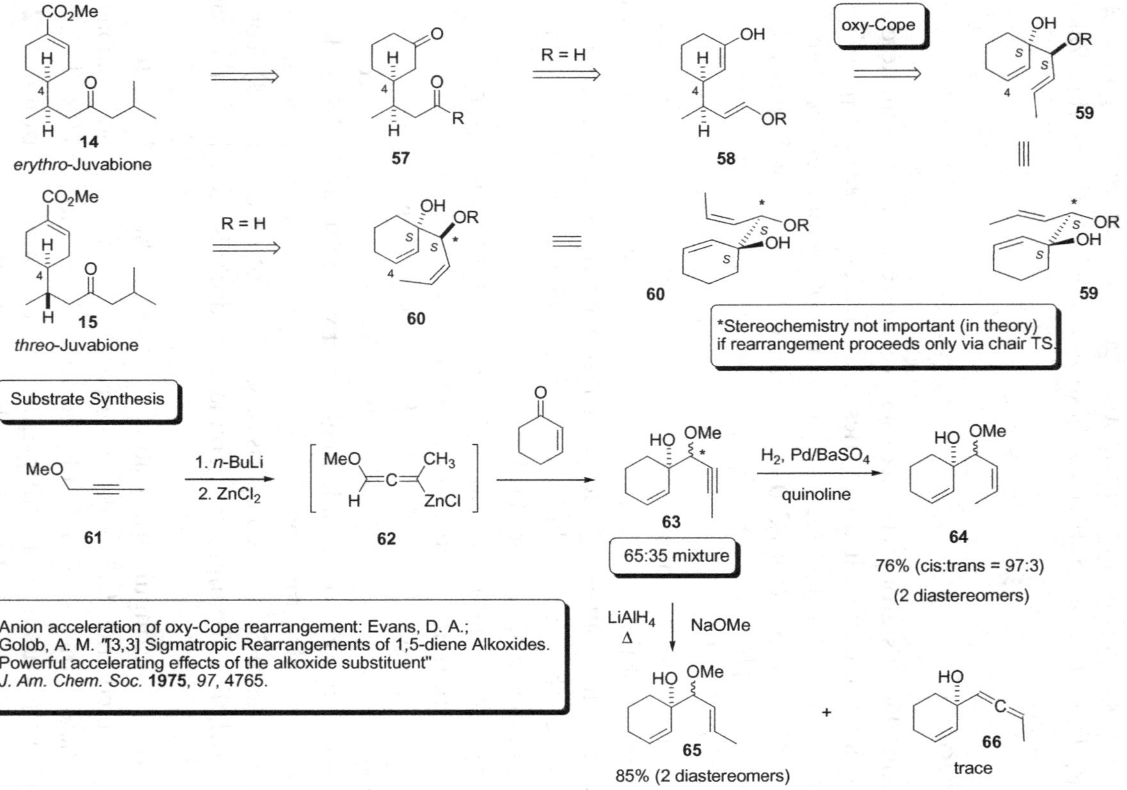

Juvabione-8

Juvabione-8

Recall our discussion of the Claisen rearrangement. This usually takes place via a well-defined chair-like transition state. We have already seen that this can translate to high transfer of chirality (Prostaglandins-24). It can also translate to excellent control of vicinal stereochemistry in an C_{sp3}-C_{sp3} bond-forming reaction. If the olefin geometry of the allyl and vinyl portions of the rearrangement substrate are well defined, this translates to excellent control over vicinal stereochemistry. The same is true for the Cope rearrangement, the [3.3]-sigmatropic rearrangement of a 1,5-hexadiene.[4]

In 1980 the Evans group reported a synthesis of *erythro*-juvabione (**14**) during the course of studies designed to reveal the nature of transition states in the anion accelerated version of the *oxy*-Cope rearrangment. The idea was that **14** could be prepared from an intermediate of type **57**. We have already seen this endgame in action during the Ficini synthesis of **14** (Juvabione-7). Compound **57** is a 1,6-dicarbonyl compound. Any 1,6-dicarbonyl compound can, in principle, be derived from the corresponding *bis*-enol tautomer (or a derivatives thereof). Thus, one precursor of **57** (where R = H) would be substituted 1,5-hexadiene **58**, which could be derived in turn from an *oxy*-Cope rearrangement of isomeric 1,5-hexadiene **59**. The key point is that the rearrangement of **59** to **58** establishes the vicinal stereochemical relationship of the juvabiones. On the other hand, the exocyclic *Z*-olefin (**60**) would provide the vicinal stereochemical relationship present in *threo*-juvabione (**15**). Note that (in theory), stereochemistry at the center marked with an asterisk should not play a major role in determining the stereochemical course of the rearrangments of **59**. In other words, **59** and its C* diasteromer should both serve as precursors to **14**, while **60** and its C* diastereomer should both serve as precursors to **15**. We will come back to this point.

The synthesis of the *oxy*-Cope substrates was accomplished in a direct, but non-stereoselective manner. Thus, lithiation of **61** and transmetallation to provide presumed allenylzinc reagent **62**, followed by reaction with cyclohexenone, gave **63** as a mixture of diastereomers. Semihydrogenation of **63** gave *Z*-olefin **64** (compare with **60**), and reduction of **63** with lithium aluminum hydride gave *E*-olefin **65** (compare with **59**) along with a trace of allene **66**.[5] Both sets of diastereomers were separated by chromatography over silica gel impregnated with silver nitrate, and their rearrangements were examined.

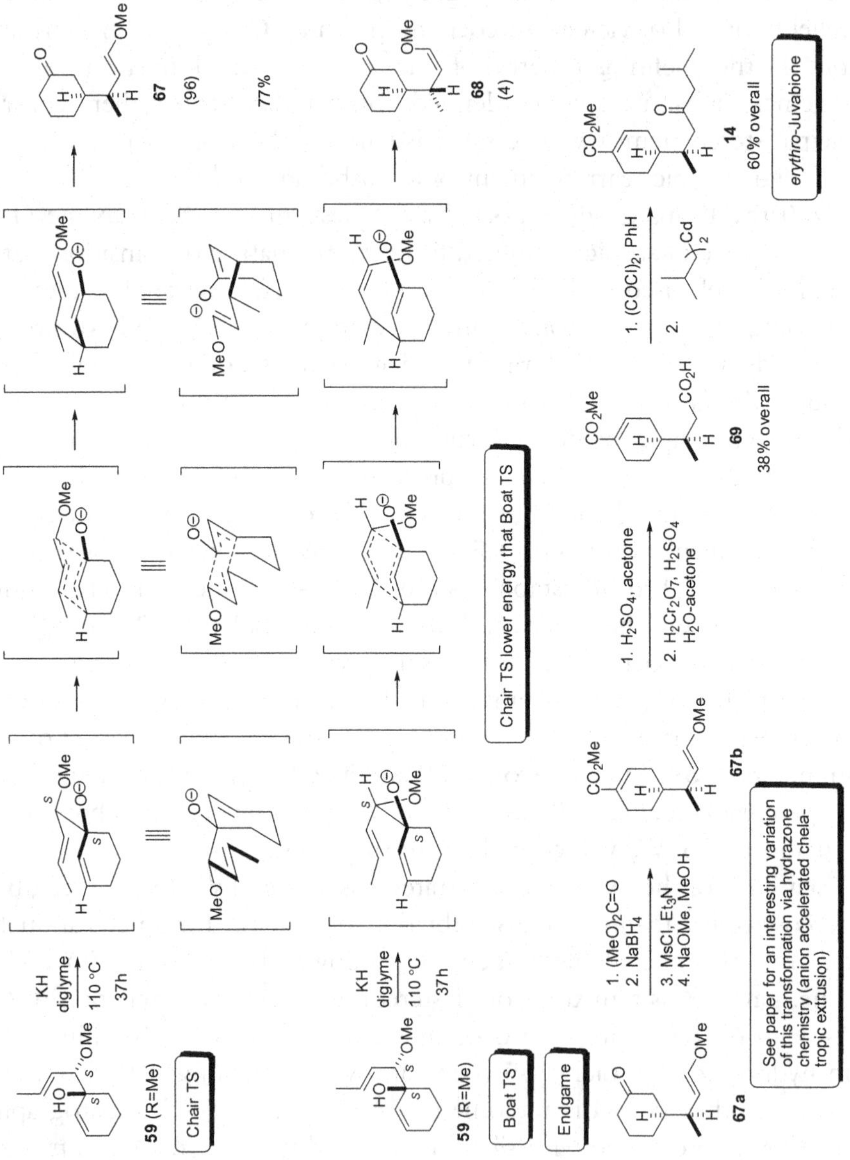
59 (R=Me)
Chair TS
KH
diglyme
110 °C
37h
67
(96)
77%
59 (R=Me)
Boat TS
68
(4)
Chair TS lower energy that Boat TS
Endgame
67a
1. (MeO)2C=O
2. NaBH4
3. MsCl, Et3N
4. NaOMe, MeOH
67b
1. H2SO4, acetone
2. H2Cr2O7, H2SO4
H2O-acetone
69
38% overall
1. (COCl)2, PhH
2.
14
60% overall
erythro-Juvabione
See paper for an interesting variation of this transformation via hydrazone chemistry (anion accelerated chelatropic extrusion)

Juvabione-9

Juvabione-9

Let's begin with **59** (R=Me). Treatment of this rearrangement substrate with KH provided a 96:4 ratio of **67** + **68**. These diastereomers differ in terms of vicinal stereochemistry *and* enol ether geometry. The major product (**67**) presumably was formed via rearrangement of the intermediate alkoxide via a chairlike transition state as anticipated. The minor product (**68**) was presumed to arise from rearrangement via a competing boat-like transition state. This nicely explains the differences in stereochemistry between the observed products (vicinal stereochemistry and olefin geometry).

The synthesis was completed by installation of the unsaturated ester (**67a** → **67b**), hydrolysis of the vinyl ether, oxidation of the intermediate aldehyde to acid **69**, and installation of the rest of the sidechain.

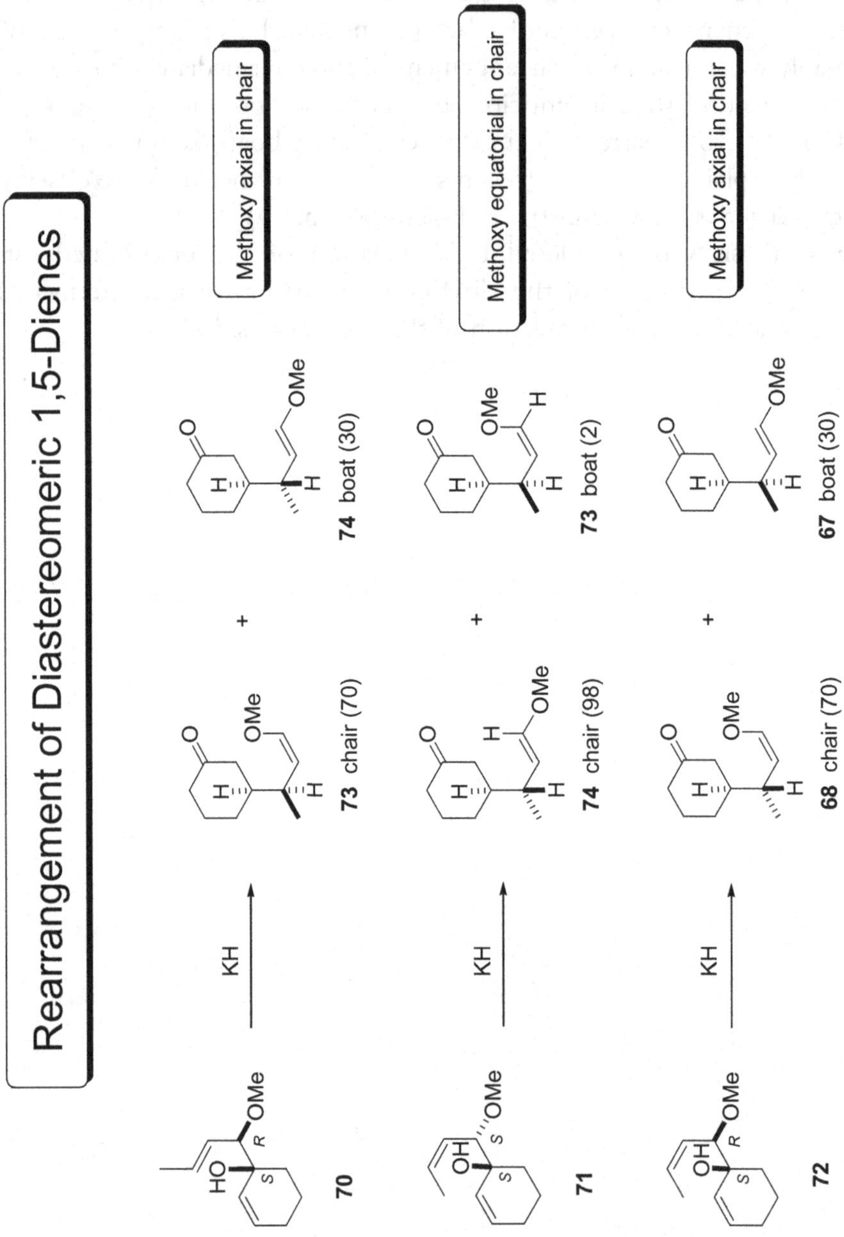

Juvabione-10

Juvabione-10

The remaining three diastereomeric rearrangement substrates [*epi*-**59** (R=Me) and the two diastereomers of **65**] were also subjected to rearrangement conditions. The epimer of **59** (**70**) gave **73**, with the vicinal stereochemistry required for *erythro*-juvabione (**14**), as the major product. This material was accompanied, however, by significant amounts of **74**, with the vicinal stereochemistry of *threo*-juvabione (**15**). The origin of these products can be explained exactly as shown for **59** (R=Me) [Juvabione-9]. The erosion in stereoselectivity presumably reflects a boat TS competing more successfully with a chair TS. Note that the methoxy group now occupies an "axial" site in the chair TS and is no longer sterically as differentiated from the boat TS. In accord with this observation substrate **71** (one of the two diastereomers of **65**) rearranges with high selectivity (the methoxy group occupies an equatorial site in the chair TS) and **72** (the other diastereomer of **65**) suffers considerable erosion of stereocontrol (the methoxy group occupies an axial site in the chair TS).

This study provided insight into mechanistic aspects of the anion accelerated oxy-Cope rearrangment. It also illustrates how control of terminal olefin stereochemistry in [3.3]-sigmatropic rearrangements can be used to establish vicinal stereochemistry. Note that both **59** (R=Me) and **71** give superb control. Only when there are overriding steric factors (**70** and **72**) does erosion of stereocontrol begin to occur.

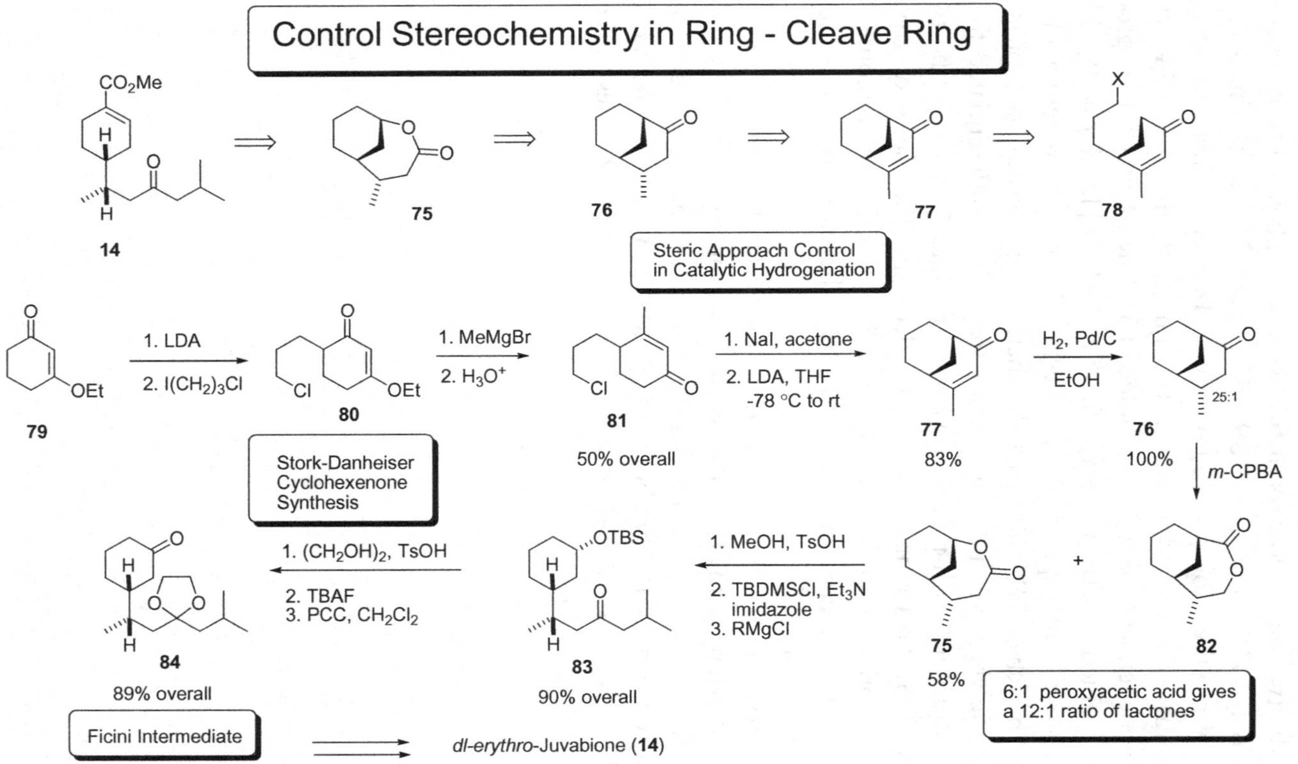

Schultz, A. G.; Dittami, J. P. "The Bicyclo[3.3.1]nonane Solution to the Problem of Vicinal Stereochemical Control at a Substituted Cyclohexane ring. A Total Synthesis of *dl-erythro*-Juvabione" *J. Org. Chem.* **1984**, *49*, 2615

Juvabione-11

Juvabione-11

Here is another synthesis that relies on setting stereochemistry in a ring and then opening the ring. The strategy was to prepare sterically biased bicyclic olefin **77**. It was imagined that reduction of the olefin would occur selectively from the face flanked by the methano bridge, to give **76**. Baeyer-Villiger oxidation of the ketone was expected to occur in the normal manner, with migration of the most highly substituted carbon, to provide lactone **75**. There are a number of reasonable ways to proceed from **75** to **14**. Execution of the plan first called for a synthesis of **77**. This was to be accomplished by an intramolecular alkylation of a compound of type **78** (where X = some leaving group).

In the forward direction, chloride **81** was assembled using a reliable method for the preparation of substituted cyclohexenones. Alkylation of **79** (notice the differentiation of electrophilic carbons by use of different halides), followed by introduction of the methyl group, and hydrolysis–dehydration of the intermediate β-hydroxyketone derivative, gave **81**. Conversion of the chloride to a better leaving group and kinetic enolate generation using LDA provided **77**. Hydrogenation of **77** took place with 25:1 diastereoselectivity to provide **76**. Baeyer-Villiger oxidation with *m*-CPBA gave the desired lactone along with the regioisomeric oxidation product **82**. Regioselectivity improved to 12:1 when peroxyacetic acid was used as the oxidant. Hydrolysis of lactone **75** was followed by protection of the secondary alcohol as a TBDMS ether. Treatment of the ester with isobutylmagnesium chloride under controlled conditions gave ketone **83**. The ketone was protected as an acetal. The alcohol was liberated with TBAF and oxidized with PCC to afford **84**, an intermediate in the Ficini synthesis of *erythro*-juvabione (**14**).

Before moving on, let's compare the Schultz (this work) and Ficini approaches. Whereas both of these syntheses use the "set stereochemistry in a ring and then open the ring" approach, the Schultz synthesis places oxidation in the side chain precisely where it is needed and the Ficini approach does not. Both syntheses, however, add the rest of the sidechain (4 carbons in the Schultz synthesis and 5 carbons in the Ficini synthesis) in a single reaction. My own bias is that the Schultz plan is nicer with respect to the sidechain, but the Ficini synthesis does accomplish the task. It is notable that up to now, the Ficini synthesis is the only synthesis we have considered that does not place the sidechain oxygen directly in the desired location. Of course, the Ficini approach leads to an examination of ynamine chemistry whereas the other approaches would not. There is give and take in the planning of any synthesis.

A Recent Approach to the Problem

Pearson, A. J.; Paramahamsan, H.; Dudones, J. D. "Vicinal Stereocontrol during Nucleophilic Addition to Arene Chromium Tricarbonyl Complexes: Formal Synthesis of *dl-erythro*-Juvabione" *Org. Lett.* **2004**, *6*, 2121-2124.

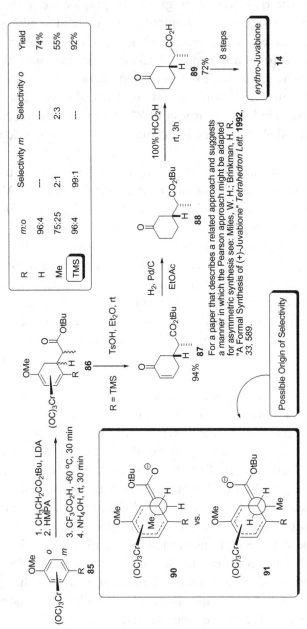

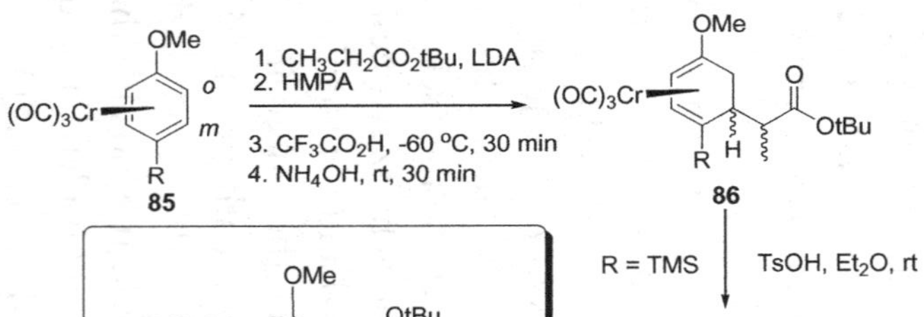

R	*m:o*	Selectivity *m*	Selectivity *o*	Yield
H	96:4	---	---	74%
Me	75:25	2:1	2:3	55%
TMS	96:4	99:1	---	92%

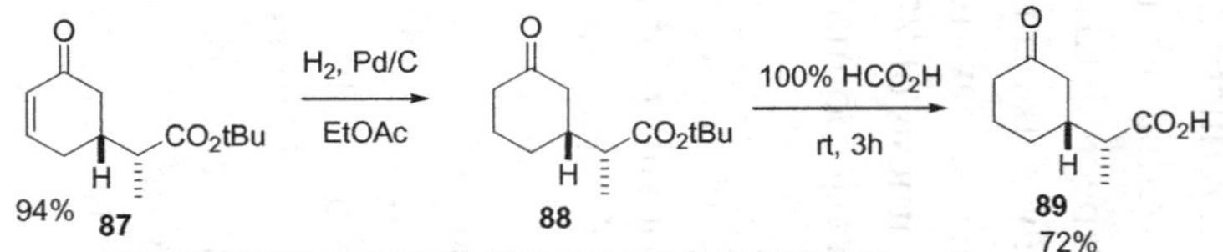

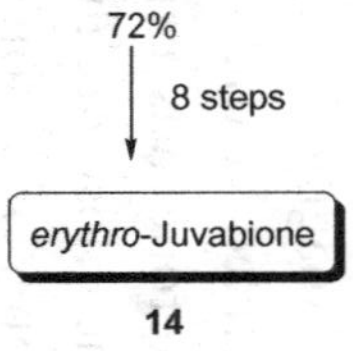

For a paper that describes a related approach and suggests a manner in which the Pearson approach might be adapted for asymmetric synthesis see: Miles, W. H.; Brinkman, H. R. "A Formal Synthesis of (+)-Juvabione" *Tetrahedron Lett.* **1992**, *33*, 589.

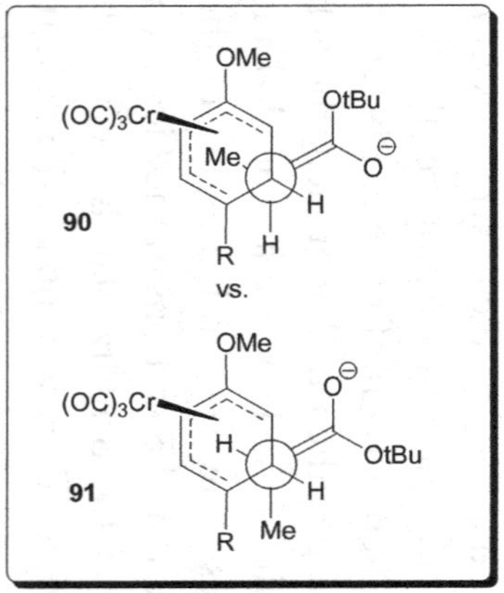

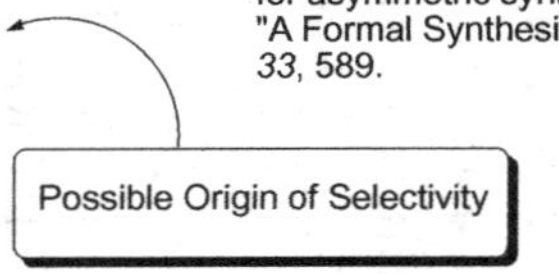

Total synthesis can serve as a template for reaction development. If we look back this was the case for this synthesis and the Evans synthesis (and several that will follow). Asymmetric synthesis is also worth considering at the point. All of the preceeding syntheses afforded racemic material. At what point could you introduce asymmetry into each of these syntheses? Which strategies would be most amenable to asymmetric synthesis? Of course this problem has been addressed (not necessarily within the context of the strategies we have seen thus far) and we will now turn to this issue.

Juvabione-12

Let's now look at a recent approach to the problem. It is actually a study wherein the "problem" was used to learn something about the stereochemical course of a new reaction. The Pearson group was studying reactions of ester enolates with arene chromium tricarbonyl complexes. As part of this study, the reaction of anisole derivatives (**85**) with the enolate derived from *tert*-butyl propionate was examined. The result was formation of a chromium tricarbonyl complex of a 1-methoxy-1,3-cyclohexadiene of type **86**. Conversion of the chromium-diene complex into a cyclohexenone would provide material that could serve as an intermediate in a juvabione syntheses if the reaction took place with good diastereoselectivity.

There are both regiochemical and stereochemical issues here. The enolate (presumably largely of *E*-geometry based on the work of Ireland) could add "ortho" or "meta" to the methoxy group, "syn" or "anti" to the metal, and provide either "erythro" or "threo" vicinal stereochemistry. A series of compounds of type **85** were examined. It was well known that nucleophiles add anti to the metal in such complexes. Good "meta" selectivity was achieved with R=H or R=TMS. Excellent diastereoselectivity was observed for the meta adduct in the case of R=TMS. Whereas regioselectivity was good when R=H, diastereoselectivity was poor. The stereochemistry of **86** (R=TMS) was established by conversion to **89**, an intermediate in the Ficini synthesis. Thus, metal decomplexation and protonolysis of the TMS group was achieved using *p*-toluenesulfonic acid in ether. The resulting enone **87** was hydrogenated to give **88**, and the *tert*-butyl ester was cleaved using acid to provide **89**. It was suggested that the observed stereoselectivity reflected a minimzation of steric interactions between the methyl group of the enolate and the R-group of **85** in a transition state that leads to a staggered array of substituents along the forming sp^3-sp^3 bond. This is consistent with lower selectivity observed in the meta adduct for **85** where R=H or Me.

The Pearson synthesis is a nice example of total synthesis serving as a template for reaction development. If we look back, this was also the case for the Ficini and Evans syntheses (and several that will follow). Asymmetric synthesis is also worth considering at this point. All of the preceeding syntheses afforded racemic material. At what point could you introduce asymmetry into each of these syntheses? Which strategies would be most amenable to asymmetric synthesis? Of course this problem has been addressed (not necessarily within the context of the strategies we have seen thus far) and we will now turn to this issue.

Terpenes as Starting Materials

Pawson, B. A.; Cheung, H.-C.; Gurbaxani, S.; Saucy, G. "Stereospecific Synthesis and Absolute Stereochemistry of Natural (+)-Juvabione" *J. Chem. Soc., Chem. Commun.* **1968**, 1057-1058.

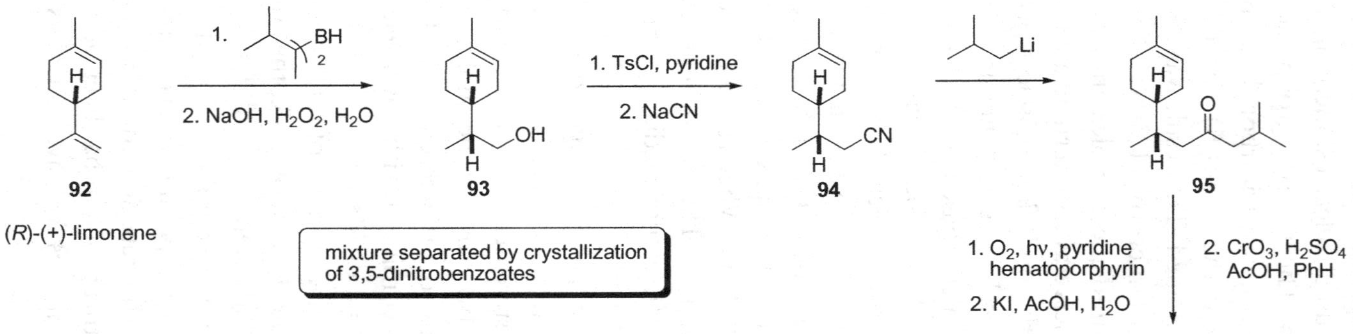

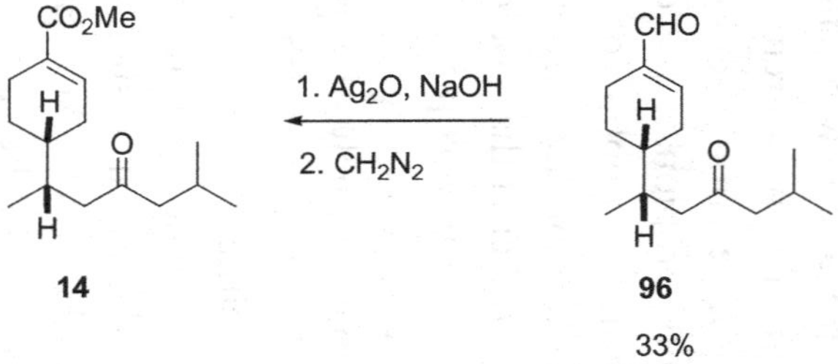

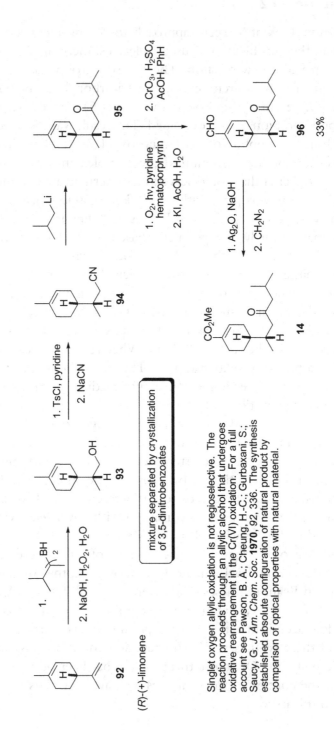

Singlet oxygen allylic oxidation is not regioselective. The reaction proceeds through an allylic alcohol that undergoes oxidative rearrangement in the Cr(VI) oxidation. For a full account see Pawson, B. A.; Cheung, H.-C.; Gurbaxani, S.; Saucy, G. *J. Am. Chem. Soc.* **1970**, *92*, 336. The synthesis established absolute configuration of natural product by comparison of optical properties with natural material.

Juvabione-13

Juvabiones-13

One approach is to begin with a single enantiomer of a chiral starting material. This is the "chiral pool" approach (Prostaglandins-15). The Hoffman-LaRoche group began their synthesis with (*R*)-limonene (**92**). The challenges in moving from limonene to juvabione involved (1) introduction of the sidechain with control of stereochemistry at the exocyclic stereogenic center and (2) oxidation of the methylcyclohexene to the unsaturated ester. The first issue was not addressed with control over stereochemistry. Hydroboration-oxidation of **92** using a hindered borane gave a mixture of alcohols that were separated after derivatization. Once pure **93** was in hand, the rest of the sidechain was introduced in a straightforward manner (**93** → **94** → **95**). The methylcyclohexene was then oxidized using singlet oxygen, followed by oxidation of the mixture of intermediates. This reaction was not regioselective, but did give unsaturated aldehyde **96** in a modest yield. Oxidation of the aldehyde and esterification completed the synthesis of **14**. This synthesis may not seem elegant, but it established the absolute configuration of the natural product.

Fuganti, C.; Serra, S.; "Baker's Yeast mediated enantio selective synthesis of the bisabolene sesquiterpenes (+)-epijuvabione and (-)-juvabione "*J. Chem. Soc., Perkin Trans. 1*, 2000, 97-101.

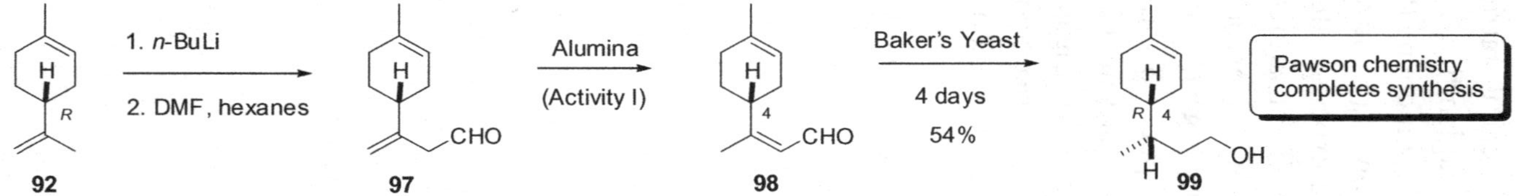

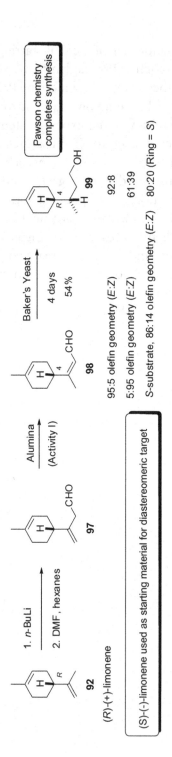

Fujii, M.; Aida, T.; Yoshinara, M.; Ohno, A. "NAD(P)$^+$ - NAD(P)H Models. 71. A Convenient Route to the Synthesis of Juvabione" *Bull. Chem. Soc. Jpn.* **1990**, 63, 1255-1257.

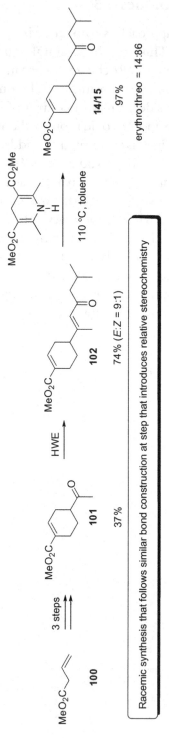

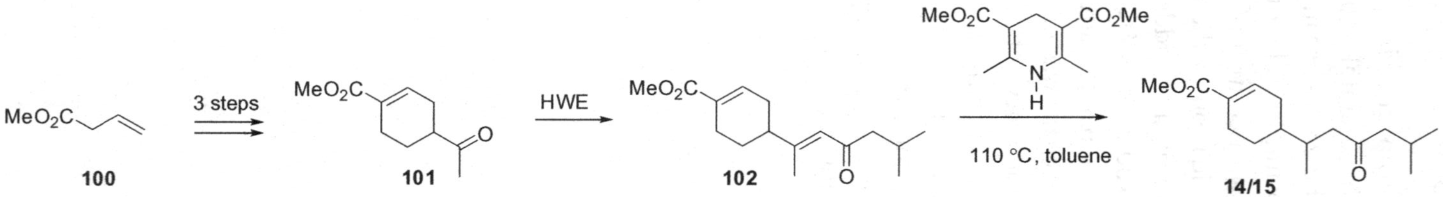

Juvabione-14

Juvabione-14

Here is another approach from *R*-limonene (**92**). The idea was to use reagent control (see Prostaglandins-12) to establish the exocyclic stereogenic center. Substrate **98** was prepared from **92** via formylation of an intermediate allylic lithium reagent, and conjugation of intermediate enal **97**. The chiral reagent in this case was a collection of enzymes secreted by fermenting Baker's yeast. Thus, when the *E*:*Z* olefin ratio in **98** was 95:5, the reduction gave **99** as a 92:8 mixture of diastereomers (at the exocyclic center). Alcohol **99** has the stereochemistry required for *erythro*-juvabione (**14**). When the *E*:Z olefin ratio in **98** was 5:95, the reduction was less selective, affording a 61:39 mixture of stereoisomers with **99** as the major isomer. Finally, if *S*-limonene was the starting point for the synthesis, and the *E*:*Z* olefin ratio was 86:14, an 80:20 mixture of the C_4 diastereomer of **99** and *ent*-**99** was obtained. The bottom line is that the enzymes responsible for the reduction were capricious. They were reasonably selective when *E*-**98** was the substrate, but showed lower selectivity (or more sensitivity to the C_4 stereogenic center) with other substrates. Nonetheless the approach is interesting. A racemic synthesis that follows a similar bond construction in the step that introduces relative stereochemistry is shown here without much comment. It is a very direct synthesis, but it lacks selectivity obtained in many of the syntheses we have previously examined.

Trost, B. M.; Tamaru, Y. "2-Methylthioacetic Acid and Diethyl Malonate as Acyl Anion Equivalents. Synthesis of Juvabione" *Tetrahedron Lett.* **1975**, 3797-3800

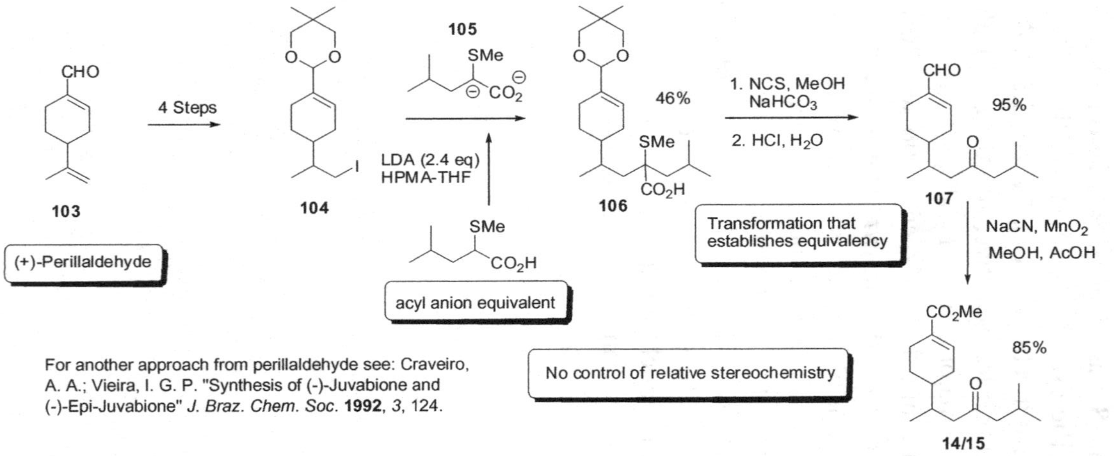

For another approach from perillaldehyde see: Craveiro, A. A.; Vieira, I. G. P. "Synthesis of (-)-Juvabione and (-)-Epi-Juvabione" *J. Braz. Chem. Soc.* **1992**, *3*, 124.

Negishi, E.; Sabanski, M.; Katz, J.-J.; Brown, H. C. "An Efficient Synthesis of Juvabione and Todomatuic Acid via Hydroboration-Oxidation" *Tetrahedron* **1976**, *32*, 925-926.

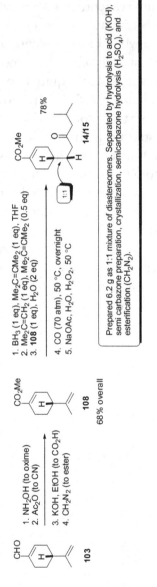

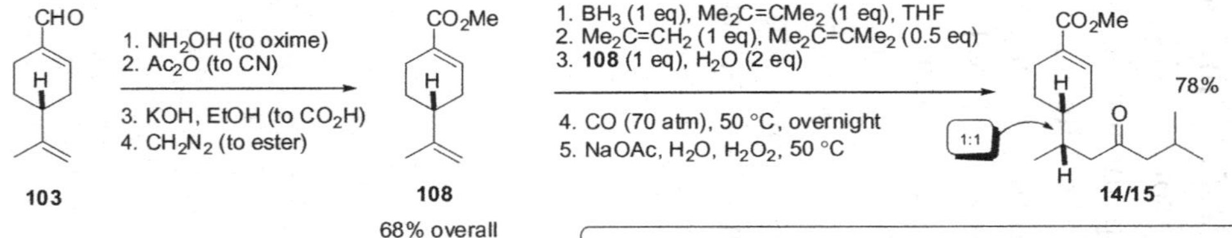

Prepared 6.2 g as 1:1 mixture of diastereomers. Separated by hydrolysis to acid (KOH), semi carbazone preparation, crystallization, semicarbazone hydrolysis (H_2SO_4), and esterification (CH_2N_2).

Juvabione-15

One shortcoming of limonene as a starting material is that the cyclohexene substituent (methyl group) is in the wrong oxidation state, and the allylic oxidation procedure that was used lacked regioselectivity. Perillaldehyde (**103**) does not have this oxidation state problem. Several groups have used this terpene as a point of departure in syntheses of the juvabiones. Two examples are presented here. The Trost synthesis illustrated the development of a new acyl anion equivalent (**105**). The Brown synthesis presented methodology for "stitching" alkenes together to produce unsymmetrically substituted ketones (**108** + isobutylene + CO → **14/15**). Neither synthesis addresses the vicinal stereochemistry problem. The Brown synthesis, however, is very efficient and provided substantial amounts of the natural products as a mixture.

Kawamura, M.; Ogasawara, K. "Stereo- and Enantio-controlled Synthesis of (+)-Juvabione and (+)-Epijuvabione from (+)-Norcamphor" *J. Chem. Soc., Chem. Commun.* **1995**, 2403-2404.

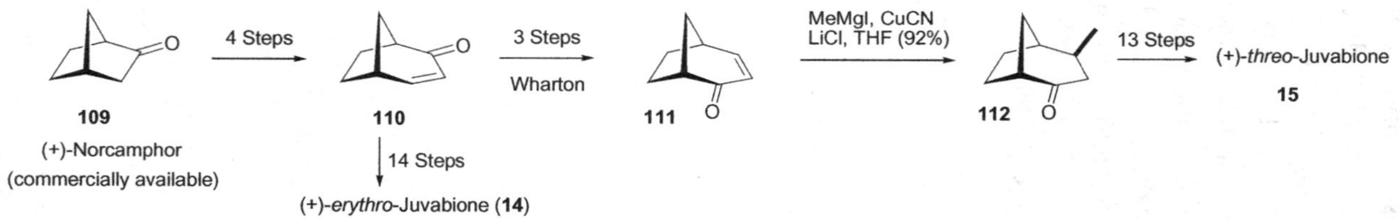

For an equally tortuous route from racemic norcamphor that involves a lipase-mediated resolution of the alcohols derived from the enones shown above see: Nagata, H.; Taniguchi, T.; Kawamura, M.; Ogasawara, K. "A Lipase-mediated Route to (+)-Juvabione and (+)-Epijvabione from Racemic Norcamphor" *Tetrahedron Lett.* **1999**, *40*, 4207-4210.

Nagano, E.; Mori, K. "Synthesis of (+)-Juvabione, a Compound with Juvenile Hormone Activity" *Biosci. Biotech. Biochem.* **1992**, *56*, 1589-1591

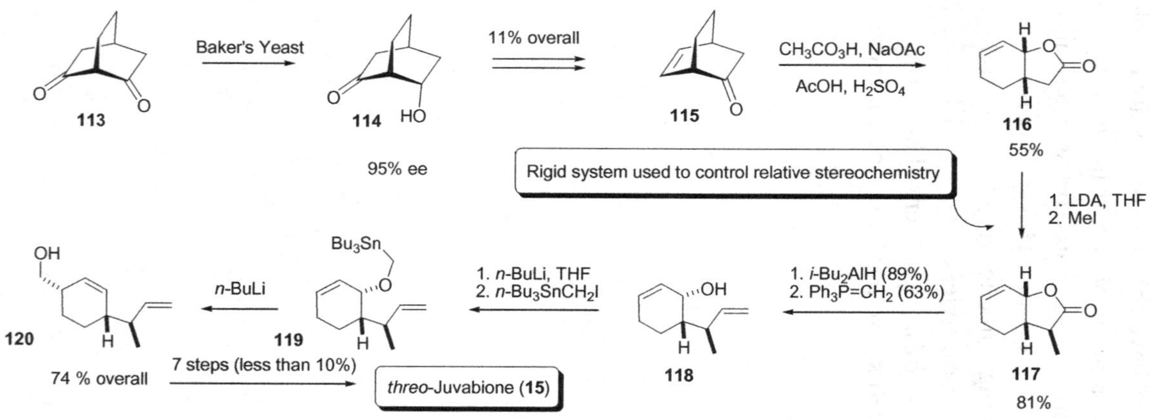

Juvabione-16

Juvabione-16

Two more enantioselective syntheses are described here. The first begins with norcamphor. This served as a starting material for the synthesis of **110** and its enantiomer **111**. These compounds were used to prepare both **14** and **15** using an organocuprate conjugate addition to set vicinal stereochemistry. A Baeyer-Villiger oxidation was used to "liberate" the sidechain. A weak point in the choice of **109** is that a 5-membered ring must be expanded to a 6-membered ring. That is one reason why this approach is lengthy.

The second synthesis began with a nifty reduction of diketone **113** using Baker's yeast. Conversion of **114** to **115** was followed by Baeyer-Villiger oxidation and rearrangment to lactone **116** (compare with the Corey approach to prostaglandins; Prostaglandins-6). Alkylation of the enolate derived from **116** gave **117** and established the vicinal stereochemical relationship. Reduction of the lactone to a lactol was followed by a Wittig olefination to give **118**. A [2,3]-sigmatropic rearrangement was then used to prepare **120**, which was carried on to *threo*-juvabione.

Both of these approaches use the now-familiar "set stereochemistry in a ring, then open the ring" strategy. They are effective in establishing vicinal stereochemistry, but are long because of the choice of starting material.

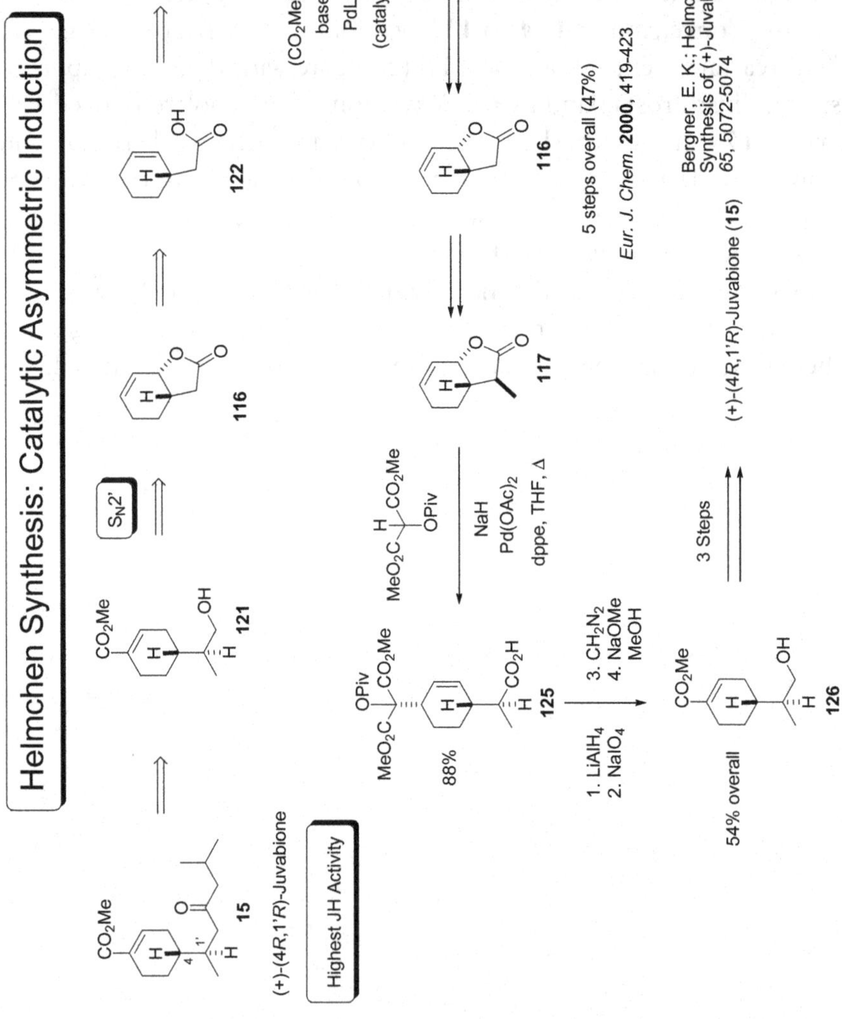

Juvabione-17

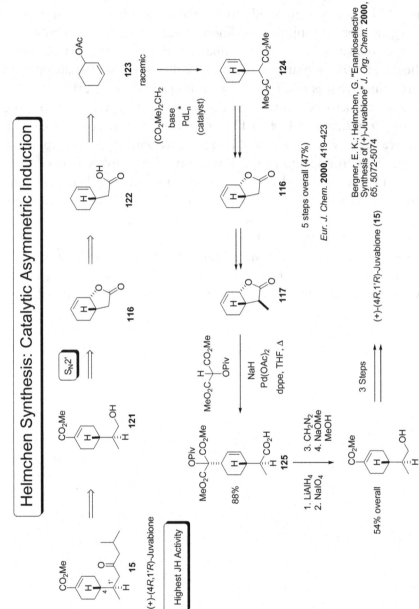

Juvabione-17

Another enantioselective route to **116**, that features catalytic asymmetric induction, is shown here (see Prostaglandins-12 for another example). This synthesis begins with racemic allylic acetate **123**. A palladium-mediated allylation of dimethyl malonate in the presence of chiral ligands (for the Pd) provided **124** with excellent enantioselectivity. This material was converted to **116**. Alkylation as per the Mori synthesis (Juvabione-16) gave **117**, which was converted to **125** using another Pd-mediated malonate allylation. Malonate **125** was converted to **126** via an intermediate tetraol. The synthesis of *threo*-juvabione (**15**) was then completed using a short reaction sequence.

This synthesis was designed to showcase organopalladium chemistry developed in the Helmchen laboratories. The stereochemistry of the ring stereogenic center was handled using reagent-based control. The bowl-shaped nature *cis*-bicyclo[4.3.0]nonane was used to control stereochemistry of what becomes the exocyclic stereogenic center of the juvabiones.

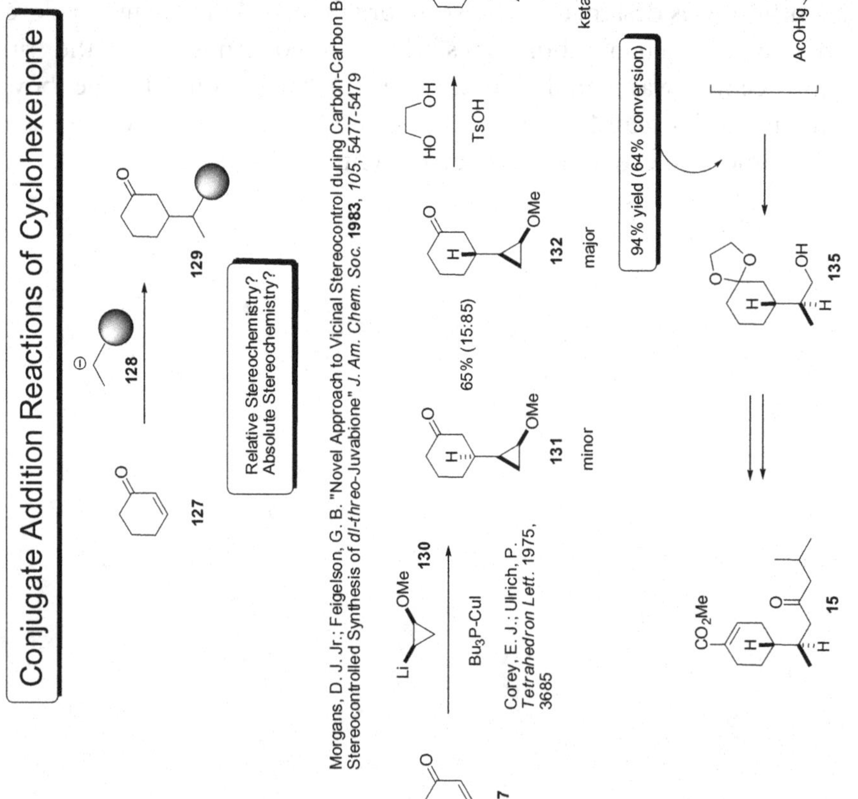

Juvabione-18

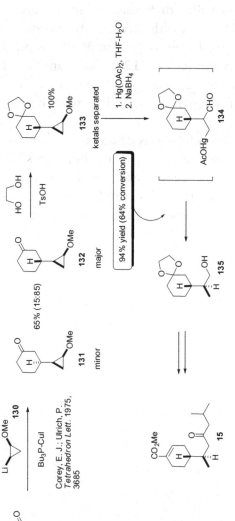

Juvabione-18

A conceptually different approach to the juvabiones would be to develop reactions of cyclohexenone (**127**) with nucleophiles of type **128** that afford conjugate adducts of type **129** with good control over relative (vicinal) stereochemistry and absolute stereochemistry. This plan is conceptually related to the Pearson approach (Juvabione-12) but uses the electrophile needed to directly connect with the now familiar 3-substituted cyclohexanone endgame for juvabione synthesis. We will look at three different approaches that adapt this general strategy.

Morgans examined reactions of **127** with cuprates derived from **130**. The choice of **130** as a "real" version of **128** was based (in part) on the hope that the stereogenic center adjacent to the carbon-lithium bond might influence the stereochemical course of the conjugate addition. It is notable that this is really a "relative asymmetric induction" approach to the problem. If the solution worked in a relative sense, then all that would be needed to accomplish an enantioselective synthesis would be to start with a single enantiomer of **130**. After surveying a number of organometallic derivatives of *rac*-**130**, conversion to racemic **131** and **132** was accomplished with a reasonable degree of diastereoselectivity. The mixture of ketones was protected and the major ketals were separated to provide **133**. Treatment of **133** with mercuric acetate, followed by reduction of the presumed intermediate organomercurial **134**, gave **135**. This material was carried through to *threo*-juvabione (**15**) in the usual manner.

Tokoroyama, T.; Pan, L.-R. "Efficient Stereoselective Synthesis of Both *dl*-Juvabione and *dl-epi*-Juvabione by New Extracyclic Stereocontrol Methodology" *Tetrahedron Lett.* **1989**, *30*, 197.

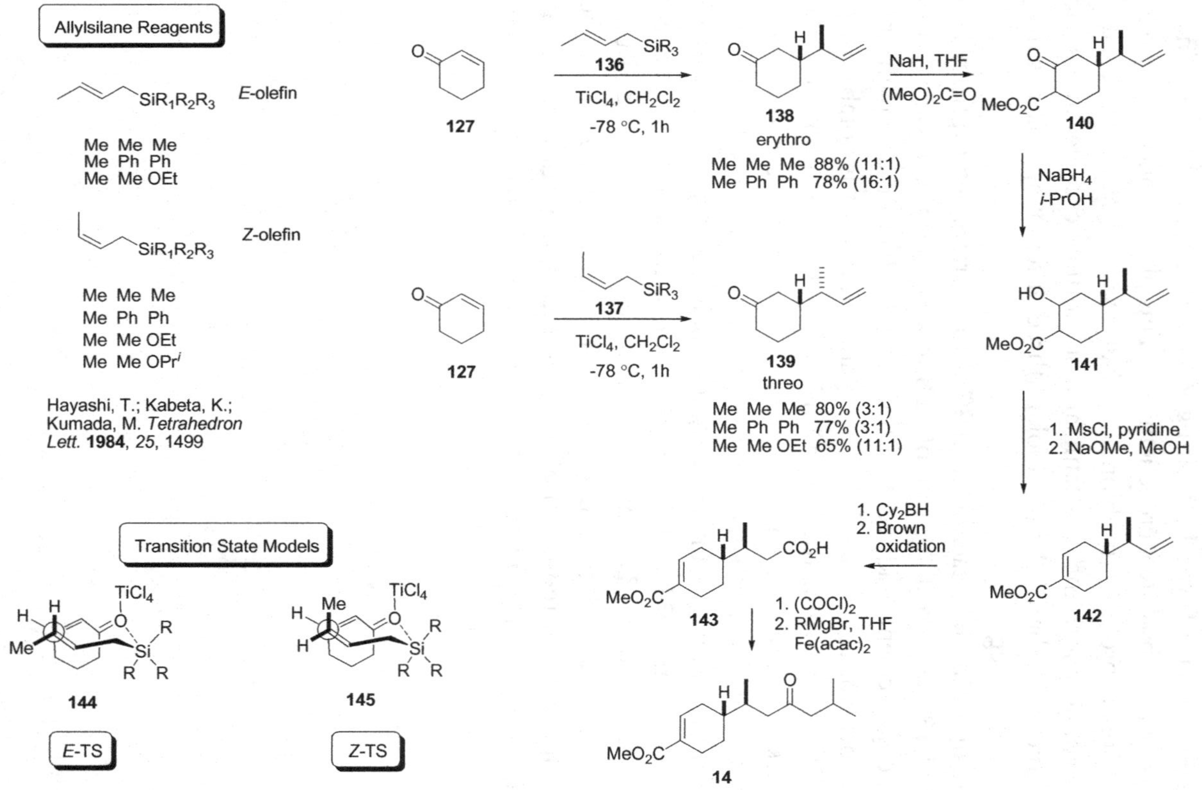

Juvabione-19

Juvabione-19

The addition of allyl silanes to α,β-unsaturated ketones to give δ,ε-unsaturated ketones is known as the Sakurai reaction. Thus, reaction of cyclohexenone (**127**) with allylsilanes of type **136** and **137** provides conjugate adducts of type **138** and **139**. The relationship of these stereoisomers to *erythro*- and *threo*-juvabiones (**14** and **15**, respectively) is clear. The Tokoroyama group surveyed a series of *E*-crotylsilanes (**136**) and *Z*-crotylsilanes (**137**) in this process. They determined that *E*-crotylsilanes provided the *erythro* adduct (**138**) as the major diastereomer, where as the *Z*-crotylsilanes gave the *threo* adduct (**139**) as the major diastereomer. Selectivities were very good with proper selection of silicon substituents. Both conjugate adducts could be transformed to the respective juvabiones. For example, the unsaturated ester was introduced to **138** using the standard acylation-reduction-dehydration sequence. The terminal olefin of **142** was then chain-extended via acid **143** to provide *erythro*-juvabione **14**. These stereochemical observations led to the proposal of transition-state models for the reaction of the *E*- and *Z*-crotylsilanes as shown in structures **144** and **145**, respectively. This is another nice example of synthesis being used to learn something about a specific reaction. Of course, whether or not this mechanistic proposal is correct would need to be determined by further studies. The nice feature of this study is that it clearly demonstrates a relationship between reagent olefin geometry and reaction diastereoselectivity.

Watanabe, H.; Shimizu, H.; Mori, K. "Synthesis of Compounds with Juvenile Hormone Activity. XXXI. Stereocontrolled Synthesis of (+)-Juvabione from a Chiral Sulfoxide" *Synthesis* **1994**, 1249. For relevant preliminary work see: Hua, D. H.; Venkataraman, S.; Coulter, M. J.; Sinai-Zingde, G. *J. Org. Chem.* **1987**, *52*, 719. Binns, M. R.; Haynes, R. K.; Katsifis, A. G.; Schober, P. A.; Vonwiler, S. C. *J. Am. Chem. Soc.* **1988**, *110*, 5411.

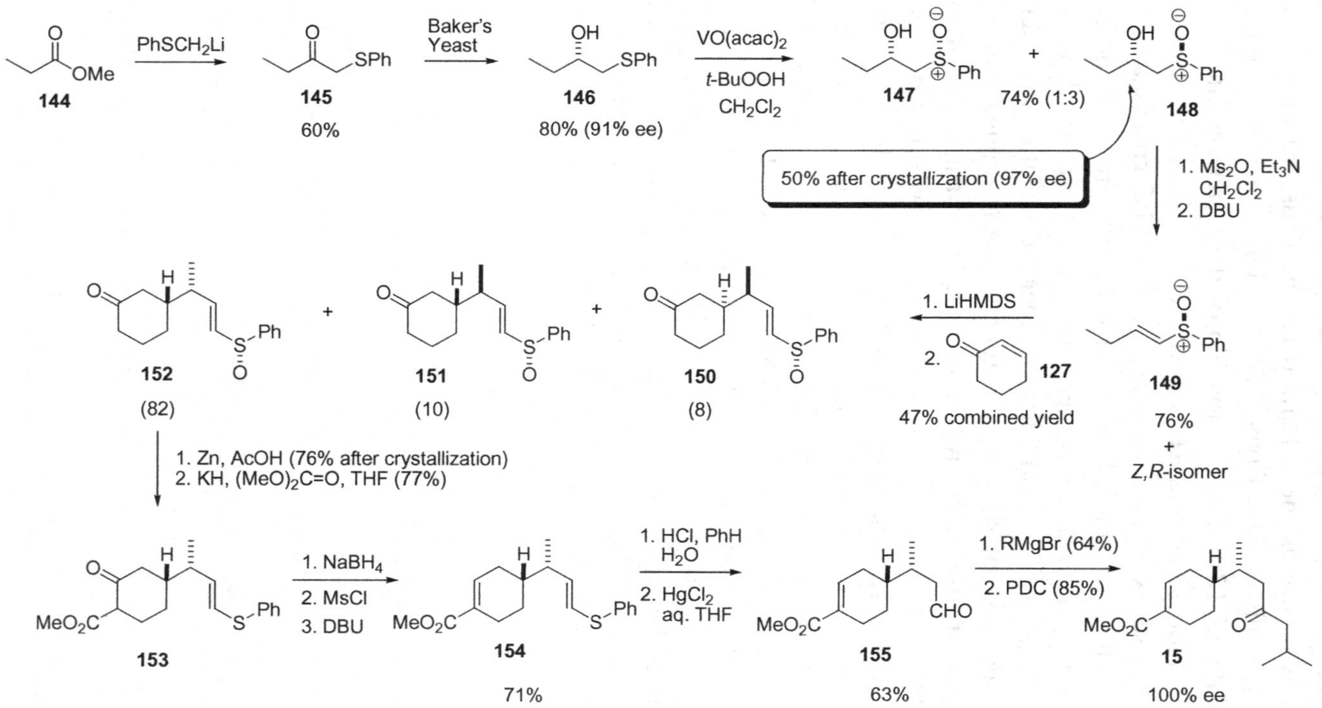

Juvabione-20

Juvabione-20

Mori was clearly intrigued by juvabione. Nearly 25 years after his first synthesis, his group described the third "conjugate addition" approach we will consider. This synthesis addresses both relative and absolute stereochemistry problems. The plan was to prepare the anion derived from vinyl sulfoxide **149** and examine its reaction with cyclohexenone (**127**). There was precedence for the γ-carbon of this type of allylic anion to behave as the nucleophile in conjugate additions (Hua). The hope was that this addition would take place with good diastereoselectivity and that the sulfoxide would influence the absolute stereochemistry of the process.

The first task was to prepare the chiral sulfoxide. The synthesis began with the conversion of methyl propionate (**144**) to keto-sulfide **145**. Enzymatic reduction of the ketone using Baker's Yeast gave **146** with decent enantioselectivity. A "directed oxidation" of the sulfide provided an unequal mixture of sulfoxides **147** and **148** (and presumably minor amounts of material derived from the 4–5% of *ent*-**146** present in the starting material) from which **148** could be isolated in 50% yield. Dehydration of the alcohol provided **149** (along with some of the *Z*-isomer). Notice that Mori decided to place the alcohol *beta* to the sulfoxide in the precursor of **149**. There might be a number of reasons for this, but one is that it facilitated the elimination reaction (dehydration) because of the electron-withdrawing properties of the sulfoxide.

The planned conjugate addition gave adducts **150–152**. Both the vicinal diastereoselection (**150**+**152**:**151** = 9:1) and diastereoselection relative to the sulfoxide (**152**:**150** = 10:1) were good. Vinyl sulfoxide **152** was carried on to unsaturated ester **154** using a straightforward reaction sequence. Vinyl sulfide **154** was hydrolyzed to aldehyde **155**, establishing the aforementioned homoenolate anion equivalency. The synthesis was completed in two steps to provide a single enantiomer of *threo*-juvabione (**15**). A comparison of Mori's first synthesis (Juvabione-4) and this synthesis provides an indication of how synthesis changed over this period of time.

A Completely Different Approach

Soldermann, N.; Velker, J.; Vallat, O.; Stoekli-Evans, H.; Neier, R. "Application of the Novel Tandem Process Diels-Alder Reaction/Ireland-Claisen Rearrangement to the Synthesis of *rac*-Juvabione and *rac*-Epijuvabione" *Helv. Chim. Acta* **2000**, *83*, 2266-2276.

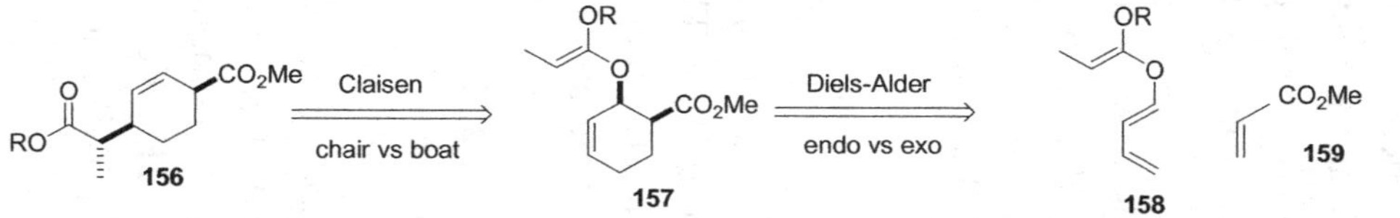

This paper presents a nice "overview" of other approaches (13 key intermediates). The "plan" involves an interesting disconnection that is related to the Evans solution to the vicinal stereochemistry problem. The plan is great, but lack of selectivity in the orbital symmetry controlled reactions are problematic.

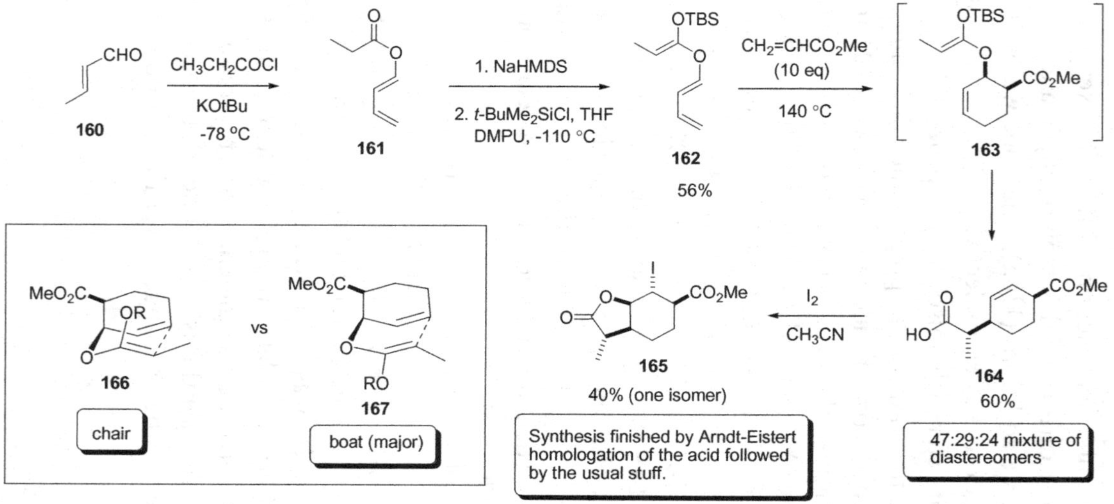

Juvabione-21

Juvabione-21

The last synthesis we will consider follows a completely different plan. The idea was that a compound of type **156** would serve as a late intermediate in the synthesis. The vicinal stereochemistry was to be set using a Claisen rearrangement of an allyl vinyl ether of type **157**. Being a cyclohexene, **157** was to be prepared by a Diels-Alder reaction between a diene of type **158** and a methyl acrylate (**159**). Given that Claisen and Diels-Alder reactions are thermally mediated, it was hoped that this might all happen in one pot merely by heating the diene and dienophile. For this plan to succeed, the Diels-Alder reaction would have to show *endo*-selectivity. This is the normal expectation for such a Diels-Alder reaction. The Claisen rearrangement would also have to occur via a well-defined transition state (**166** or **167**). This was a reasonable expectation based on the Evans-Nelson studies of a related Cope rearrangement (Juvabione-8).

The required dienophile was prepared from crotonaldehyde (**160**). *O*-Acylation of the derived dienolate gave **161**. Enolate generation and silylation gave **162**. The tandem Diels-Alder/Claisen sequence gave **164** in 60% yield as a mixture of diastereomers. The structure of the major diastereomer was established by conversion of the mixture to a mixture of iodolactones from which **165** was isolated and analyzed by X-ray crystallography. Working backwards, it was deduced that the major stereoisomer came from an *endo*-Diels-Alder followed by Claisen rearrangement from boat-like TS **167**. Perhaps this is not surprising given the steric hinderance present in the chair-like conformation **166**. From the standpoint of the synthesis, the mixture of acids **164** was carried on to the end to provide a mixture of juvabiones **14** and **15**. This was a great plan, but it just did not deliver the stereoselectivity of other syntheses.

References

1. For early history of the topic see Trost, B. M.; "The Juvenile Hormone of Hyalophoria Cecropia" *Acc. Chem. Res.* **1970**, *3*, 120 and references cited therein. See also, Rees, H. H.; Goodwin, T. W. "Molting Hormones" *Biochem. Soc. Trans.* **1974**, *2*, 1027–1032. Karlson, P.; Sekeris, C. E. "Ecdysone, an Insect Steroid Hormone, and its Mode of Action" *Recent Progress in Hormone Research* **1966**, *22*, 473–495.
2. Giguere, R. J. "Nonconventional Reaction Conditions: Ultrasound, High Pressure, and Microwave Heating in Organic Synthesis in "*Organic Synthesis: Theory and Applications*" **1989**, *1*, 103–172.
3. Corey, E. J. "Catalytic Enantioselective Diels-Alder Reactions: Methods, Mechanistic Fundamentals, Pathways, and Applications" *Angew. Chem. Int Ed.* **2002**, *41*, 1560–1567. Fringuelli, F.; Piermatti, O.; Pizzo, F.; Vaccaro, L. "Recent Advances in Lewis Acid Catalyzed Diels-Alder Reactions in Aqueous Media" *Eur. J. Org. Chem.* **2001**, 439–455. Evans, D. A.; Johnson, J. S. "Diels-Alder Reactions" *Comprehensive Asymmetric Catalysis* **1999**, *3*, 1177–1235.
4. Vittorelli, P.; Hansen, H. J.; Schmid, H. "Kinetics and Stereochemical Course of the Thermal Rearrangement of the Four Stereoisomers of Propenyl But-2′-enyl Ether" *Helv. Chim. Acta* **1975**, *58*, 1293–1309.
5. Corey, E. J.; Katzenellenbogen, J. A.; Posner, G. H. "New Stereospecific Synthesis of Trisubstituted Olefins. Stereospecific Synthesis of Farnesol" *J. Am. Chem. Soc.* **1967**, *89*, 42454247. Molloy, B. B.; Hauser, K. L. "Effects of Metal Alkoxides on the Lithium Aluminum Hydride Reduction of Substituted Prop-2-ynyl Alcohols" *J. Chem. Soc., Chem. Commun.* **1968**, 1017–1019.

Problems

1. Outline a series of reactions that shows how the Wicha strategy could be used to prepare the C_{20} epimer of **5**. (Juvabione-1)
2. Consider the following reactions. Predict the major product of each reaction. Which reactions you expect to take place with good diastereoselectivity or poor diastereoselectivity? Explain the basis for your predictions. (Juvabione-3)

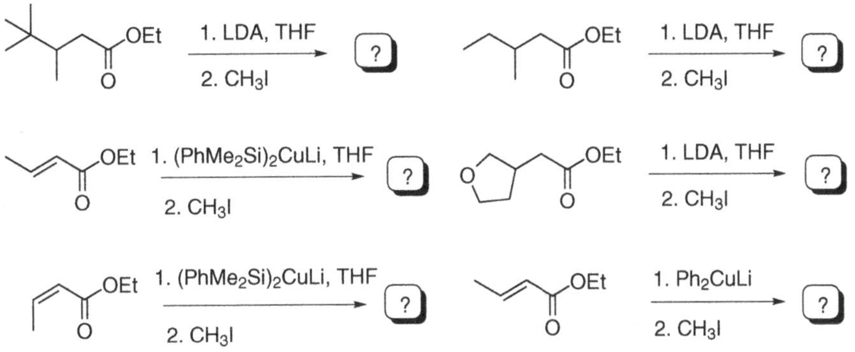

For lead references see: Yamamoto, Y.; Maruyama, K. "The Opposite Diastereoselectivity in Alkylation and Protonation of Enolates" *J. Chem. Soc., Chem. Commun.* **1984**, 904–905. Fleming, I.; Lewis, J. J. "A Paradigm for Diastereoselectivity in Electrophilic Attack on Trigonal Carbon Adjacent to a Chiral Center: The Methylation and Protonation of Some Open-Chain Enolates" *J. Chem. Soc., Chem. Commun.* **1985**, 149–151. McGarvey, G. J.; Williams, J. M. "Stereoelectronic Controlling Features of Allylic Asymmetry. Application to Ester Enolate Alkylations" *J. Am. Chem. Soc.* **1985**, *107*, 1435–1437. Hart, D. J.; Krishnamurthy, R. "Investigation of a Model for 1,2-Asymmetric Induction in Reactions of α-Carbalkoxy Radicals: A Stereochemical Comparison of Reactions of α-Carbalkoxy Radicals and Ester Enolates" *J. Org. Chem.* **1993**, *57*, 4457–4470.

3. What isomers of **23** would you expect to be present following the Birch reduction-hydrolysis sequence? (Juvabione-4)
4. What are the structures of the diastereomeric semicarbazones derived from **27**? (Juvabione-4)
5. Suggest a mechanism for the conversion of **44** → **45**. (Juvabione-6)
6. Suggest some tactics that might (1) improve diastereoselectivity in the Diels-Alder cycloaddition and/or (2) provide a stereoselective

alternative to the cyanohydrin-dehydration route from **40** (or a related compound) to either **14** or **15**. (Juvabione-6)

7. Propose a modification of the Ficini synthesis that might provide *threo*-juvabione (**15**) [the $C_{1'}$-epimer of **14**] in a stereoselective manner. (Juvabione-7)
8. Explain the regioselectivity of the acylation of **55**. (Juvabione-7)
9. Predict the stereochemical course of the following reactions. (Juvabione-10)

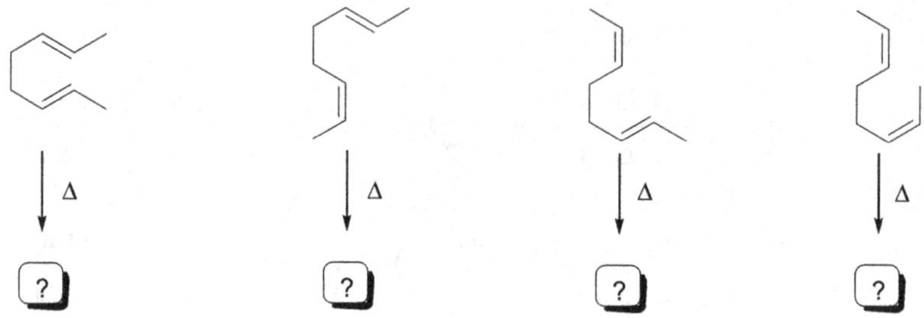

Doering, W. v. E.; Roth, W. R. "The Overlap of Two Allyl Radicals or a Four-Centered Transition State in the Cope Rearrangement" *Tetrahedron* **1962**, *18*, 67–74.

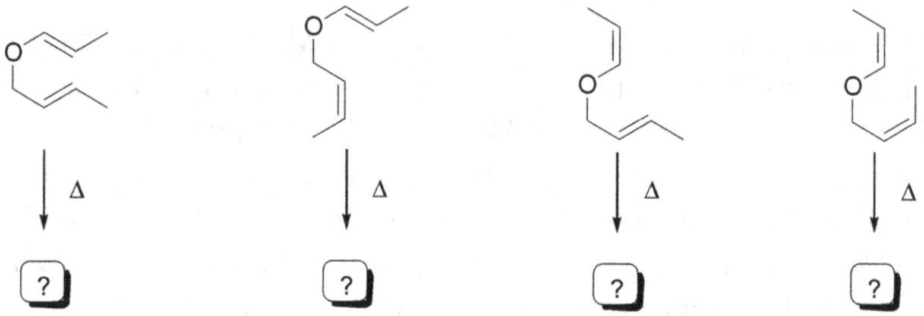

Vittorelli, P.; Hansen, H. J.; Schmid, H. "Kinetics and Stereochemical Course of the Thermal Rearrangement of the Four Stereoisomers of Propenyl But-2'-enyl Ether" *Helv. Chim Acta* **1975**, *58*, 1293–1309.

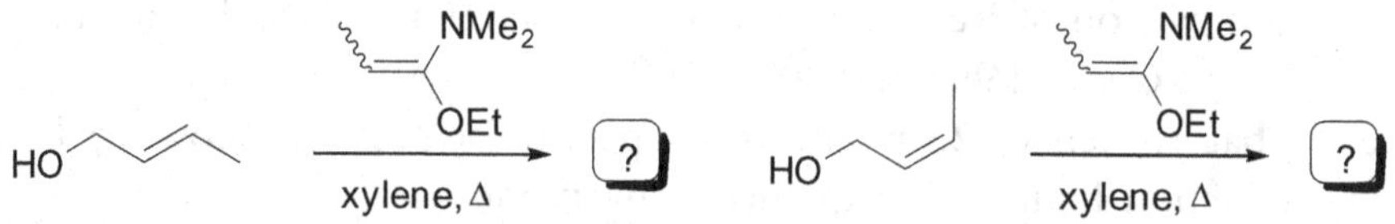

Sucrow, W.; Richter, W. "Stereochemistry of the Claisen Rearrangement with 1-Dimethlamino-1-Methoxy-1-Propene" *Chem. Ber.* **1971**, *104*, 3679–3688.

10. Explain why the Eschenmoser-Claisen rearrangement and the Bartlett variation of the Eschenmoser-Claisen shown below give different stereochemical results. (Juvabione-10)

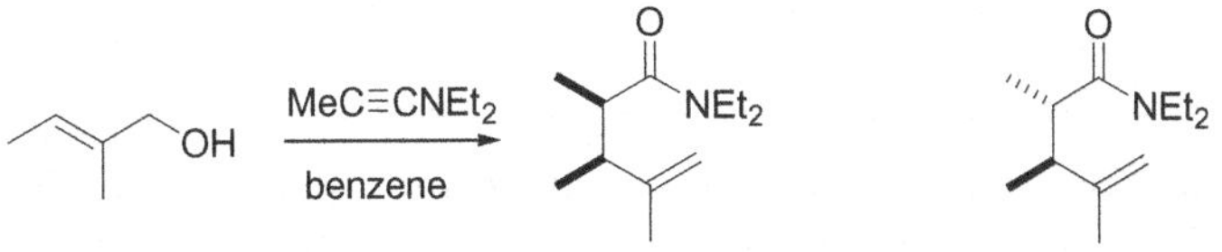

add catalytic BF_3-Et_2O to alcohol; stir 4 days at rt	10	(50%)	1
add ROH to ynamine over 18 h period	1	(56%)	10

Bartlett, P. A.; Hahne, W. F. "Stereochemical Control of the Ynamine-Claisen Rearrangement" *J. Org. Chem.* **1979**, *44*, 882–883.

11. Design a synthesis of pilocarpine that relies on [3.3]-sigmatropy to establish vicinal stereochemistry. (Juvabione-10)

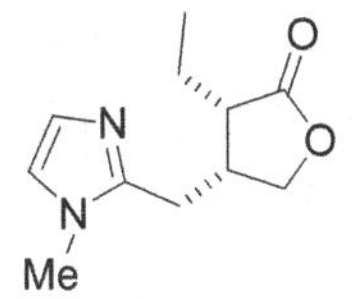

pilocarpine

12. Based on the following observation, propose a synthesis of *threo*-juvabione (**15**). How might your plan be modified to afford *erythro*-juvabione (**14**)? (Juvabione-11)

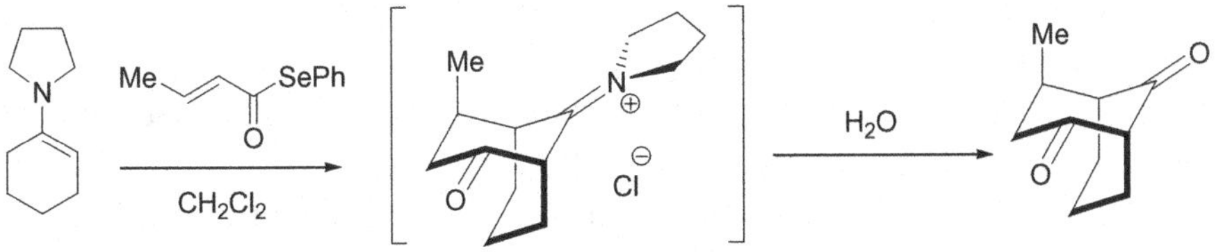

References: Byeon, C.-H.; Hart, D. J.; Lai, C.-S.; Unch, J. "Reactions of cyclohexanone enamines with α,β-unsaturated thioesters and selenoesters" *Synlett* **2000**, 119–121. Hickmott, P. W.; Miles, G. J.; Sheppard, G.; Urbani, R.; Yoxall, C. T. "Enamine chemistry. XVII. Reaction of α,β-unsaturated acid chlorides with enamines. Further mechanistic investigations. Effect of triethylamine on the reaction path." *J. Chem. Soc., Perkin Trans. 1* **1973**, 1514–1519. Harding, K. E.; Clement, B. A.; Moreno, L.; Peter-Katalinic, J. "Synthesis of some polyfunctionalized bicyclo[3.3.1]nonene-2,9-diones and bicyclo[4.3.1]decane-1,10-diones." *J. Org. Chem.* **1981**, *46*, 940–948. Gelin, R.; Gelin, S.; Dolmazon, R. "Acylation of 1-morpholinecyclohexene by ethylenic acid chlorides." *Bull Soc. Chim. Fr.* **1973**, 1409–1416.

13. Consider the Birch, Ficini, Evans, Schultz and Pearson syntheses and suggest how each synthesis might be modified to address enantioselectivity issues. (Juvabione-12)
14. Suggest a mechanism for the conversion of **95** → **96**. (Juvabione-13)
15. Suggest tactics that could be used to proceed from **100** → **102** via **101**. (Juvabione-14)
16. Compare **111** (Juvabione-16) and **77** (Juvabione-11) as starting points for juvabione synthesis. (Juvabione-16)
17. Propose an alternative synthesis of **115**. (Juvabione-16)
18. Suggest a mechanism for the conversion of **119** → **120**. (Juvabione-16)
19. Suggest a reaction sequence that will convert **124** → **116**. (Juvabione-17)
20. Provide a reaction sequence that will convert **116** to the $C_{1'}$ diastereomer of **117**. (Juvabione-17)
21. Discuss how the malonic acid derivative used to convert **117** → **126** serves as a "carbomethoxy anion equivalent". Suggest other tactics for accomplishing this transformation (other than those used in Juvabione-16). (Juvabione-17)
22. Provide a mechanism for the conversion of **133** → **135**. (Juvabione-18)
23. The reaction of **149** with **127** might be a direct conjugate addition. An alternative mechanism would be a 1,2-addition to the enone from the α-carbon of the metallated sulfoxide followed by an anion accelerated oxy-Cope rearrangment. Evaluate this possibility using a stereochemical analysis of each step of the reaction. (Juvabione-20)
24. Discuss (in general terms) how the anion derived from **149** served as a homoenolate anion equivalent. (Juvabione-20)
25. Propose a reaction sequence that would convert **164** to a mixture of **14** and **15**. Juvabione-21)
26. Recall the geminally activated cyclopropane chemistry we saw in Chapters 3 and 4. Propose diastereoselective syntheses of **14** and **15** that revolve around this methodology. (Juvabione-21)

CHAPTER 6

Functional Group Reactivity Patterns and Difunctional Relationships

General Comments on Target-Oriented Synthesis

Target-oriented organic synthesis is a complex exercise that requires at least the following steps: (1) Selection of the target (2) Development of a synthetic plan (commonly called the strategy) (3) choice of reagents for accomplishing the plan (commonly called tactics) (4) execution of the plan in the laboratory (a process that frequently involves revisiting the selected tactics).

The first step can be quite personal. Compounds that are biologically important, but in scarce supply, are frequently selected as targets for synthesis. The notion that a given compound will have valuable properties (biological or otherwise) can serve as an impetus for synthesis. For example, process research groups are often faced with the synthesis of compounds that are likely to have practical importance. On occasion synthesis still serves as a means for proving the structure of a natural product. An interest in testing a particular type of chemistry can dictate target selection. A fascinating structure (which is personal opinion more than fact) can draw one to attempting a synthesis. The bottom line, however, is that this exercise is target-oriented and thus, a specific target or small family of targets must be selected before one can develop a synthetic strategy.

The second step can be approached in a number of ways. A process commonly called “retrosynthetic analysis” is often used. In this process, one begins with the target structure and systematically works backwards through (usually) less complex intermediates that one feels can be moved to the desired targets by application of certain tactics. In the last four chapters we have seen a number of examples of this process as applied to steroids, prostaglandins, pyrrolizidine alkaloids, and juvabione. It is obvious that this process (retrosynthetic analysis) can generate a series of pathways from simpler materials to the final target. Development of a given strategy can be a function of the practioner's imagination, knowledge of tactics, understanding of mechanistic and stereochemical principles. A quite different approach to stategy development involves “starting material recognition”. The syntheses of juvabione that begin with limonene and perillaldehyde are good examples of such strategies. Recognition of whether or not a synthesis can be accomplished in a covergent manner (rather than stepwise manner) is also important when developing (or comparing) strategies. For example the brevity of the three-component approach to the prostaglandins is a result of its convergent nature.

I will not comment in detail on the third and fourth steps. These are largely practitioner dependent. An understanding of functional group compatiblities, however, is clearly an important aspect of these steps.

Difunctional Relationships-1

Difunctional Relationships-1

Target-oriented organic synthesis is a complex exercise that requires at least the following steps: (1) selection of the target (2) development of a synthetic plan (commonly called the strategy) (3) choice of reagents for accomplishing the plan (commonly called tactics) (4) execution of the plan in the laboratory (a process that frequently involves revisiting the selected tactics). A discussion of these four points is presented on page 204 (Functional Groups-1)

Since strategy development is presumably logical, one goal of a number of research efforts has been to develop a set of guidelines that, if followed, will allow one to generate plans for the synthesis of any target. The efforts of the Corey group to develop computer-aided methods for strategy development are perhaps most famous in this regard. "The Logic of Chemical Synthesis" is a readable book that presents a distillation of many of the ideas developed in the Corey group. A number of other authors have written books (or articles) that address the development of synthetic strategies and some of these (that DJH has found particularly interesting, entertaining and useful) are shown on the next page. The Warren books are particularly well-organized and include a detailed discussion of difunctional relationships, the topic of this chapter.

Although books are clearly useful, opinions and ideas expressed by teachers (mentors/advisors/colleagues/students) also are bound to have an effect on how a given practitioner approaches strategy development. For example, I was strongly influenced by Richard G. Lawton (University of Michigan), William G. Dauben (UC Berkeley) and David A. Evans (Caltech) and I am very grateful to them for the insights they shared with me, many of which I hope I am passing along to you.

With these general comments, I want to now move to the topic of functional group reactivity patterns and difunctional relationships. An understanding of the first topic is essential to synthesis. The second topic can be a helpful consideration when developing synthetic strategies, and is revealing when analyzing published syntheses. For example, we will see how such an analysis shows why cyclohexenes and cyclopentenes can serve as precursors of 5- and 6-membered rings, respectively (recall the Johnson approach to progesterone). We will start with some general considerations and then move to examples from syntheses we have already seen.

Useful Books

Ireland, R. E. "Organic Synthesis", Prentice-Hall, 1969 (147 pages)

Fleming, I. "Selected Organic Syntheses: A Guidebook for Organic Chemists", John Wiley and Sons, 1973 (227 pages)

Warren, S.; "Organic Synthesis: The Disconnection Approach", John Wiley and Sons, 1982 (391 pages)

Corey, E. J.; Cheng, X.-M. "The Logic of Chemical Synthesis", John Wiley and Sons, 1989 (436 pages)

Nicolaou, K. C.; Sorensen, E. J. "Classics in Total Synthesis: Targets, Strategies, Methods", Wiley-Verlag, 1996 (798 pages)

Functional Groups and Charge

Since most carbon-carbon bond forming reactions are polar in nature, it is useful to classify functional groups with regard to the polarity they impose on carbon. For example, a halogen imposes partial positive charge on the carbon to which it is bonds. So does a hydroxyl group in the form of a mesylate (or many other derivatives). So does an oxygen that is doubly bonded to carbon (carbonyl compounds). These "groups" are all more electronegative than carbon.

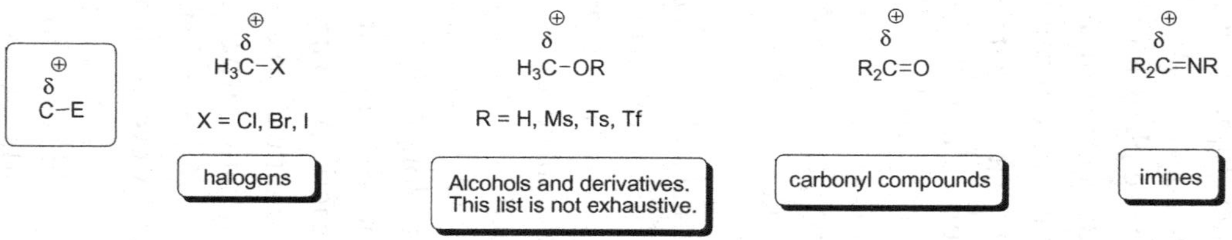

Thus one can generate a set of functional groups that impart partial positive charge to carbon. Using terminology introduced by David Evans, I will call these E-functions (E for Electrophile). It is sometimes important to consider the protonated versions of E-functions to fully appreciate the charge they impart on carbon.

Difunctional Relationships-2

Difunctional Relationships-2

When I first started to learn organic chemistry, I searched for ways to visualize bond-forming reactions. It seemed to me that most reactions were much like bringing magnets together. Oppositely charged atoms formed bonds and like-charged atoms did not, just like the different poles of magnets.* Although many reactions are not best classified in terms of polar intermediates (pericyclic reactions, free radical reactions, some transition metal mediated reactions), many carbon-carbon bond forming reactions are polar in nature. Therefore it is useful to classify functional groups in regard to the polarity they impose on carbon. For example a halogen imposes partial positive charge to the carbon on which it resides. It renders the carbon "electrophilic". So does a hydroxyl group in the form of a mesylate (or many other derivatives). So does an oxygen that is doubly bonded to carbon (carbonyl compounds). These groups are all more electronegative than carbon. One can generate a set of functional groups that impart partial positive charge to carbon. Using terminology related to that introduced by David Evans, we will call these groups E-functions (E for Electrophilic).[2] Note that it is sometimes important to consider the protonated versions of E-functions, or their complexes with other Lewis acids, to fully appreciate the charge they impose on carbon.

*In the mid-1970s I spent two years as a postdoctoral fellow with David Evans' group, then at the California Institute of Technology where I was exposed to a well-organized way to think about polar coupling reactions. The conceptual framework that I will use was developed by Evans, as was much of the terminology used in this chapter. Whereas this presentation is much less detailed than the concepts developed by Evans, I hope it will be clear enough to students of organic synthesis to be useful. Any shortcomings in this presentation are due to me and not the concept. I also note that a somewhat related analysis of functional group behavior and relationships can be found in Stuart Warren's book (*vide supra*).

Some functional groups impart negative characteristics to a carbon (N-functions for Nucleophilic, not to be confused with nitrogen). These groups are largely Group IA and IIA metals (groups that are less electronegative that carbon).

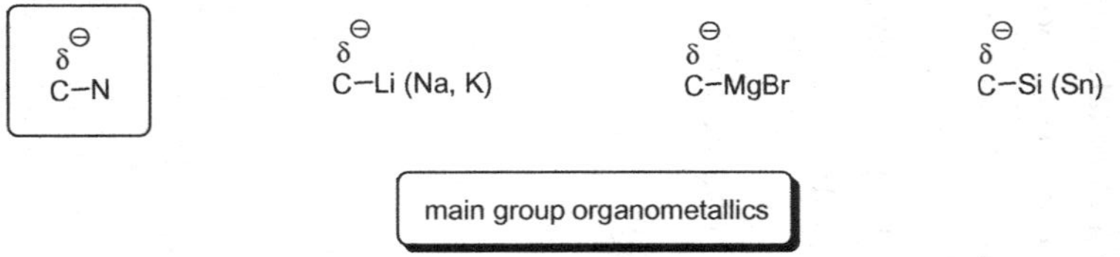

The majority of functional groups, however, can impose either E or N characteristics to carbon. These can be classifed as A-functions (for Alternate). A simple example of an A-function is triply bonded nitrogen:

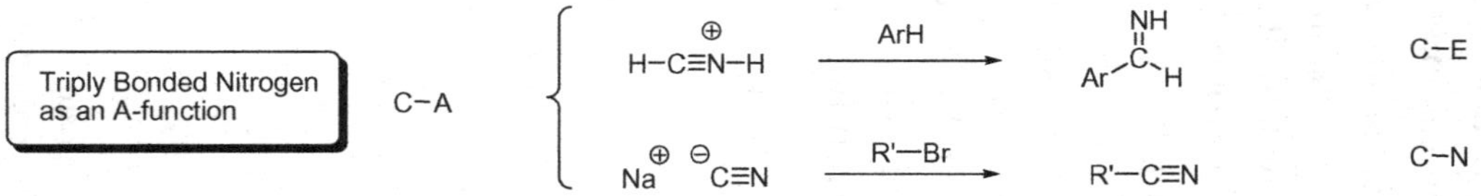

This seems simple, but is worth recognizing and can be helpful when deciding how to construct certain bonds. Here are some other A-functions. Note that it is important to consider tautomers of some functional groups, as well as their conjugate acids and conjugate bases, to appreciate the polarity that they impart on carbon.

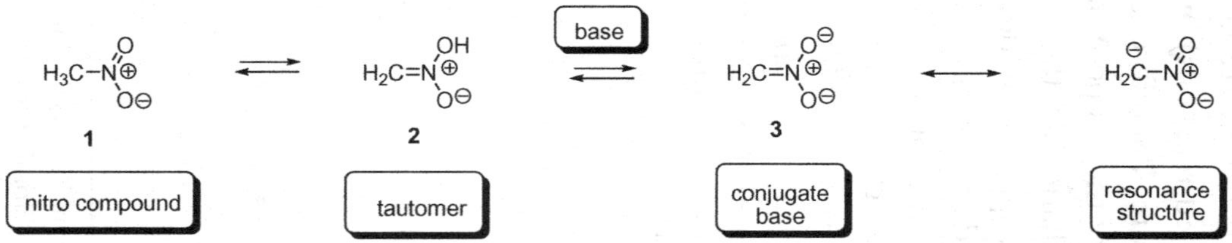

Difunctional Relationships-3

Difunctional Relationships-3

Some functional groups impart negative characteristics to carbon (N-functions for Nucleophilic, not to be confused with nitrogen). These groups are largely Group IA and IIA metals (groups that are less electronegative than carbon), for example organolithium reagents, Grignard reagents, and even some silanes and stannanes.

The majority of functional groups, however can impose either E or N characteristics to carbon. These can be classified as A-functions (for Alternate). An example is triply bonded nitrogen. Consider hydrogen cyanide (H-C≡N). It can behave as an electrophile when protonated. One example would be the Gattermann reaction in which HCN can be used to introduce a formyl group to an aromatic ring via an electrophilic aromatic substitution reaction.[3] On the other hand, the triply bonded nitrogen is a strong enough electron-withdrawing group that HCN can be easily deprotonated to generate cyanide, which is a good nucleophile. This seems simple, but is worth recognizing and can be helpful when deciding how to construct certain bonds.

Here are some other A-functions. Consider nitroalkanes with nitromethane (**1**) as an example.[4] The nitro group is a good electron-withdrawing group (note the positive charge on the nitrogen) and thus can behave as an E-function. Nitroalkanes with α-hydrogens can tautomerize to the corresponding nitronic acids (**2**). This tautomer is electrophilic (note the presence of the iminium ion within **2**). The Nef reaction involves the conversion of nitroalkanes to the corresponding aldehydes or ketones (with aqueous acid).[5] This transformation is a direct result of the electrophilic nature of carbon bonded to a nitro group. Deprotonation of **1** (or **2**) gives the corresponding conjugate base (**3**). This structure places negative charge on the α-carbon and illustrates how the nitro group can behave as an N-function. The Henry reaction, wherein the conjugate base of a nitroalkane behaves as a nucleophile toward an aldehyde or ketone, illustrates the nucleophilic properties imparted on carbon by the nitro group.[6]

Note that oximes and hydrazones have the potential for inverting the "normal" electophilic nature of a carbonyl carbon. Recall the following "unusual" reaction encountered in the steroid syntheses.

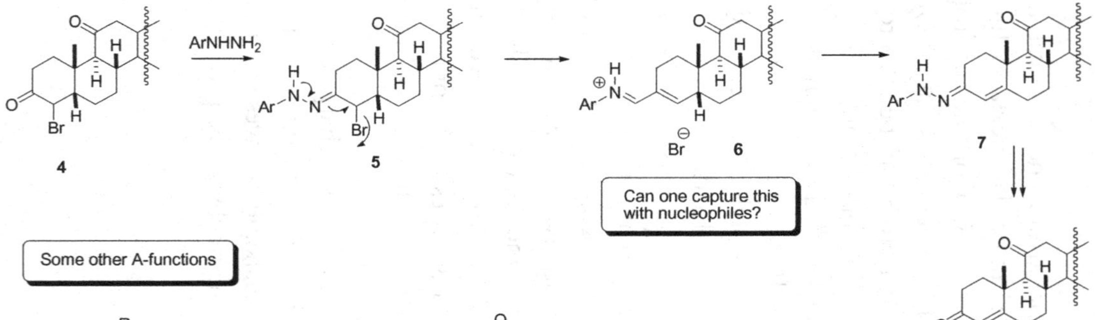

Some other A-functions

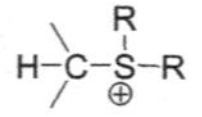

sulfonium salts

$\overset{\oplus}{C}$: alkyating agents

$\overset{\ominus}{C}$: conjugate base (S-ylid) is nucleophile

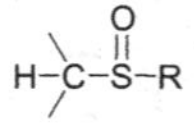

sulfoxides

$\overset{\ominus}{C}$: conjugate base is nucleophile

$\overset{\oplus}{C}$: protonation generates electrophile (Pummerer intermediate)

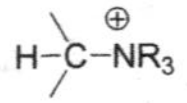

quarternary ammonium salts

$\overset{\ominus}{C}$: conjugate base is nucleophile

$\overset{\oplus}{C}$: alkylating agents (recall Holton PG synthesis)

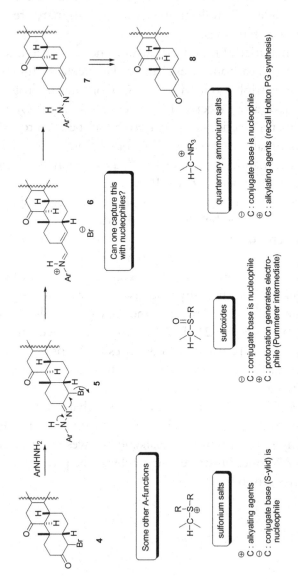

Difunctional Relationships-4

Difunctional Relationships-4

Oximes and hydrazones have the potential for inverting the "normal" electrophilic nature of a carbonyl carbon. Consider the two resonance structures of an oxime (or oxime ether if R = alkyl) shown here. One implies that the carbonyl carbon should be an electrophile and the other implies that it might behave as a nucleophile. The same goes for hydrazones. Consider a diazoalkane and its conjugate acid. One is nucleophilic at carbon and the other electrophilic. Work the problems suggested for this section for examples where =NOR and $=NNR_2$ and diazonium groups behave as E-functions and/or N-functions. An old-fashioned but interesting method for dehydrohalogenation of an α-haloketone relies on the "charge reversal" or "umpolung" or "charge affinity inversion" that occurs upon converting a ketone to a hydrazone derivative (**4** → **8**).

Some additional A-functions include sulfonium salts, sulfoxides and quaternary ammonium salts. For example, sulfonium salts can be used to generate sulfur ylids (N-function) or as leaving groups in substitution reactions (E-function). Sulfoxides can be deprotonated to generate nucleophilic carbon (N-function) or activated by electrophiles to generate electrophilic carbon after loss of an α-proton (E-function). Quaternary ammonium salts are good leaving groups (E-function) and can be deprotonated to generate nucleophilic nitrogen ylids (N-function).

Difunctional Relationships

Much of organic synthesis consists of establishing relationships between two functional groups (difunctional relationships). Such relationships can be classified as "odd" or "even".

1,2-Difunctional Relationship | 1,3-Difunctional Relationship | Even Difunctional Relationship | Odd Difunctional Relationship

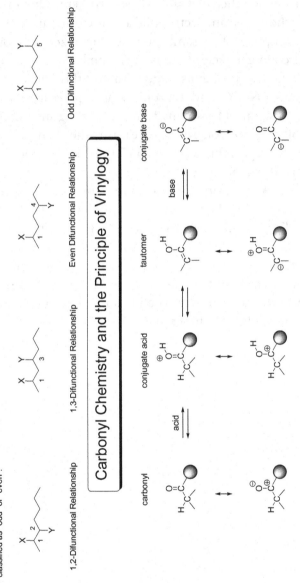

Carbonyl Chemistry and the Principle of Vinylogy

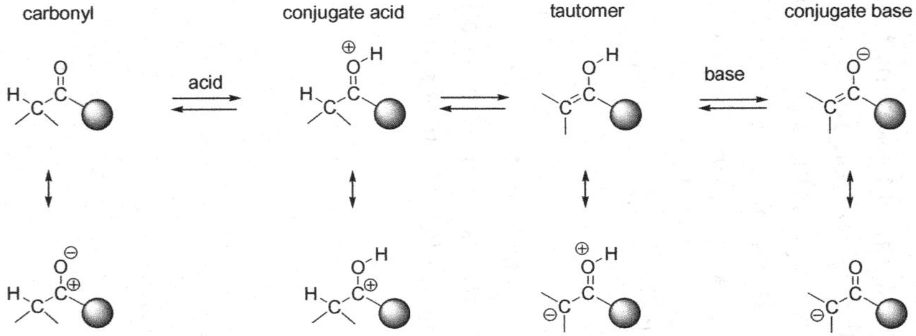

The natural polarity imparted by carbon is "carbonyl carbon positive" and "α-carbon negative". Also note that carbonyl groups can be converted into almost any other functional group, or can be prepared from almost any other functional group. They are extremely versatile and important in synthesis.

Difunctional Relationships-5

Difunctional Relationships-5

Given this introduction to the polarity that functional groups impart on carbon, let's consider difunctional relationships. Consider a compound with two functional groups. They can be on adjacent carbons (1,2-difunctional relationship) or separated by a carbon (1,3-difunctional relationship). Any difunctional relationship can be classified as either even (1,2 or 1,4 or 1,6) or odd (1,3 or 1,5 or 1,7).*

Since most functional groups can be derived from a carbonyl compound (or converted to a carbonyl compound) we will focus on this functional group. The natural polarity imparted by a carbonyl group on proximal carbons is "carbonyl carbon positive" and "α-carbon negative". This is a natural consequence of the acid-base properties of carbonyl compounds as illustrated here: carbonyl carbons and their conjugate acids are electrophilic at the C=O carbon; the tautomers of carbonyl compounds and their conjugate bases are nucleophilic at the α-carbon.

*Evans has refered to even and odd difunctional relationships as "dissonant" and "consonant" for reasons that will become apparent.

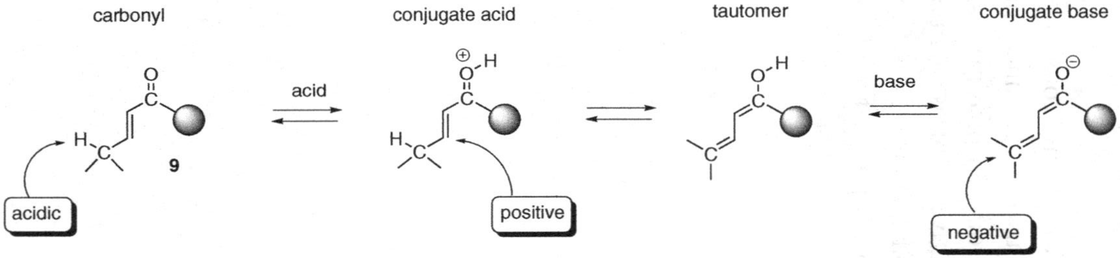

If one allows for conjugation, the natural polarity imparted by a carbonyl group on a carbon chain is:

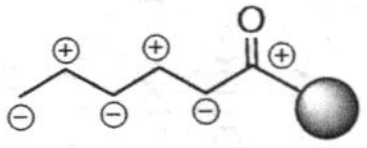

10

How will two carbonyl groups interact in terms of polarity?

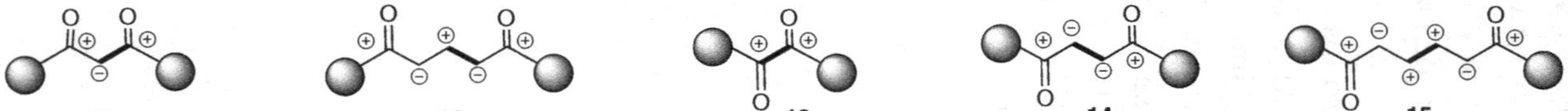

11	**12**	**13**	**14**	**15**
1,3	1,5	1,2	1,4	1,6

Odd difunctional relationships "reinforce" one another.

Even difunctional relationships "interfere" with one another.

A consequence of this relationship is that odd difunctional relationships can be constructed using the normal polarity of the carbonyl group. Construction of an even difunctional relationship will require the use of an A-function.

Difunctional Relationships-6

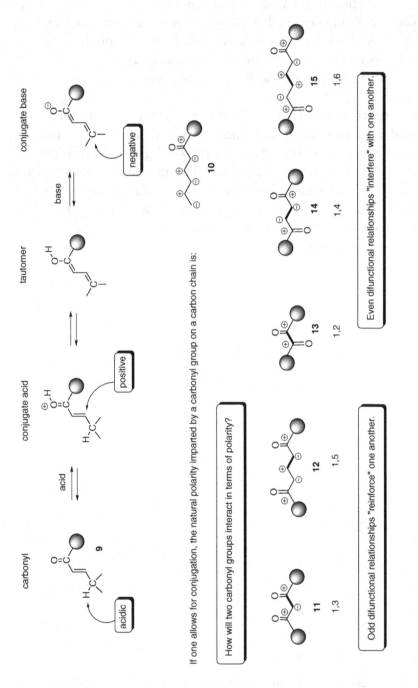

Difunctional Relationships-6

What is the principle of vinylogy? Consider the acid-base behavior of carbonyl compounds. When a vinyl group is inserted between the carbonyl group and α-carbon of a carbonyl compound, the acidity of the α-CH bond is extended to the γ-CH bond (if not overridden by stereoelectronic considerations).* In addition, the β-carbon of the unsaturated charge associated with the C=O carbon is extended to the γ-carbon. This is simply conjugation. The vinyl group allows the α-carbon (now gamma in **9**) to "talk" with the carbonyl. It wires them together. From the standpoint of charge distribution imparted by a carbonyl group to surrounding atoms, a conjugated π-bond extends the alternating charge pattern by two carbons. A second π-bond would extend the alternating charge pattern by two more carbons as shown in structure **10**.

How will two carbonyl groups interact in terms of polarity? Odd difunctional relationships "reinforce" one another.** Even difunctional relationships "interfere" with one another (see bold bonds in **13–15**).*** A consequence of this relationship is that odd difunctional relationships can be constructed using the normal polarity of the carbonyl group (see bold bonds in **11** and **12**). Construction of an even difunctional relationship will require use of an A-function.

*The C-H bond must still be able to align with the carbonyl π-bond. In other words, stereoelectronic considerations still apply to unsaturated systems.

**Evans described these relationships as "consonant".

***Evans described these relationships as "dissonant".

Construction of a 1,3-Difunctional Relationship

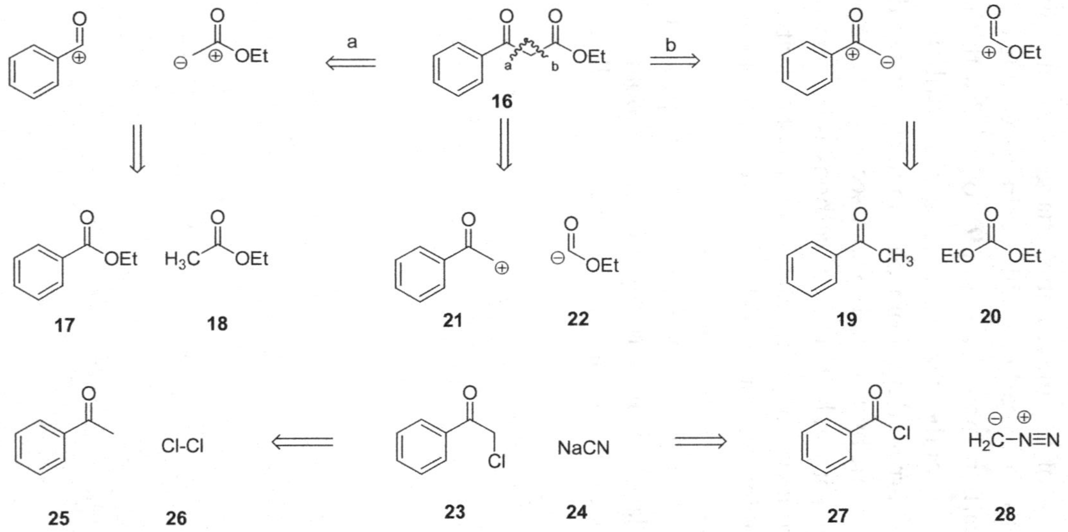

1,2-difunctional compound

Whereas you can construct a 1,3-difunctional relationship in any number of ways, it will always be possible to construct this relationship using the normal polarity of carbonyl groups (no A-functions needed). This is a valuable consideration, BUT NOT A RULE. Sometimes the use of an A-function (in a complex setting) will have advantages.

Difunctional Relationships-7

Difunctional Relationships-7

Let's consider possible syntheses of compound **16** from monofunctional compounds. Polar disconnection of bond "a" leads to an "acyl cation" and an "enolate". Real world versions of these polar species would be benzoate **17** and ester **18** (which would be deprotonated to provide the enolate). Thus, a logical way to contruct **16** might use a crossed Claisen condensation of **17** and **18**. Disconnection of bond "b" leads to a crossed condensation between acetophenone (**19**) and diethyl carbonate (**20**). Both of these approaches to construction of the 1,3-difunctional relationship in **16** rely on the normal polarity of carbonyl groups. No A-function is needed.

We could have disconnected bond "b" to a cation such as **21** and acyl anion **22**. This disconnection, however, does not rely on the normal polarity of the carbonyl group. This does not mean that the synthesis could not be done this way. Chloroketone **23** and cyanide (**24**) are real world versions of **21** and **22**, respectively. A displacement reaction followed by ethanolysis of the α-cyanoketone might provide **16**. Chloroketone **23**, however, is a difunctional compound. Two methods for preparing **23** are shown. The synthesis from ketone **25** makes use of the normal polarity of the carbonyl group. The synthesis from carbon fragments **27** and **28** makes use of an A-function.

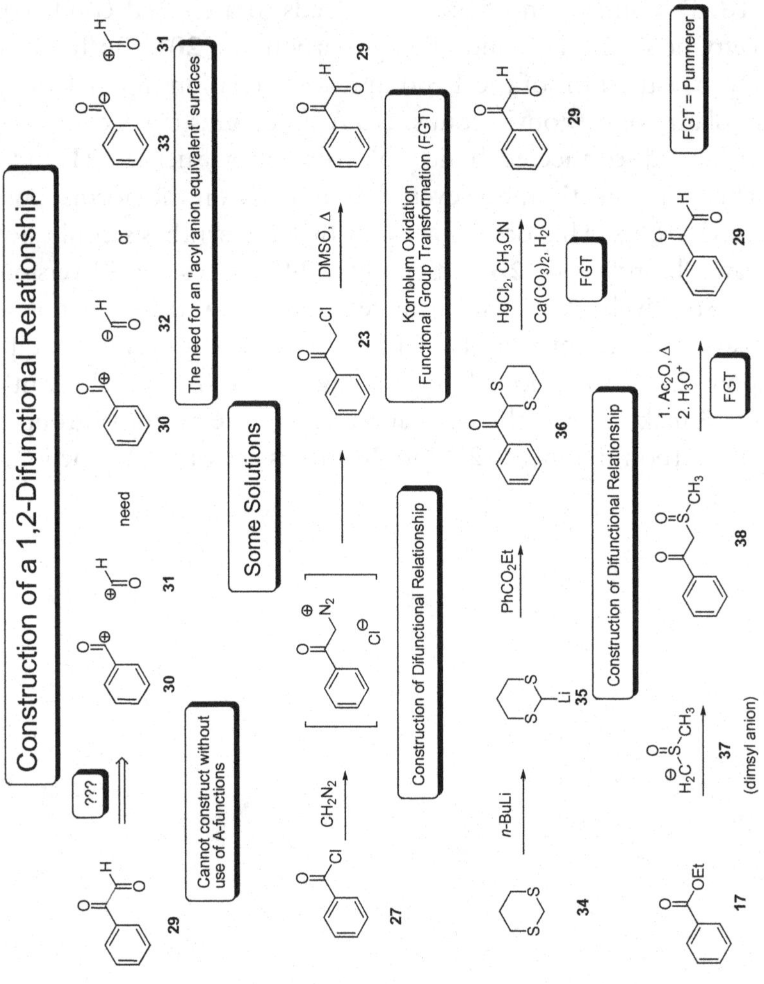

Difunctional Relationships-8

Difunctional Relationships-8

Let's look at construction of a 1,2-dicarbonyl compound within the context of ketoaldehyde **29**. There is no simple way to construct the bond between the carbonyls without use of an A-function. An analysis of this problem reveals the need for acyl anion equivalents of type **32** or **33**. Several possibilities are shown here. We have already seen that a monofunctional compound (**27**) can be converted to a 1,2-difunctional compound (**23**) using diazomethane (A-function chemistry). The problem now simply becomes one of oxidation states. The chloride has to be oxidized to the aldehyde oxidation state. One method that could be used is the Kornblum oxidation (carbon is oxidized and sulfur is reduced).[7] This is an example of a functional group transformation (FGT). It does *not* involve construction of a difunctional relationship. That work was done in the first step of the sequence (**27** → **23**). It is the FGT that establishes the *equivalency* of diazomethane with the "unreal" formyl anion **31** (see Steroids-18 and Prostaglandins-3, 6 and 15).

Lithiated dithiane **35** is another example of a formyl anion equivalent.[8] Acylation would provide **36** and thioacetal hydrolysis would provide **29**. The thioacetal hydrolysis is the FGT that establishes the equivalency. Acylation of dimsyl anion **37** (from metallation of DMSO) would afford β-ketosulfoxide **38**. Once again an oxidation is needed to establish the desired equivalency. This can be accomplished using a Pummerer rearrangement wherein oxidation occurs at carbon with reduction at sulfur.[9] Hydrolysis of the resulting S,O-acetal would complete the required FGT.

This is a small sampling of potential tactics for accomplishing the desired transformation. In fact, **29** is a very reactive compound because it has two adjacent carbonyl carbons (repelling magnets). Whereas it is easy to draw **29** on the page, it is more difficult to put it in a bottle because of the reactivity of the carbonyl groups with nucleophiles (including water). Nonetheless this exercise illustrates what must be considered if the task is construction of a 1,2-difunctional relationship.

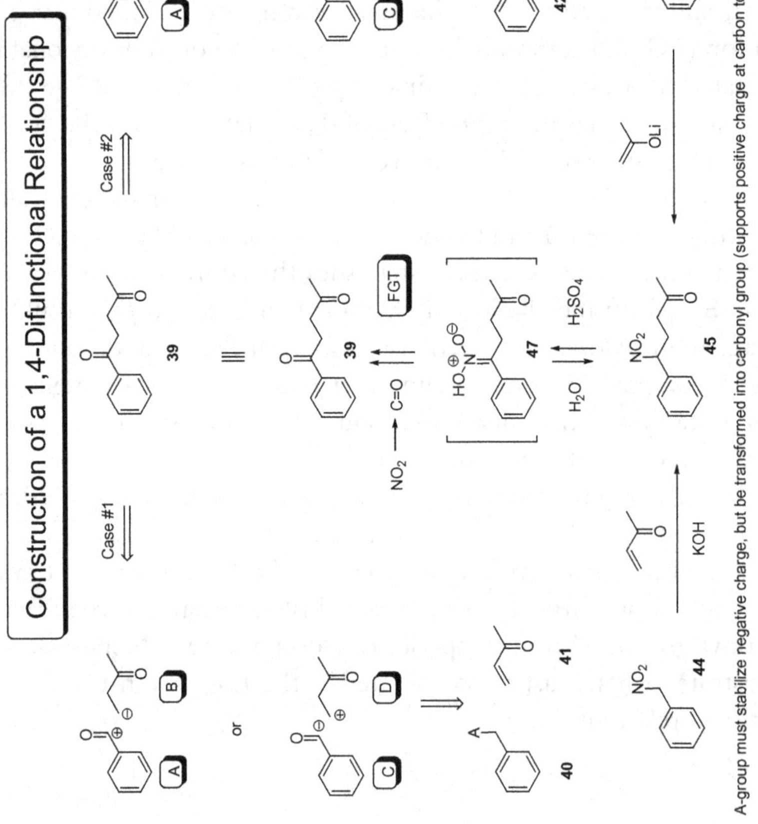

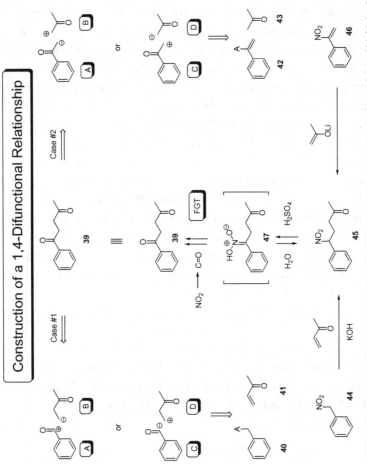

Difunctional Relationships-9

Difunctional Relationships-9

Now let's consider 1,4-dicarbonyl compound **39**. One polar disconnection leads to charged fragments **A** + **B** or **C** + **D** (Case #1). The **C** + **D** pair can easily be related to an acyl anion equivalent and methyl vinyl ketone (by application of the principle of vinylogy). The acyl anion equivalent can be generalized by structure **40** in which an A-function behaves as a N-function.

An alternative disconnection is shown as Case #2. In this case it is a coin toss between **A** + **B** or **C** + **D**. Both require an enolate (from a ketone) and an α-acylcation equivalent. We will consider the use of an enolate derived from acetone (**43**) and a generalized α-ketocation equivalent **42**, in which an A-function must behave as an N-function, rendering the terminal olefin electrophilic.

What A-function might one use? One choice could be a nitro group. Working through Case #1 in the forward direction, deprotonation of **44** followed by conjugate addition to **41** might give **45**. For Case #2, reaction of the enolate of acetone (or an appropriate derivative) with nitroalkene **46** might be expected to provide **45**. A Nef reaction would then accomplish the FGT needed to convert **45** to **39**.

More Solutions

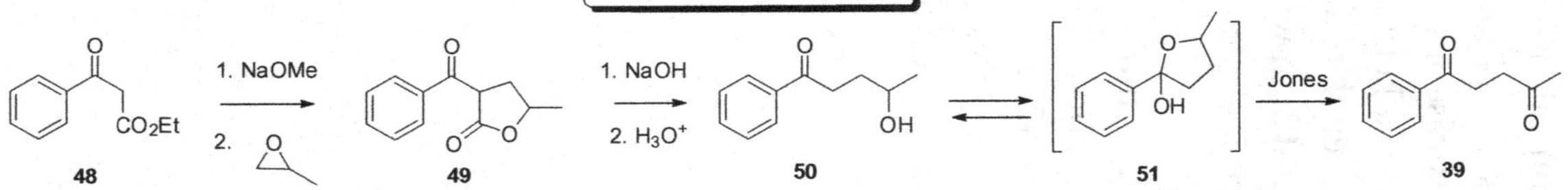

Here is another solution to the Case #2 approach. It is not a particularly good solution relative to other approaches, but it can be used to illustrate some points: (1) Ketone activation can be useful. In this case if PhCOMe (acetophenone) was used as the nucleophile, aldol condensations would surely compete with the desired epoxide opening because the epoxide opening will surely be slow. (2) You can go up and down in oxidation state along your reaction pathway (alcohol to ketone). (3) An epoxide can be regarded as a 1,2-difunctional compound. This is the actual origin of the ultimate even (1,4) difunctional relationship in the target. (4) Epoxides are derived from alkenes. Thus, alkenes are wonderful precursors of 1,2-difunctional relationships. So here is a problem. What other tactics can you suggest that would get you from **48** to **39**?

Construction of a 1,5-Difunctional Relationship

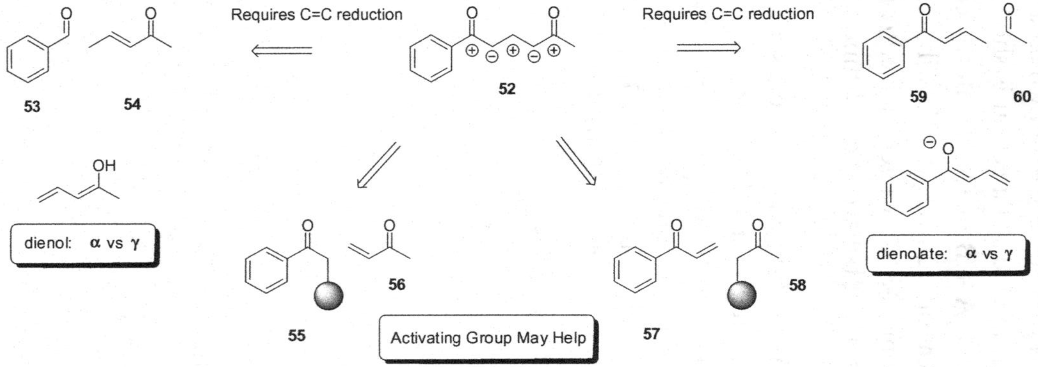

Difunctional Relationships-10

Difunctional Relationships-10

Here is another solution to the Case #2 approach. It is not necessarily a good solution relative to other approaches, but it can be used to illustrate some points: (1) Ketone activation can be useful. In this case if PhCOMe (acetophenone) were used as the nucleophile, aldol condensations would surely compete with the desired epoxide opening because the epoxide opening is slow. (2) You can go up and down in oxidation state along the reaction pathway (alcohol to ketone). (3) An epoxide can be regarded as a 1,2-difunctional compound. *This is the actual origin of the ultimate even (1,4) difunctional relationship in the target.* (4) Epoxides are derived from alkenes. Thus, alkenes are wonderful precursors of 1,2-difunctional relationships. So here is a problem. What other tactics can you suggest that would get you from **48** to **39**?

The normal polarity imparted by carbonyl groups, in conjunction with the principle of vinylogy, can be used to construct 1,5-difunctional relationships. Disconnection of each of the four bonds connecting the carbonyl groups in **52** affords fragments that could (in principle) be coupled to give the target structure. For example, the dienol or dienolate derived from **54** might react with a carbonyl compound of type **53** to afford **52**. Dienols tend to react with electrophiles at the γ-carbon so this transformation should be possible. The same comments apply to **59** and **60**. Another possibility would be the conjugate addition of **55** (or **58**) to enone **56** (or **57**). Since both partners in these reactions could behave as both nucleophile or electrophile, the use of an activating group might help control the chemistry (*vide supra*). The bottom line is that construction of this 1,5-difunctional relationship can be accomplished without the use of A-functions through use of the normal polarity of carbonyl compounds.

Application to Cyclohexanones

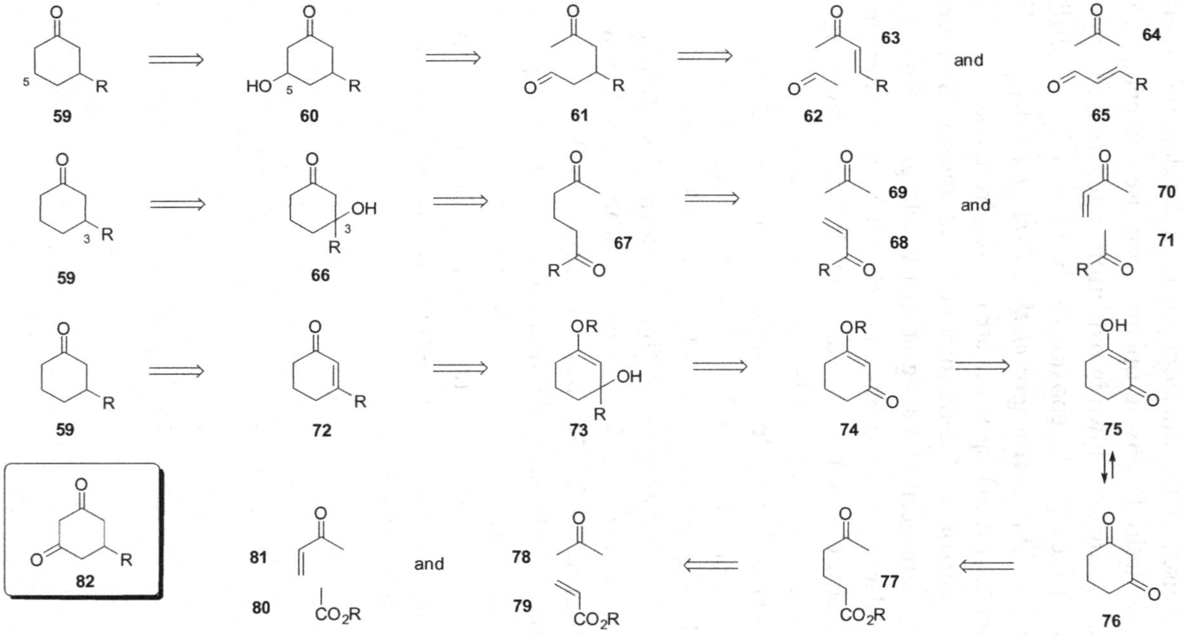

These disconnections describe annulation (annelation) routes to cyclohexanone derivatives. They are not all-inclusive. Notice that the acyclic precursors are 1,5-dicarbonyl compounds and that the ring-forming reaction constructs a 1,3-difunctional relationship.

Difunctional Relationships-11

Difunctional Relationships-11

Let's apply these concepts to the synthesis of 3-alkylcyclohexanones (**59**) from acyclic precursors. Cyclohexanone **59** is a monofunctional compound. It could be prepared from aldol **60** (a difunctional compound) by dehydration and reduction of the intermediate enone. Aldol **60** could be prepared from **61** by an intramolecular aldol condensation. Diketone **61** contains a 1,5-difunctional relationship. Thus it can be prepared using the normal polarity of the carbonyl group, for example by conjugate addition of **62** to enone **63** or conjugate addition of **64** to enal **65**. Enal **65** could be prepared, in turn, by aldol-dehydration chemistry. A critical point in the "development" of this plan is placement of the second functional group, working backwards from **59** at a position that generates an odd (consonant) difunctional relationship. It is this choice that leads to a plan that relies on the polarity of the carbonyl group.

Placement of the second functional group at the 3-position of **59** takes one back through 1,5-dicarbonyl **67** to enone **68** and ketone **69**, or enone **70** and ketone **71**. Another plan that works back through odd (consonant) difunctional relationships passes through vinylogous ester **74** and 1,3-diketone **76** to ketone **78** and unsaturated ester **79** or ester **80** and unsaturated ketone **81**. See if you can work back from **59** through 1,3-diketone **82** in a similar manner before moving on with this chapter.

All of these disconnections describe annulation (annelation) routes to cyclohexanone derivatives. There are tactical issues involved with executing the strategies in the lab that we will not discuss here. Of course these are not the only ways to construct simple cyclohexanone derivatives. The problems explore several other approaches. Nonetheless these are versatile strategies that have been widely used in organic synthesis. One point to take away from this exercise is that 6-membered rings can be prepared without resorting to use of A-functions. Although we have seen this only for carbocycles, it turns out this is true for heterocycles as well (see the problems).

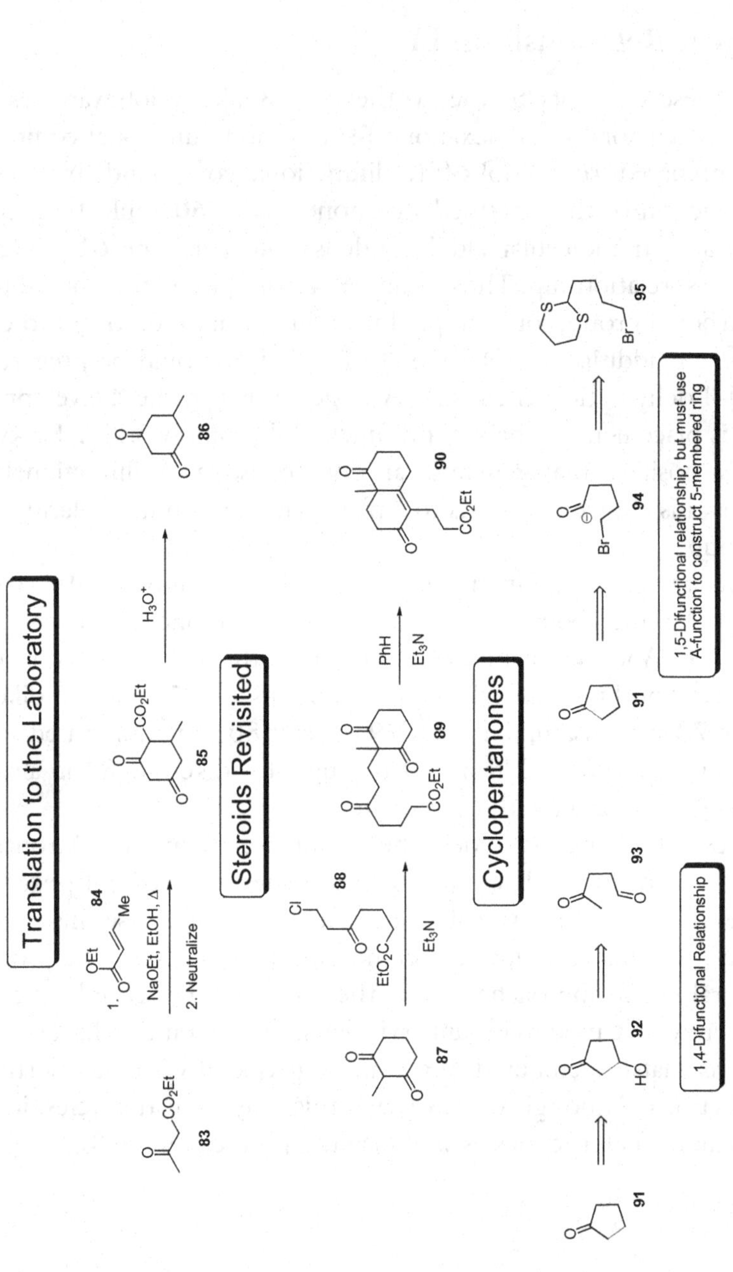

Difunctional Relationships-12

Difunctional Relationships-12

Let's see how this can translate to the laboratory. The reaction of ethyl acetoacetate (**83**) with ethyl crotonate (**84**) provides **85** in 55% yield.[10] This sequence involves initial formation of the enolate of **83**. Enolate formation is insured by the use of an "activating group" that is removed in a later operation. Enolate formation is followed by a conjugate addition to **84**. A subsequent proton transfer and (ester enolate to ketone enolate) is followed by a nucleophilic acyl substitution whose driving force is presumably formation of a stable 1,3-diketone enolate. Protonation, ester hydrolysis and decarboxylation complete the synthesis of **86**. This synthesis is a variation of **78** + **79** → **77** → **76** (Functional Groups-11).

Let's revisit the Wieland approach to steroids (Steroids-7). The reaction of **87** with the enone derived from *in situ* dehydrohalogenation of **88** gave **89**. This was followed by an intramolecular aldol-dehydration to give **90**. This is variation of **68** + **69** → **67** → **66** (followed by hydration to a cyclohexenone) as seen in Functional Groups-11. There are many more tactics that have been developed to accomplish this fundamental strategy in the laboratory.

How about the synthesis of cyclopentanones from acyclic precursors? This could also be accomplished by aldol-dehydration strategies. For example **93** will undergo an intramolecular aldol to give **92**, and dehydration, followed by reduction of the resulting olefin, would give **91**. This strategy starts with a 1,4-difunctional relationship. We have already seen that construction of this relationship will require the use of an A-function. We could prepare the cyclopentanone from a 1,5-difunctional compound. For example metallated **95** might function as an equivalent of **94** and provide **91** via an intramolecular alkylation. It is interesting to note that this transformation also requires the use of an A-function. This is because the 1- and 5-positions of a 1,5-dicarbonyl compound are both inherently positive. The polarity at one carbon must be reversed to facilitate bond construction. A useful lesson that falls out of this is that the construction of 5-membered rings generally requires the use of A-functions. It is also interesting to note that aldol-dehydration routes to 5-membered rings frequently purchase (rather than construct) the critical 1,4-difunctional relationship!

Progesterone Revisited

This synthesis addresses problems associated with construction of the D-ring in the synthesis of *dl*-16,17-dehydroprogesterone. Can one terminate the polyolefin cyclization to directly afford a 5-membered ring?

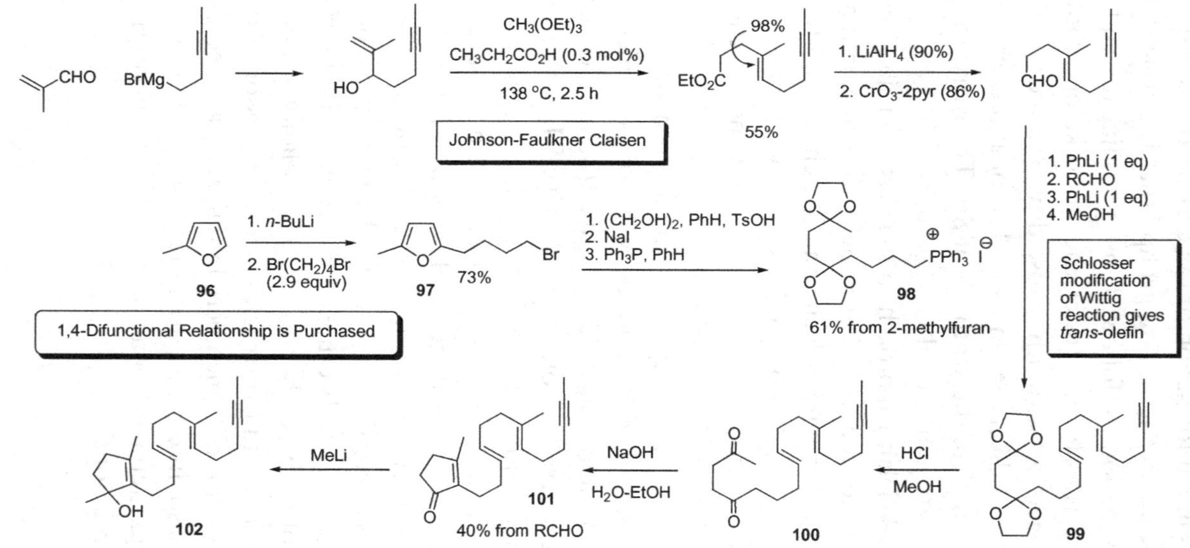

Johnson, W. S.; Gravestock, M. B.; McCarry, B. E. "Acetylenic Bond Participation in Biogenetic-Like Olefinic Cyclizations. II. Synthesis of *dl*-Progesterone" *J. Am. Chem. Soc.* **1971**, *93*, 4332-4334

Difunctional Relationships-13

Consider the Johnson synthesis of progesterone. The cyclization precursor (**102**) was derived from cyclopentenone **101**, which was prepared by an intramolecular aldol-dehydration of 1,4-diketone **100**. Johnson did not construct the 1,4-difunctional relationship present in **100**. It was purchased in the form of 2-methylfuran (**96**). This simple heterocyclic compound is at the same oxidation state as the 1,4-dicarbonyl compound, released by hydrolysis of the furan, first in a protected form (**98** and **99**) and finally ready for use in the aldol dehydration (**99** → **100**).

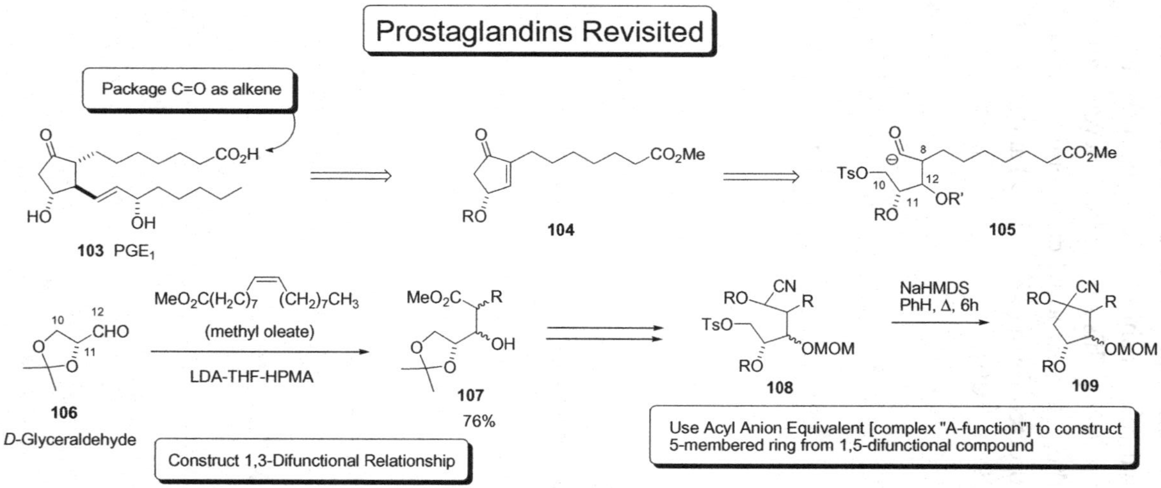

Stork, G.; Takahashi, T. "Chiral Synthesis of Prostaglandins (PGE_1) from *D*-Glyceraldehyde" *J. Am. Chem. Soc.* **1977**, *99*, 1275-1276

Difunctional Relationships-14

Difunctional Relationships-14

The prostaglandins contain a 5-membered ring. In the syntheses we examined, how was this ring constructed within the context of our current discussion? Several of the syntheses purchase the 5-membered ring (Corey lactone approach, Sih synthesis, the three-component coupling syntheses, Holton's approach, the Stork free radical cyclization approach). The Stork "glyceraldehyde" approach to PGE_1 (**103**) proceeded through cyclopentenone **104**. The enone was introduced by dehydration of a β-alkoxycyclopentanone. This 1,3-difunctional relationship was established by addition of the enolate of methyl oleate to glyceraldehyde (**106** → **107**). The 5-membered ring was established from a 1,5-difunctional compound making use of a metallated cyanohydrin as an A-function (**108** → **109**). Note that the 1,2-difunctional relationships, required by intermediates such as **107** and **108**, were purchased (as **106**).

The Importance of Alkenes

Alkenes are latent 1,2-difunctional compounds. Cyclohexenes are latent 1,6-difunctional compounds (and thus 5-membered rings). Cyclopentenes are latent 1,5-difunctional compounds (and thus 6-membered rings).

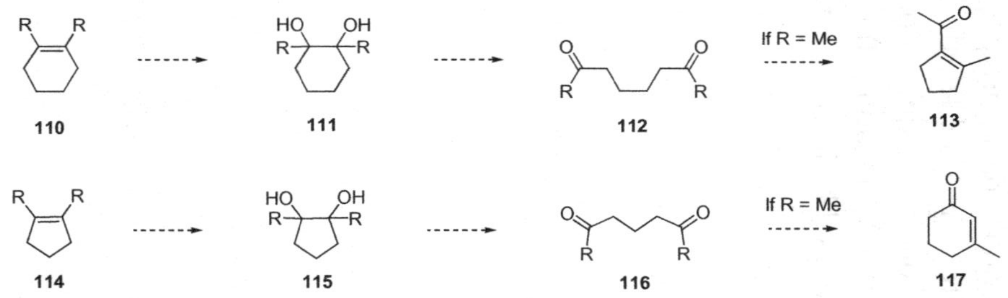

Prostaglandins Revisited

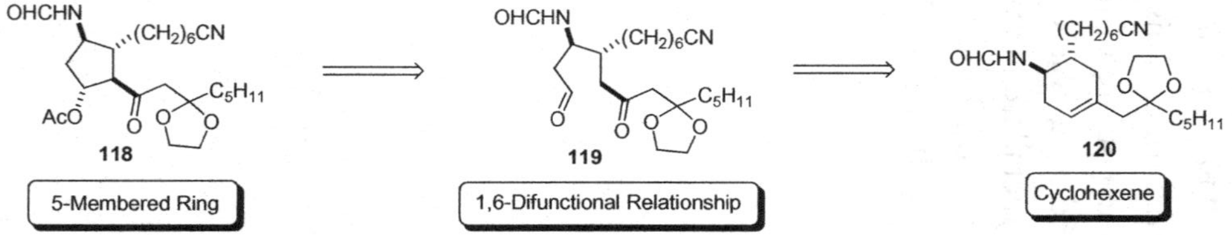

Corey, E. J.; Andersen, N. H.; Carlson, R. M.; Paust, J.; Vedejs, E.; Vlattas, I.; Winter, R. E. K. "Total Synthesis of Prostaglandins. Synthesis of the Pure *dl*-E_1,-$F_{1\alpha}$, $F_{1\beta}$, A_1, and B_1 Hormones" *J. Am. Chem. Soc.* **1968**, 90, 3245-3247.

Difunctional Relationships-15

Difunctional Relationships-15

Alkenes are latent 1,2-difunctional compounds. Cyclic alkenes are latent odd (consonant) or even (dissonant) dicarbonyl compounds depending on the their ring size. For example a cyclohexene is a latent 1,6-dicarbonyl, and a cyclopentene is a latent 1,5-dicarbonyl. It now becomes clear why cyclohexenes can be transformed to 5-membered rings and cyclopentenes to 6-membered rings. We first saw these transformations in the area of steroid synthesis. We can also see that the first Corey syntheses of prostaglandins and the Woodward steroid synthesis use the same fundamental strategy to construct the 5-membered rings (illustrated in Functional Groups-15 with the Corey PG synthesis). A Diels-Alder reaction was used to prepare a cyclohexene (**120**). This was followed by oxidative cleavage of the olefin to give a 1,6-dicarbonyl compound (**119**), that was then converted to a 5-membered ring (**118**), relying on the normal behavior of the carbonyl group.

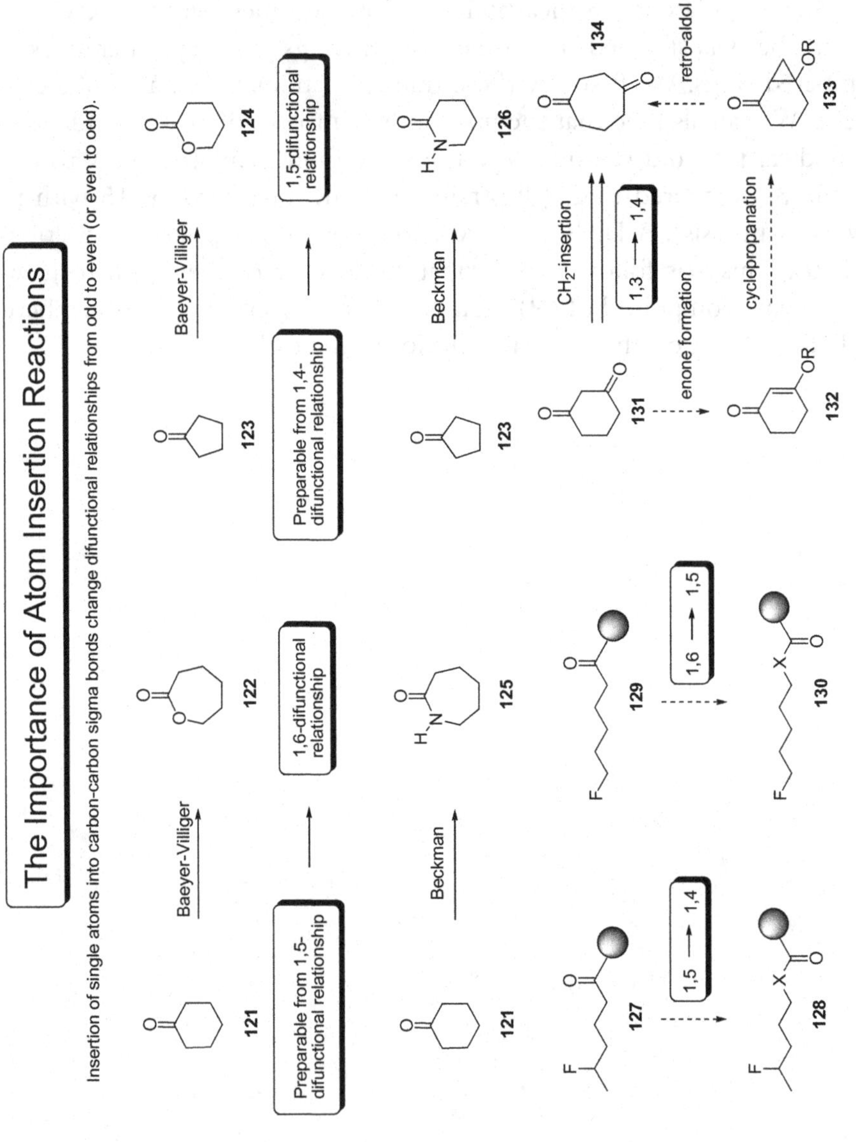

Difunctional Relationships-16

Difunctional Relationships-16

Atom insertions constitute another important family of reactions. Insertion of single atoms into carbon-carbon σ-bonds change difunctional relationships from odd to even (or even to odd). Cyclohexanones (for example **121**) are preparable from 1,5-dicarbonyls (see Functional Groups-11) or 1,7-dicarbonyls. Baeyer-Villiger and Beckman rearrangments convert cyclohexanones into 1,6-difunctional compounds. Conversely, cyclopentanones (for example **123**) can be prepared from 1,4-dicarbonyls or 1,6-dicarbonyls (*vide supra*) and can be similarly converted to 1,5-difunctional compounds. This is also true for acyclic compounds (**127** → **128** and **129** → **130**). Carbon insertion reactions accomplish the same change of difunctional relationships. For example insertion of a methylene between the carbonyl groups of **131** converts this 1,3-dicarbonyl compound to a 1,4-dicarbonyl compound. This relationship can be useful from time to time.

In concluding this section I want to emphasize that consideration of difunctional relationships is only one way to think about designing a synthesis. Entirely different approaches can also be useful. Nonetheless this is clearly one analytical "tool" that can be used to guide decision-making when designing a synthesis of a given target molecule.

References

1. For an introduction to "retrosynthetic analysis" see Corey, E. J.; Cheng, X-M. "The Logic of Chemical Synthesis", John Wiley and Sons, 1989 (pages 5–16). See also Wyatt, P.; Warren, S. "Organic Synthesis — Strategy and Control" John Wiley and Sons, 2007.
2. Evans, D. A.; Andrews, G. C. "Allylic Sulfoxides: Useful Intermediates in Organic Synthesis" *Acct. Chem. Res.* **1974**, *7*, 147–155.
3. Gattermann, L. *Chem. Ber.* **1898**, *31*, 1149–1152.
4. Seebach, D.; Colvin, E. W.; Lehr, F.; Weller, T. "Nitroaliphatic Compounds — Ideal Intermediates in Organic Synthesis?" *Chimia* **1979**, *33*, 1–18. Henning, R.; Lehr, F.; Seebach, D. "α,β-Doppeldeprotoierte Nitroalkane: Super-Enamine?" *Helv. Chim. Acta* **1976**, *59*, 2213–2217.
5. Nef, J. U. "Ueber die Constitution der Salze Nitroparaffine" *Annalen* **1894**, *280*, 263. Noland, W. E. "The Nef Reaction" *Chem. Rev.* **1955**, *55*, 137–155. Pinnick, H. W. "The Nef Reaction" *Organic Reactions* **1990**, *38*, 655–792. Ballini, R.; Petrini, M. "Recent Synthetic Developments in the Nitro to Carbonyl Conversion (Nef Reaction)" *Tetrahedron* **2004**, *60*, 1017–1047.
6. Henry, L. "Nitro-alcohols" *Compt. Rend.* **1895**, *120*, 1265–1268.
7. Kornblum, N.; Powers, J. W.; Anderson, G. J.; Jones, W. J.; Larson, H. O.; Levand, O.; Weaver, W. M. "A New and Selective Method of Oxidation" *J. Am. Chem. Soc.* **1957**, *79*, 6562.
8. Groebel, B. T.; Seebach, D. "Umpolung of the Reactivity of Carbonyl Compounds through Sulfur-Containing Reagents" *Synthesis* **1977**, 357–402. Seebach, E.; Corey, E. J. "Generation and Synthetic Applications of 2-Lithio-1,3-dithianes" *J. Org. Chem.* **1975**, *40*, 231–237.
9. Pummerer, R. "Ueber Phenylsulfoxy-essigsaure. II." *Chem. Ber.* **1910**, *43*, 1401–1412. Russell, G. A.; Mikol, J. G. "Acid-catalyzed Rearrangements of Sulfoxides and Amine Oxides. The Pummerer and Polonovski Reactions" *Mech. Mol. Migrations* **1968**, *1*, 157–207.
10. Blanchard, J. P.; H. L. Goering "5-Methyl-2-cyclohexenone" *J. Am. Chem. Soc.* **1951**, *73*, 5863–5864.

Problems

1. Provide mechanisms for the following transformations. (Difunctional Relationships-4)

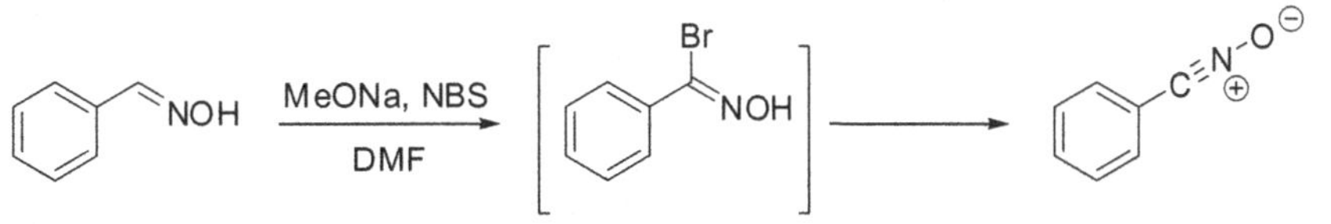

Grundmann, C.; Richter, R. "Nitrile Oxides. X. Improved Method for the Preparation of Nitrile Oxides from Aldoximes" *J. Org. Chem.* **1968**, *33*, 476–478.

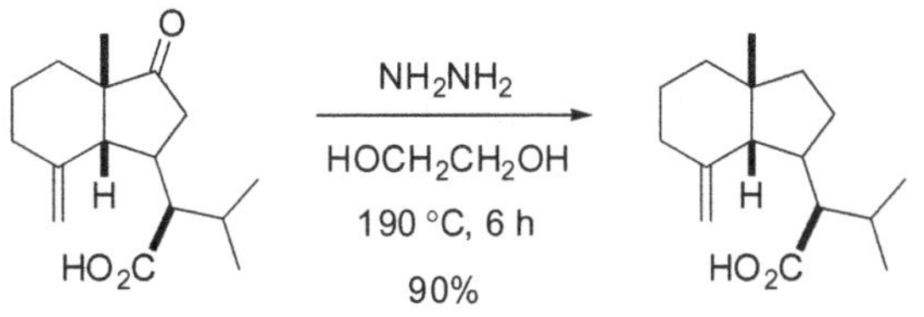

Piers, E.; Yeung, B. W. A.; Rettig. S. J. "Methylenecyclohexane Annulation. Total Synthesis of (±)-Axamide-1, (±)-Axisonitrile-1, and the Corresponding C-10 Epimers" *Tetrahedron* **1987**, *43*, 5521–5535.

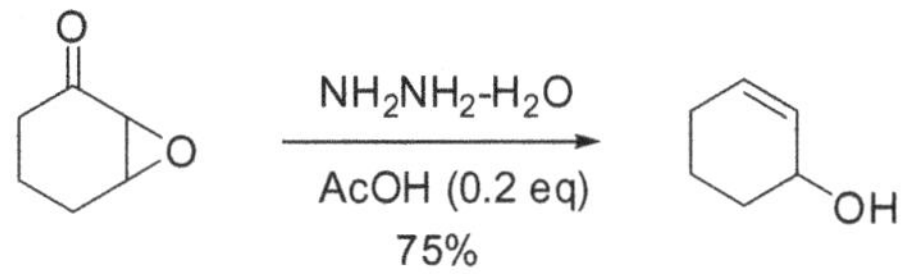

Wharton, P. S.; Bohlen, D. H. "Hydrazine Reduction of α,β-Epoxy Ketones to Allylic Alcohols" *J. Org. Chem.* **1961**, *26*, 3615–3616. For a mechanistic study see Stork, G.; Williard, P. G. "Five- and Six-Membered-Ring Formation from Olefinic α,β-Epoxy Ketones and Hydrazine" *J. Am. Chem. Soc.* **1977**, *99*, 7067–7068. For a recent application of the Wharton rearrangement see Liu, J.; Hsung, R. P.; Peters, S. D. "Total Syntheses of (+)-Cylindricines C-E and (−)-Lepadiformine through a Common Intermediate Derived from an *aza*-Prins Cyclization and Wharton's Rearrangement" *Org. Lett.* **2004**, *6*, 3989–3992.

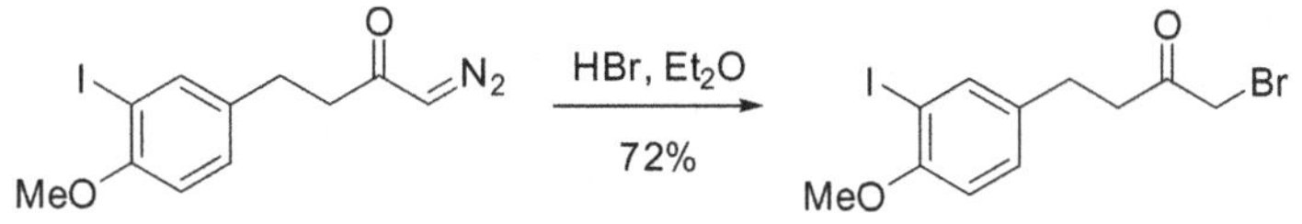

Hart, D. J.; Hong, W. P.; Hsu, L. Y. "Total Synthesis of (±)-Lythrancepine II and (±)-Lythrancepine III" *J. Org. Chem.* **1987**, *52*, 4665–4673.

2. Provide a mechanism for the following reaction that illustrates how sulfonium and oxosulfonium groups behave as both *N*-functions and an *E*-functions in the following reactions. (Difunctional Relationships-4)

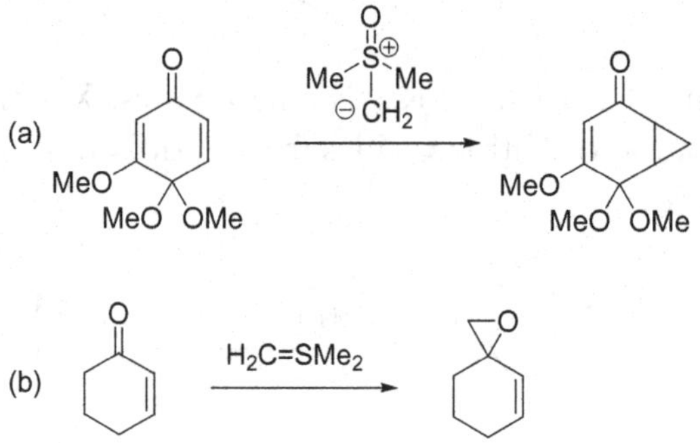

3. Provide mechanisms for the following reactions and discuss in terms of how sulfinyl [S(=O)R] or trialkylammonium groups influence charge on carbon. (Difunctional Relationships-4)

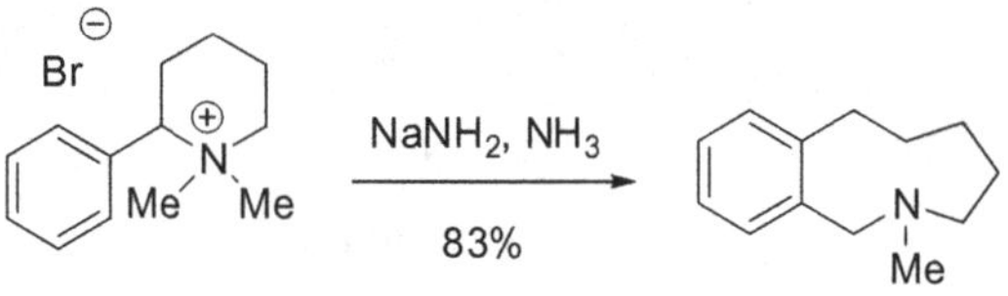

Lednicer, E.; Hauser, C. R. "A Novel Ring Enlargement Involving the Ortho Substitution Rearrangement by Means of Sodium Amide in Liquid Ammonia" *J. Am. Chem. Soc.* **1957**, *79*, 4449–4451.

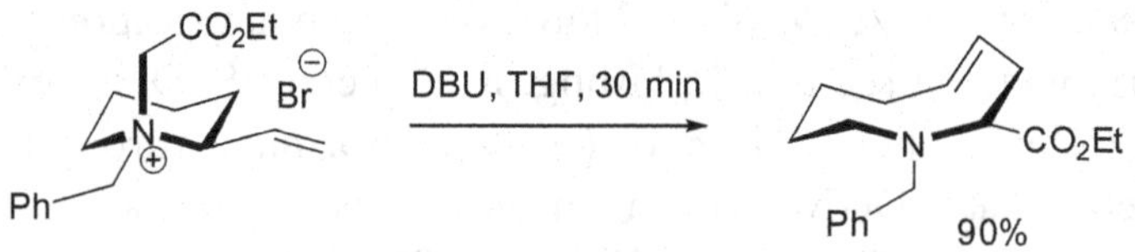

Vedejs, E.; Arco, M. J.; Powell, D. W.; Renga, J. M.; Singer, S. P. "Ring Expansion of 2-Vinyl Derivatives of Thioane, N-Benzylpiperidine, and Thiepane by [2,3] Sigmatropic Shift" *J. Org. Chem.* **1978**, *43*, 4484–4485.

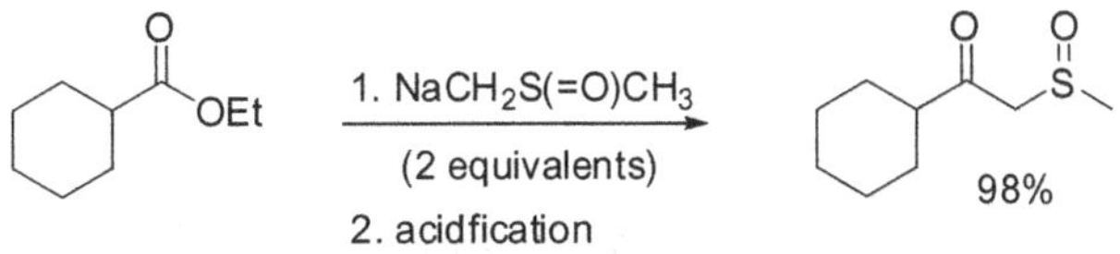

Corey, E. J.; Chaykovsky, M. "New Synthesis of Ketones" *J. Am. Chem. Soc.* **1964**, *86*, 1639–1640.

4. The terms "charge reversal", "umpolung" and "charge affinity inversion" have been used to describe processes wherein (1) a group is operated upon in a manner that reverses the normal polarity of the group (2) the operation that inverts the normal polarity of the group can be reversed to reveal the original functionality. For this process to have any practical importance, the derivative in which the normal polarity must be usable in some bond-forming reaction. Predict the products of the following reaction sequences and identify the individual steps that consitute the aforementioned process. (Difunctional Relationships-4)

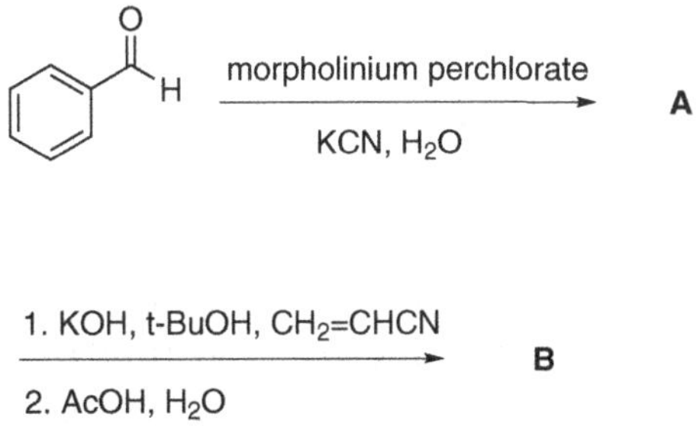

Leete, E.; Chedekel, M. R.; Bodem, G. B. "Synthesis of Myosmine and Nornicotine using an Acylcarbanion Equivalent as an Intermediate" *J. Org. Chem.* **1972**, *56*, 4465–4466.

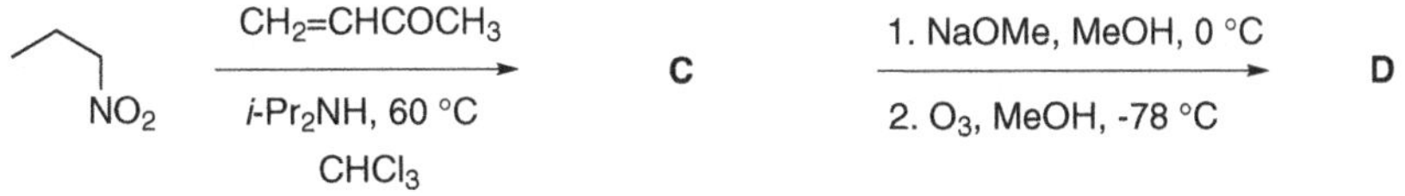

McMurry, J. E.; Melton, J. "Conversion of Nitro to Carbonyl by Ozonolysis of Nitronates: 2,5-Heptanedione" *Organic Syntheses* **1977**, *56*, 36–39.

Corey, E. J.; Seebach, D. "Carbanions of 1,3-Dithianes. Reagents for C-C Bond Formation by Nucleophilic Displacement and Carbonyl Addition" *Angew. Chem. Int. Ed.* **1965**, *4*, 1075–1077.

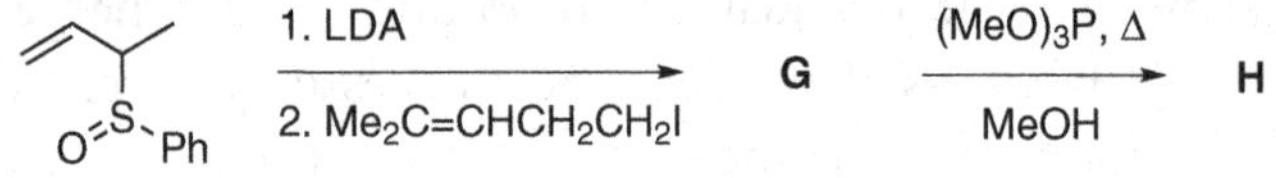

Evans, D. A.; Andrews, G. C.; Fujimoto, T. T.; Wells, D. "Stereoselective Synthesis of Trisubstituted Olefins" *Tetrahedron Lett.* **1973**, *15*, 1389–1394.

5. Explain why **A** can be deprotonated by lithium diisopropylamide (LDA) where as **B** cannot. Explain why this is not a breakdown of the principle of vinylogy. (Difunctional Relationships-6)

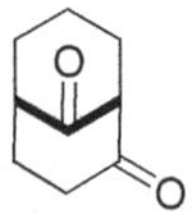

A

B

6. Compound **C** is commonly called a vinylgous acid. As the name implies, it has a pK_a close to that of a carboxylic acid (5.25). Explain using the principle of vinylogy. Provide the structure of the vinylogous methyl ester, vinylogous acid chloride, vinylogous amide that can be derived from **C**. Suggest methods for their preparation (reason by analogy with the chemistry of carboxylic acids). (Difunctional Relationships-6)

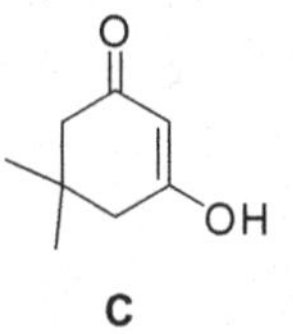

C

Fehnel, E. A.; Paul, A. P. "Thiapyran Derivatives. V. The Monosulfinyl and Monosulfonyl Analogs of Phlorogulcinol" *J. Am. Chem. Soc.* **1955**, *77*, 4241–4244.

7. Which of the two approaches to the triketone shown below is likely to give the higher yield? Why? (alkylations with α-bromoacetone vs propargyl bromide) (Difunctional Relationships-7)

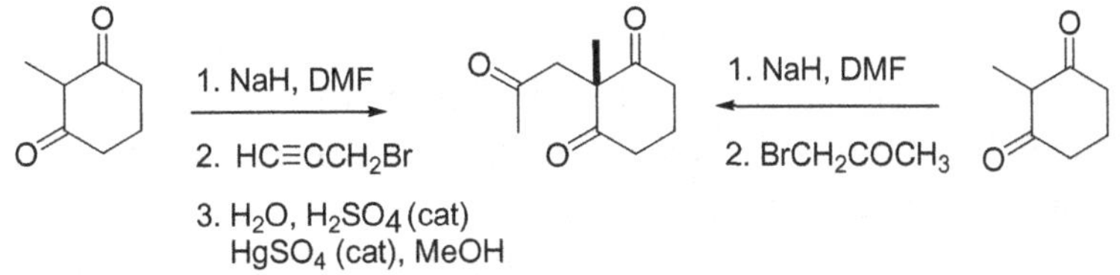

8. Provide the structure of intermediates and mechanism for the conversion of **38** to **29**. (Functional Groups-8)
9. Show how a phensulfinyl group [PhS(=O)] might play the role of the A-group in the approaches to **39** outlined in Functional Groups-9.
10. The benzoin condensation is an example of a reaction that proceeds via acyl anion equivalent intermediates. Provide mechanisms for the following reactions. (Functional Groups-10)

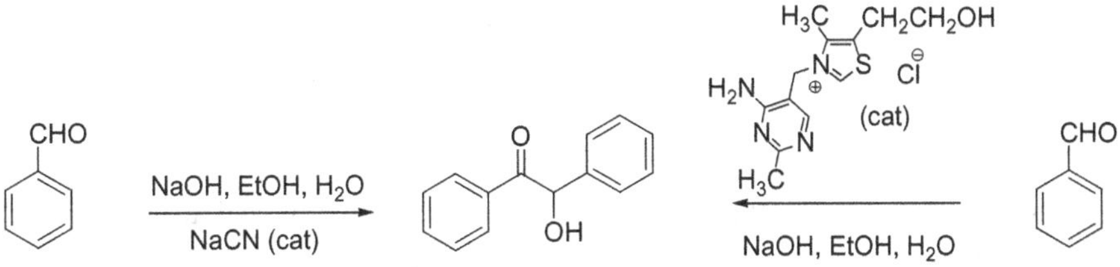

11. Treatment of **D** with **E** affords 1,4-dicarbonyl compound **F** in 77% yield. Provide a mechanism for this reaction. (Functional Groups-10)

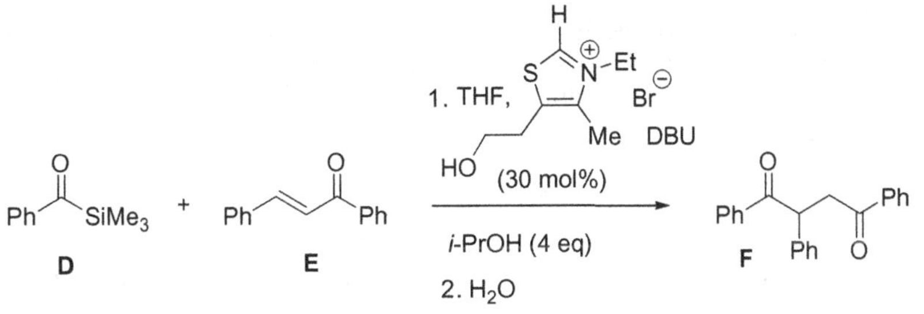

Mattson, A. E.; Bharadwaj, A. R.; Scheidt, K. A. "The Thiazolium-Catalyzed Sila-Stetter Reaction: Conjugate Addition of Acylsilanes to Unsaturated Esters and Ketones" *J. Am. Chem. Soc.* **2004**, *126*, 2314–2315.

12. Diketones **76** and **82** are almost aromatic. To what aromatic compound would addition of one mole of hydrogen provide **76** or **82**? (Functional Groups-11)
13. Provide an analysis of 2-substituted cyclohexanone synthesis that is related to Functional Groups-11. Repeat the exercise for 4-substituted cyclohexanones, 2,4-disubstituted cyclohexanones, and so on. (Functional Groups-11)
14. Propose intermediates for the following reaction sequences and think about difunctional relationships along the transformation pathways. (Stevens pyridine, Danishefsky steriods) (Functional Groups-11)

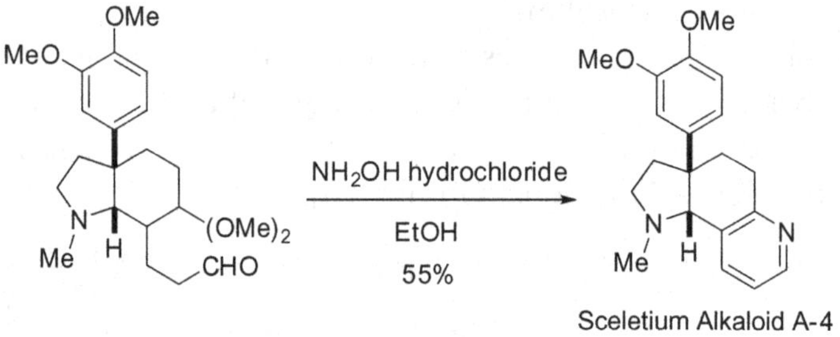

Stevens, R. V.; Lesko, P. M.; Lapalme, R. "General Methods of Alkaloid Synthesis. XI. Total Synthesis of the Sceletium Alkaloid A-4 and an Improved Synthesis of (±)-Mesembrine" *J. Org. Chem.* **1975**, *40*, 3495–3498.

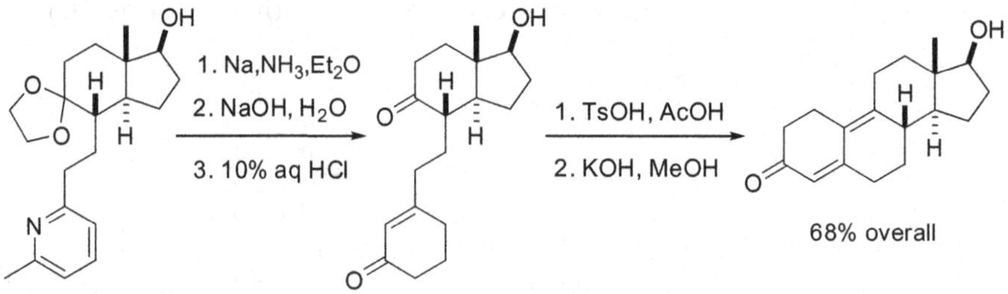

Danishefsky, S.; Cain, P. "Optically Specific Synthesis of Estrone and 19-Norsteroids from 2,6-Lutidine" *J. Am. Chem. Soc.* **1976**, *98*, 4975–4983.

15. A synthesis of furans from 1,3-dicarbonyl derivatives is shown below. Provide the structure of possible intermediates and discuss in terms of difunctional relationship transformations. [Garst, M. E.; Spencer, T. A. "General Method for the Synthesis of 3- and 3,4-substituted Furans. Simple Syntheses of perillene and Dendrolasin" *J. Am. Chem. Soc.* **1973**, *95*, 250–252.] (Difunctional Relationships-13)

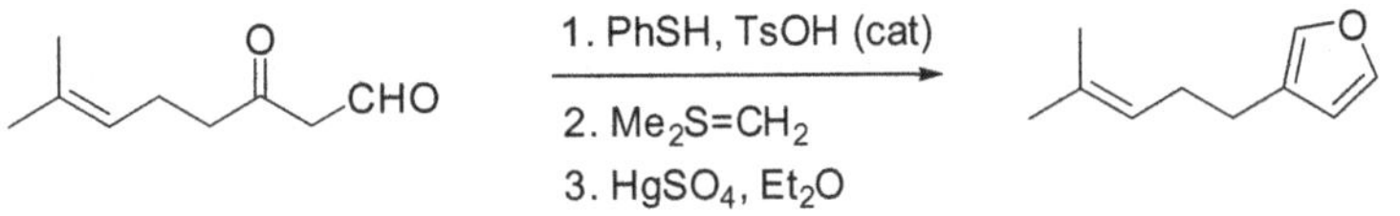

Garst, M. E.; Spencer, T. A. "General Method for the Synthesis of 3- and 3,4-Substituted Furans. Simple Syntheses of Perillene and Dendrolasin" *J. Am. Chem. Soc.* **1973**, *95*, 250–252.

16. Propose a synthesis of the following compounds from acyclic starting materials. (Difunctional Relationships-15)

Bartlett, P. A.; Green, F. R., III "Total Synthesis of Brefeldin A" *J. Am. Chem. Soc.* **1978**, *100*, 4858–4865.

CHAPTER 7

Some Unnatural Products — Twistane and Triquinacene

Twistane Analysis: One-Bond Disconnections

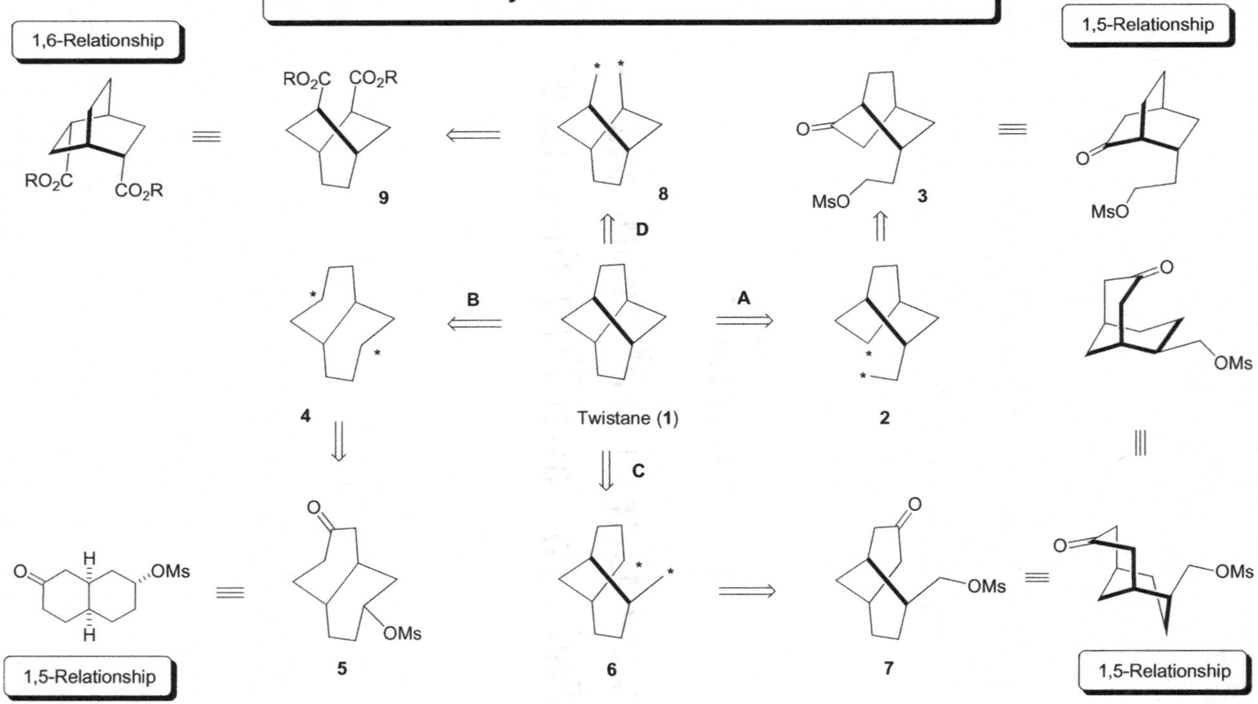

For a nice paper that includes multiple-bond disconnections see Hamon, D. P. G.; Young, R. N. "The Analytical Approach to Synthesis: Twistane" *Aust. J. Chem.* **1976**, *29*, 145-161.

Unnatural Products-1

In Chapter 6 we saw that an analysis of difunctional relationships within a synthetic target could be used to guide development of a synthetic strategy. In this chapter we will examine syntheses of two unnatural products, with an eye on difunctional relationships.

We will first focus on twistane (**1**), a hydrocarbon of theoretical interest because it contains only six-membered rings, fused such that each ring is constrained in a twist-boat conformation. Twistane is quite symmetrical, but is also chiral, and thus enantioselective synthesis is also an issue with this molecule.

Twistane has no functional groups. Development of a synthetic strategy requires passing through functionalized intermediates, and thus we are faced with decisions regarding where to place functionality as we develop a plan. We will restrict our "retrosynthetic analysis" of twistane to one-bond disconnections. For a more thorough analysis of this target that includes multiple-bond disconnections, the reader should consult the excellent paper by Harmon and Young cited in Unnatural Products-1.[1]

Twistane has four different sigma bonds and thus, there are four possible one-bond disconnections that can be used to convert it to a difunctional intermediate. These disconnections are labelled as Paths A-D in Unnatural Products-1. Path A requires construction of a bond between the two carbons marked with asterisks in structure **2**. We can imagine making this bond in a number of ways, one of which involves passing through 1,5-difunctional intermediate **3**. An enolate alkylation might be used to construct the required C–C bond. Compound **3** is easily recognized as a compound that could be constructed by a Diels-Alder reaction. Path B requires the bond construction required by structure **4**. Once again this might be accomplished via a 1,5-difunctional compound of type **5**. The bond disconnection depicted in Path C (**1** → **6**) can also be accomplished from a compound containing a 1,5-difunctional relationship (**7**). Bond disconnection via Path D does not directly lead to a 1,5-dicarbonyl compound. Instead it suggests construction via a 1,6-difunctional compound of type **9**. Let's now look at syntheses that follow each of these paths.

First Synthesis of Twistane

Whitlock, H. W. Jr. "Tricyclo[4.4.0.0^{3,8}]decane" *J. Am. Chem. Soc.* **1962**, *84*, 3412.

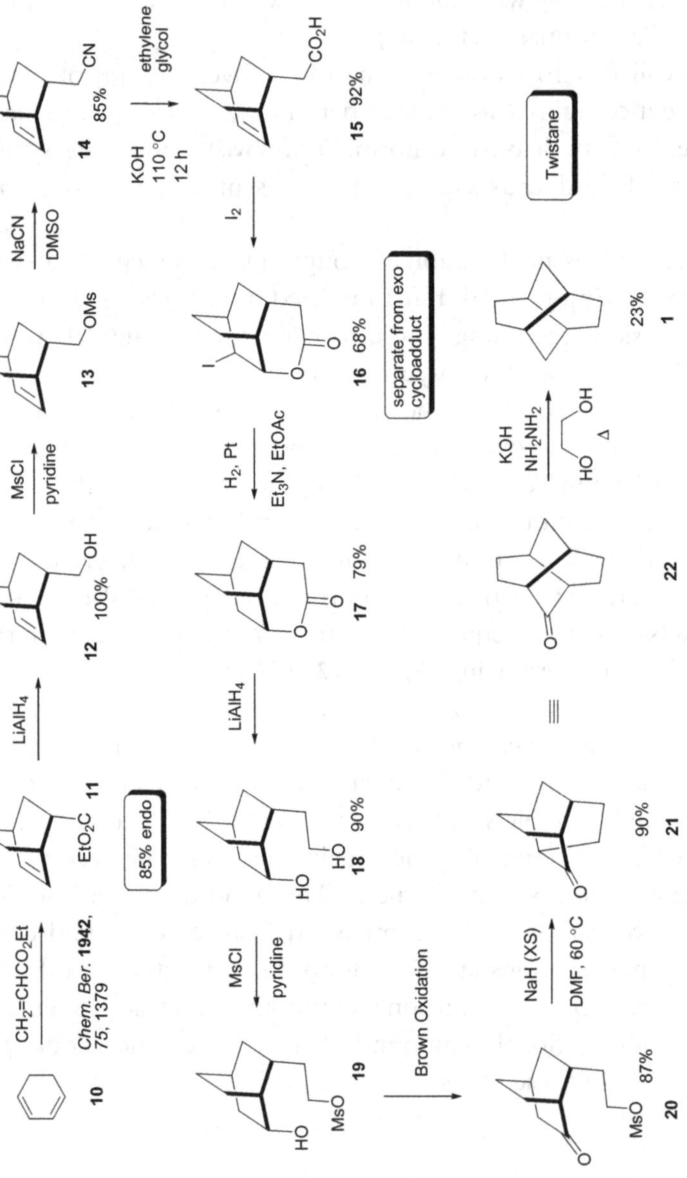

Unnatural Products-2

The first synthesis of twistane, reported by Howard Whitlock (University of Wisconsin) followed Path A. A Diels-Alder reaction between ethyl acrylate and 1,3-cyclohexadiene (**10**) gave a mixture of *endo* and *exo* cycloadducts **11**. A standard homologation sequence was used to convert **11** to carboxylic acid **15** (and its *exo* isomer). An iodolactonization reaction, with six-membered ring lactone formation predominating over seven-membered ring lactone formation, gave **16** and established the desired 1,5-difunctional relationship.[2] In addition, **16** was easily separated from the unreactive *exo* isomer of **15**. A series of adjustments in oxidation state gave keto-mesylate **20**. An intramolecular alkylation gave **22** and a Wolf-Kishner reduction provided twistane (**1**).

Twistane Analysis: One-Bond Disconnections

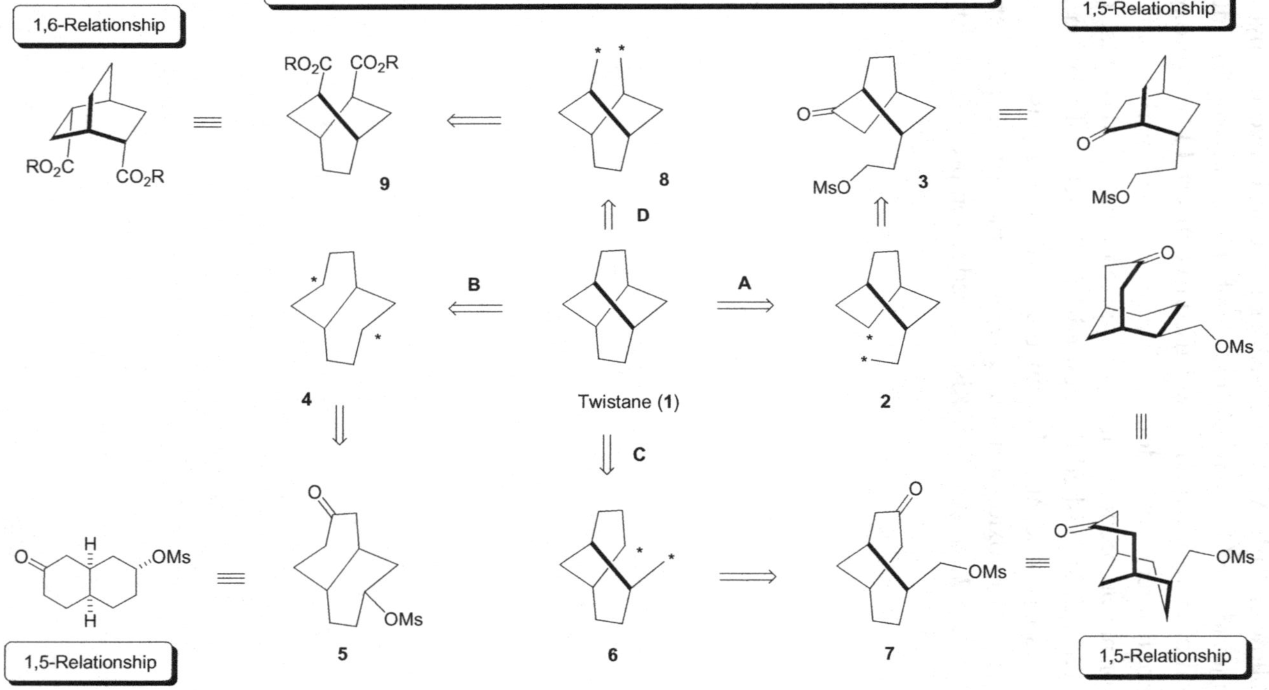

For a nice paper that includes multiple-bond disconnections see Hamon, D. P. G.; Young, R. N. "The Analytical Approach to Synthesis: Twistane" *Aust. J. Chem.* **1976**, *29*, 145-161.

Unnatural Products-3

Unnatural Products-3

One-bond disconnection via Path B leads a symmetrical species of type **4**. The strategy introduced in Chapter 6 suggests that making one of the carbons in the disconnected bond an electrophile (place a leaving group on the carbon), and the other a nucleophile (place a carbonyl adjacent to the carbon). After stereochemical considerations (stereoelectronic requirements of an S_N2 reaction), keto mesylate **5** can be proposed as a 1,5-difunctional intermediate that might undergo the required bond construction.

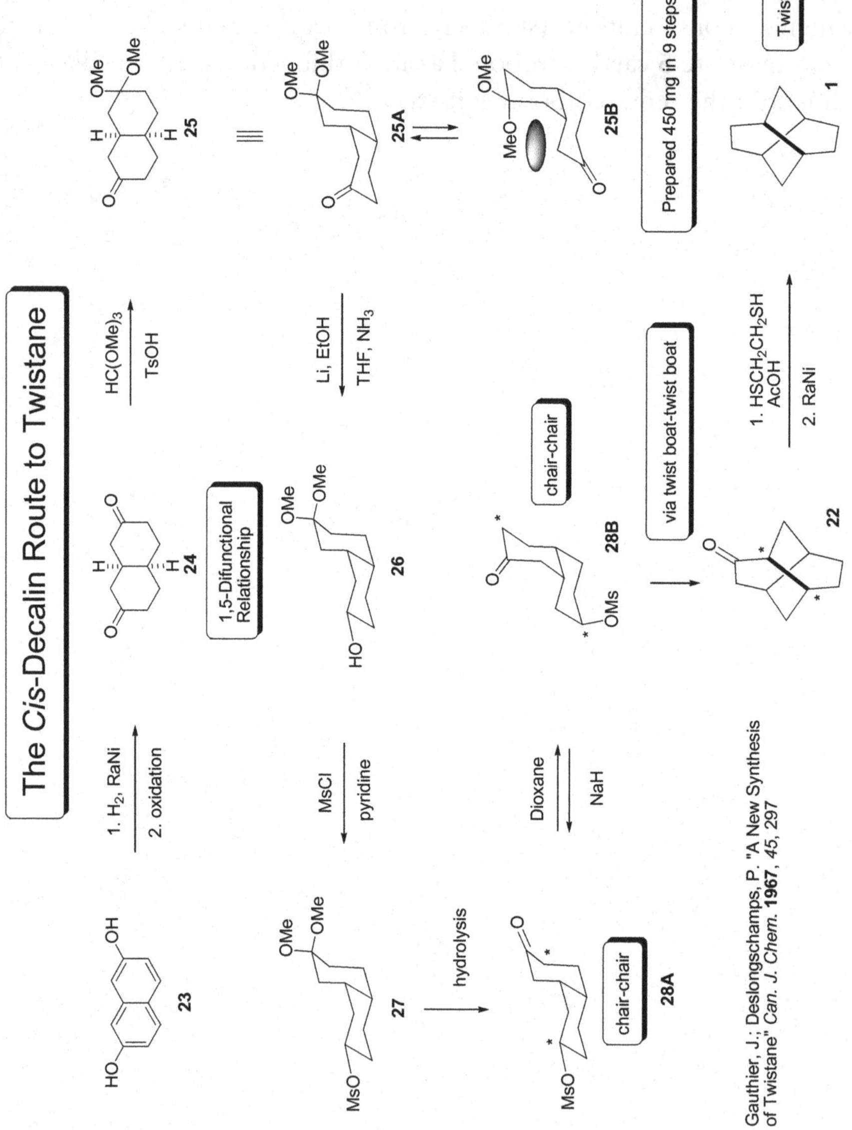

Gauthier, J.; Deslongschamps, P. "A New Synthesis of Twistane" *Can. J. Chem.* **1967**, *45*, 297

Unnatural Products-4

Unnatural Products-4

Delongschamps was the first to report a synthesis via a difunctional intermediate of type **5**. The synthesis began with naphthol **23**. Catalytic hydrogenation of the naphthalene, followed by oxidation of intermediate alcohols, gave 1,5-diketone **24** and established the *cis*-decalin ring fusion required for the synthesis. Monoprotection of diketone **24** gave acetal **25**. Dissolving metal reduction of **25** provided equatorial alcohol **26**. This tactic is known to afford largely equatorial alcohols from rigid cyclohexanone substrates.[3] Since acetal **25** presumably prefers conformation **25A** over **25B** for steric reasons, the reduction led to formation of **26** rather than the epimeric alcohol. Alcohol **26** was converted to **28** via mesylate **27**, and the enolate derived from **28** underwent the desired intramolecular S_N2 reaction, presumably via a twist-boat/twist-boat conformation derived from **28B**, to give ketone **22**. Reduction of **22**, by desulfurization of an intermediate dithiane, completed the synthesis of twistane (**1**).

Asymmetric Synthesis of Twistane

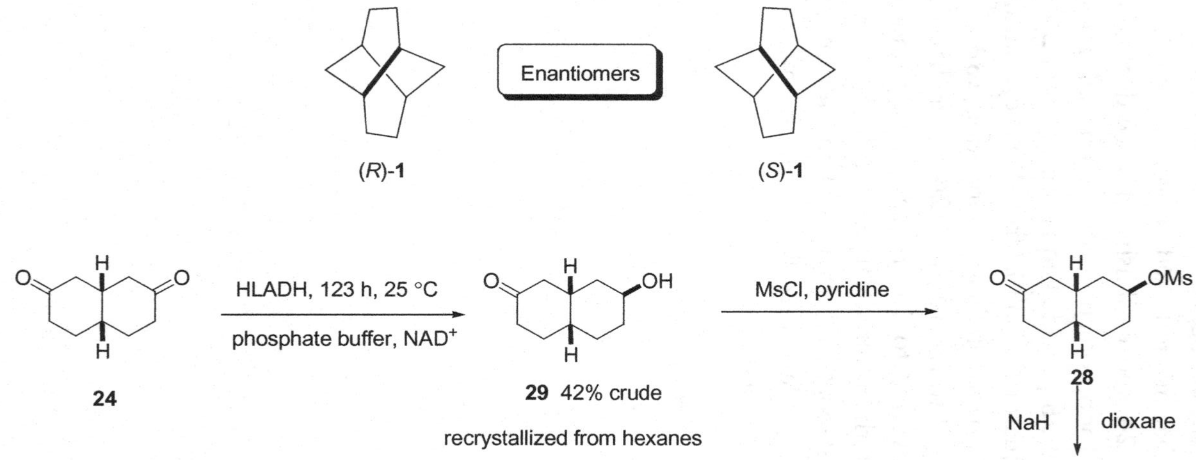

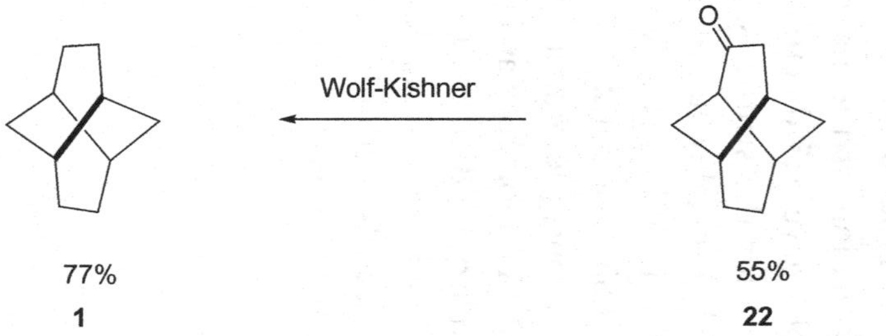

Nakazaki, M.; Chikamatsu, H.; Tarriguchi, M. "Horse Liver Alcohol Dehydrogenase (HLADH) Mediated Chemicoenzymatic Asymmetric Synthesis of (+)-Twistane from *cis*-Decalin-2,7-dione" *Chem. Lett.* **1982**, 1761-1764

Unnatural Products-5

Unnatural Products-5

Fifteen years after the Delongschamps group completed their synthesis, an enantioselective synthesis of (*R*)-twistane was reported using the same key bond-construction. Diketone **24** was enzymatically reduced (horse liver alcohol dehydrogenase) to give ketol **29**. The synthesis was completed in a straightforward manner.

Another Asymmetric Synthesis of Twistane

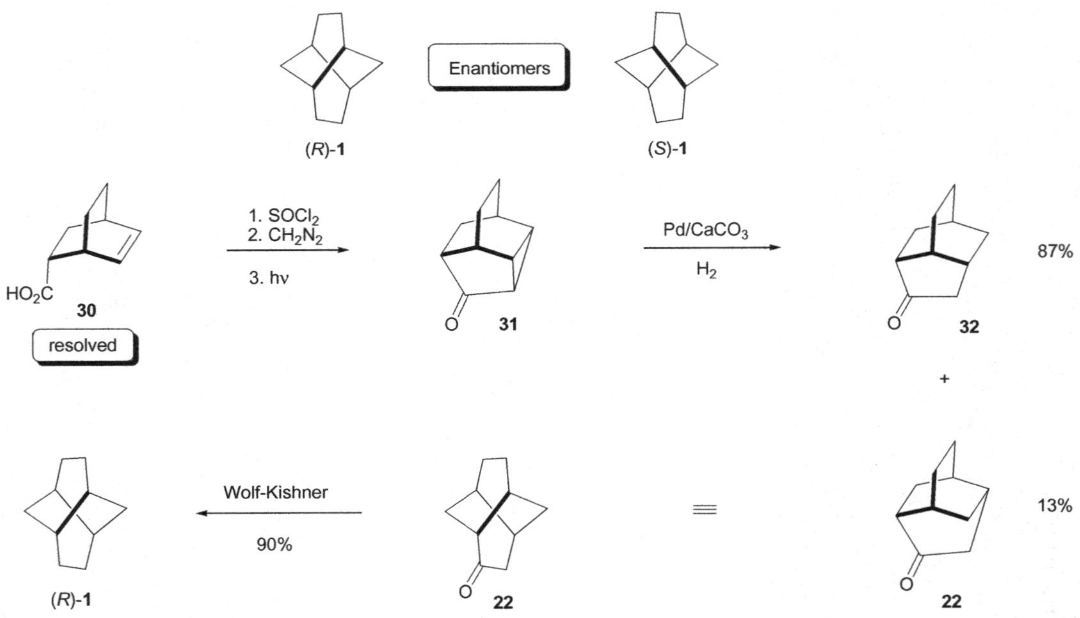

Tichy, M. "On the Absolute Configuration of Tricyclo[4.4.0.0^{3,8}]decane" *Tetrahedron Lett.* **1972**, 2001-2004

Unnatural Products-6

Unnatural Products-6

Another enantioselective synthesis of (R)-twistane was reported by Tichy. This synthesis began with Diels-Alder cycloadduct **30** which was resolved and then converted to cyclopropylketone **31** via an intramolecular carbene addition reaction. Hydrogenolysis of **31** provided ketones **32** and **22**. Wolf-Kishner reduction of the minor product (**22**) gave twistane (**1**). The hydrogenolysis of **31** occurred at the cyclopropane σ-bond best alligned with π-bond of the carbonyl group. Other than poor regioselectivity in the hydrogenolysis, this was a very direct synthesis of twistane.

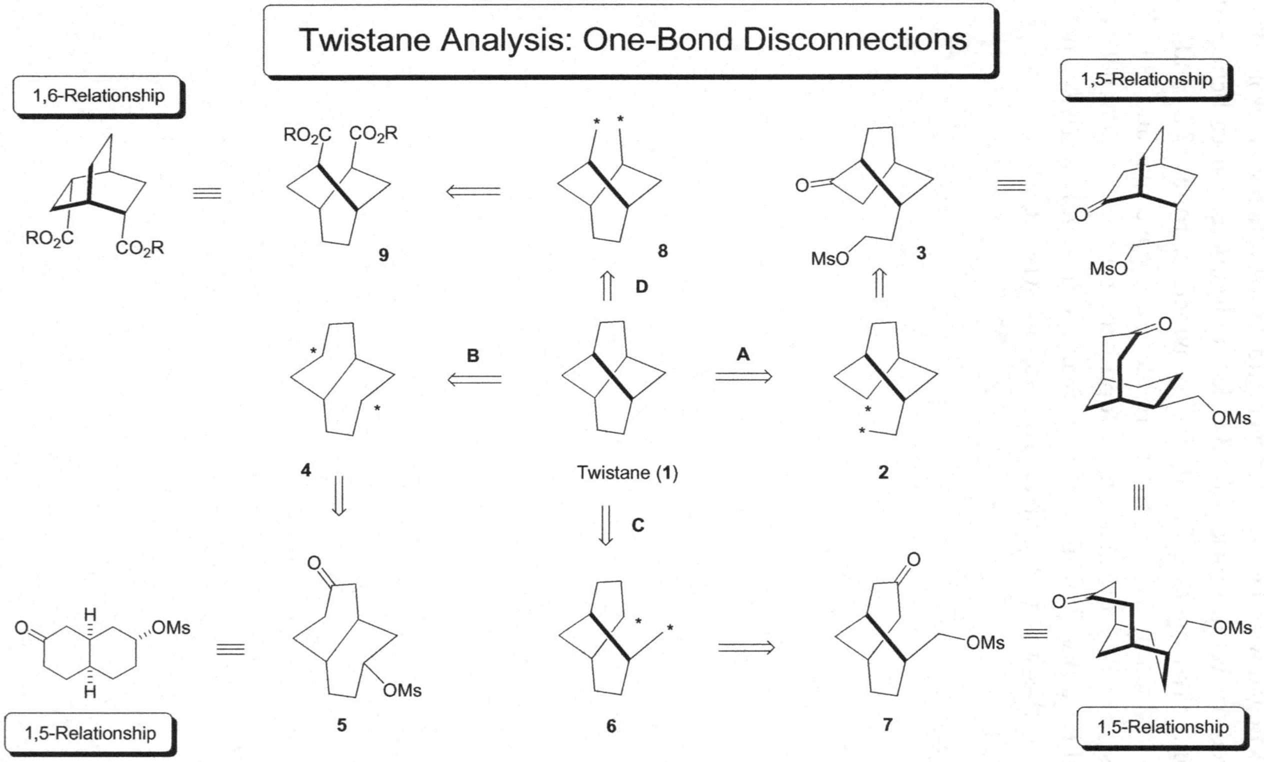

For a nice paper that includes multiple-bond disconnections see Hamon, D. P. G.; Young, R. N. "The Analytical Approach to Synthesis: Twistane" *Aust. J. Chem.* **1976**, *29*, 145-161.

Unnatural Products-7

One-bond disconnection via Path C generates bicyclo[3.3.1]nonane intermediates. Keto-mesylate 7 is a 1,5-difunctional intermediate one might convert to twistane via yet a third intramolecular alkylation. This is the pathway pursued by Hamon and Young.[1]

The Bicyclo[3.3.1]nonane Route to Twistane

Hamon, D. P. G.; Young, R. N. "The Analytical Approach to Synthesis: Twistane" *Aust. J. Chem.* **1976**, *29*, 145-161.

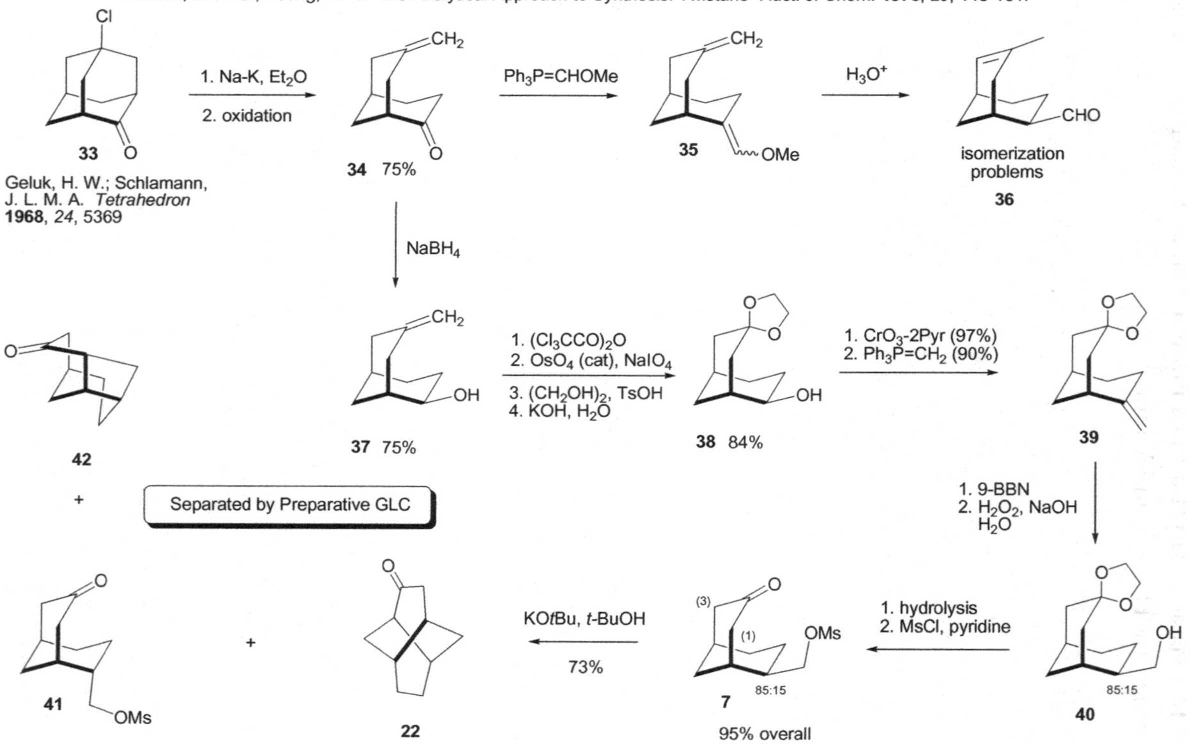

Unnatural Products-8

Unnatural Products-8

The Hamon and Young synthesis began with reductive fragmentation of the known adamantanone derivative **33** to provide **34**. Conversion of **34** to key intermediate **7** required a bit of work. This transformation required that the ketone of **34** be converted to a hydroxymethyl group via a reductive homologation reaction, and that the olefin be oxidatively cleaved to a ketone. In practice, Wittig olefination of **34** provided enol ether **35**, but hydrolysis of the enol ether was complicated by isomerization of the olefin to the more stable endocyclic isomer **36**. Thus it was decided to reverse the homologation and oxidative cleavage operations. Reduction of **34** gave alcohol **37**. Protection of the alcohol (through esterification), oxidative cleavage of the exocyclic methylene, protection of the resulting ketone as an acetal, and removal of the trichloroacetate protecting group, gave **38**. Oxidation of the alcohol was followed by Wittig olefination to provide **39**. Hydroboration-oxidation of **39** gave alcohol **40** as the major stereoisomer. Hydrolysis of the ketal and mesylate formation completed the synthesis of key intermediate **7** contaminated with 15% of stereoisomer **41**.

Treatment of **7** with potassium *tert*-butoxide provided a separable mixture of materials that included **41** (unreacted from the impure starting material), and a 2:1 ratio of **22**:**42**. We have already witnessed the conversion of **22** to twistane. It is notable that **7** must undergo energetically uphill conformational changes to meet the stereoelectronic requirements of the S_N2 reactions leading to either **22** or **42**. Apparently there is little energy difference between the transition states leading to the intramolecular alkylation products.

Despite the problems associated with the last step, this is an interesting approach to twistane that provides one with the opportunity to think of alternatives for construction of the final carbon-carbon bond. I hope the problems will stimulate you in this regard.

Twistane via a 1,6-Difunctional Relationship

Tichy, M.; Sicher, J. "Synthesis and Absolute Configuration of Tricyclo[4.4.0.0^{3,8}]dec-4-ene (Twistene)" *Tetrahedron Lett.* **1969**, 4609

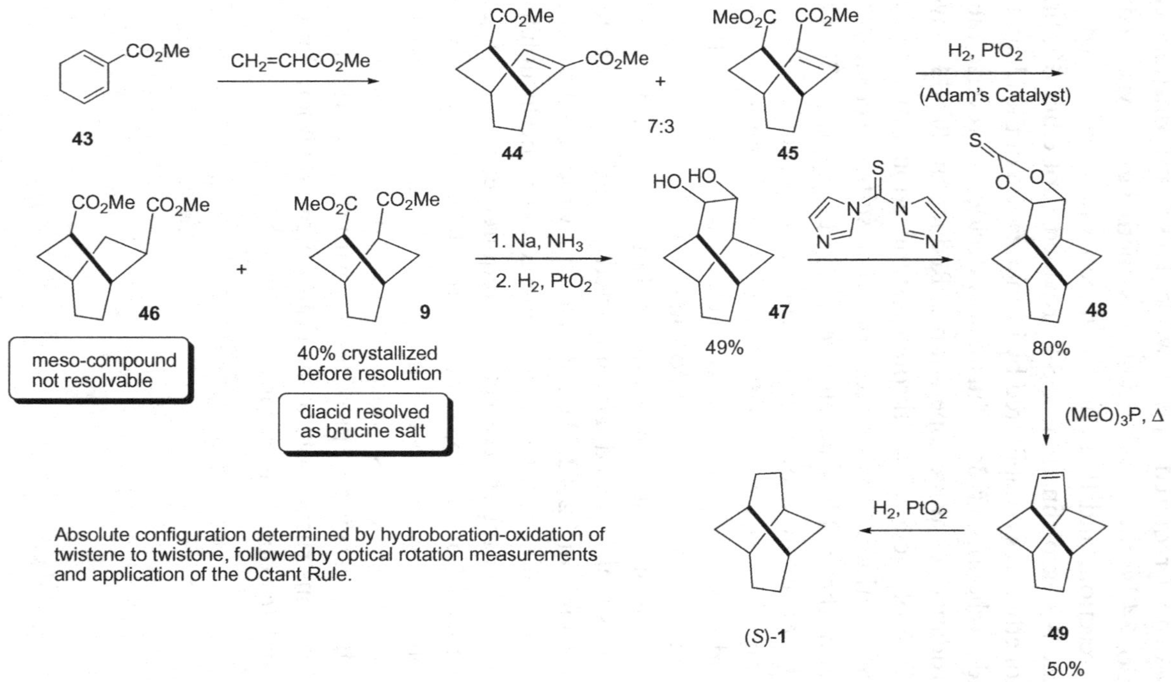

Absolute configuration determined by hydroboration-oxidation of twistene to twistone, followed by optical rotation measurements and application of the Octant Rule.

Unnatural Products-9

Unnatural Products-9

The final twistane synthesis we will consider originates from the one-bond disconnection indicated by Path D in Unnatural Products-1. Just as with Path A, the synthesis uses a Diels-Alder reaction to establish the bicyclo[2.2.2]octane substructure of key intermediate **9**. The cycloaddition of cyclohexadiene **43** and methyl propiolate afforded a mixture of regioisomeric cycloadducts **44** and **45**. Catalytic hydrogenation of the mixture occured from the sterically most accessible face of the olefin to afford *meso*-compound **46** and its diastereomer **9**. Diester **9** was subjected to an acyloin condensation, and catalytic hydrogenation of the resulting α-hydroxyketone gave diol **47**. A Corey-Winter reaction was used to convert **47** to **49** via thionocarbonate **48**.[4] Catalytic hydrogenation completed the synthesis.

Woodward Synthesis of Triquinacene

Woodward, R. B.; Fukunaga, T.; Kelly, R. C. "Triquinacene" *J. Am. Chem. Soc.* **1964**, *86*, 3162

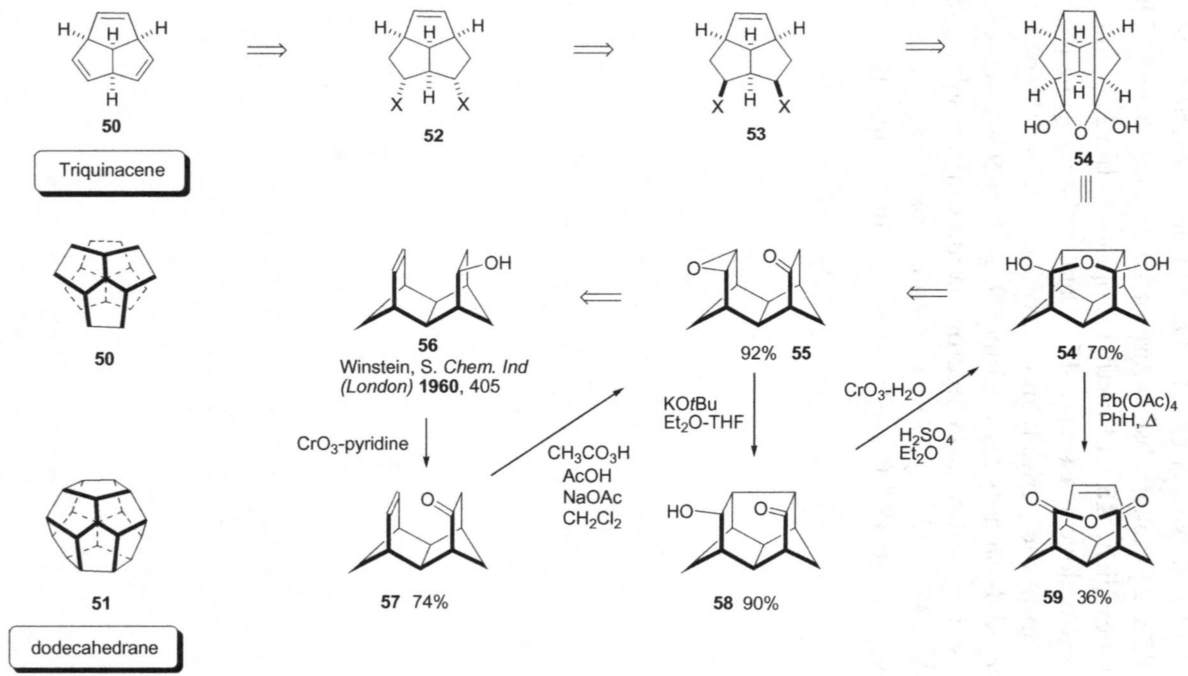

Unnatural Products-10

Unnatural Products-10

Triquinacene (**50**) is another hydrocarbon that has long been of interest. One reason is that an appropriate dimerization of triquinacene would provide dodecahedrane (**51**). From the standpoint of synthesis, triquinacene forces one to address the problem of five-membered ring synthesis. In addition, it is interesting to examine the difunctional relationships used in syntheses by practitioners in the field.

I will start with the Woodward synthesis. Triquinacene is symmetrical and thus, it is not surprising that the synthesis (plan) passes through a series of symmetrical intermediates. For example, it was felt that triquinacene could be prepare by elimination of two moles of H–X from an intermediate such as **52** or **53**. At this point it is difficult to know what provided Woodward with the insight that led back to **54** as a projected intermediate. Perhaps it was recognition that the known compound **56**, derived from norbornadiene and cyclopentadiene, was only one carbon-carbon bond away from containing the triquinacene carbon skeleton as a substructure.

The synthesis began with **56**, which had been described by the Winstein group at UCLA.[5] Oxidation to ketone **57**, and epoxidation from the sterically most excessible face of the olefin, provided **55**. Base mediated intramolecular opening of the epoxide established the final carbon-carbon bond of the triquinacene skeleton, as alluded to above. Oxidation of **58** to the diketone was accompanied by hydrate formation to provide **54**. It was now necessary to break two carbon-carbon bonds to reveal the tricyclic nucleus of triquinacene. This was accomplished by treatment of **54** with lead tetraacetate to reveal anhydride **59**, with the first of three double bonds present in triquinacene in place.

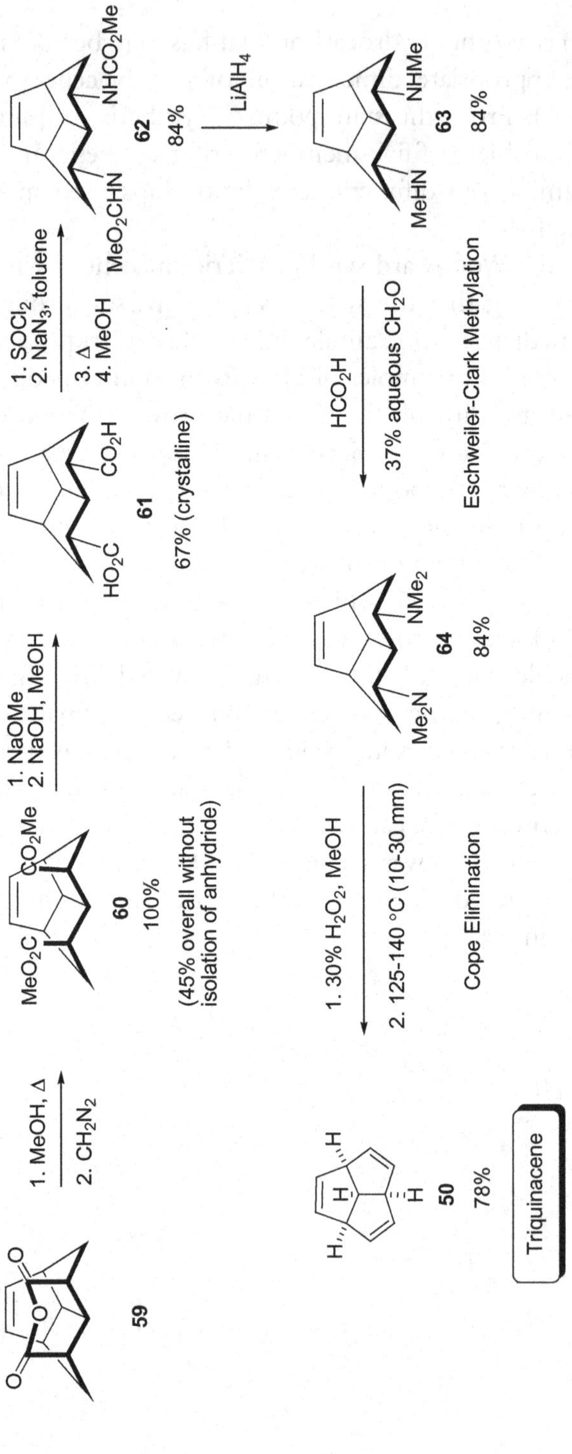

Unnatural Products-11

Unnatural Products-11

Anhydride **59** was next converted to diester **60**. Epimerization to the more stable diester, with the carbomethoxy groups on the convex face of the triquinacene nucleus, followed by saponification, gave diacid **61**. The next task was to convert the carboxyl groups to groups suitable for introducing the remaining double bonds using an appropriate elimination reaction. It was decided to use a *bis*-Cope elimination and thus, diamine **64** became the next target. A "double" Curtius rearrangement of **61** provided *bis*-carbamate **62**. Reduction of the carbamate to diamine **63** was followed by Eschweiler-Clark methylation to give the target diamine **64**. Oxidation of **64** to the *bis*-amine oxide was followed by thermal *syn*-elimination of *N,N*-dimethylhydroxylamine to provide triquinacene (**50**).

The origin of the three 5-membered rings in this synthesis is notable. Two of the rings were purchased in the form of starting materials (cyclopentadiene and norbornadiene). The final ring was constructed from a 1,6-difunctional intermediate, keto-epoxide **55**.

Deslongschamps Synthesis of Triquinacene

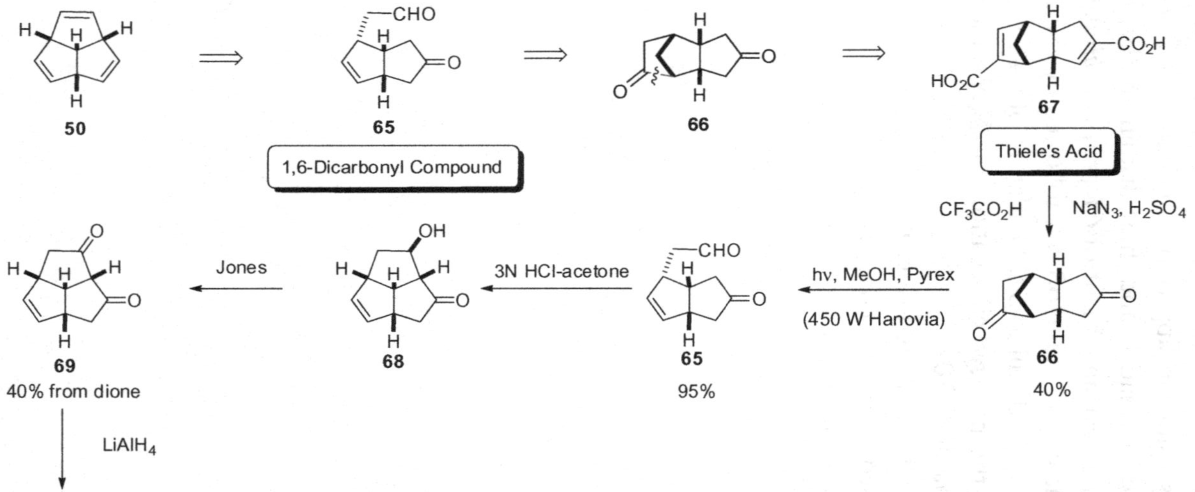

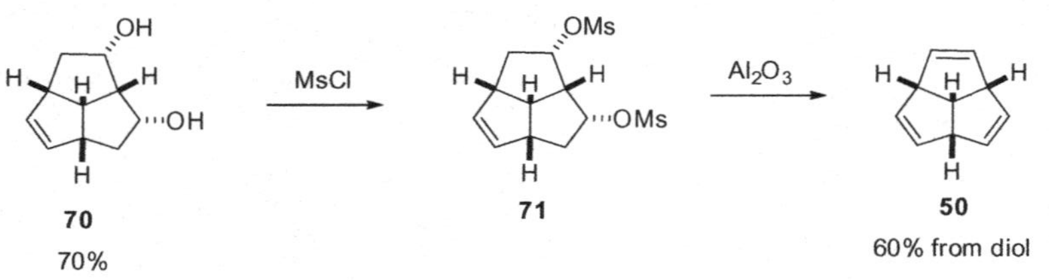

Mercier, C.; Soucy. P.; Rosen, W.; Deslonschamps, P. "A Convenient Synthesis of Triquinacene" *Syn. Commun.* **1973**, 3, 161-164. See also Russo, R. Lambert, Y.; Deslongschemps, P. *Can. J. Chem.* **1971**, *49*, 531. For derivatives and variations on this theme see Delongschamps, P.; Cheriyan, U. O.; Lambert, Y.; Mercier, J.-C.; Ruest, L.; Russo, R.; Soucy, P. "Synthesis of Triquinacene and some of its derivatives" *Can. J. Chem.* **1978**, *56*, 1687.

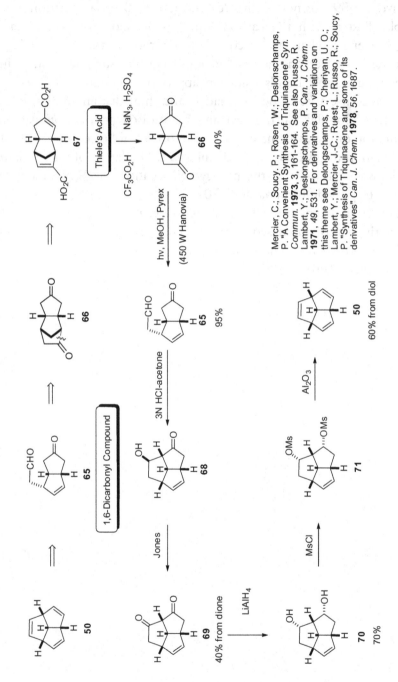

Unnatural Products-12

Unnatural Products-12

Ten years after appearance of the Woodward synthesis, a second synthesis was reported by the Delongschamps group (University of Sherbrooke in Canada). The Delongschamps strategy involved some classical disconnections along the line of carbonyl group chemistry. Thus, it was imagined that triquinacene (**50**) could be prepared from **65** (a 1,6-dicarbonyl compound) via an intramolecular aldol condensation, followed by some oxidation state adjustments to introduce the final two double bonds. It was recognized that **65** might be prepared from **66** by cleavage of the indicated bond. Diketone **66** was to be derived from Thiele's acid (**67**), the Diels-Alder dimer of 1,3-cyclopentadiene 2-carboxylic acid.[6]

Moving forward from Thiele's acid (**67**), degradation to diketone **66** was followed by photochemical conversion of **66** to **65**.[7] An acid-promoted aldol condensation was used to construct the third five-membered ring (**65** → **68**). The alcohol was then oxidized to a ketone (**69**) and reduction of the diketone, with hydride delivery from the convex face of the molecule, provided diol **70**. Formation of the *bis*-mesylate (**71**) and a double elimination reaction completed the synthesis of triquinacene.

Once again it is interesting to see that two of the five-membered rings were purchased (cyclopentadiene) and the third ring was constructed from a 1,6-difunctional compound (**65**). Once again, the latter stages of the synthesis make good use of symmetry.

Cook Synthesis of Triquinacene

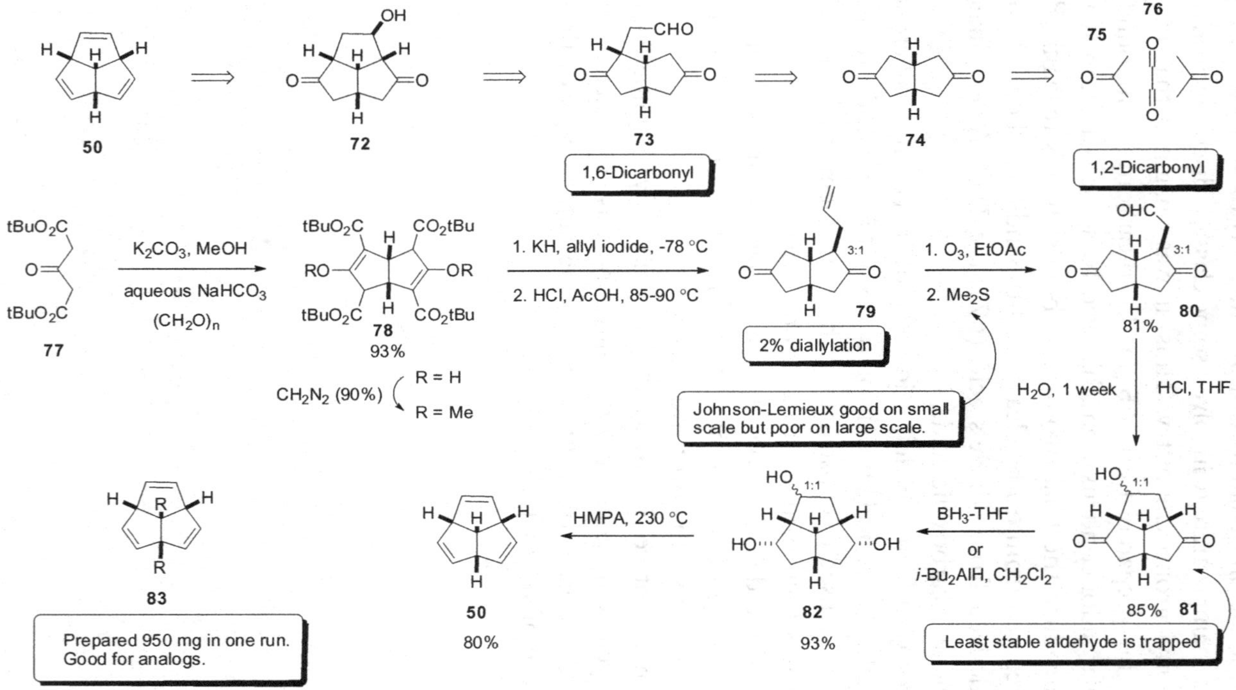

Gupta, A. K.; Lannoye, G. S.; Kubiak, G.; Schkeryantz, J.; Wehrli, S.; Cook, J. M. "General Approach to the Synthesis of Polyquinenes. 8. Syntheses of Triquinacene, 1,10-Dimethyltriquinacene, and 1,10-cyclohexanotriquinacene" *J. Am. Chem. Soc.* **1989**, *111*, 2169-2179. Jim Cook was DJH's teaching assistant for an undergraduate laboratory course at the University of Michigan. Thanks Jim!

Unnatural Products-13

Fifteen years after the Delongschamps synthesis, Jim Cook's group (University of Wisconsin-Milwaukee) described a versatile route to triquinacene. The synthesis revolved around a three-component reaction known as the Weiss reaction. The "plan" was to prepare **50** from tricyclic intermediate **72** which was to be prepared from 1,6-dicarbonyl compound **73**. The precursor of **73** was to be diketone **74**, to be prepared in turn from $(CHO)_2$ (**76**) and acetone (**75**) or a derivative thereof. It is notable that the union of **75** with **76** involves the construction of only 1,3- and 1,5-difunctional relationships. It is the 1,2-difunctional relationship present in glyoxal (**76**) that facilitates construction of the two five-membered rings. The conversion of **74** to **73** would necessarily involve construction of a 1,4-difunctional relationship. Finally, it is clear that stereochemistry must be considered when introducing the acetaldehyde unit going from **74** to **73**. Only the endo isomer (shown) is disposed properly for the intramolecular aldol condensation (**73** → **72**). Note that this was also an issue in the Delongschamps synthesis, who handled the problem by opening a ring constrained with the required stereochemistry (**66** → **65**).

In the laboratory, acetone equivalent **77** was treated with glyoxal and base to afford **78** (R=H), followed by conversion to the *bis*-enol ether (R=Me) using diazomethane. Compound **78** (R=Me) is a vinylogous malonate and thus, reaction with base and an alkylating agent (allyl iodide), followed by ester hydrolysis and decarboxylation, gave **79** as a 3:1 mixture of diastereomers. The enolate derived from the monoalkylated intermediate is presumably more hindered than the enolate derived from **78** and thus, polyalkylation of **78** is not a major problem. Introduction of the acetaldehyde residue was completed by ozonolysis of **79** to provide **80**, once again as a mixture of isomers. Treatment of **80** with acid promoted the desired aldol condensation to give **81** in 85% yield. Clearly the *exo*-isomer of **80** can be epimerized to the *endo*-isomer needed for the aldol. Thus, incorporation of the "extra carbonyl group", compared to the Delongschamps synthesis, removed the need to establish stereochemistry at the sidechain stereogenic center. The rest of the synthesis is straightforward. Reduction of the ketones provided **82** and a triple-dehydration completed the synthesis of triquinacene (**50**).

This approach is excellent from the standpoint of analog synthesis. For example using 2,3-butanedione as the 1,2-dicarbonyl compound in the Weiss reaction ultimately provided **83** where R=Me.

Paquette's Remarkable Synthesis

Wyvratt, M.; Paquette, L. A. "Domino Diels Alder Reactions. II. A Four-Step Conversion of Cyclopentadienide to Triquinacene" *Tetrahedron Lett.* **1974**, 2433-2436

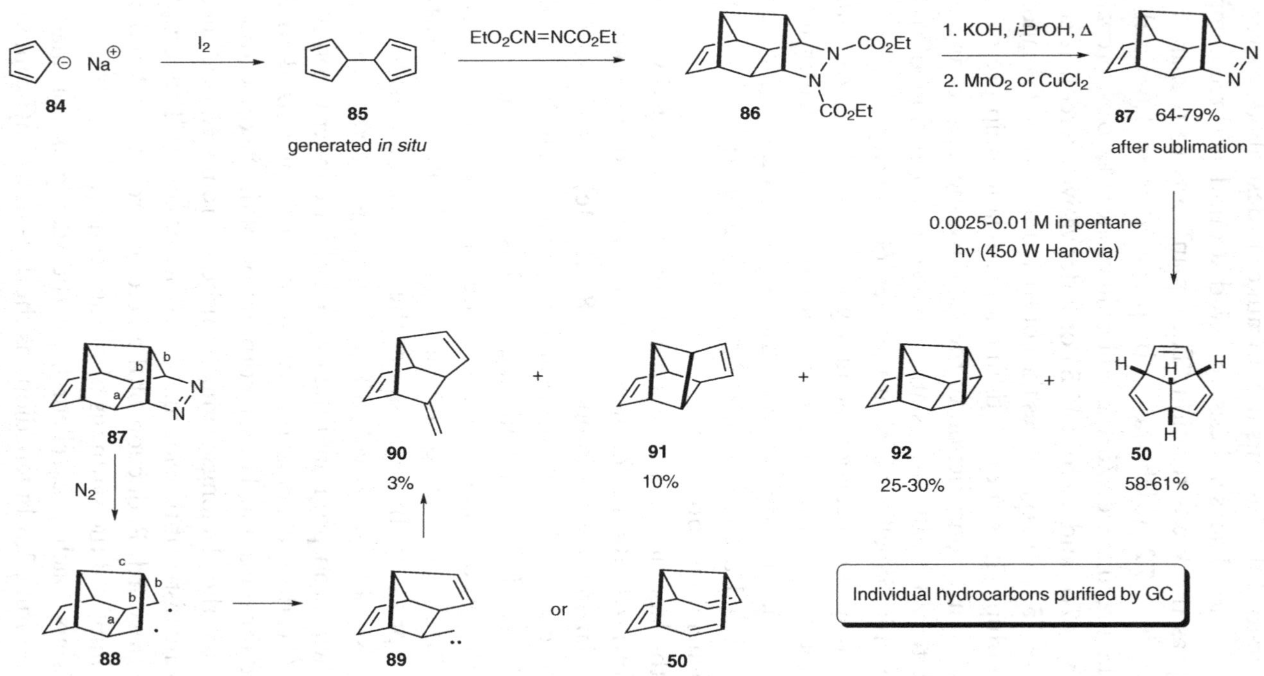

Unnatural Products-14

Unnatural Products-14

The final synthesis I want to present is a short and imaginative one described by Leo Paquette. Let's jump right in. The synthesis begins with oxidative coupling of sodium cyclopentadienide (**84**) to afford **85**. This is a highly reactive compound that undergoes reaction with diethyl azodicarboxylate to provide **86** via two sequential Diels-Alder reactions. The initial cycloaddition occurs largely from the face of the cyclopentadiene opposite the 1,3-cyclopentadien-5-yl group (recall the Corey approach to prostaglandins), setting up the second Diels-Alder reaction. Hydrolysis of the carbamates and oxidation of the intermediate hydrazine provided azo compound **87**.

Examination of the azo compound (**87**) reveals that homolysis of the C–N bonds (with loss of dinitrogen) would provide a diradical (**88**) that would, upon homolysis of the a-sigma bond, result in formation of triquinacene (**50**). In the lab, irradiation of **87** accomplished this transformation in good yield. A number of other products reveal that this mechanistic pathway to triquinacene is reasonable. For example, formation of minor product **92** can be rationalized by intramolecular coupling of biradical **88**. Homolysis of one of the b-sigma bonds in **88** would provide carbene **89** which could lead to **90** (C-H insertion). A 1,2-shift of the c-sigma bond gives **91**. This synthesis is remarkable in its brevity and its use of symmetry.

I hope that this chapter has served as a pleasant diversion from natural products synthesis, and has also helped (to a degree) place the discussion of difunctional relationships in a practical context.

References

1. Hamon, D. P. G.; Young, R. N. "The Analytical Approach to Synthesis: Twistane" *Aust. J. Chem.* **1976**, *29*, 145–161.
2. Baldwin, J. E. "Rules for Ring Closure" *J. Chem. Soc., Chem. Commun.* **1976**, 734–736. Baldwin, J. E. "Approach Vector Analysis: A Stereochemical Approach to Reactivity" *J. Chem. Soc., Chem. Commun.* **1976**, 738–741.
3. Huffman, J. W.; Charles, J. T. "The Metal-Ammonia Reduction of Ketones" *J. Am. Chem. Soc.* **1968**, *90*, 6486–6492.
4. Corey, E. J.; Winter, R. A. E. "A New, Stereospecific Olefin Synthesis from 1,2-Diols" *J. Am. Chem. Soc.* **1963**, *85*, 2677–2678.
5. Bruck, P.; Thompson, D.; Winstein, S. "Dechlorination of Isodrin and Related Compounds" *Chem. Ind.* (London) **1960**, 405–406.
6. Thiele, J. "Ueber Abkömmlinge des Cyclopentadiens" *Chem Ber.* **1901**, *34*, 68–71.
7. Matsui, T. "On the Mechanism of the Photolysis of Strained Cycloalkanones" *Tetrahedron Lett.* **1967**, *9*, 3761–3765.
8. Gupta, A. K.; Fu, X.; Snyder, J. P.; Cook, J. M. "General Approach for the Synthesis of Polyquinenes via the Weiss Reaction" *Tetrahedron* **1991**, *47*, 3665–3710.

Problems

1. Compound **20** is a 1,5-difunctional compound. Therefore one can design a synthesis that proceeds via intermediates that only contain odd difunctional relationships. Propose such a synthesis and indicate potential problems with your approach. If you can't get started, have a look at the following reference: Ballester, P.; Costa, A.; Raso, A. G.; Gomez-Solivellas, A.; Mestres, R. "Dienediolates from Unsaturated Carboxylic Acids. Michael Addition of Dilithium 1,3-Butadiene-1,1-diolate (from Crotonic Acid) to Unsaturated Ketones" *J. Chem. Soc., Perkin 1: Organic and Bio-Organic Chemistry* **1988**, 1711–1717. (Unnatural Products-2)
2. Propose another synthesis of **1** via the Path A disconnection and a 1,5-difunctional intermediate that differs from **20** in regard to placement of the functional groups. (Unnatural Products-2)
3. Suggest a 1,6-difunctional intermediate, and a reaction of your proposed intermediate, that might be used to construct the bond disconnected in Path B. (Unnatural Products-3)
4. Suggest an alternative bicyclo[3.3.1]nonane derivative that might be used to "connect" the carbons marked by asterisks in structure **6**. Suggest tactics that will transform your proposed intermediate into twistane. (Unnatural Products-7)
5. Provide a mechanism for the conversion of **33** to **34**. (Unnatural Products-8)
6. Propose an alternative synthesis of **7** that passes through intermediates with "odd" difunctional relationships. (Unnatural Products-8)
7. Explain the stereoselectivity associated with the conversion of **39** → **40**. (Unnatural Products-8)
8. Review olefin metathesis and the Ramburg-Bachlund reaction (see examples below). Suggest alternative tactics for converting **9** → **49** based on these reactions. (Unnatural Products-9)

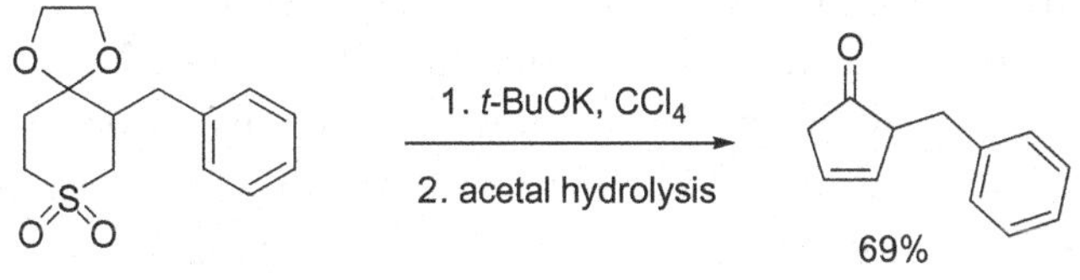

Matsuyama, H.; Miyazawa, Y.; Takei, Y. "Tetrahydrothiopyran-4-one. A Useful 5-Carbon Synthon for the Synthesis of 3-Cyclopentenones"

Chem. Lett. **1984**, 833–836. For another variation of the Ramburg-Backlund reaction see Matusuyama, H.; Miyazawa, Y.; Kobayashi, M. "Regioselective Alkylation and the Ramberg-Baecklund Type Reaction of α-(*p*-tolylsulfonyl)thiane *S,S*-Dioxide. A New Route to the Synthesis of 3-Alkyl-3-Cyclopentenones" *Chem. Lett.* **1986**, 433–436.

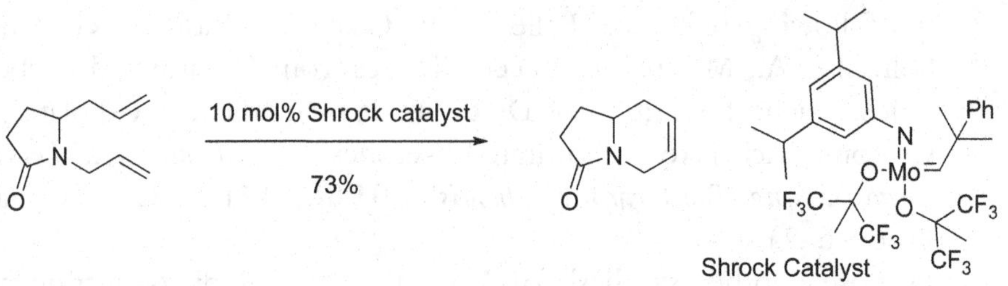

Martin, S. F.; Chen, H-J.; Courtney, A. K.; Liao, Y.; Patzel, M.; Ramser, M.; Wagman, A. S. "Ring-Closing Olefin Metathesis for the Synthesis of Fused Nitrogen Heterocycles" *Tetrahedron* **1996**, *52*, 7251–7264. For a review and other catalysts see Grubbs, R. H.; Chang, S. "Recent Advances in Olefin Metathesis and its Application to Organic Synthesis" *Tetrahedron* **1998**, *54*, 4413–4450.

9. Outline a synthesis of **56** from 1,3-cyclopentadiene and norbornadiene. (Unnatural Products-10)
10. Suggest a mechanism for the conversion of **54** → **59**. (Unnatural Products-10)
11. Provide the mechanism of the reactions used to convert **59** → **60**. (Unnatural Products-11)
12. Provide the structure of the products from each reaction along the path from **60** to **62**. (Unnatural Products-11)
13. Suggest a mechanism for the Eschweiler-Clarke methylation (**63** → **64**). (Unnatural Products-11)
14. Illustrate how 1,3-cyclopentadiene 2-carboxylic acid behaves as a "2-hydroxy-1,3-cyclopentadiene equivalent" in the synthesis of **67**. Suggest intermediates along the path from **67** → **66** under the conditions cited in Unnatural Products-12? (Unnatural Products-12)
15. Suggest a mechanism for the conversion of **66** → **65**. Suggest a reason for the observed regioselectivity (reaction of only one of the ketones) based on your mechanism. (Unnatural Products-12)
16. Provide a mechanism for the conversion of **77** to **78** (R=H). (Unnatural Products-13)

17. Provide the structure of the triquinacene derivative that could be prepared from cyclohexan-1,2-dione using this strategy. (Unnatural Products-13)
18. Predict the triquinacene one might prepare from PhC(=O)CHO using this strategy. (Unnatural Products-13)
19. Suggest a mechanism for the formation of **85** from **84**. Provide the structure of the intermediate in the conversion of **85** → **86**. Provide the structure of an "unproductive" cycloadduct obtained from the reaction of **85** with diethyl azodicarboxylate. (Unnatural Products-14)
20. Suggest a mechanism for the formation of **92** from photolysis of **87**. (Unnatural Products-14)

17. Provide the structure of the ... derivative ... from ... [illegible]
problems 18)

18. ... the ... and ... prepare ... (CO) ...
... (Chapter Problems 18)

... a mechanism ... structure ... 98. Provide the structure ... velocity ... of 58 ...

20. ... the formation of ... 3 ...
... [illegible]

CHAPTER 8

Alkaloids — Difunctional Relationships and the Importance of the Mannich Reaction

Alkaloids: Importance of the Mannich Reaction

Corey, E. J.; Balanson, R. D. "Total Synthesis of *dl*-Porantherine" *J. Am. Chem. Soc.* **1974**, *96*, 6516.

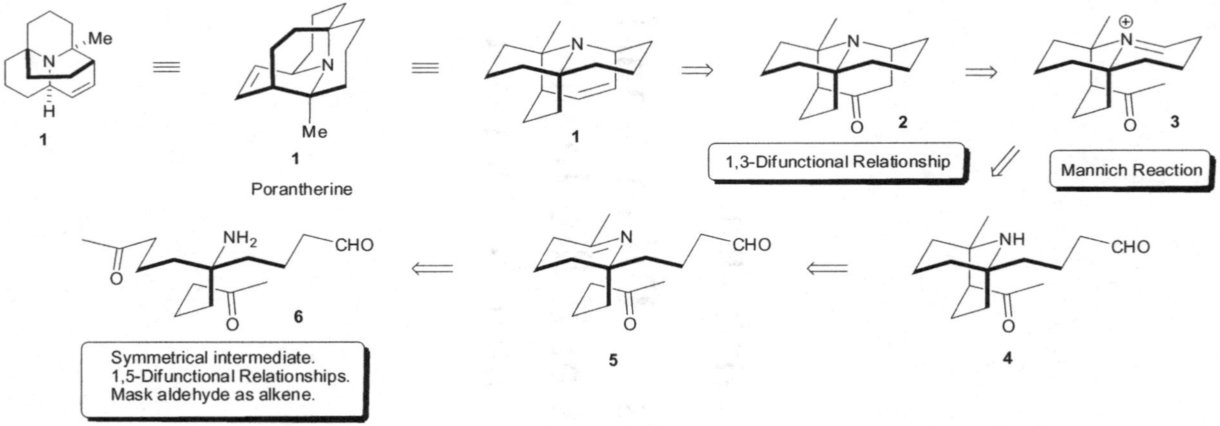

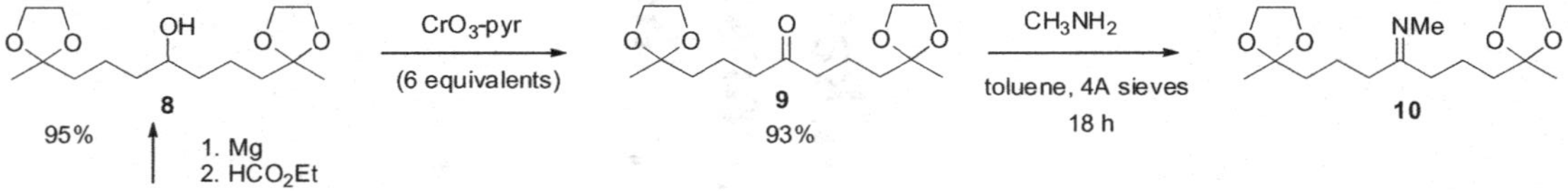

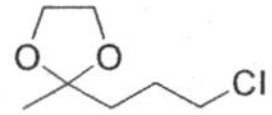

7

In this case, symmetry and the availability of a 1,4-difunctional starting material makes the use of a γ-alkylation equivalent workable.

Alkaloids-1

Alkaloids-1

In this chapter we will focus on alkaloid total synthesis. We will start with several syntheses that rely on the addition of an iminium ion (or imine) to an enol (or enolate). This process is commonly refered to as the Mannich reaction.[1] It is really the nitrogen counterpart of the aldol condensation. The Mannich reaction is important not only in the laboratory synthesis of many alkaloids, but also in the biosynthesis of many alkaloids. The fact that nature assembles alkaloids using this process is clearly what renders it useful in developing synthetic strategies for alkaloid synthesis.

We will start with porantherine (**1**). This alkaloid is a difunctional molecule. A simple "last step" in any projected synthesis of porantherine would be introduction of the olefin from a ketone. There are two ketone precursors that could be considered. Ketone **2** is one of these. It is also a difunctional compound with a 1,3-difunctional relationship between the ketone and amine functional groups. Note that in the "other ketone", the ketone and amine functional groups would have a vicinal or 1,2-difunctional relationship. One can imagine **2** being available from **4** (via **3**) via an intramolecular Mannich reaction. Compound **4** might be prepared from **5** via another intramolecular Mannich reaction. And **5** would come from **6** via intramolecular imine formation. Going in the forward direction there would be several points to address: (1) Regioselective imine formation going from **6** → **5** would be an issue. This could be addressed by masking the aldehyde in some manner. (2) The stereochemistry of the acetyl group in **4** might be an issue in moving forward to **2**. For example, the epimer of **2** would not be disposed to undergo the intramolecular Mannich reaction. It is notable that **6** contains only 1,5-difunctional relationships and thus, should be available via traditional carbonyl group chemistry. It is also notable that this plan makes good use of symmetry.

This plan has been executed in two somewhat related manners by the Corey and Stevens groups. The Corey-Balanson synthesis began with the synthesis of symmetrical ketone **9** from 1,4-difunctional chloro-ketal **7**. Imine formation provided **10**.

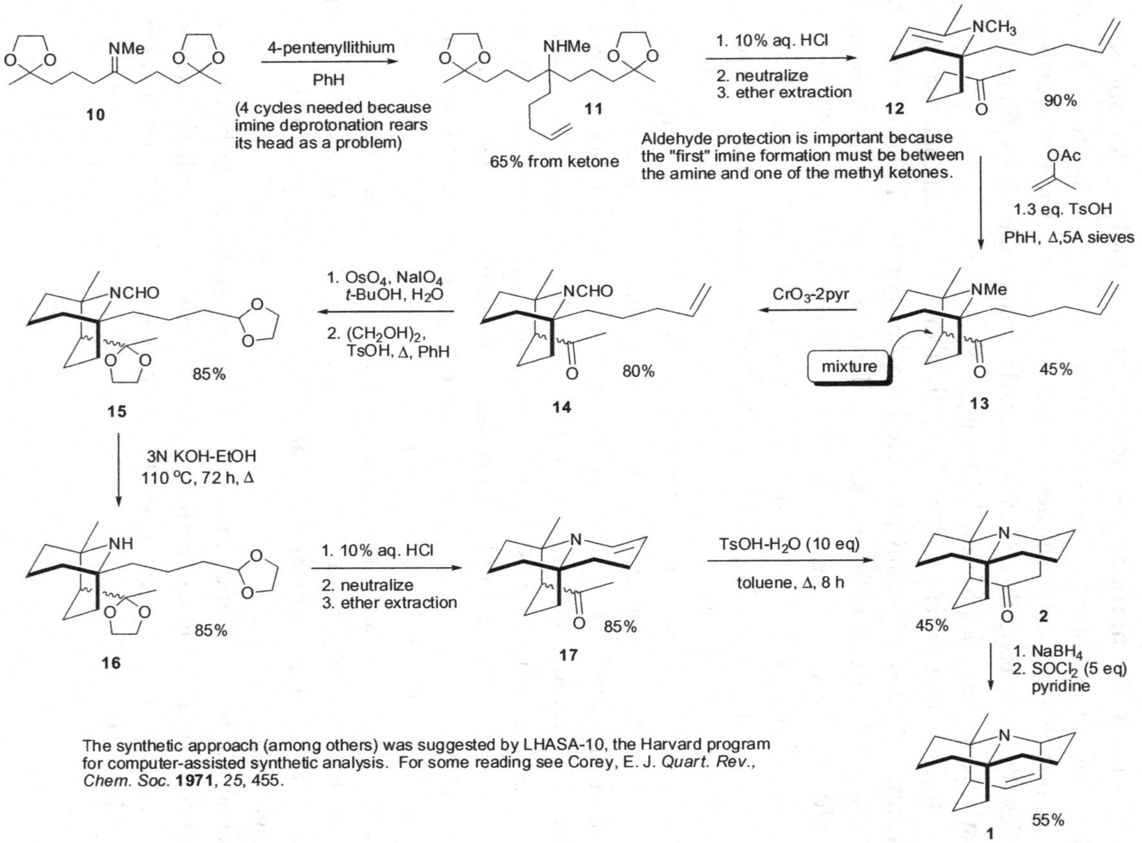

The synthetic approach (among others) was suggested by LHASA-10, the Harvard program for computer-assisted synthetic analysis. For some reading see Corey, E. J. *Quart. Rev., Chem. Soc.* **1971**, *25*, 455.

Alkaloids-2

Reaction of **10** with 4-pentenyllithium gave **11**. Enolization of **10** competed with imine addition and thus, "recycling" of this material was needed to achieve a respectable overall yield of **11** from ketone **9**. Acid hydrolysis of **11** furnished enamine **12** after neutralization. An acid-mediated Mannich reaction converted **12** to a mixture of epimeric ketones **13**. Oxidation of the *N*-methyl group provided formamide **14**.

The next stage of the synthesis called for conversion of the terminal olefin to an aldehyde. This was accomplished using a Johnson-Lemieux oxidation. The resulting ketoaldehyde was then converted to *bis*-ketal **15**. Basic hydrolysis of the formamide provided **16** and hydrolysis of the acetals afforded a mixture of diastereomeric enaminoketones **17**. An acid-promoted intramolecular Mannich reaction gave **2**. Presumably epimerization of the ketone allowed both diastereomers of **17** to follow the path to **2**. Reduction of ketone **2** with sodium borohydride provided an alcohol, which underwent formal dehydration upon treatment with thionyl chloride and pyridine, to give porantherine (**1**) as a racemic mixture.

Overall this synthesis is quite direct. It is notable that the approach was suggested by a computer-assisted synthetic analysis developed at Harvard. It is also notable that the choice of 1,3-difunctional aminoketone **2** as a late intermediate may have been responsible for recognition of this "Mannich reaction" strategy.

Related Approach to Porantherine

Porantherine is a Euphorbaceae alkaloid isolated from the woody shrub *Poranthera corymbosa* Brogn. It is known to poison cattle wherever it grows. It came to light for poisoning cattle in New South Wales and Queensland, Australia. Some other members of this family of alkaloids are shown below.

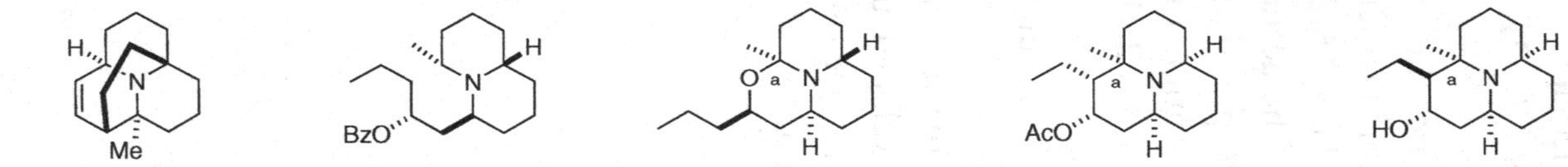

Porantherine **1** — Porantherilidine, Bz = benzoyl, **18** — Porantheridine **19** — Porantheriline **20** — Porantherícine **21**

Ryckman, D. M.; Stevens, R. V. "Stereorational Total Synthesis of *dl*-Porantherine" *J. Am. Chem. Soc.* **1987**, *109*, 4940-4948. The critical analysis is contained in Stevens' synthesis of coccinelline and precoccinelline: Stevens, R. V.; Lee, A. W. M. *J. Am. Chem. Soc.* **1979**, *101*, 7032.

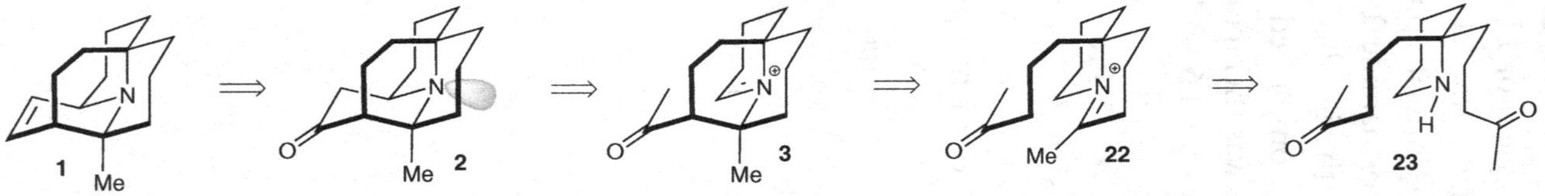

Stevens' notion was that enol (or enolate) addition to an iminium ion (or imine) occurs such that the developing carbon-carbon bond and nitrogen lone pair have an anti-periplanar relationship (a stereoelectronic arguement). Stevens' approach is a variation of the Corey-Balanson synthesis and it takes advantage symmetry.

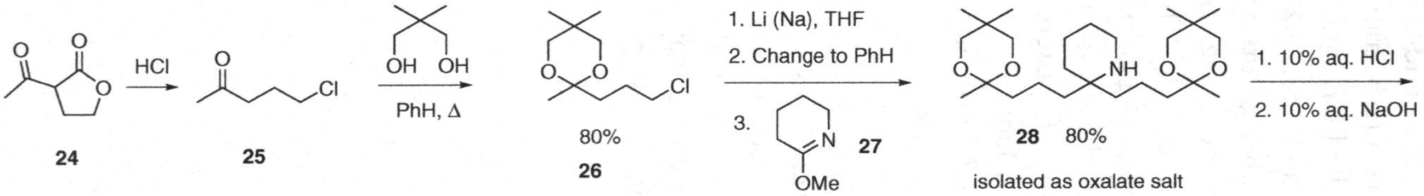

Alkaloids-3

Porantherine is one member of a family of structurally (and presumably biosynthetically) related alkaloids shown here. One might imagine that a late step in the synthesis (or biosynthesis) of alkaloids **19–21** could also involve capture of an iminium ion by an appropriate nucleophile (to construct the bonds marked "a"). Even porantherilidine (**18**) contains the 1,3-difunctional relationship needed to develop a synthetic strategy that takes advantage of the Mannich reaction.

A related approach to porantherine was reported by the Stevens group about 10 years after publication of the Corey synthesis. The Stevens synthesis emphasized the notion that the stereoelectronic requirement for addition of a nucleophile to an imine (or iminium ion) was development of an *anti*-periplanar relationship between the nucleophile and the developing lone pair on nitrogen.[2,3] This notion also leads to the disconnection of **2** → **3** → **22**, and iminium ion **22** was to be prepared by oxidation of piperidine **23**. Thus, the major difference between the Stevens and Corey approaches is the choice of a surrogate for the aldehyde required for generation of iminum ion **3**. In the forward direction, the lithium reagent derived from **26** was treated with iminoether **27** to provide **28**. Hydroysis of **28** gave enamine **29**.

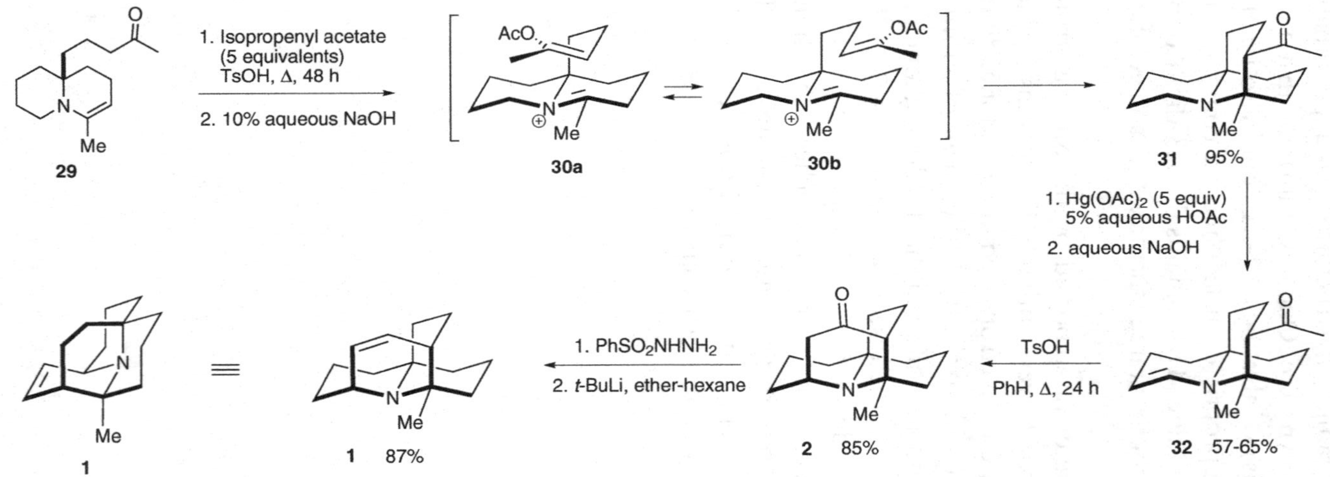

David, M.; Dhimane, H.; Vanucci-Bacque, C.; Lhommet, G. "Efficient Total Synthesis of Enantiopure (-)-Porantheridine" *J. Org. Chem.* **1999**, *64*, 8402-8405.

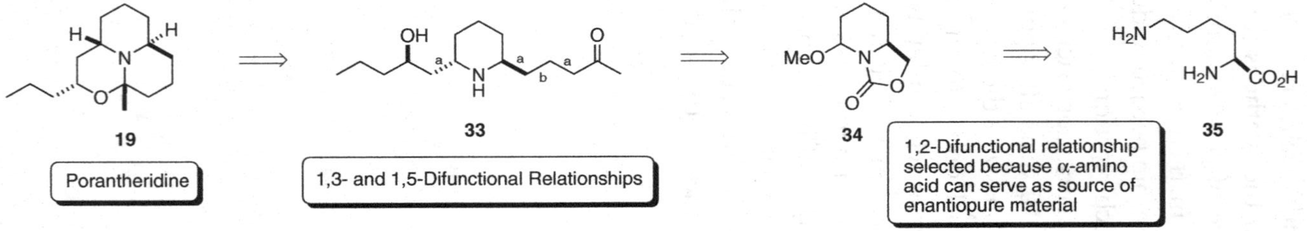

Alkaloids-4

Alkaloids-4

As in the Corey synthesis, an intramolecular Mannich reaction converted **29** to **31**. Mercuric acetate oxidation of **31** to the corresponding iminium ion, followed by basification, gave enaminoketone **32**. This intermediate does not have the proper stereochemistry for an intramolecular Mannich reaction, but under acidic conditions, epimerization is followed by cyclization to provide **2**. The synthesis of porantherine (**1**) was completed using a Bamford-Stevens reaction to introduce the olefin.[4]

Let's briefly look at a synthesis of porantheridine (**19**) reported by the Lhommet group. Disconnection of the *N,O*-acetal provided **33** as a key intermediate. The notion was that **33** would "collapse" to provide the target alkaloid. There are stereochemical issues associated with this process that we will address in time. Compound **33** has only 1,3- and 1,5-difunctional relationships and thus, should be straightforward to prepare via intermediates that also contain such relationships. For example, the "a" bonds in **33** could all be constructed using carbonyl addition or Mannich-type reactions. Indeed this was the strategy followed for introduction of the 2-hydroxypentyl sidechain. Disconnection of the "b" bond in structure **33** generated a 1,2-difunctional compound. This disconnection did, however, have an advantage. It suggested lysine (**35**) as an enantiopure starting material.

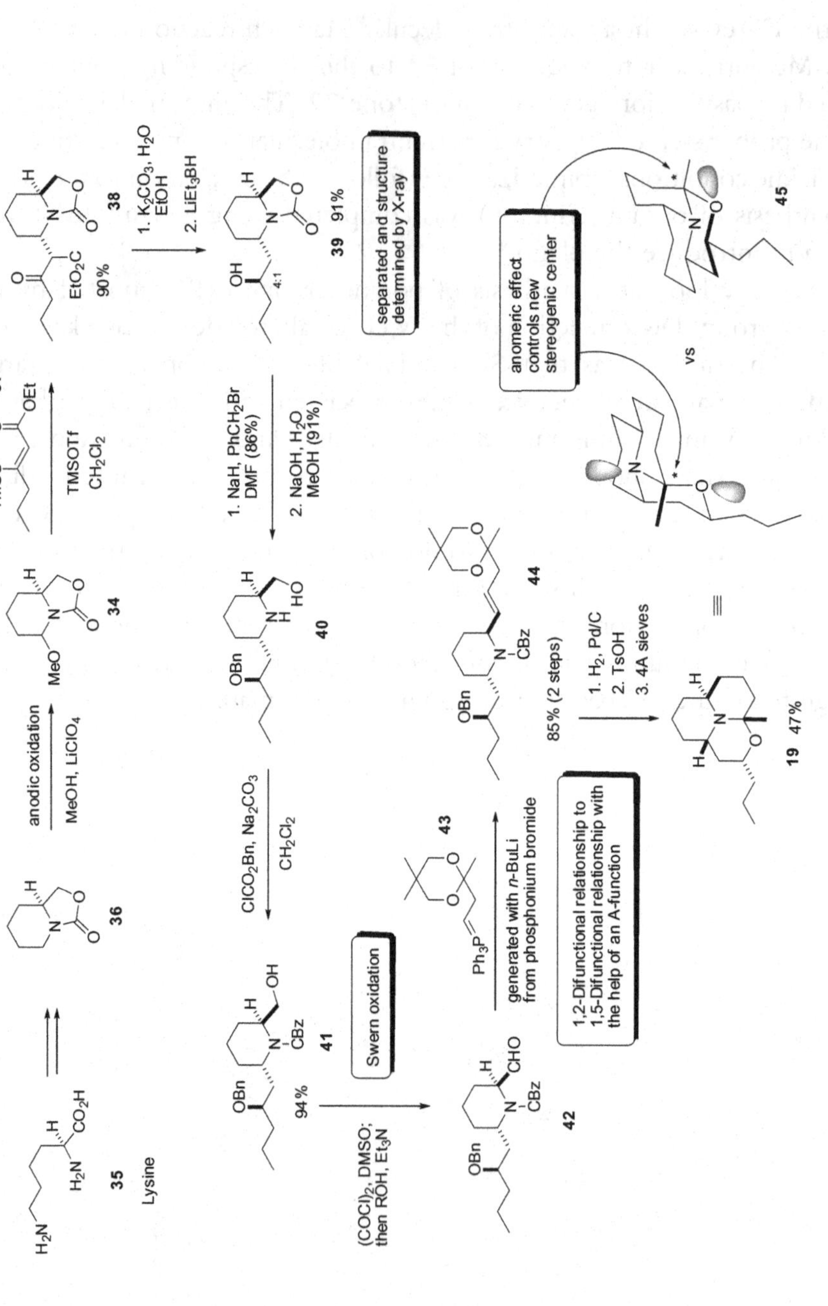

35
Lysine
36
anodic oxidation
MeOH, LiClO4
34
TMSOTf
CH2Cl2
37
90%
38
1. K2CO3, H2O EtOH
2. LiEt3BH
4:1
39 91%
separated and structure determined by X-ray
1. NaH, PhCH2Br DMF (86%)
2. NaOH, H2O MeOH (91%)
40
ClCO2Bn, Na2CO3
CH2Cl2
41
94%
Swern oxidation
(COCl)2, DMSO; then ROH, Et3N
42
43
generated with n-BuLi from phosphonium bromide
1,2-Difunctional relationship to 1,5-Difunctional relationship with the help of an A-function
44
85% (2 steps)
1. H2, Pd/C
2. TsOH
3. 4A sieves
19 47%
anomeric effect controls new stereogenic center
vs
45

Alkaloids-5

Alkaloids-5

Lysine (**35**) was converted to lactam **36**. Anodic oxidation of the lactam provided *N,O*-acetal **34**, chemistry pioneered by the Shono group in Japan.[5] An acid-promoted intermolecular Mannich reaction between **34** and **37** provided *β*-ketoester **38**. The stereochemical course of this reaction, at the new stereogenic center in the piperidine ring, follows from the stereoelectronic analysis suggest by Stevens for iminium ion addition reactions (Alkaloids-3). Hydrolysis and decarboxylation of the *β*-ketoester, followed by reduction of the resulting ketone, gave **39** as a separable mixture of diastereomers. The alcohol was protected as a benzyl ether and the cyclic carbamate was hydrolyzed to give amino alcohol **40**. The amine was protected with a Cbz group (**40** → **41**) and the alcohol was then oxidized to aldehyde **42**. Wittig olefination with *β*-acyl anion equivalent **43**, provided **44**. Catalytic hydrogenation of the olefin followed by hydrolysis of the acetal in the presence of a "water sponge" gave porantheridine (**19**) with excellent control of stereochemistry. From the standpoint of steric effects, **19** and its *N,O*-acetal diastereomer (**45**) look similar. But **19** enjoys double anomeric stabilization [nitrogen lone pair *anti*-periplanar C–O bond and oxygen lone pair *anti*-periplanar to C–N bond], whereas **45** enjoys only "single" anomeric stabilization. It is this feature that most likely controls stereochemistry in the final step of the synthesis.[6]

Biomimetic Synthesis Revisited

van Tamelen, E. E.; Foltz, R. L. "The Biogenetic-Type Synthesis of *dl*-Sparteine" *J. Am. Chem. Soc.* **1960**, *82*, 2400

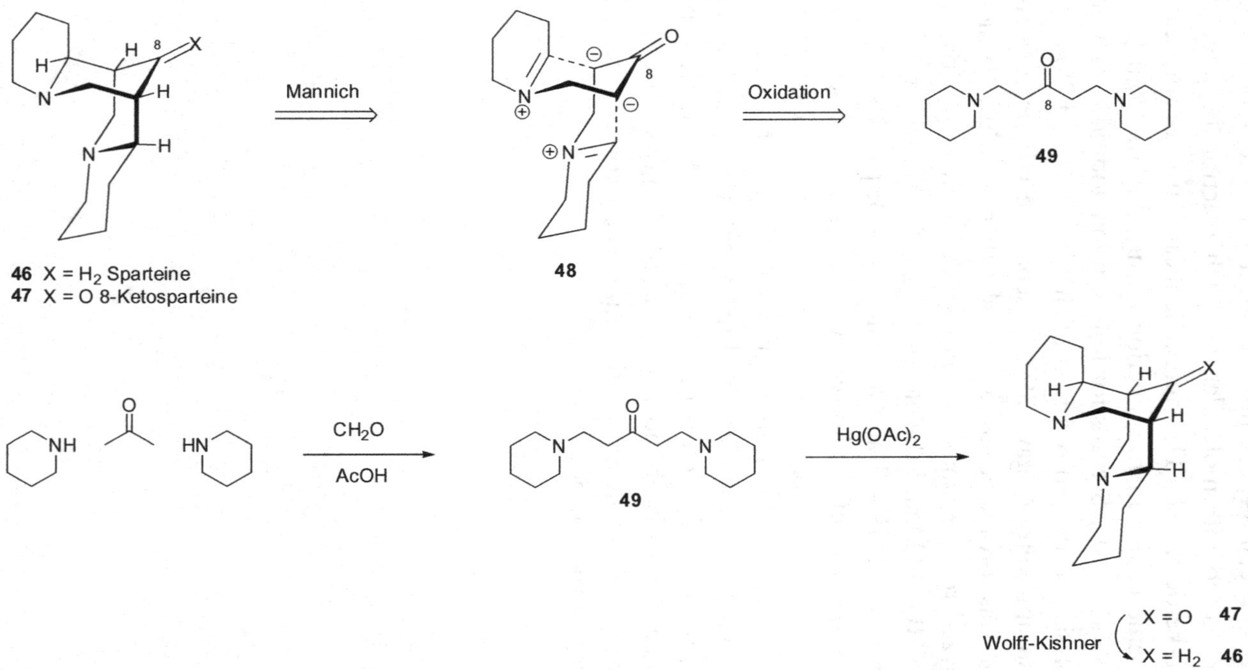

Alkaloids-6

Alkaloids-6

The notion of "biomimetic synthesis" was introduced in Chapter 2 within the context of polyolefin cyclizations. It was suggested early in this chapter that alkaloid syntheses that employ Mannich reactions may also incorporate (in part) a biomimetic strategy. The synthesis of sparteine shown here is an elegant example of such a strategy.

The idea was that sparteine (**46**) might arise from the double intramolecular Mannich reaction suggested by structure **48** (followed by reduction of the 8-keto group). A precursor of **48** might be **49**. In the laboratory, **49** was assembled from piperidine, acetone and formaldehyde via a double intermolecular Mannich reaction. Mercuric acetate oxidation of **49** gave 8-ketosparteine (**47**), presumably via intermediates resembling **48** (stepwise oxidation-Mannich events seem most likely).[7] A Wolff-Kishner reduction of **47** completed the synthesis of **46**. Yields were not given for this synthesis and thus, the efficiency of this route (in terms of yield) cannot be evaluated. Efficiency in terms of the number of steps cannot be denied.

First Total Synthesis of Luciduline

Luciduline is a member of the *Lycopodium* family of alkaloids. For a short review see: Inubushi, Y.; Harayama, T. "Total Synthesis of Lycopodium Alkaloids" *Heterocycles* **1981**, *15*, 611-635.

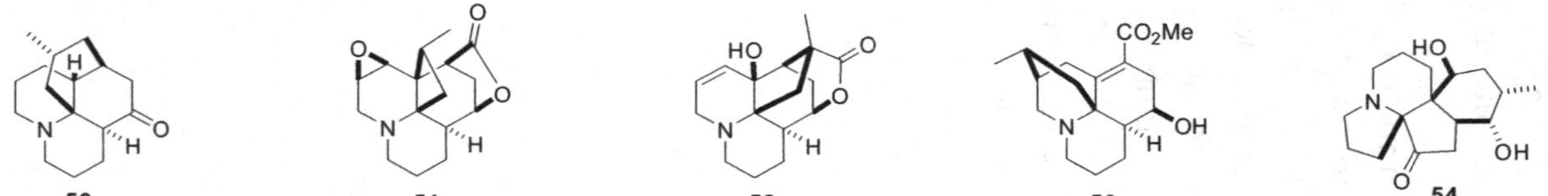

50	**51**	**52**	**53**	**54**
Lycopodine	Annotinine	Annotine	Annopodine	Serratinine

Scott, W. L.; Evans, W. A. "The Total Synthesis of dl-Luciduline" *J. Am. Chem. Soc.* **1972**, *94*, 4779.

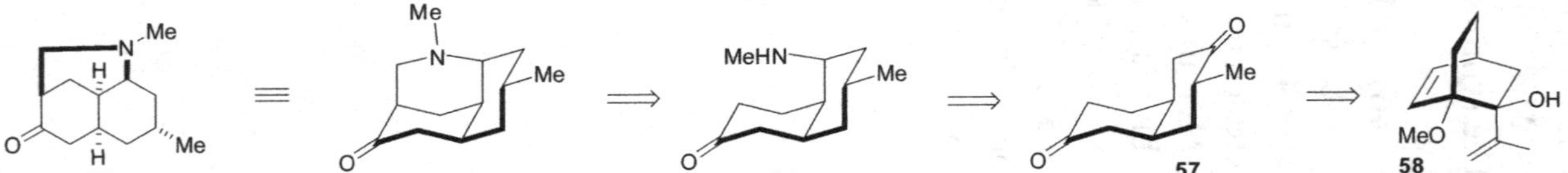

55 Luciduline; **55**; **56**; **57**; **58**

1,6-Dicarbonyl ⟹ Oxy-Cope

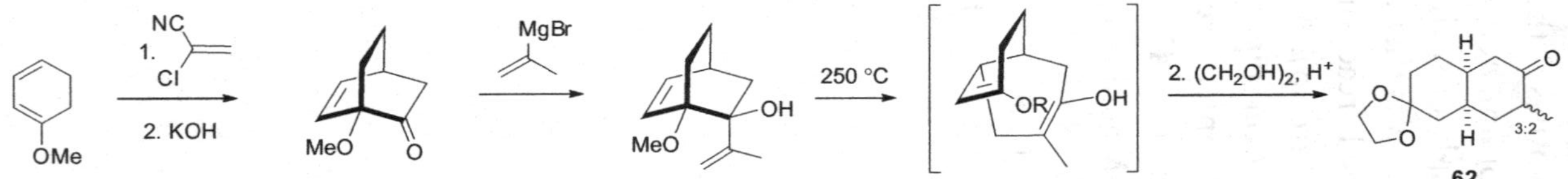

59 → **60** → **58** → [**61**, R = Me] → **62**, 65% from ketone

Alkaloids-7

Let's look at another alkaloid whose structure screams "Mannich reaction" for an endgame. Luciduline is one member of the *Lycopodium* family of alkaloids. This is a large family of natural products. A few structures are shown here (**50–54**). Luciduline (**55**) is a β-aminoketone. This is precisely the difunctional relationship that results from the Mannich reaction. Thus, a strategy that passes through **56** *en route* to **55** seems likely to succeed. Aminoketone **56** is a 1,5-difunctional compound and, in principle, should be available using "normal" carbonyl chemistry. We will see an example of this later, but the first synthesis of luciduline approached **56** from 1,6-difunctional intermediate **57**. One versatile method for the preparation of 1,6-dicarbonyl compounds is the oxy-Cope rearrangement. We saw this in the Evans synthesis of juvabione (Chapter 5) and indeed, this is the methodology used by Evans and Scott in their synthesis of luciduline (**58** $\rightarrow$ **57**).

The synthesis began with the preparation of rearrangement substrate **58**. A Diels-Alder reaction between α-chloroacrylonitrile (a ketene equivalent) and cyclohexadiene **59** gave **60** after "hydrolysis" of the intermediate cycloadduct (recall the Corey synthesis of prostaglandins). Reaction of ketone **60** with isopropenylmagnesium bromide provided **58**. Thermal rearrangement of **58** gave the ketone derived from tautomerization of **61**. Regioselective ketal formation provided **62**, the enol ether reacting faster that the ketone.

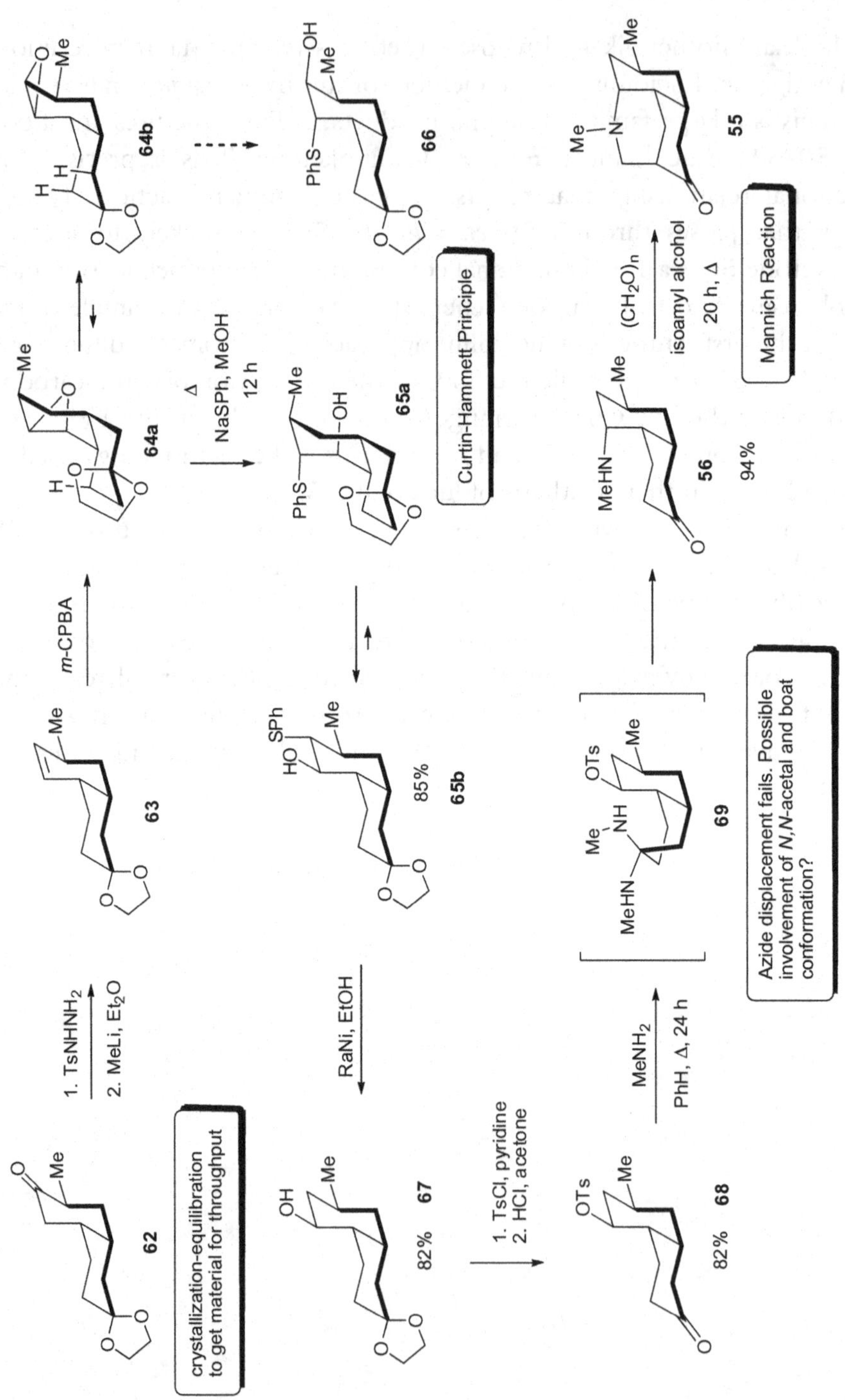
62
crystallization-equilibration to get material for throughput
1. TsNHNH2 2. MeLi, Et2O
63
m-CPBA
64a
64b
Δ NaSPh, MeOH 12 h
65a
65b
85%
66
Curtin-Hammett Principle
RaNi, EtOH
67
82%
1. TsCl, pyridine 2. HCl, acetone
68
82%
MeNH2 PhH, Δ, 24 h
69
Azide displacement fails. Possible involvement of N,N-acetal and boat conformation?
56
94%
(CH2O)n isoamyl alcohol 20 h, Δ
55
Mannich Reaction

Alkaloids-8

Alkaloids-8

The diastereomers of **62** were separated by crystallization, and equilibration of the mother liquors provided more material. It was next required to move the ketone functionality to the adjacent carbon to establish the required 1,5-difunctional relationship. Whereas how this was done is very interesting, this adjustment is sort of the bottle-neck in the synthesis.

A Bamford-Stevens reaction was used to convert **62** to olefin **63**. Epoxidation of **63** occurred from the convex face of the *cis*-octalin to provide **64**. The epoxide was then opened by sodium thiophenoxide to provide **65** and Raney-Ni was used to reduce the C–S bond and provide **67**. The epoxide opening is interesting and illustrates how mechanistic thinking must be an integral part of synthesis planning. It is well-known that epoxides are opened by nucleophiles to initially provide an *anti*-relationship between the nucleophile and the developing alcohol. In cyclohexene oxide this means that products are born with a *trans*-diaxial relationship (the Furst-Plattner rule).[8] Epoxide **64** is likely to exist as an equilibrating mixture of conformations **64a** and **64b**. One might imagine that **64b** would predominate for steric reasons (note the axial ketal oxygen-CH_2 interaction in **64a**). *Anti*-periplanar opening of **64b**, however, would provide **66** and this was not observed. On the other hand, *anti*-periplanar opening of **64a** would provide the observed product, born as **65a** but sure to prefer conformation **65b** once formed. An inspection of models shows that attack of thiophenoxide on **64b** results in an interaction of the nucleophile with the two axial C–H bonds indicated in the structure. On the other hand, only one axial C–H provides cause for concern as thiophenoxide approaches **64a**. Thus, the rate of reaction of the less stable conformation appears to be faster than the rate of reaction of the more stable conformation. This is a lovely example of the Curtin-Hammett principle in operation.[9]

Recall that the goal was to arrive at aminoketone **56**. The projected Mannich reaction requires that the amino group be on the concave face surface of the *cis*-decalin ring system. Notice that the alcohol stereochemistry in **67** is set such that an S_N2 reaction at the carbinol center would establish the required stereochemistry in **56** [We will see another approach to establishing this stereochemistry shortly]. Tosylate formation followed by acetal hydrolysis provided **68**, but treatment of this material with azide failed to give any of the desired S_N2 product. Treatment of **68** with methylamine, however, gave **56** in excellent yield. Given the results with azide, it is probable that this displacement occurs with intramolecular delivery the nucleophile via involvement of an *N,N*-acetal (**69**). The final Mannich reaction proceeded as anticipated to provide luciduline (**55**).

Oppolzer's Nitrone Cycloaddition Route to Luciduline

Oppolzer, W.; Petrzilka, M. "A Total Synthesis of *dl*-Luciduline by a Regioselective Intramolecular Addition of an N-Alkenylnitrone" *J. Am. Chem. Soc.* **1976**, *98*, 6722-6723. Oppolzer, W.; Petrzilka, M. "An Enantioselective Total Synthesis of Natural (+)-Luciduline" *Helv. Chim. Acta* **1978**, *61*, 2755.

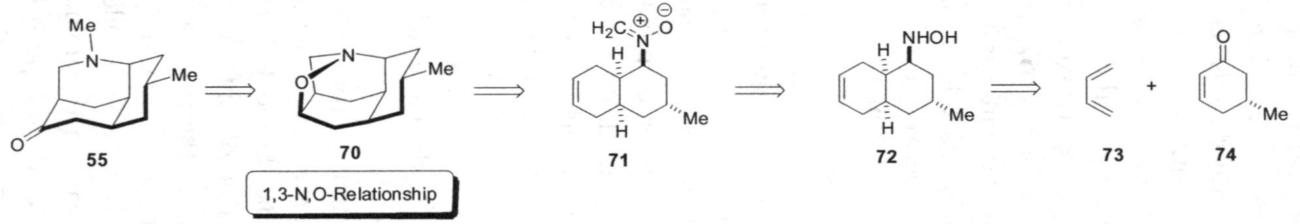

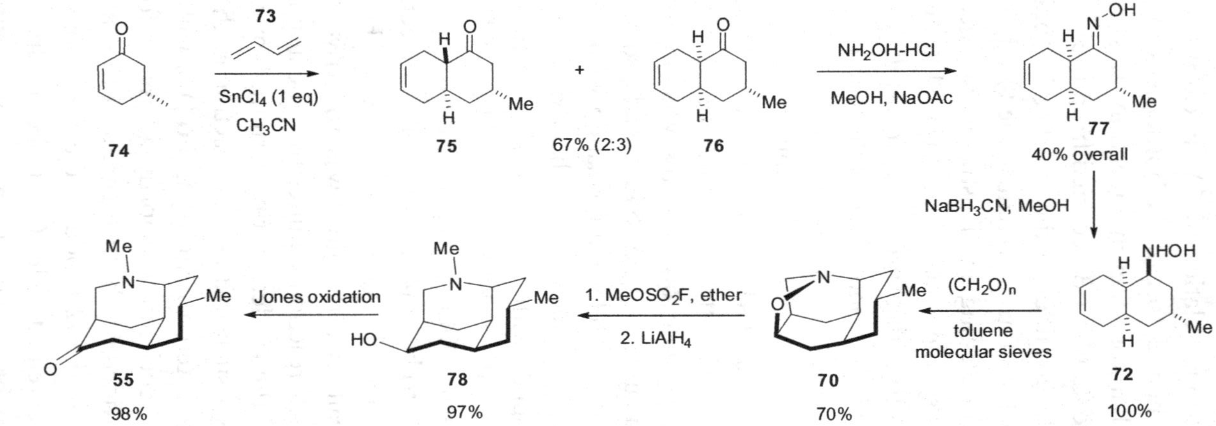

Alkaloids-9

Alkaloids-9

I hope it is clear by now that 1,3-*N,O* relationships appear in many alkaloids because of the manner in which they are made by nature. Since the Mannich reaction generates a 1,3-*N,O* relationship, it is a very useful reaction for alkaloid synthesis. Taking this a step further, it is probable that *any* reaction that establishes a 1,3-*N,O* relationship may find broad use in alkaloid synthesis. If we return to the pyrrolizidine alkaloid syntheses we briefly visited in Chapter 4, it is apparent that nitrone-alkene cycloadditions represent one such reaction. Let's see how this was applied to a synthesis of luciduline by the Oppolzer group.

The plan was to prepare **55** from **70**, which would be derived from an intramolecular nitrone cycloaddition of **71**. *Cis*-octalin **71** was to be prepared from hydroxylamine **72** and formaldehyde, a standard method for nitrone preparation. The cycloaddition required that the hydroxylamine reside on the concave face of the *cis*-octalin. It was felt that this stereochemistry could be established by hydride reduction of an oxime, which would be derived in turn from the corresponding ketone. A cycloaddition between 1,3-butadiene (**73**) and cyclohexenone **74** was to serve as the starting point of the synthesis.

The cycloaddition proceeded smoothly, when promoted by acid, to provide ketones **75** and **76**. Whereas **76** was surely the kinetic product, acid promoted enolization-isomerization provide **75** as well. Treatment of the mixture with hydroxylamine gave the desired oxime (**77**) and Borch reduction from the convex face provided **72**.[10] The rest of the synthesis proceeded as expected. The intramolecular nitrone cycloaddition gave **70**. Methylation of the amine and reduction of the weak N–O bond provided **78**. Jones oxidation (Cr^{VI} in acid) of **78** completed the synthesis of luciduline (**55**). When a single enantiomer of **74** was used, a single enantiomer of luciduline was obtained.

Reserpine

Reserpine was isolated in 1952 from the Indian snake-root, *Rauwolfia serpentina* Benth. The structure was determined by 1955 and the compound quickly became important for the treatment of nervous and mental disorders. The first total synthesis of reserpine was reported a year later and the compound has remained a popular target for synthesis throughout the past half century. We will examine six syntheses that span this time period. We will begin with the Woodward group synthesis: Woodward, R. B.; Bader, F. E.; Bickel, H.; Frey, A. J.; Kierstead, R. W. "The Total Synthesis of Reserpine" *J. Am. Chem. Soc.* **1956**, *78*, 2023-2025. Woodward, R. B.; Bader, F. E.; Bickel, H.; Frey, A. J.; Kierstead, R. W. "A Simplified Route to a Key Intermediate in the Total Synthesis of Reserpine" *J. Am. Chem. Soc.* **1956**, *78*, 2657. Woodward, R. B.; Bader, F. E.; Bickel, H.; Frey, A. J.; Kierstead, R. W. "The Total Synthesis of Reserpine" *Tetrahedron* **1958**, 2, 1-57.

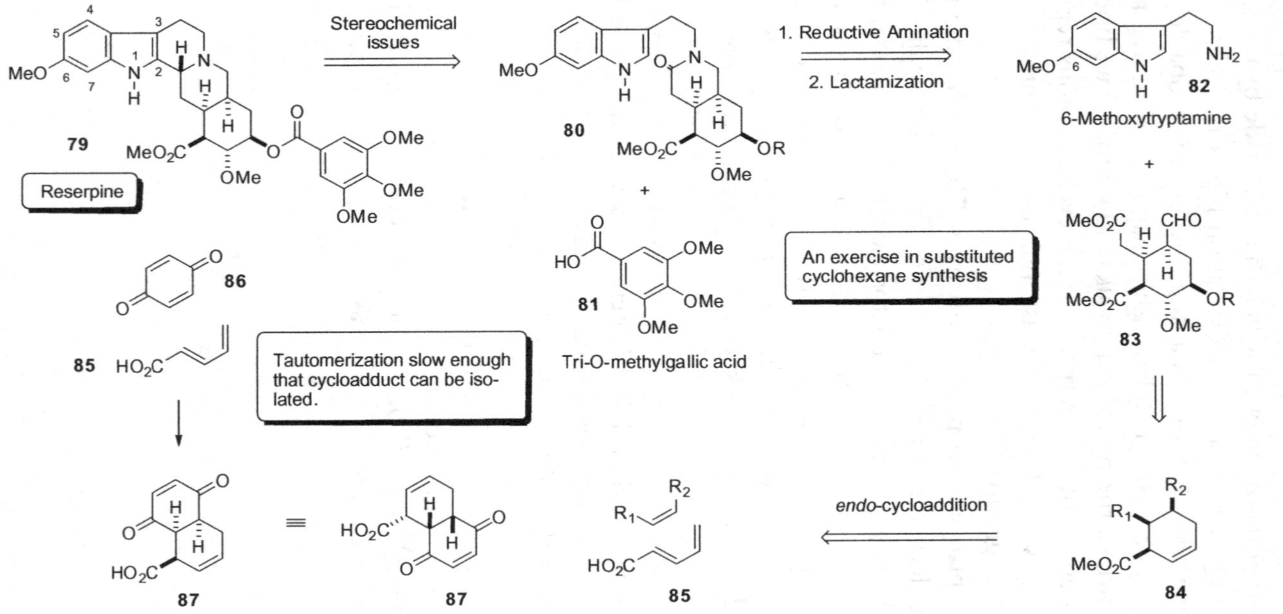

Alkaloids-10

Alkaloids-10

I will next move to reserpine. One reason is that, in all approaches to this alkaloid, the final reaction used to assemble the pentacyclic skeleton is formally a Mannich-type reaction (a Bischler-Napieralski reaction or a Pictet-Spengler reaction).[11,12] A more honest reason is that this alkaloid is a classical target in natural product synthesis that has retained the interest of synthetic organic chemists for more than a half-century, including myself. As we will see, there are some good lessons to be learned from approaches to this alkaloid.

Reserpine (**79**) was isolated in 1952 from the Indian snake-root, Rauwolfia serpentina Benth. The structure was determined by 1955 and the compound quickly became important for the treatment of nervous and mental disorders. The first total synthesis of reserpine was reported a year later and the compound has remained a popular target for synthesis throughout the past half century. We will examine six syntheses that span this time period. We will begin with the Woodward group synthesis. It is notable that this synthesis was being conducted about the same time that this group was developing their approach to steroids. In fact, the steroid work and the reserpine synthesis share some features, particularly in the early stages.

The plan was to disconnect reserpine into indole **80** and tri-*O*-methylgallic acid (**81**). Indole **80** was to be converted to reserpine using a variant of the well-known Bischler-Napieralski synthesis of isoquinoline derivatives. This transformation (**80** → **79**) generates a stereogenic center and thus, there are some stereochemical issues associated with this portion of the synthesis. They are rather fascinating and, in time, we will see how various groups addressed this issue. Lactam **80** was projected to come from 6-methoxytryptamine (**82**) and aldehyde **83**. It was anticipated that reductive amination of the aldehyde would be followed by cyclization of the intermediate aminoester to provide the lactam. Aldehyde **83** is a densely functionalized cyclohexane. Disconnection of the vicinal OR groups in **83** led to cyclohexene **84** as a possible intermediate, which was to be prepared by a Diels-Alder reaction between pentadienoic acid **85** and an appropriate *cis*-1,2-disubstituted ethylene. Whereas there might be regiochemical issues associated with the Diels-Alder reaction, the expected *endo*-cycloaddition would generate the proper relative stereochemistry at the three contiguous stereogenic centers in **84**. Furthermore, the regiochemical issue disappeared when *p*-benzoquinone (**86**) was selected as the dienophile. As anticipated, the initial cycloaddition proceeded smoothly to provide **87**. The next stages of the synthesis were two-fold in nature: (1) the olefin had to be converted to a vicinal diol derivative with control over stereochemistry and regiochemistry and (2) the enedione

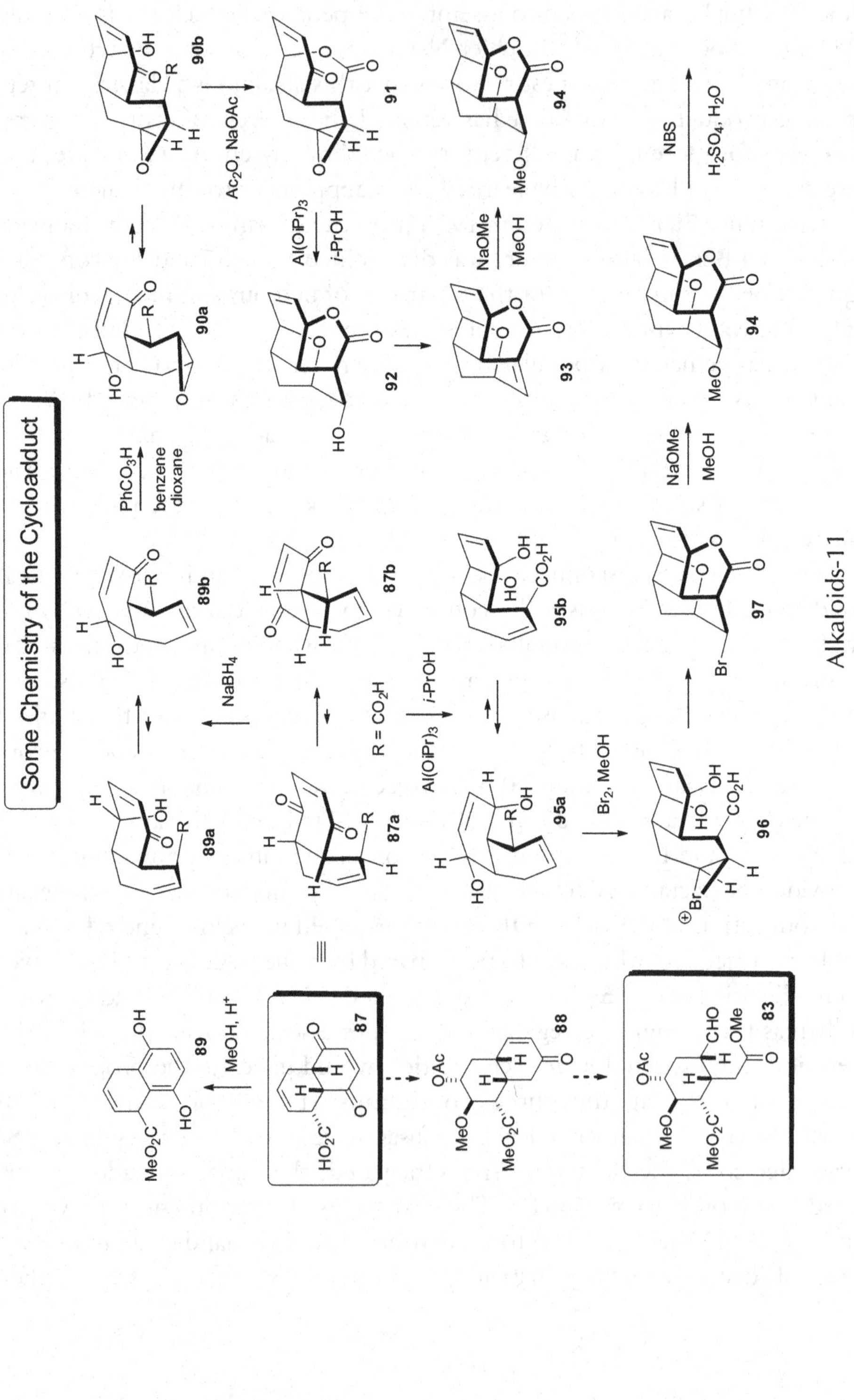
Some Chemistry of the Cycloadduct
MeOH, H+
89
87
88
83
89a
NaBH4
89b
PhCO3H
benzene
dioxane
90a
90b
Ac2O, NaOAc
91
87a
87b
R = CO2H
Al(OiPr)3
i-PrOH
92
93
94
NaOMe
MeOH
95a
95b
Br2, MeOH
96
97
NBS
H2SO4, H2O
Alkaloids-11

substructure had to be degraded in an unsymmetrical manner to reveal the aldehydo-ester needed for reaction with 6-methoxytryptamine (**82**).

Alkaloids-11

The plan for accomplishing these two tasks, and partial execution of the plan, is outlined here. It was hoped that **87** could be converted to **88**, which has the vicinal diol problem solved and is poised to reveal the aldehydo-ester in **83** (R = Ac) via oxidative cleavage of the enone.

Enedione **87** was a sensitive compound. It is at the same oxidation state as *p*-hydroquinone **89** and indeed, acid promoted this transformation. Reduction of **87** with sodium borohydride, however, differentiated the two carbonyl groups to provide **89** (R = CO_2H). The stereochemistry of the reduction can be rationalized by attack of hydride from the convex face of **87**. The regiochemistry can be rationalized by reduction of the sterically more accessible carbonyl group, if one assumes that reduction occurs from the presumably more stable conformation **87b**. Introduction of the differentiated vicinal diol came next. This transformation required regioselective *anti*-periplanar addition of two oxygens to the more electron-rich olefin in **89**. One oxygen had to be delivered from the convex face of **89**, and the other from the concave face. This was accomplished by epoxidation of **89** to provide **90**, stereochemistry being controlled by the bowl-shape of **89**. The next task was to regioselectively open the epoxide with "hydroxide" from the concave face of **90**. This required several steps, but was eventually accomplished by intramolecular delivery of the nucleophilic oxygen as follows (recall the conversion of **68** to **56** in the luciduline synthesis). Dehydration of **90** gave δ-lactone **91**. Reduction of the ketone in **91**, from the only accessible face of the carbonyl group was followed by γ-lactone formation. This froze the substrate in a conformation related to **90b**, in which the hydroxyl group was properly disposed for *anti*-periplanar opening of the epoxide to provide **92**. Base-promoted elimination of the alcohol gave unsaturated ester **93**, and treatment of **93** with sodium methoxide introduced the methoxyl group with proper stereochemistry (addition of the nucleophile to the convex face of unsaturated ester).

A streamlined route from **87** to **94** was also developed. Thus, Meerwein-Pondorf-Verley reduction of **87** provided diol **95**. Treatment of **95** with bromine in methanol gave β-bromolactone **97**, presumably via intermediate bromonium ion **96**. Reaction of **97** with sodium methoxide gave **94**. Note that this route takes advantage of the same stereochemical principles followed by the epoxide route to **94**.

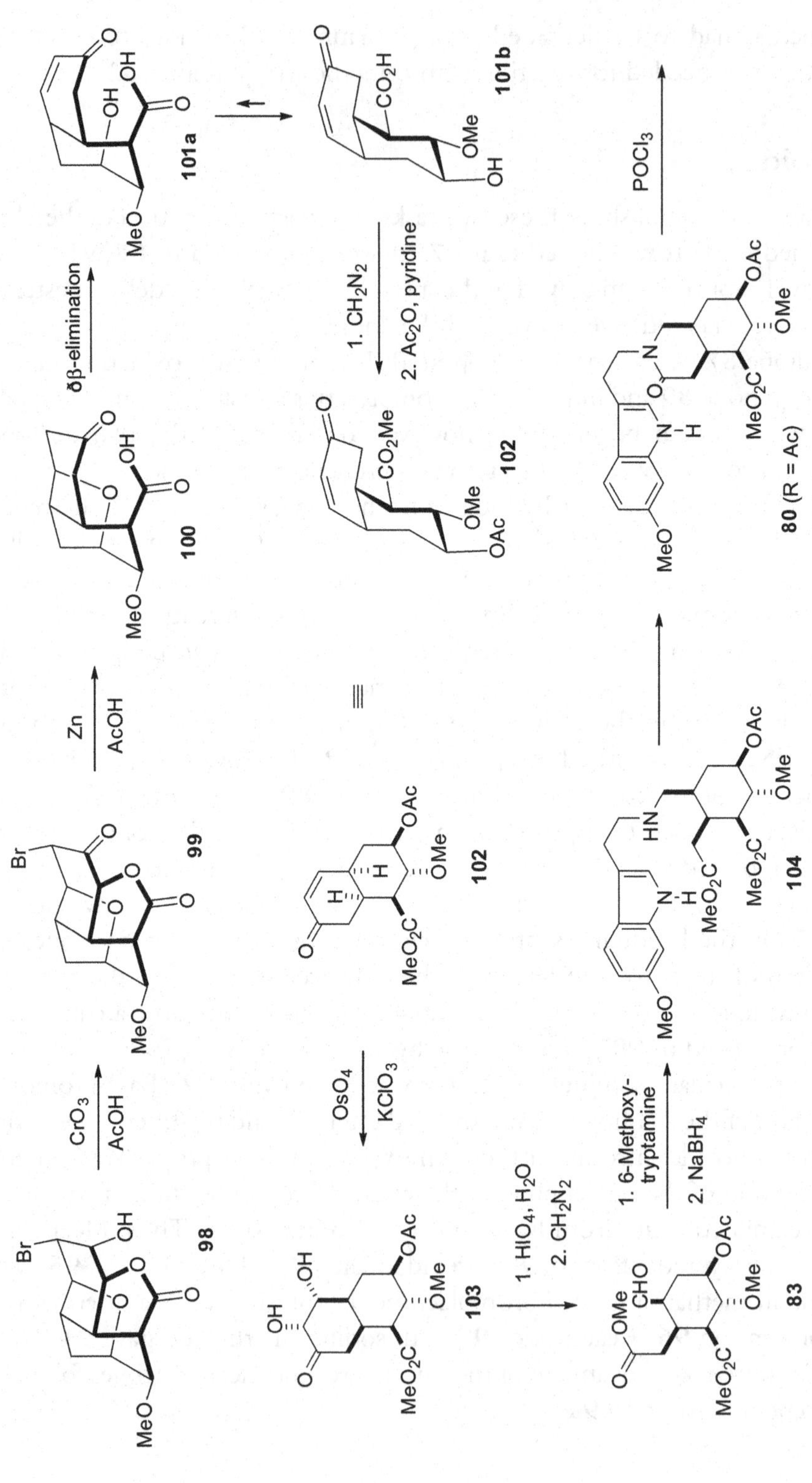

Alkaloids-12

Alkaloids-12

The next task was to convert **94** to unsaturated ester **102** (same as **88**) in preparation for the oxidative cleavage reaction to provide **83**. This transformation began by treating **94** with aqueous bromine to provide bromohydrin **98**. The regiochemistry and stereochemistry of this transformation can once again be rationalized using a variation of the Furst-Plattner rule.[8] *Anti*-periplanar opening of the bromonium ion derived from addition to the convex face of **94**, provides **98**. Oxidation of the alcohol, followed by reductive cleavage of the α-C–O and C–Br bonds in **99**, followed by an acid-promoted β-elimination from **100**, provided **101**. Notice that this sequence springs **99** open to **101a**, most likely the least stable conformation of **101**. Conversion of the acid to a methyl ester (with diazomethane) and esterification of the secondary alcohol completed the synthesis of **102** (same as **88**).

Oxidative cleavage of the enone was accomplished in two steps (diol formation followed by periodate cleavage of **103**) and esterification of the resulting acid gave **83**. The synthesis continued as expected. 6-Methoxytryptamine reacted with **83** to provide the expected imine, which was reduced to provide lactam **80**, presumably via amino ester **104**.

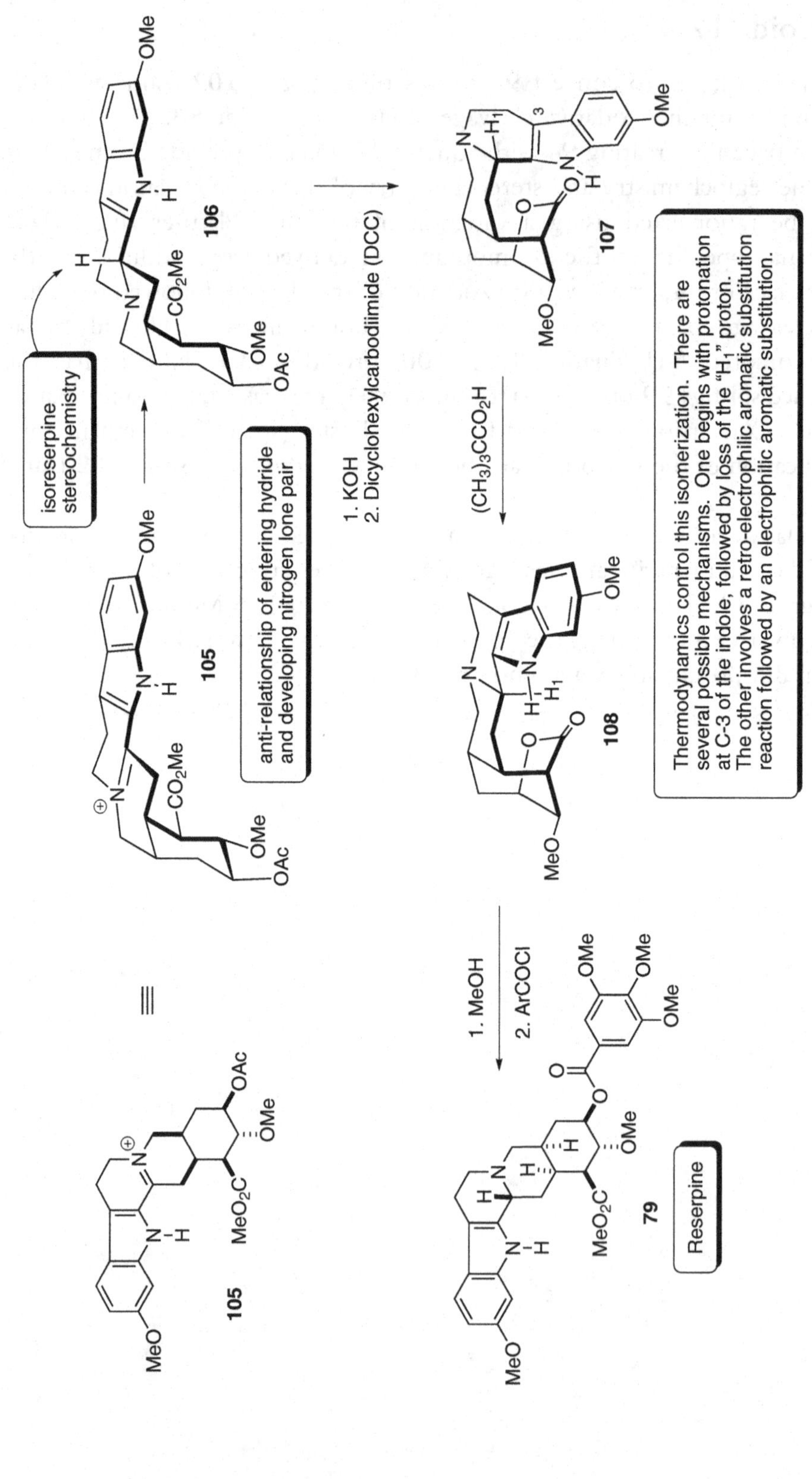

Alkaloids-13

Alkaloids-13

Now we come to the critical Bischler-Napieralski cyclization. Treatment of **104** with $POCl_3$ gave iminium ion **105** via amide activation and an intramolecular electrophilic aromatic substitution. Reduction of this iminium ion with sodium borohydride gave **106**. This material has the opposite stereochemistry (isoreserpine stereochemistry), at the new stereogenic center, to that required by reserpine. There are at least two explanations for this observation: (1) the reduction proceeds from the convex face of **105** and (2) the stereoelectronics of the iminium ion addition favor this reduction (recall the Stevens synthesis of porantherine). How was this stereochemical "mistake" fixed? Hydrolysis of the esters and lactonization of the resulting hydroxy acid converted **106** to **107** (note that conformations available to *cis*-decahydroisoquinolines can be treated in the same way that one analyzes conformations of *cis*-decalins). The transformation generates some serious 1,3-diaxial interactions that can be relieved by epimerization of **107** to **108** (see "H_1" hydrogen). One would expect **108** to be *thermodynamically* more stable than **107**. It turns out that this epimerization was accomplished by heating **107** with an acid (pivalic acid). There are several possible mechanisms one can imagine for this transformation. One begins with protonation at C_3 of the indole, followed by loss of the "H_1" proton. The other involves a retro-electrophilic aromatic substitution reaction followed by an aromatic substitution reaction. Opening of the lactone with methanol, followed by esterification of the secondary alcohol, completed the synthesis of reserpine (**79**).

Another Route to the Functionalized Cyclohexane

Pearlman, B. A. "A Method for Effecting the Equivalent of a deMayo Reaction with Formyl Acetic Ester" *J. Am. Chem. Soc.* **1979**, *101*, 6398. Pearlman, B. A.; "A Total Synthesis of Reserpine" *J. Am. Chem. Soc.* **1979**, *101*, 6404.

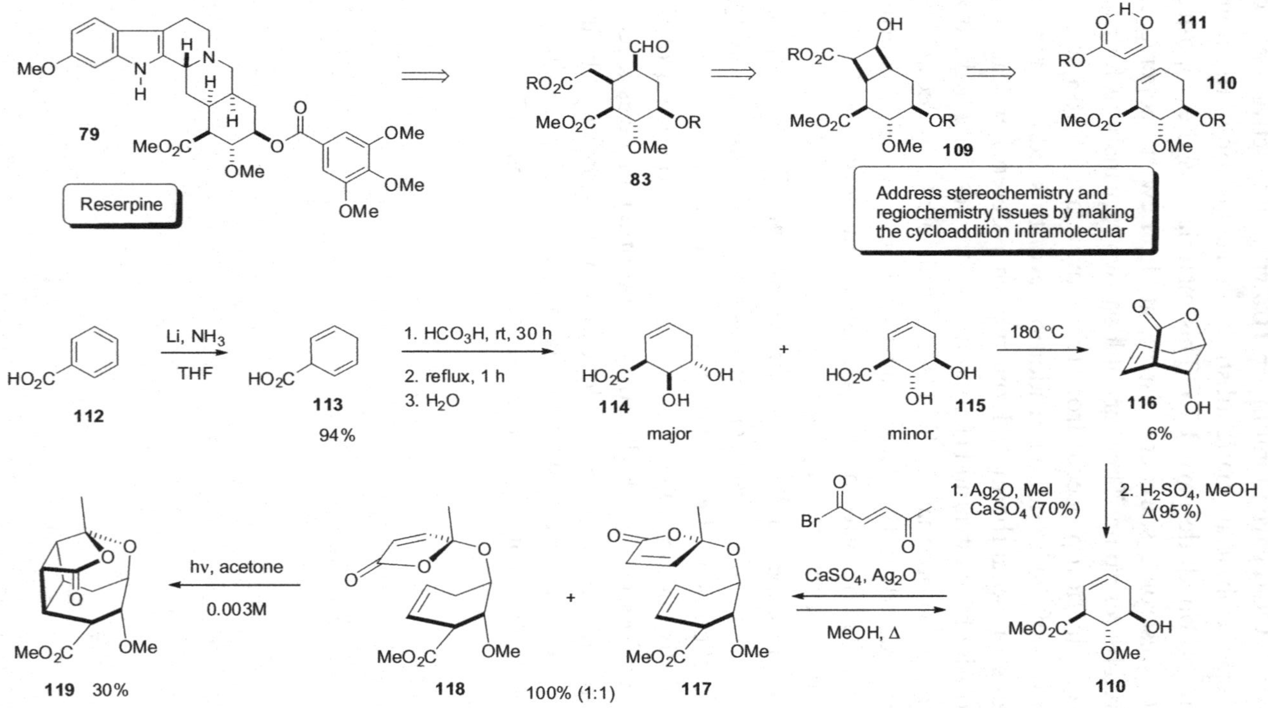

Alkaloids-14

Alkaloids-14

We will now look at several other syntheses of reserpine, some in more detail than others. Here we see an approach reported by Pearlman. The plan was related to the Woodward approach at a late stage, as **83** was projected to be an intermediate. The approach to **83**, however, was very different. The plan was to prepare strained β-hydroxyester **109** and then spring it loose to **83** by a retro-aldol reaction. The cyclobutane was to be prepared by a photocycloaddition between cyclohexene **110** and enol **111**.[13] The stereochemical and regiochemical issues associated with the cycloaddition were to be handled by making the photocycloaddition an intramolecular process. It is notable that this is an example of a synthesis that was designed to develop and illustrate new synthetic methodology, in this case a method for the "directed" *syn*-addition of two functionalized carbons across a carbon-carbon π-bond.

The synthesis began with reduction of benzoic acid (**112**) to provide **113**. Epoxidation of **113**, followed by opening, provided diols **114** (major) and **115** (minor). Heating the mixture of hydroxy acids provided lactone **116** in low yield. Alkylation of the free hydroxyl group followed by methanolysis of the lactone gave **110**, the desired photocycloaddition substrate. The desired course of the photocycloaddition was "cis to the hydroxyl group with formyl group vicinal to the hydroxyl group". This is demanding (does not work) without the use of intramolecular direction. In the end, **110** was converted to a mixture of diastereomeric acetals **117** and **118**, the latter of which was disposed to undergo the desired photocycloaddition to provide **119**.

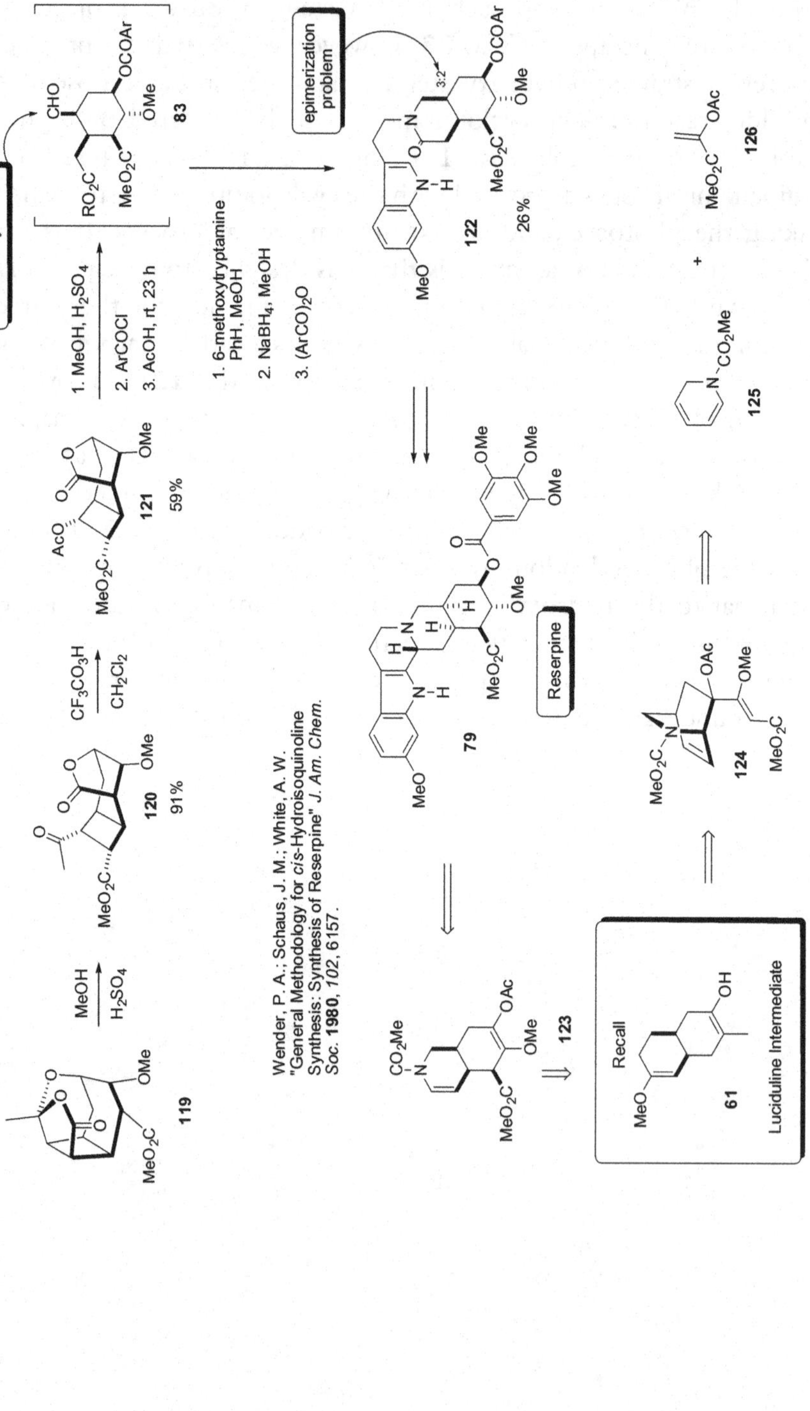

Wender, P. A.; Schaus, J. M.; White, A. W. "General Methodology for *cis*-Hydroisoquinoline Synthesis: Synthesis of Reserpine" *J. Am. Chem. Soc.* **1980**, *102*, 6157.

Alkaloids-15

Alkaloids-15

The conversion of **119** to retro-aldol substrate **121** was accomplished by hydrolysis of the acetal and esterification of the resulting acid to give **120**. This process was accompanied by epimerization of both the ester and the resulting methyl ketone, presumably to relieve strain. Baeyer-Villiger oxidation of **120** completed the transformation to **121**.[14] The sensitive aldehyde was then converted to **122** using chemistry analogous to that developed by the Woodward group. The bottom line is that this is a creative plan that had trouble when it came to the details (tactics).

The Wender group (then Harvard and now Stanford) reported another approach. The idea was to prepare reserpine from an enamide of type **123**. Modification of the *N*-carbethoxy group and protonation of the enamine/enamide would afford the required pentacyclic skeleton. Reduction of the cyclohexene, with control of stereochemistry, would also be required to establish the five contiguous stereogenic centers of reserpine. Notice that **123** is a heterocyclic analog of luciduline intermediate **61**. It is a 1,5-hexadiene. This is a recognition key for a Cope rearrangment. Working backward leads to azabicyclo[2.2.2]octane **124**, which was to be prepared by a Diels-Alder reaction between N-acyldihydropyridine **125** and dienophile **126**. Let's see what happened in the lab.

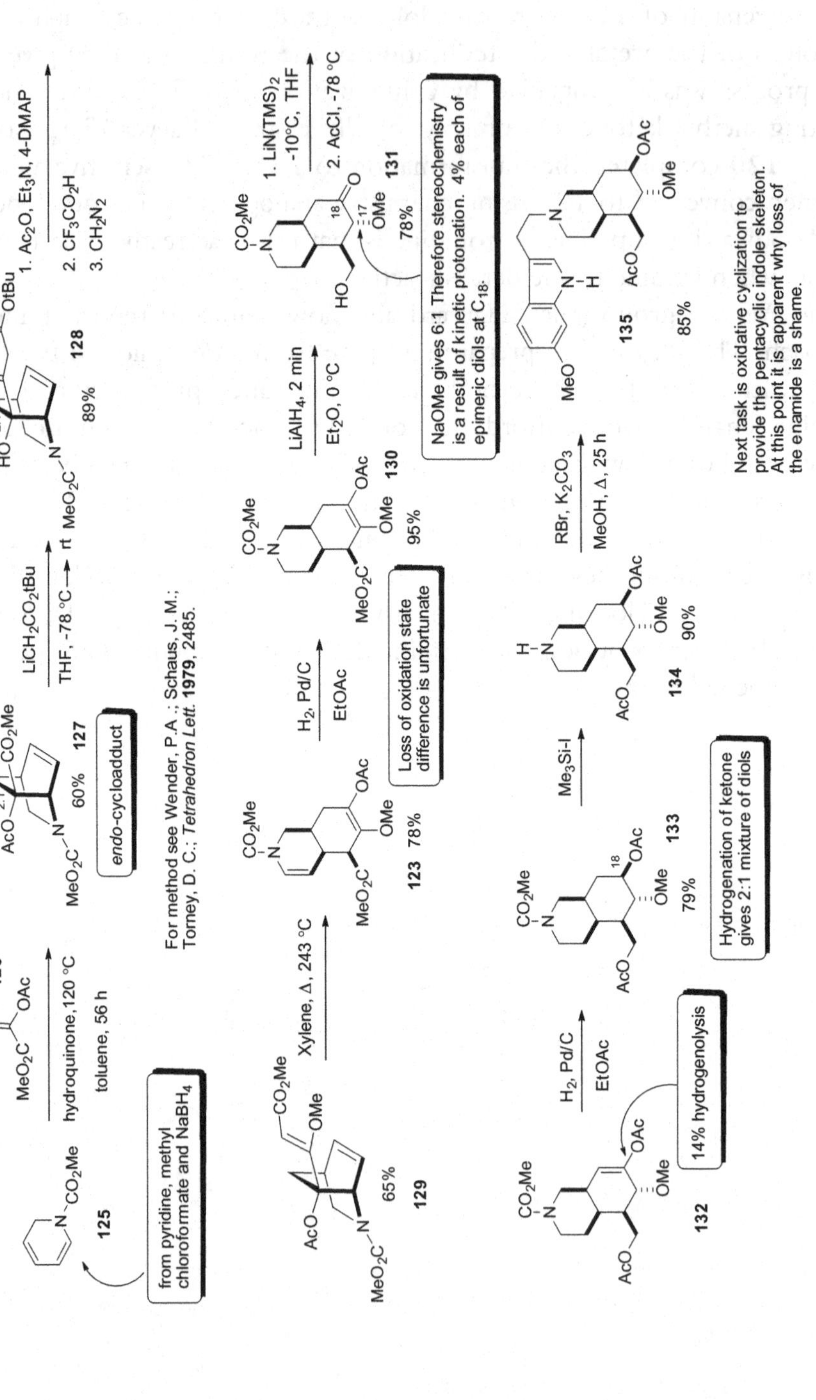

Alkaloids-16

Alkaloids-16

The initial Diels-Alder reaction worked well to provide a mixture of endo (**127**) and exo (not shown) cycloadducts. A crossed Claisen condensation provided **128**, and a series of functional group manipulations provided Cope substrate **129**. The Cope rearrangement proceeded well to give **123** in excellent yield. Unfortunately a number of problems were encountered thereafter.

It was not possible to reduce the enediol derivative without also reducing the enamide. For example, catalytic hydrogenation of **123** provided **130**. The enamide is what differentiated oxidation states at the carbons adjacent to nitrogen and thus, we will see, it is unfortunate that differentiation was lost at this point of the synthesis. Continuing from **130**, lithium aluminum hydride reduction converted the carbomethoxy group to a primary alcohol, and cleaved the enol acetate to afford an enolate. Kinetic protonation of the enolate afforded **131**, although thermodynamics also favored **131** over its C_{17} diastereomer. Catalytic hydrogenation of **131** gave a 2:1 mixture of alcohols at C_{18}. On the other hand, conversion of **131** to enol acetate **132**, followed by catalytic hydrogenation, provided **133** with the desired C_{18} stereochemistry.

The next stage of the synthesis required introduction of the indolylethyl group and oxidative cyclization to afford the pentacyclic skeleton of reserpine. First the carbamate was removed from decahydroisoquinoline **133** to give **134**. Alkylation then gave **135**. The next task was oxidative cyclization to provide the pentacyclic indole skeleton. At this point it becomes clear why loss of the enamide was unfortunate.

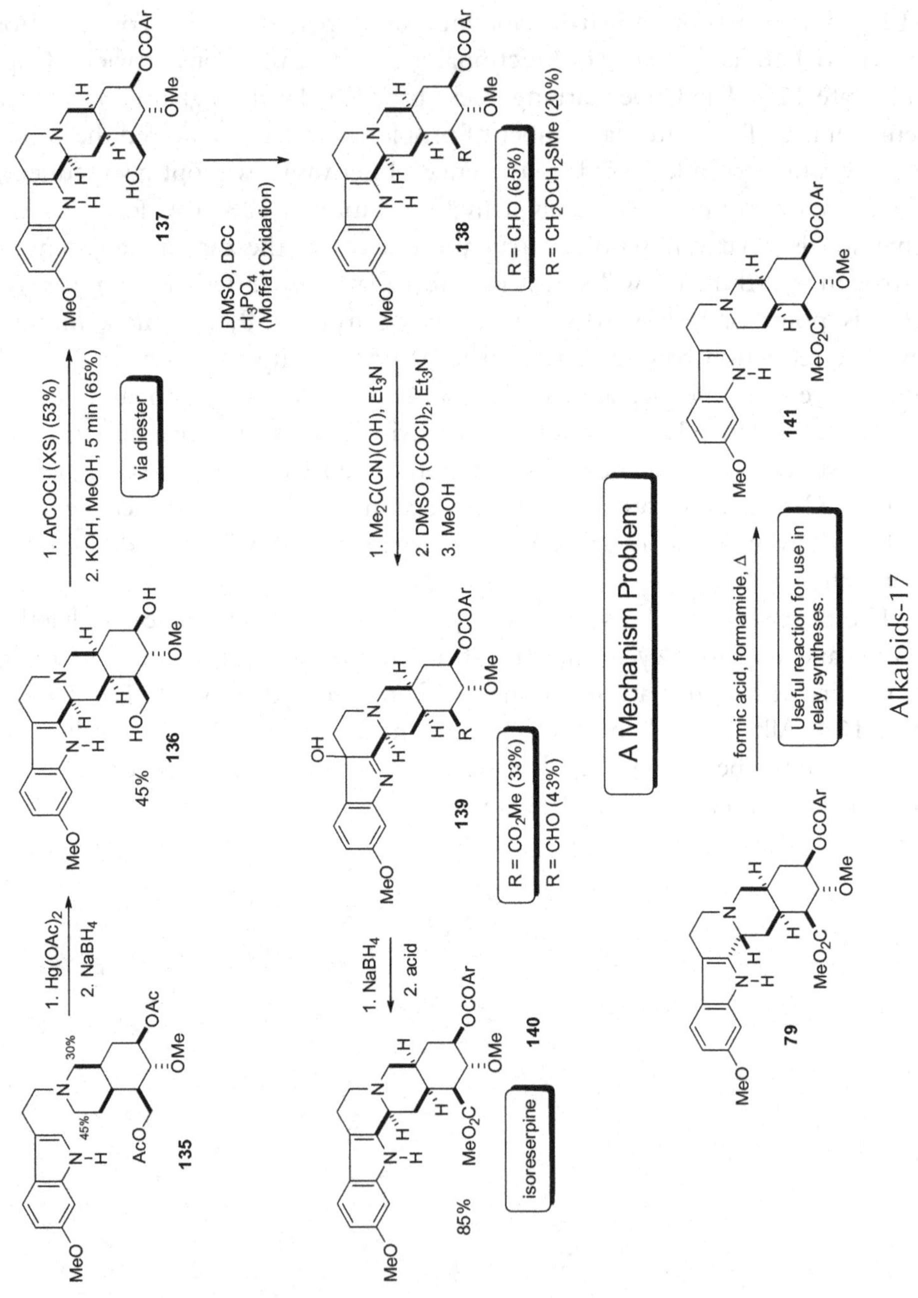
1. Hg(OAc)$_2$
2. NaBH$_4$
135
45%
30%
136
1. ArCOCl (XS) (53%)
2. KOH, MeOH, 5 min (65%)
via diester
137
DMSO, DCC
H_3PO_4
(Moffat Oxidation)
138
R = CHO (65%)
R = CH_2OCH_2SMe (20%)
1. $Me_2C(CN)(OH)$, Et_3N
2. DMSO, $(COCl)_2$, Et_3N
3. MeOH
139
R = CO_2Me (33%)
R = CHO (43%)
1. NaBH$_4$
2. acid
140
85%
isoreserpine
A Mechanism Problem
79
formic acid, formamide, Δ
Useful reaction for use in relay syntheses.
141
Alkaloids-17

Alkaloids-17

Oxidative cyclization of **135** occurred with little regioselectivity. Oxidation at C_3 afforded isoreserpine analog **136**, and a nearly equal amount of material derived from oxidation at C_{21} was obtained (so-called inside isomers). Based on earlier work, it is probable that the sequence of events leading to **136** involve oxidation of **135** to an iminium ion, cyclization via an electrophilic aromatic substitution reaction, oxidation of the resulting cyclized products to an iminium ion related to **105** (see Alkaloids-13) and finally, reduction ($NaBH_4$) of that iminium ion to provide **136**. Conversion of **136** to isoreserpine (**140**) was relatively straightforward. I will leave the details to you (see problems for guidance). Since isoreserpine had previously been converted to reserpine (see Wooward synthesis for example), this constituted a synthesis of **79**.

Intramolecular Diels-Alder Route to Reserpine

Martin, S. F.; Rueger, H.; Williamson, S. A.; Grzejszczak, S. "General Stategies for the Synthesis of Indole Alkaloids. Total Syntheses of *dl*-Reserpine and *dl*-α-Yohimbine" *J. Am. Chem. Soc.* **1987**, *109*, 6124-6134.

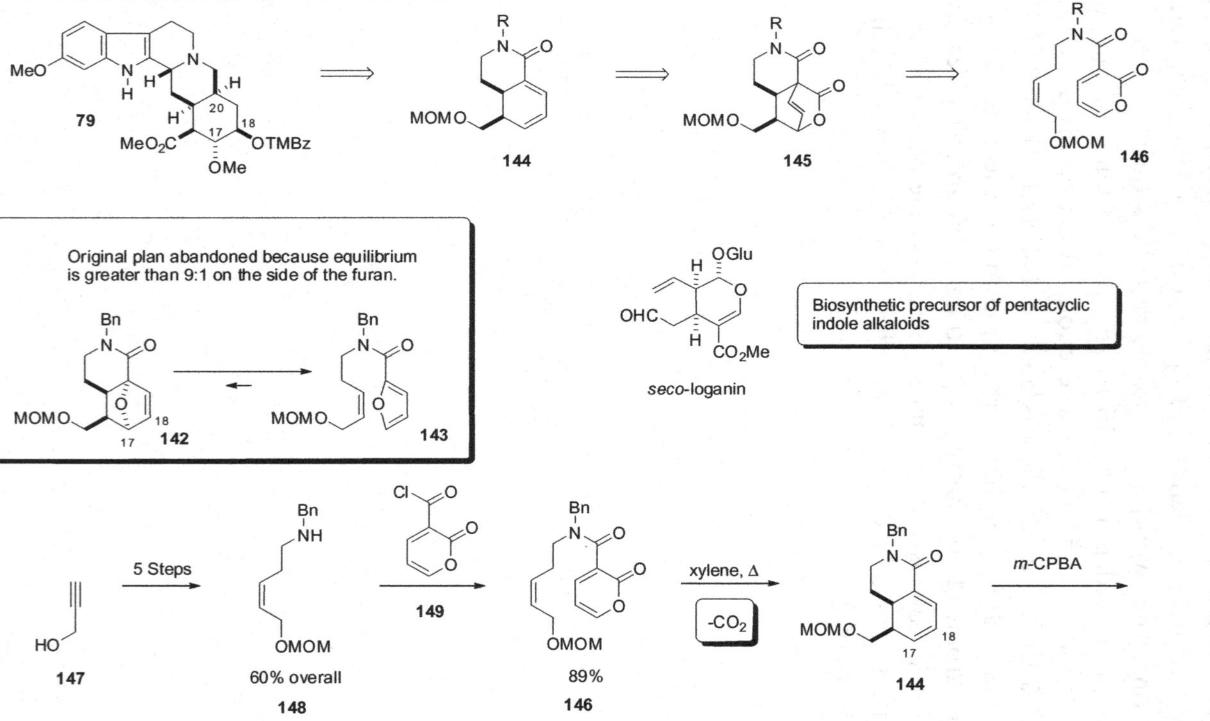

Alkaloids-18

Alkaloids-18

Martin (University of Texas) described a synthesis that followed the plan depicted here. The ultimate idea was that perhydroisoquinoline **144** could be converted to **79** using the conjugated diene as a handle for introducing the C_{17}-C_{18} diol and C_{20} stereogenic center. This diene was to come from **146** via a Diels-Alder (to provide **145**) *retro*-Diels-Alder (to provide **144**) sequence. Actually, the original plan was to prepare **142** via an intramolecular Diels-Alder of **143**. It is apparent that this plan took into account the needed substituents at C_{17} and C_{18}, but it was abandoned because the equilibrium for this reaction was well to the side of the furan (**143**).

The synthesis began with the preparation of amine **148** from propargyl alcohol (**147**). Acylation of the amine with **149** gave **146**, and the proposed Diels-Alder reaction proceeded smoothly to give **144**. The remote double bond of **144** was the most electrophilic double bond and thus, epoxidation from the sterically most accessible face gave **150** as the major product.

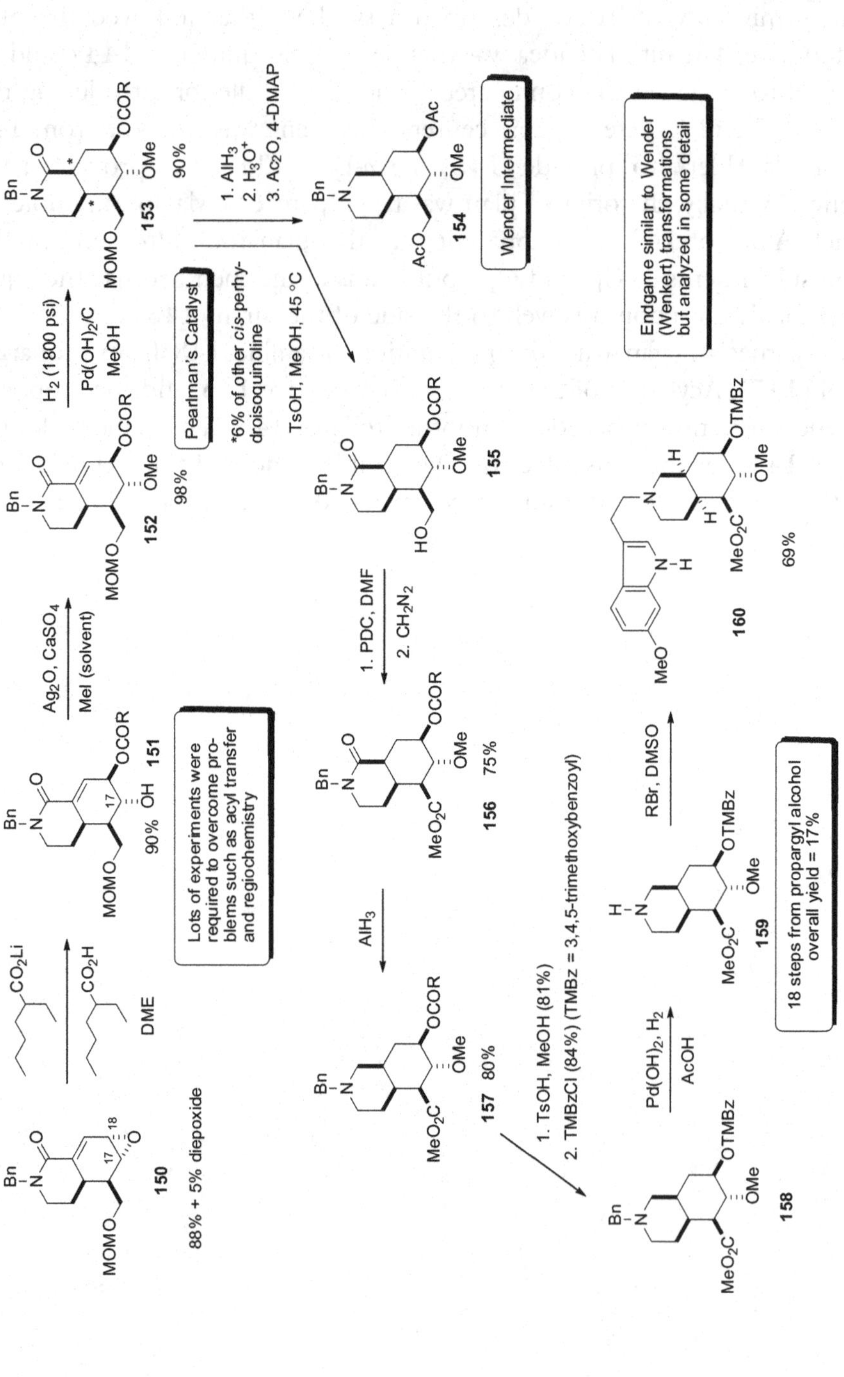

Alkaloids-19

Alkaloids-19

Epoxide **150** was then opened with lithium 2-ethylhexanoate to provide **151**. This was not the first choice of nucleophile! Lots of experiments were required to overcome problems such as acyl transfer and regioselectivity. There is no doubt that "search" is an important part of the word "research". Williamson etherification of the C_{17} alcohol gave **152**, and catalytic hydrogenation of the olefin provided **153** as the major product. This was converted to the Wender intermediate **154** (**133**) in a straightforward manner.

As an alternative, **153** was converted **157** by a series of oxidation state adjustments. The ester **157** was then converted to **158** and hydrogenolysis of the *N*-benzyl group provided **159**. Introduction of the 6-methoxytryptophyl group provided **160** to set the stage for the now-familiar endgame.

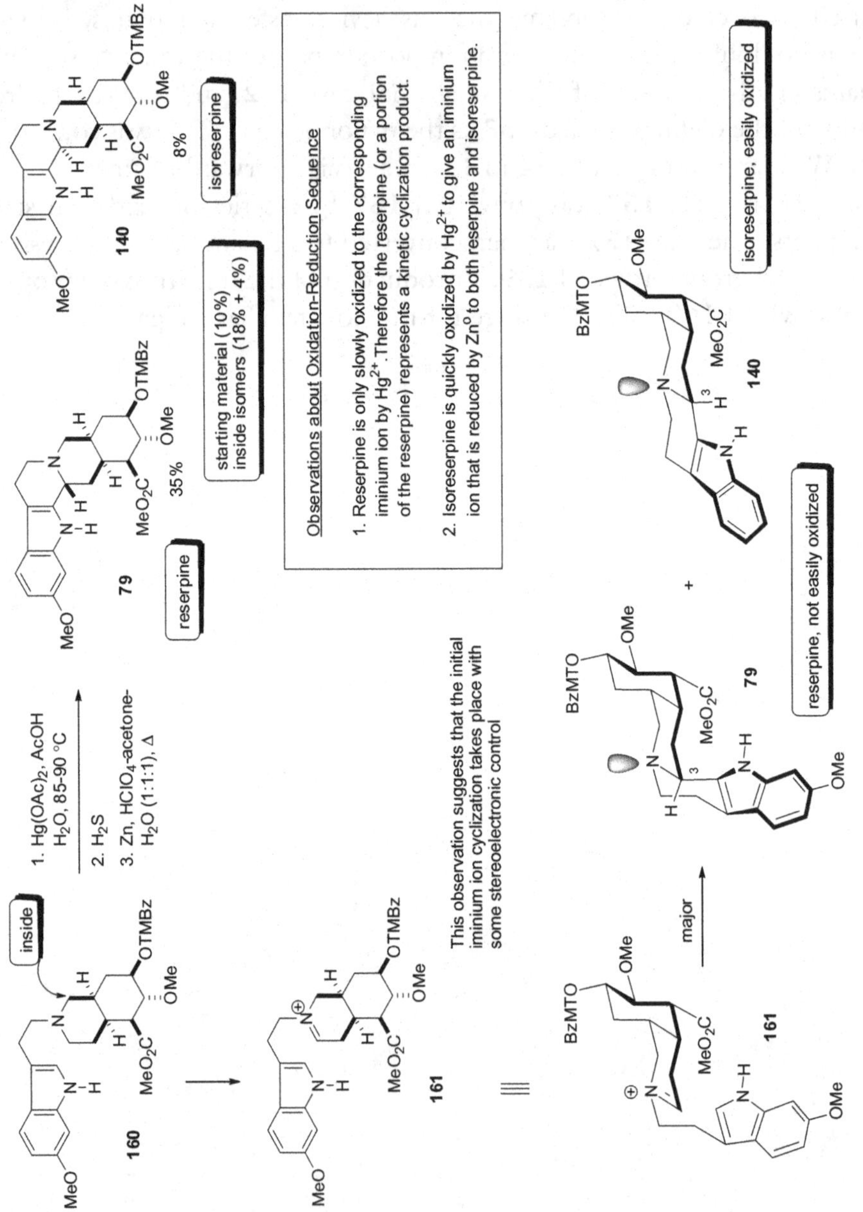

Alkaloids-20

Alkaloids-20

The Martin group used an oxidation-reduction sequence that was similar to that used by Wender. The reaction conditions, however, were slightly different and Zn-perchloric acid was used in the "reduction" step rather than sodium borohydride. The results differed from those obtained by Wender in that reserpine (**79**) was the major product and isoreserpine (**140**) the minor product. "Inside" isomers were also obtained due to lack of regioselectivity in the mercuric acetate oxidation.

Several observations were made during the course of this transformation. First, reserpine is only *slowly oxidized* to the corresponding iminium ion by mercuric acetate (see **105** on Alkaloids-13 for a related example where the TMBz group = Ac). Therefore it was proposed that the reserpine isolated from this reaction (or a portion of the reserpine) represented the kinetically prefered product formed from cyclization of iminium ion **161**. This is consistent with the stereoelectronic arguments for iminium ion-nucleophile reactions set forth by Stevens. It was also shown that isoreserpine was *quickly oxidized* by mercuric acetate to provide the corresponding iminium ion (see **105** on Alkaloids-13 for a related iminium ion), which was reduced by Zn^0 to give a mixture of both reserpine *and* isoreserpine. Examination of the most stable conformations of **79** and **140** (at least toward C_3) illustrates why one might expect to see such a difference in oxidation rates between these isomers. The C_3-H and nitrogen lone pair (the site of "attack" by the Hg^{2+}) have an *anti*-periplanar arrangement in isoreserpine, but not in reserpine! The bottom line is that the Martin synthesis begins to address the issue of directly obtaining reserpine, rather than having to rely on an isomerization of isoreserpine. We will revisit this later in this chapter.

Enantioselective Synthesis Starting from the Chiral Pool

Hannessian, S.; Pan, J.; Carnell, A.; Bouchard, H.; Lusage, L. "Total Synthesis of (-)-Reserpine Using the Chiron Approach" *J. Org. Chem.* **1997**, *62*, 465-473

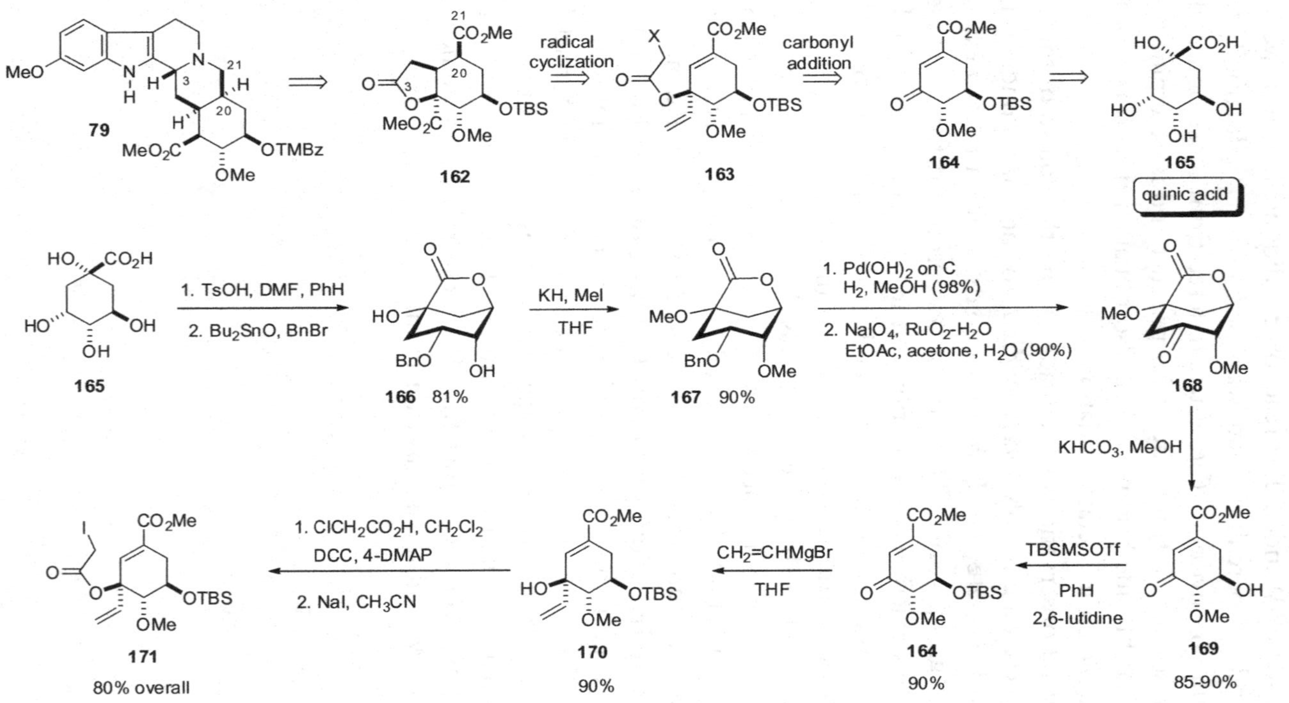

Alkaloids-21

Alkaloids-21

The syntheses we have examined thus far do not take into account enantioselectivity. The Hannesian group described a synthesis that used quinic acid (**165**) as an enantiopure starting material. The idea was to prepare reserpine (**79**) from *penta*-substituted cyclohexane **162**. The incipient C_3 was to come from an intermediate with that carbon at the aldehyde oxidation state (to be derived from reduction of the lactone). The incipient C_{21} was to be introduced at the carboxylic acid oxidation state. This plan avoids issues with formation of "inside" isomers as we saw in the Wender and Martin syntheses. Lactone **162** was to come via a radical cyclization of an intermediate of type **163**. Reduction of an intermediate radical from the convex face of the *cis*-bicyclic ring system was to control stereochemistry at C_{20}. Compound **163** was to be prepared from **164**, which was to come from quinic acid (**165**) through a series of protections and oxidation state adjustments.

The protection of quinic acid was accomplished by γ-lactone formation, benzyl ether formation at the least hindered secondary equatorial alcohol, and methylation of the remaining two alcohols. The secondary alcohol of the resulting **167** was deprotected (hydrogenolysis) and the alcohol was oxidized to provide ketone **168**. A transesterification followed by a β-elimination provided **169**. Protection of the secondary alcohol provided intermediate **164**. Axial delivery of a vinyl group to ketone **164** gave **170**. Esterification of the tertiary alcohol and a Finkelstein reaction gave free radical cyclization substrate **171**.[15]

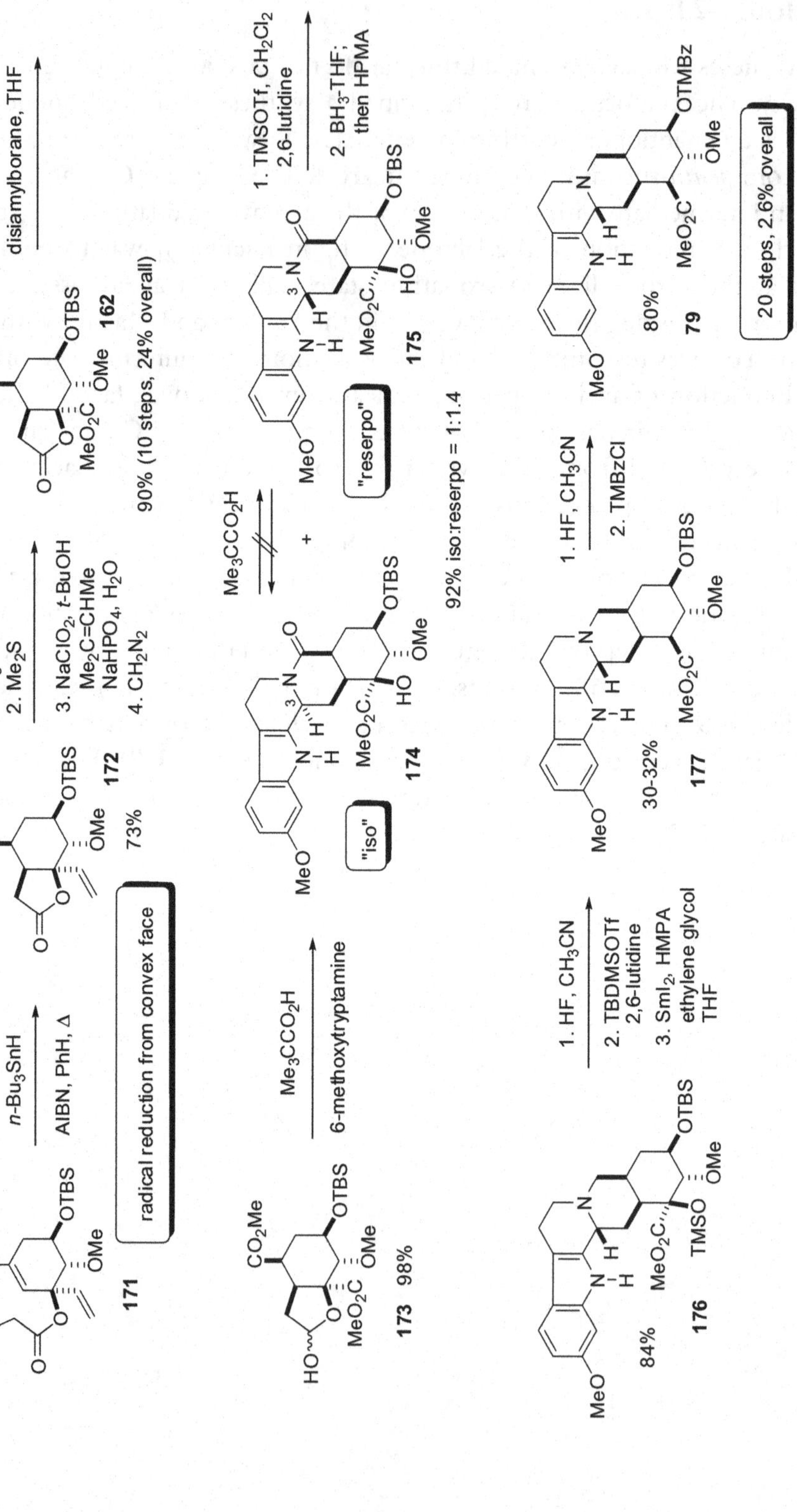

Alkaloids-22

Alkaloids-22

The projected free radical cyclization proceeded as planned to give **172**. Ozonolysis of the vinyl group, oxidation of the resulting aldehyde to an acid, and alkylation with diazomethane provided projected intermediate **162**. Reduction of the lactone provided **173**. Treatment of **173** with 6-methoxytryptamine and pivalic acid then provided a nearly equal mixture of lactams **174** (isoreserpine stereochemistry at C_3) and **175** (reserpine stereochemistry at C_3). The correct C_3 stereoisomer was moved forward to **176** (protection of the tertiary alcohol followed by reduction of the lactam). The silyl ethers were removed, the secondary ether was re-protected, and reaction with samarium iodide accomplished reduction of the α-hydroxy ester to provide **177**. Removal of the TBS group and esterification of the alcohol completed the synthesis of reserpine.

Regioselective and Stereoselective Route

Stork, G.; Tang, P. C.; Casey, M.; Goodman, B.; Toyota, M. "Regiospecific and Stereoselective Syntheses of *dl*-Reserpine and (-)-Reserpine" *J. Am. Chem. Soc.* **2005**, *127*, 16255-16262.

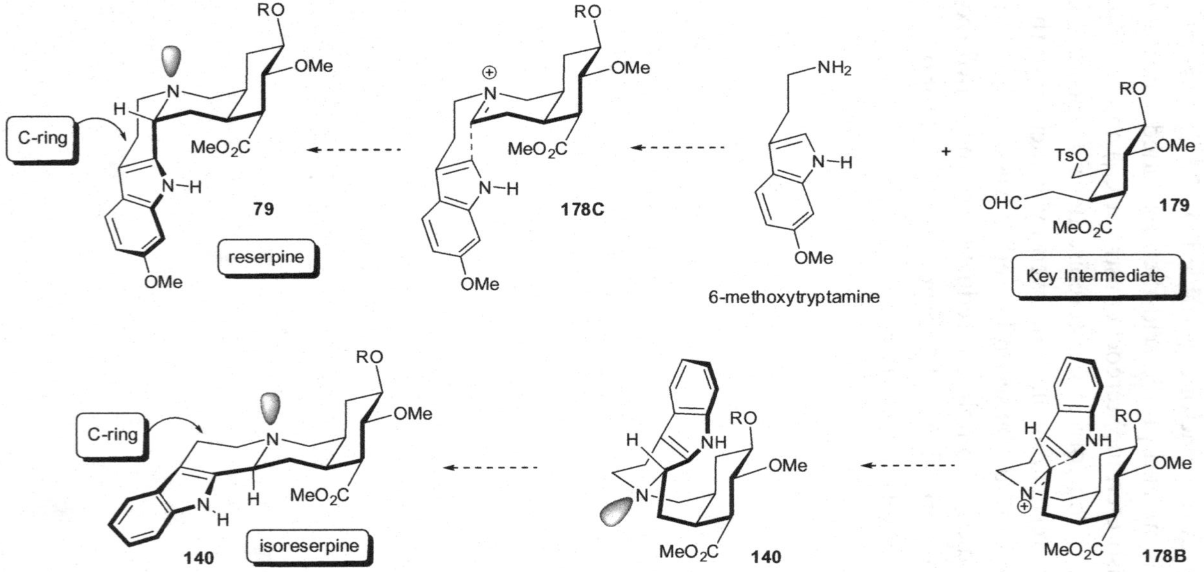

The plan revolves around the notion that the indole should kinetically add to an iminium ion such that carbon-carbon bond and lone-pair are developed with an *anti*-periplanar relationship. This can occur in two ways. If the C-ring is born as a chair (with the requisite anti-periplanar relationship), the resulting product is reserpine. If the C-ring is born as a boat, the result is isoreserpine. Although isoreserpine is more stable than reserpine, the presumed conformations in which they would be born would favor reserpine over isoreserpine. This assumes that product partitioning is controlled by kinetics (is not reversible). The "key intermediate" also deals nicely with regiochemical issues. Three syntheses of the key intermediate were described. We will examine only one of these.

Alkaloids-23

The final synthesis of reserpine (**79**) that we will examine was accomplished by the Stork group (Columbia). The plan revolves around the notion that the indole should kinetically add to an iminium ion such that the carbon-carbon bond and lone-pair are developed with an *anti*-periplanar relationship. This can occur in two ways. If the C-ring is born as a chair (with the requisite anti-periplanar relationship), the resulting product is reserpine (**178C** to **79**). If the C-ring is born as a boat, the result is isoreserpine (**178B** to **140**). Although isoreserpine is more stable than reserpine, the presumed conformations in which they would be born would favor reserpine over isoreserpine. This assumes that product partitioning is controlled by kinetics (is not reversible). Iminium ion **178** was to be generated from **179** via Schiff base formation followed by *N*-alkylation. Key intermediate **179** also deals nicely with regiochemical issues. Three syntheses of the **179** were described. We will examine only one of these.

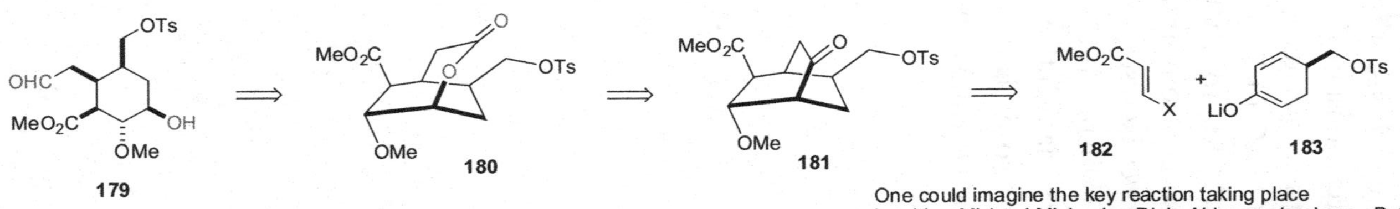

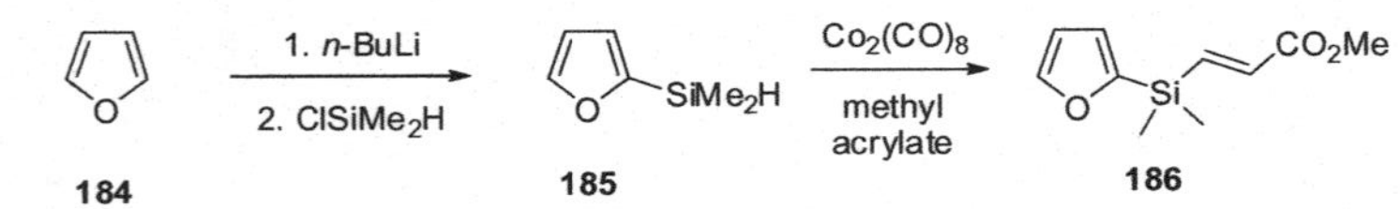

One could imagine the key reaction taking place by either Michael-Michael or Diels-Alder mechanisms. Both would be expected to occur from the face of the diene opposite the tosylate. Endo selectivity would be required for "stereochemical success". The choice of "X" is important. A methoxy group would complicate the Michael-Michael process by β-elimination, and the Diels-Alder process by reducing dienophilicity through electron donation.

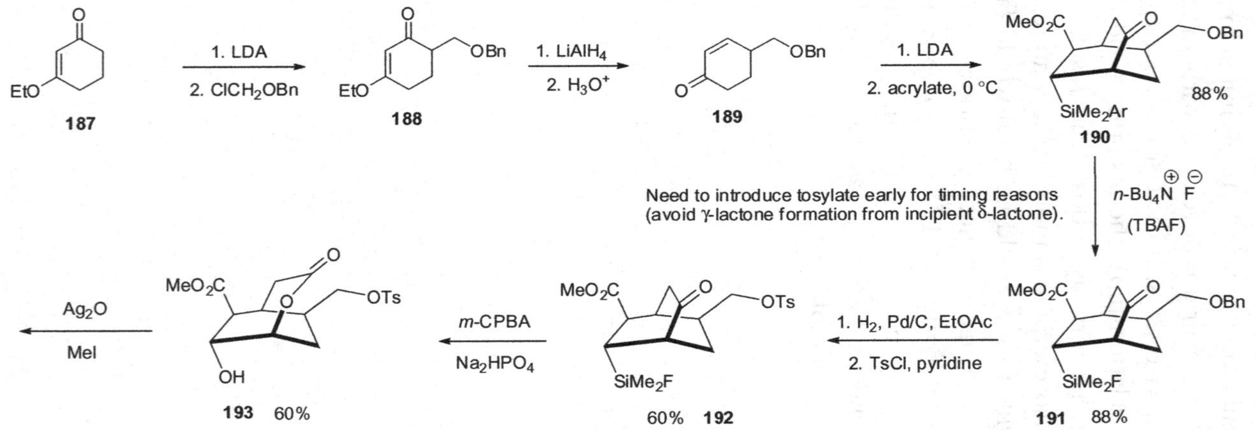

Need to introduce tosylate early for timing reasons (avoid γ-lactone formation from incipient δ-lactone).

Tamao-Fleming Oxidation is mechanistically similar to the Baeyer-Villiger Oxidation

Alkaloids-24

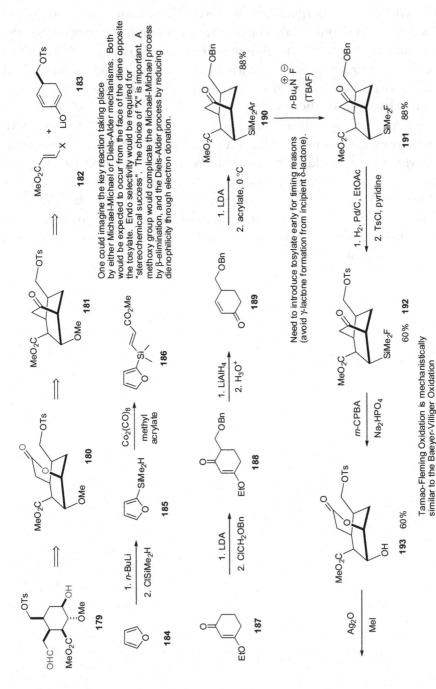

Alkaloids-24

The approach to **179** that we will examine proceeded through lactone **180**. This intermediate is a lactone reduction away from **179**. The lactone was to be generated by Baeyer-Villiger oxidation of bicyclic ketone **181** which was to come, in turn, from a formal "cycloaddition" between dienolate **183** and β-hydroxyacrylate equivalent **182** (X=OH). One could imagine the key cycloaddition taking place by either Michael-Michael or Diels-Alder mechanisms. Both would be expected to occur from the face of the diene opposite the tosylate (or an appropriate tosylate precursor). Endo selectivity would be required for "stereochemical reasons". The choice of "X" in **182** was important. A methoxy group would complicate the Michael-Michael process by β-elimination, and the Diels-Alder process by reducing the dienophilicity through electron donation.

Furan derivative **186** was ultimately selected as the "dienophile". The plan was to convert the C–Si bond to a C–O bond using a Tamao-Fleming oxidation.[16] The dienolate derived from kinetic deprotonation of cyclohexenone **189** was ultimately selected as the "diene". Cyclohexenone **189** was prepared using the classical "Stork-Danheiser synthesis" of 4-substituted cyclohexenones.[17] The reaction between the enolate of **189** and **186** was very efficient, giving **190** in excellent yield. The 2-furyl group was replaced by fluoride using tetra-*n*-butylammonium fluoride (**190** → **191**). The benzyl ether was converted to the corresponding tosylate (**192**). Baeyer-Villiger oxidation of **192** was accompanied by Tamao-Fleming oxidation of the C–Si bond to provide **193**. The timing of the conversion of the benzyl ether to the tosylate is notable. Had this been left for later in the synthesis, translactonization might have been a serious problem at the stage of intermediates such as **193**, where the OTs group is replaced by an OH group.

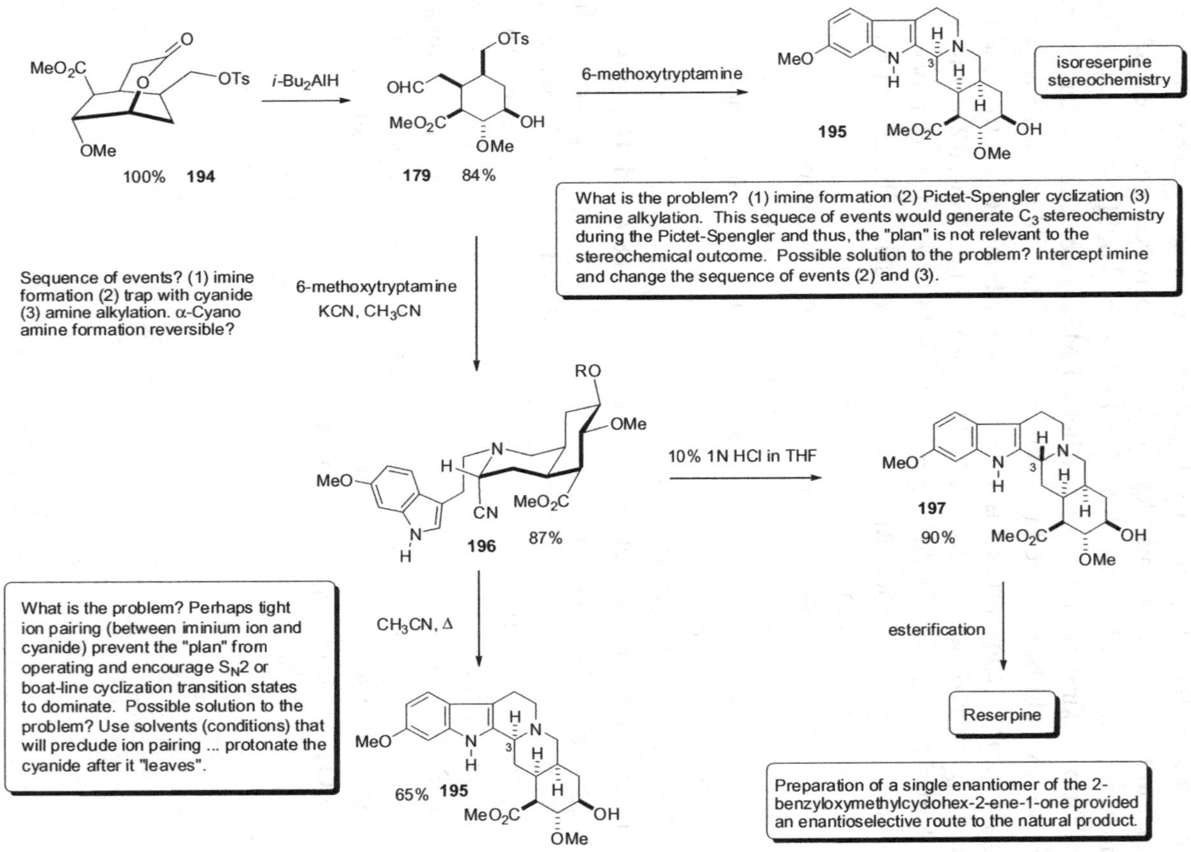

Alkaloids-25

Alkaloids-25

Etherification of **193** gave **194** and *i*-Bu_2AlH reduction of the lactone provided key intermediate **179**. Treatment of **179** with 6-methoxytryptamine provided **195** with "isoreserpo" stereochemistry at C_3! What was the problem? It was suggested that the sequence of events leading to **195** might be: (1) imine formation, (2) Pictet-Spengler cyclization and (3) *N*-alkylation of the resulting secondary amine. This sequence would generate C_3 stereochemistry during the Pictet-Spengler reaction, and thus the "plan" would not be relevant to the stereochemical outcome. What is a possible solution to the problem? Intercept the imine and change the sequence of events (2) and (3).

It was projected that cyanide might be an appropriate imine trap. Thus, when **179** was treated with 6-methoxytryptamine in the presence of KCN, α-cyanoamine **196** was obtained in excellent yield. Note that the stereochemistry of the α-cyanoamine is that predicted by "stereoelectonic considerations". It was known that α-cyanoimines could serve as precursors to iminium ions. Thus, warming **196** in a good ionizing solvent gave a cyclization product, but once again the product was **195** with the undesired C_3 stereochemistry. What was the problem? It was surmised that tight ion pairing between the iminium ion and cyanide might be preventing the "plan" from operating, and might be encouraging S_N2 or boat-like cyclization transition states to dominate. If this was the problem, it was hoped that conditions that "break up" ion pairs might give the desired stereochemical result. The tactic that was adopted was to protonate the cyanide after it "left" the α-carbon. The ploy worked and **197** was obtained as the major product in excellent yield. Esterification of the C_{19} alcohol completed the synthesis.

This synthesis was also accomplished in an enantioselective manner when a single enantiomer of **189** was prepared and moved through the synthesis. This synthesis provides a marvelous example in which mechanistic reasoning was used to "rationally" identify tactics for accomplishing a desired result. Other syntheses of reserpine have been described, but I hope this sampling has been enjoyable. We will now move to alkaloid syntheses where the focus on iminium ion chemistry is somewhat reduced, although it will not disappear completely.

References

1. For some recent reviews see: Maryanoff, B. E.; Zhang, H-C.; Cohen, J. H.; Turchi, I. J.; Maryanoff, C. A. "Cyclizations of *N*-Acyliminium Ions" *Chem. Rev.* **2004**, *104*, 1431–1628. Hajicek, J. "Mannich and Related Reactions in Total Synthesis of Alkaloids" *Chem. Listy* **2004**, *98*, 1096–1112. Arend, M.; Westermann, B.; Risch, N. "Modern Variants of the Mannich Reaction" *Angew. Chem. Int. Ed.* **1998**, *37*, 1045–1070.
2. Stevens, R. V.; Lee, A. W. M. "Stereochemistry of the Robinson-Schopf Reaction. A Stereospecific Total Synthesis of the Ladybug Defense Alkaloids Precoccinelline and Coccinelline" *J. Am. Chem. Soc.* **1979**, *101*, 7032–7035. Stevens coccinelline
3. For more examples see: Delongschamps, P. "Stereoelectronic Effects in Organic Chemistry" Pergamon Press, 1983.
4. Shapiro, R. H. "Alkenes from Tosylhydrazones" *Organic Reactions* **1976**, *23*, 405–507. Chamberlin, A. R.; Stemke, J. E.; Bond, F. T. "Vinyllithium Reagents from Arenesulfonylhydrazones" *J. Org. Chem.* **1978**, *43*, 147–154.
5. Shono, T.; Matsumura, Y.; Tsubata, K. "Anodic Oxidation of *N*-Carbomethoxy-pyrrolidine: 2-Methoxy-*N*-carbomethoxypyrrolidine" *Organic Syntheses* **1985**, *63*, 206–213.
6. For an excellent introduction to stereoelectronic effects including the anomeric effect see: Deslongschamps, P. "Stereoelectronic Effects in Organic Chemistry" Pergamon Press, 1983 (375 pages).
7. Leonard, N. J.; Hay, A. S.; Fulmer, R. W.; Gash, V. W. "Unsaturated Amines. III. Introduction of α,β-Unsaturation by Means of Mercuric Acetate: $\Delta^{1(9a)}$-Dehydroquinolizidine" *J. Am. Chem. Soc.* **1955**, *77*, 439–444.
8. Plattner, Pl. A.; Fürst, A. "Ueber Steroide und Sexualhormone: uber eine Ergiebige Method zur Herstellung des Epi-cholesterins und uber das 3α,5-Dioxy-cholestan" *Helv. Chem. Acta* **1948**, *31*, 1455–1463. Fürst, A.; Plattner, Pl. A. "Ueber Steroide und Sexualhormone: 2α,3α- und 2β,3β-Oxidocholestane; Konfiguration der 2-Oxy-cholestane" *Helv. Chem. Acta* **1949**, *32*, 275–283.
9. Seeman, J. I. "The Curtin-Hammett Principle and the Winstein-Holness Equation. New Definition and Recent Extensions to Classical Concepts" *J. Chem. Educ.* **1986**, *63*, 42–48. Winstein, S.; Holness, N. J. "Neighboring Carbon and Hydrogen. XIX. *tert*-Butylcyclohexyl Derivatives. Quantitative Conformational Analysis" *J. Am. Chem. Soc.* **1955**, *77*, 5562–5578. Curtin, D. Y. "Stereochemical Control of Organic Reactions. Differences in Behavior of Diastereoisomers. I. Ethane Derivatives. The Cis Effect" *Rec. Chem. Progr.* **1954**, *15*, 111–128. An alternative to the Curtin-Hammett argument presented on page 295 is that the epoxide reacts with thiophenoxide from conformation **64b** such that the product is born in the B-ring boat (or twist-boat) conformation of **65b**.

10. Borch, R. F.; Bernstein, M. D.; Durst, H. D. "Cyanohydridoborate Anion as a Selective Reducing Agent" *J. Am. Chem. Soc.* **1971**, *93*, 2897–2904.
11. Bischler, A.; Napieralski, B. "Zur Kenntniss einer Neuen Isochinolinsynthese" *Chem. Ber.* **1893**, *26*, 1903–1908. Whaley, W. M.; Govindachari, "The Preparation of 3,4-Dihydroisoquinolines and Related Compounds by the Bischler-Napieralski Reaction" *Organic Reactions* **1951**, *6*, 74–150.
12. Pictet, A.; Spengler, T. "Über die Bildung von Isochinolin-Derivaten durch Enwirkung von Methylal auf Phenyl-üthylamin, Phenyl-alanin und Tyrosin" *Chem. Ber.* **1911**, *44*, 2030–2036. Whaley, W. M.; Govindachari, "The Preparation of 3,4-Dihydroisoquinolines and Related Compounds by the Bischler-Napieralski Reaction" *Organic Reactions* **1951**, 6, 151–206.
13. Challand, B. D.; Hikino, H.; Kornis, G.; Lange, G.; De Mayo, P. "Photochemical Synthesis. XXV. Photochemical Cycloaddition. Some Applications of the use of Enolized *β*-Diketones" *J. Org. Chem.* **1969**, *34*, 794–806.
14. Krow, G. R. "The Baeyer-Villiger Oxidation of Ketones and Aldehydes" *Organic Reactions* **1993**, *43*, 251–798. Baeyer, A.; Villiger, V. "Ueber die Einwirkung des Caro'schen Reagens auf Ketone." *Chem. Ber.* **1900**, *33*, 858–864.
15. Finkelstein, H. "Darstellung Organischer Jodid aus den Entsprechenden Bromiden und Chloriden" *Chem. Ber.* **1910**, *43*, 1528–1532.
16. Fleming, I. "Silyl-to-Hydroxy Conversion in Organic Synthesis" *Chemtracts: Organic Chemistry* **1996**, *9*, 1–64. Tamao, K.; Ishida, N.; Kumada, M. "(Diisopropoxymethylsilyl)methyl Grignard Reagent: A New, Practically Useful Nucleophilic Hydroxymethylating Agent" *J. Org. Chem.* **1983**, *48*, 2120–2122.
17. Stork, G.; Danheiser, R. L. "Regiospecific Alkylation of Cyclic *β*-Diketone Enol Ethers. General Synthesis of 4-Alkylcyclohexenones" *J. Org. Chem.* **1973**, *38*, 1775–1776.

Problems

1. What is the "other ketone" that could be converted to porantherine (**1**) in the same manner as ketone **2**? (Alkaloids-1)
2. Outline a synthesis of **9** that passes through only "odd" difunctional intermediates. (Alkaloids-1)
3. Suggest mechanisms for the cyclization of **12**→**13** and for the oxidation of **13** → **14**. (Alkaloids-2)
4. Suggest alternative tactics for the demethylation of **13**. Do your alternative tactics call for a change in the sequence of steps leading from **13** → **15**? (Alkaloids-2)
5. Suggest mechanisms for the oxidation of **31** → **32** and for the conversion of **2** → **1**. (Alkaloids-4)
6. Suggest alternative tactics for the oxidation of **31** → **32**. (Alkaloids-2)
7. Apply the stereoelectronic principles suggested by Stevens (Alkaloids-3) to the conversion of **34** → **38**. Show how this notion rationalizes the stereochemical course at the newly-formed stereogenic center in the piperidine ring of **38**. (Alkaloids-5)
8. Explain why oxidation of the nitrogen was not problematic in the last step of this synthesis. (Alkaloids-9)
9. Predict the product expected from the following reactions (Bischler-Napieralski, Pictet-Spengler and related reactions). (Alkaloids-10)

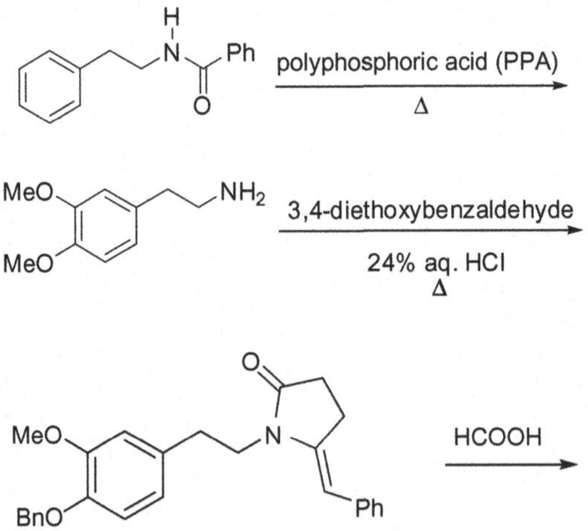

Whaley, W. M.; Hartung, W. H. "Synthesis of Isoquinoline Derivatives" *J. Org. Chem.* **1949**, *14*, 650–654. Weinbach, E. C.; Hartung, W. H.

"Synthesis of Tetrahydroisoquinoline Derivatives" *J. Org. Chem.* **1950**, *15*, 676–679. Burnett, D. A.; Hart, D. J. "Conformational Effects on the Oxidative Phenolic Coupling of Benzyltetrahydroisoquinolines to Morphinan and Aporphine Alkaloids" *J. Org. Chem.* **1987**, *52*, 5662–5667.

10. Draw the two chair-chair conformations for each *cis*-decalin shown below and indicate which conformation will be more stable. (Alkaloids-11)

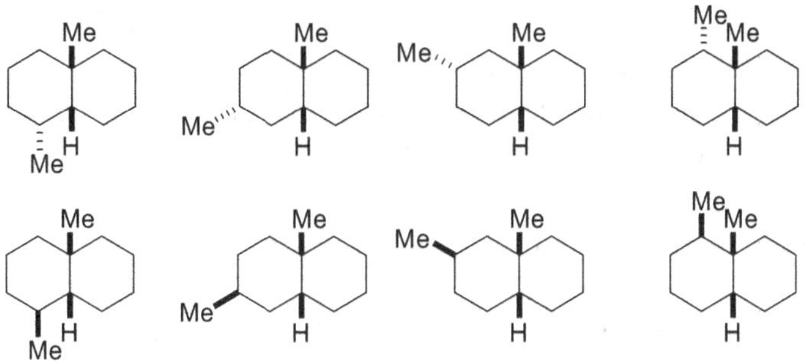

11. Rationalize the stereochemical course of the addition of methoxide to unsaturated ester **93** using stereoelectronic arguments. (Alkaloids-11)
12. Provide a mechanism for the conversion of **99** → **100**. (Alkaloids-12)
13. Outline a synthesis of 6-methoxytryptamine. (Alkaloids-12)
14. Provide mechanistic details for the conversion of **107** → **108** by each of the mechanisms suggested in Alkaloids-13. (Alkaloids-13)
15. Provide a mechanism for the conversion of **112** → **113**. (Alkaloids-14)
16. Can you provide a mechanistic rationale for the stereochemical course of the conversion of **113** → **114** + **115**? for the conversion of **110** → **117** + **118**? (Alkaloids-14)
17. Provide a mechanistic rationale for the stereochemical course of the protonation in the conversion of **130** → **131**. Explain why thermodynamics favor **131** over its C_{17} diastereomer. (Alkaloids-16)
18. A Moffat oxidation was used to convert **137** → **138** (R=CHO). A minor product in this reaction was **138** (R=CH_2OCH_2SMe). Provide a mechanistic explanation for formation of this product. (Alkaloids-17)
19. Provide structures of intermediates in the conversion of **138** → **139**. (Alkaloids-17)
20. Propose a mechanism for the conversion of **139** → **140**. (Alkaloids-17)

21. Propose a reaction sequence that might convert **142** (were it available) to the following compound. (Alkaloids-18)

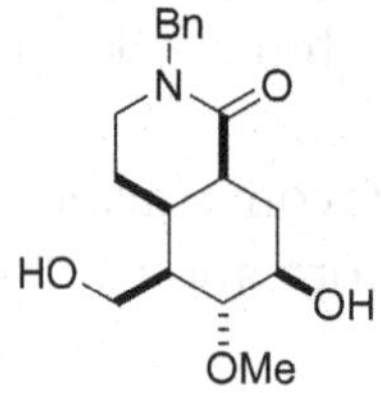

22. 2-Ethylhexanoic acid is a commodity chemical (very inexpensive). Propose a synthesis of this acid that you think might be amenable to synthesis on a ton-scale. (Alkaloids-19)
23. Note that the sequence going from racemic **150** → **157** necessarily will involve diastereomeric mixtures if racemic 2-ethylhexanoic acid is used in the conversion of **150** → **151**. Explain. Note that this is also true even if one enantiomer of 2-ethylhexanoic acid is used in this step. Explain. (Alkaloids-19)
24. Provide the products along the way from **153** → **154**. (Alkaloids-19)
25. Lactones have some interesting properties relative to their their acyclic cousins (esters). Why is the lactone reduced faster than the esters in the reduction of **162**? Why is ethyl acetate water insoluble, but γ-butyrolactone is completely water miscible? (Alkaloids-21)
26. Suggest a mechanism (or several mechanisms) for the conversion of **173** → **174** + **175**. (Alkaloids-22)
27. Suggest a mechanism for the Tamao-Fleming portion of the conversion of **192** → **193**. (Alkaloids-24)
28. Elaborate on the use of "stereoelectronic considerations" to explain the conversion of **179** → **196**. Discuss the possibility that the gauche effect also plays a role in the stereochemical course of this transformation. (Alkaloids-25)

CHAPTER 9

Alkaloids from "Dart-Poison" Frogs

Alkaloids from "Dart-Poison" Frogs

One prolific source producer (or maybe not ... read on) of alkaloids are dart-poison frogs, a family of colorful and petite amphibians native to Costa Rica, Panama, Ecuador, and Colombia. The intense coloration of these frogs serves as a warning to potential predators who would have an unpleasant experience if they tried to make a meal of one of these tiny creatures. This is because the frogs secrete toxic alkaloids from glands on the surface of their skin. These secretions are so toxic that they have been used by locals to poison blowgun darts used in hunting, hence the name dart-poison frogs. It has recently been shown that, in many cases, the frogs actually injest the alkaloids from a dietary source (ants) before secreting them as a defense substance. It has also been shown that frogs raised in captivity do not secrete these alkaloids.

Scientists from the National Institutes of Health have isolated and determined the structures of many of the compounds present in these skin secretions using the techniques of mass spectrometry and NMR spectroscopy. So far the structures of well over 200 different alkaloids have been determined. The most potent of these toxins is batrachotoxin (from *Phyllobates terribilis*), an example of a steroidal alkaloid. Other toxins include histrionicotoxin (from *Dendrobates histrionicus*), pumiliotoxin C (from *Dendrobates pumilio*), and gephyrotoxin.

Because many of these compounds are produced in only microgram quantites by the frogs, laboratory syntheses have often been developed to confirm structures and also to provide a supply of material for pharmacological studies. It turns out that most of these toxins act on the nervous system by affecting the manner in which ions are transported across cell membranes. As a result of this property, several of these alkaloids are now used as research tools in the field of neuroscience. We will spend this chapter examining selected syntheses of the aforementioned toxins (with the exception of batrachotoxin). Let us begin with histrionicotoxin.

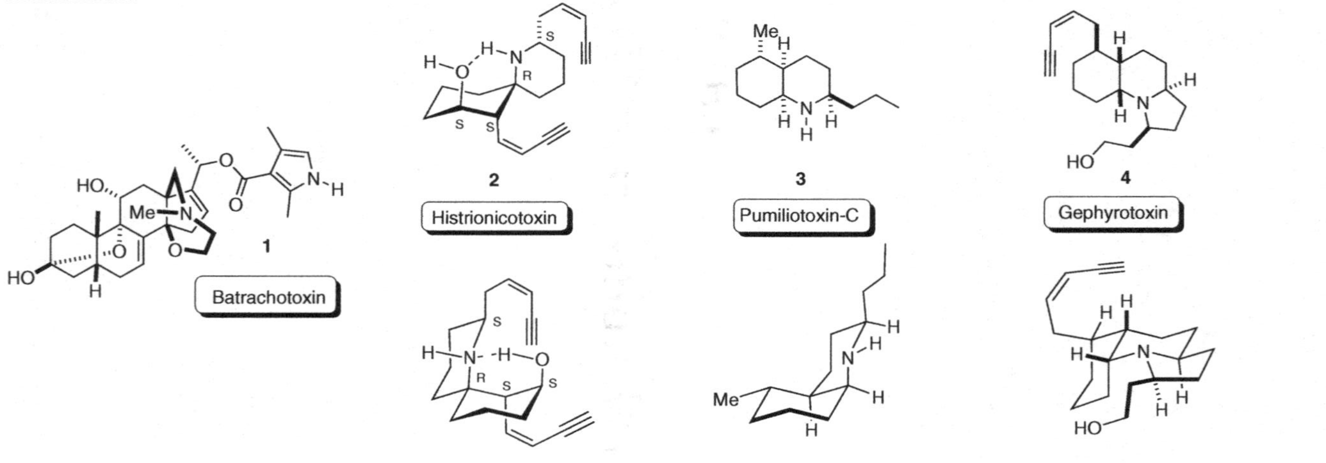

Histrionicotoxin-1

Histrionicotoxin-1

One prolific producer (or maybe not ... read on) of alkaloids are dart-poison frogs, a family of colorful and petite amphibians native to Costa Rica, Panama, Ecuador, and Colombia. The intense coloration of these frogs serves as a warning to potential predators who would have an unpleasant experience if they tried to make a meal of one of these tiny creatures. This is because the frogs secrete toxic alkaloids from glands on the surface of their skin. These secretions are so toxic that they have been used by locals to poison blowgun darts used in hunting, hence the name dart-poison frogs. It has recently been shown that, in many cases, the frogs actually injest the alkaloids from a dietary source (ants) before secreting them as a defense substance. It has also been shown that frogs raised in captivity do not secrete these alkaloids.

Scientists from the National Institutes of Health (John Daly and coworkers) have isolated and determined the structures of many of the compounds present in these skin secretions using the techniques of mass spectrometry and NMR spectroscopy. So far the structures of well over 200 different alkaloids have been determined.[1] The most potent of these toxins is batrachotoxin (**1**) (from *Phyllobates terribilis*), an example of a steroidal alkaloid. Other toxins include histrionicotoxin (**2**) (from *Dendrobates histrionicus*), pumiliotoxin C (**3**) (from *Dendrobates pumilio*), and gephyrotoxin (**4**).

Because many of these compounds are produced in only microgram quantities by the frogs, laboratory syntheses have often been developed to help assign structures and also to provide a supply of material for pharmacological studies. It turns out that most of these toxins act on the nervous system by affecting the manner in which ions are transported across cell membranes. Due to this property, several of these alkaloids are now used as research tools in the field of neuroscience. In this chapter we will examine selected syntheses of the aforementioned toxins (with the exception of batrachotoxin). We will begin with histrionicotoxin (**2**).

Histrionicotoxin

A number of histrionicotoxins have been characterized. These differ largely in the position of unsaturation on the two "side chains". Whereas the parent natural product eluded synthesis for many years, the saturated version of this compound, perhydrohistrionicotoxin, was first prepared by Kishi [Aratani, M.; Dunkerton, L. V.; Fukuyama, T.; Kishi, Y.; Kakoi, H.; Sugiura, S.; Inoue, S. "Synthetic studies on histrionicotoxins. I. A stereocontrolled synthesis of *dl*-perhydrohistrionicotoxin" *J. Org. Chem.* **1975**, *40*, 2009-2011; Fukuyama, T.; Dunderton, L. V.; Aratani, M.; Kishi, Y. "Synthetic Studies on Histrionicotoxins. II. A Practical Synthetic Route to *dl*-Perhydro- and *dl*-Octahydrohistrionicotoxin" *J. Org. Chem.* **1975**, *40*, 2011-2012] and Corey [Corey, E. J.; Arnett, J. F.; Widger, G. N. "A simple total synthesis of *dl*-perhydrohistrionicotoxin" *J. Am. Chem. Soc.* **1975**, *97*, 430-431] and indeed, by many other groups over the years. We will simply look at the Corey and Kishi approaches retrosynthetically and will only examine two syntheses of perhydrohistrionicotoxin in detail. The first will introduce the use of *N*-acyliminium ions as electrophiles and the second will revisit *N*-acylnitroso compounds as reactive species for use in alkaloid synthesis.

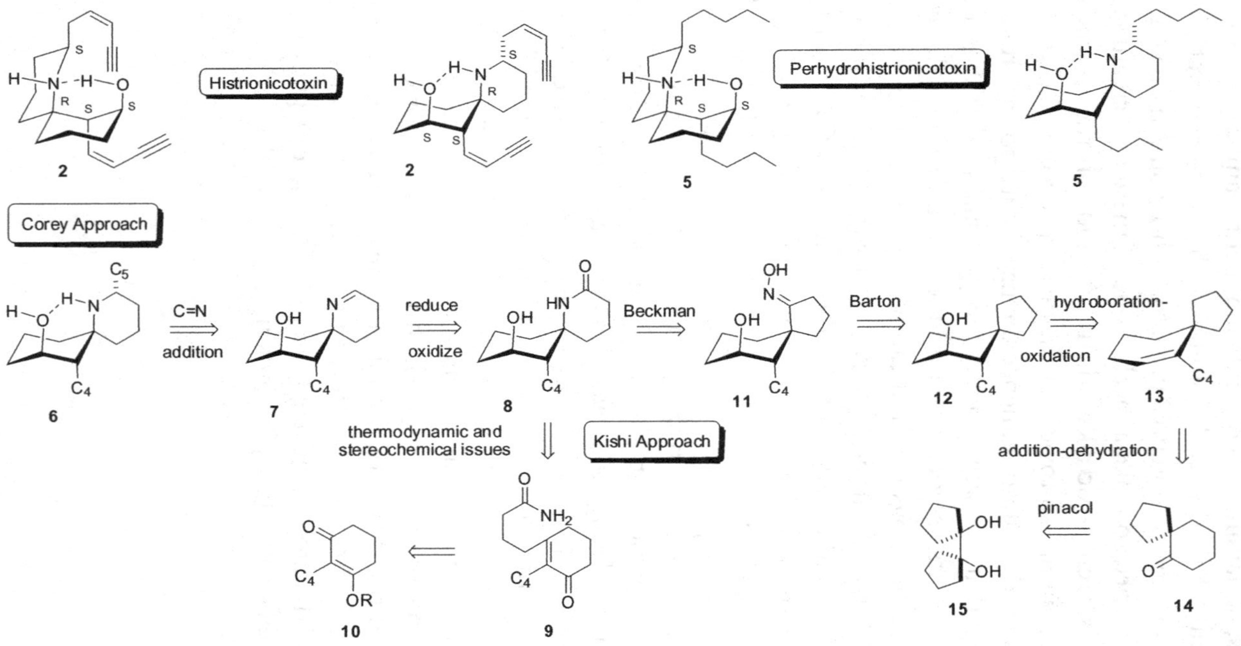

Histrionicotoxin-2

Histrionicotoxin-2

A number of histrionicotoxins have been characterized. These differ largely in the position of unsaturation on the two "side chains". Whereas the parent natural product eluded synthesis for many years, the saturated version of this compound, perhydrohistrionicotoxin (**5**), was first prepared by Kishi and Corey, and indeed, by many other groups thereafter. We will simply look at the Corey and Kishi approaches retrosynthetically and will then examine two syntheses of perhydrohistrionicotoxin in detail. The first synthesis will introduce the use of *N*-acyliminium ions as electrophiles, and the second will revisit *N*-acylnitroso compounds as reactive species for use in alkaloid synthesis.

Both the Kishi and Corey approaches to **5** proceeded through imine **7** and lactam **8**, the pentyl sidechain being introduced by addition of an organometallic reagent to the imine. Kishi approached **8** using an intramolecular conjugate addition strategy (**9** → **8**) This approach was plagued by thermodynamic and stereochemical issues, and several "corrections" were needed along the way. Cyclohexenone **9** was prepared from **10**. It is notable that this strategy passes through intermediates with 1,3- and 1,5-difunctional relationships, and relies largely on normal chemistry of the carbonyl group. Corey approached **8** via a Beckman rearrangement of oxime **11**, which came, in turn, from **12** via a Barton reaction.[2] Alcohol **12** was prepared by hydroboration-oxidation of **13**. The sequence from **13** → **12** → **11** → **8** nicely addresses the relative stereochemistry at the three contiguous stereogenic centers in perhydrohistrionicotoxin (**5**). Alkene **13** was prepared from **15** using a pinacol rearrangement to establish the spiro[4.5]decane ring system.

N-Acyliminium Ion Cyclizations

Some reactive intermediates

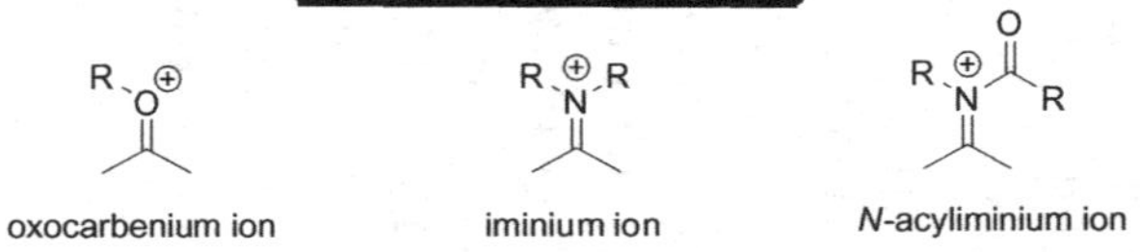

Early Results from Stork Lycopodine Synthesis

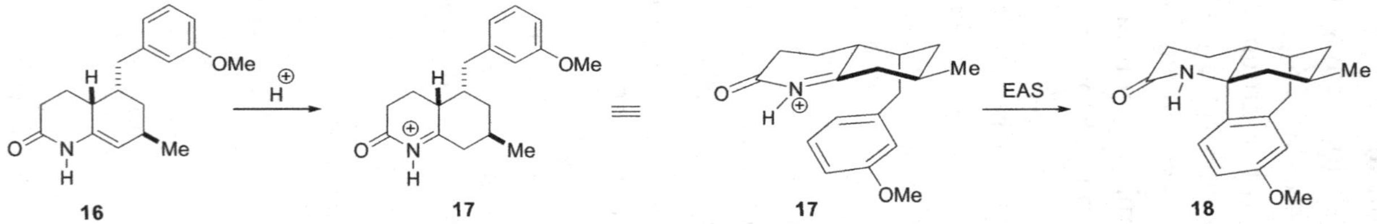

N-Acyliminium Ions as Olefin Cyclization Substrates

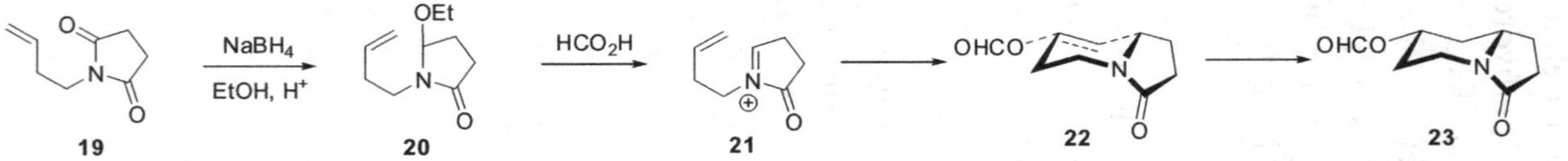

Other Reducing Agents: $NaBH_4$/MeOH or *i*-Bu_2AlH/PhMe

Initial work from group of Nico Speckamp. Looks like "anti addition" of electrophile and nucleophile across double bond. Work of Johnson served as inspiration for initial studies. Notice the construction of a 1,3-N,O relationship.

Histrionicotoxin-3

Histrionicotoxin-3

We will examine two very different approaches to perhydrohistrionicotoxin. The first one revolves around the chemistry of *N*-acyliminium ions, a field largely developed by the Speckamp group in the Netherlands.[3,4] In Chapter 8 we saw that iminium ions (the nitrogen analogs of oxocarbenium ions) are excellent electrophiles (Mannich reaction). *N*-Acyliminium ions are even more electrophilic than iminium ions. They react with both heteroatom and carbon nucleophiles as expected. An early use of an *N*-acyliminium ion in synthesis is the cyclization of enamide **16** to **18** in an electrophilic aromatic substitution reaction. This reaction was an early step in Stork's approach to lycopodine.[5] The reaction presumably proceeds through *N*-acyliminium ion **17**, generated by protonation of the enamide.

N-Acyliminium ions are "hot" and will even add to unactivated olefins. For example, *N,O*-acetal **20** reacts with formic acid to provide **23**. The stereochemistry of the process is relatively clean and can be rationalized by an *anti*-periplanar addition of electrophilic carbon and nucleophilic oxygen across the carbon-carbon double bond with a transition state that resembles a chair *N*-acylpiperidine. In reality, this process is mechanistically more complex than this simple model (for example, possible carbocation intermediates and in some cases, some underlying sigmatropic rearrangements), but the model is simple, easy to remember, and has excellent predictive value. How was this chemistry used to prepare perhydrohistrionicotoxin?

Synthesis of Perhydrohistrionicotoxin

Schoemaker, H. E.; Speckamp, W. N. "Stereocontrolled Synthesis of Functionalized 1-Azaspirans. Efficient Synthesis of Perhydrohistrionicotoxin" *Tetrahedron* **1980**, *36*, 951-958. For a related approach see Evans, D. A.; Thomas, E. W. "A formal Synthesis of *dl*-Perhydrohistrionicotoxin via α-Acylimonium Ion-Olefin Cyclizations" *Tetrahedron Lett.* **1979**, 411-414.

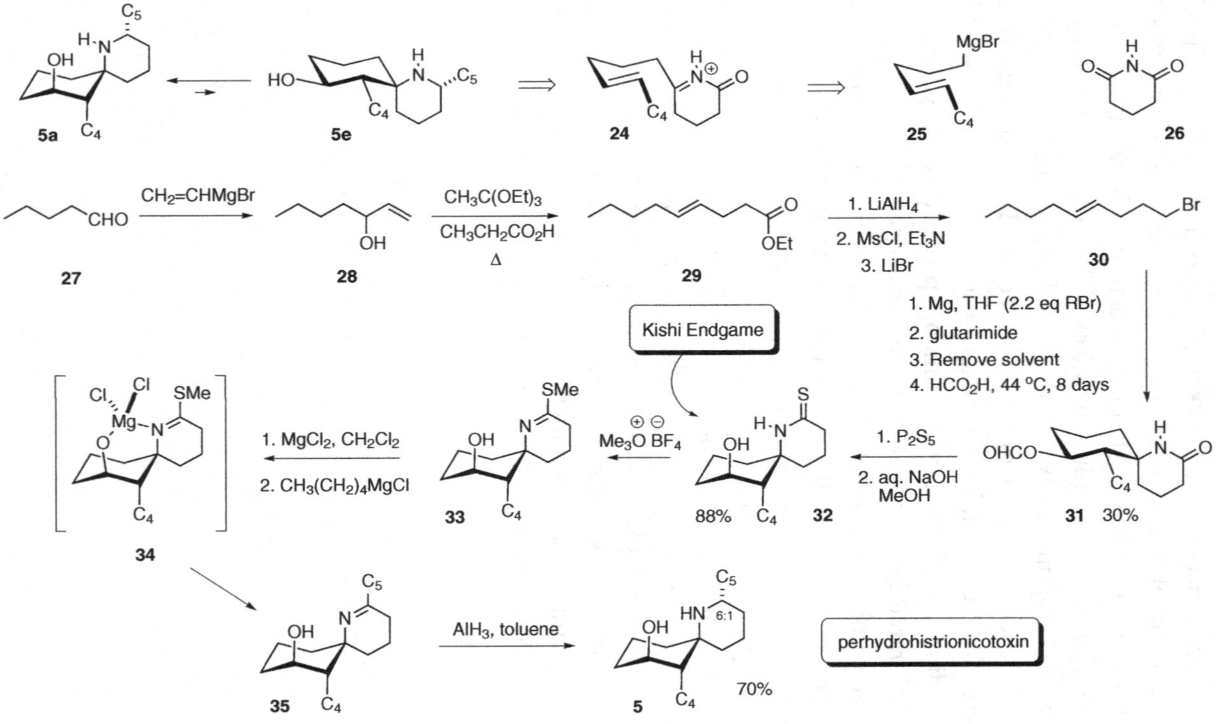

Histrionicotoxin-4

Histrionicotoxin-4

The first thing to recognize is that the *N*-acyliminium ion cyclization we just examined generates a 1,3-*N,O* relationship, precisely the kind of relationship that appears in lactam **8** (see Histrionicotoxin-2). This raises the question, can **8** be prepared by an *N*-acyliminium ion cyclization? Two chair-chair conformations are available to perhydrohistrionicotoxin (**5a** and **5e**). Examination of **5e** suggested that an *N*-acyliminium ion of type **24** might undergo cyclization to provide lactam **8** (see Histrionicotoxin-2). The key issue here is the regiochemical course of addition to the olefin. Will cyclization occur to afford a 5-membered ring or a 6-membered ring? It was felt that **24** could be obtained from Grignard reagent **25** and glutarimide (**26**).

The bromide precursor of **25** was prepared from pentanal (**27**) in a straightforward manner. Notice the use of a Claisen rearrangement to carry out an "enforced S_N2' reaction" in the preparation of γ,δ-unsaturated ester **29** (see Prostaglandins-24). Treatment of glutarimide with 2.2 equivalents of Grignard reagent **25** (one equivalent to deprotonate the imide), and treatment of the resulting mixture of materials with formic acid, gave **31** albeit in modest yield. The synthesis was completed using the endgame developed by Kishi. Conversion of **31** to thiolactam **32** was followed by transformation to thioimidate **33**. Exposure of **33** to magnesium chloride (to presumably generate complex **34**) followed by pentylmagnesium chloride gave **35**. Alane reduction of **35** completed the synthesis of **5**.

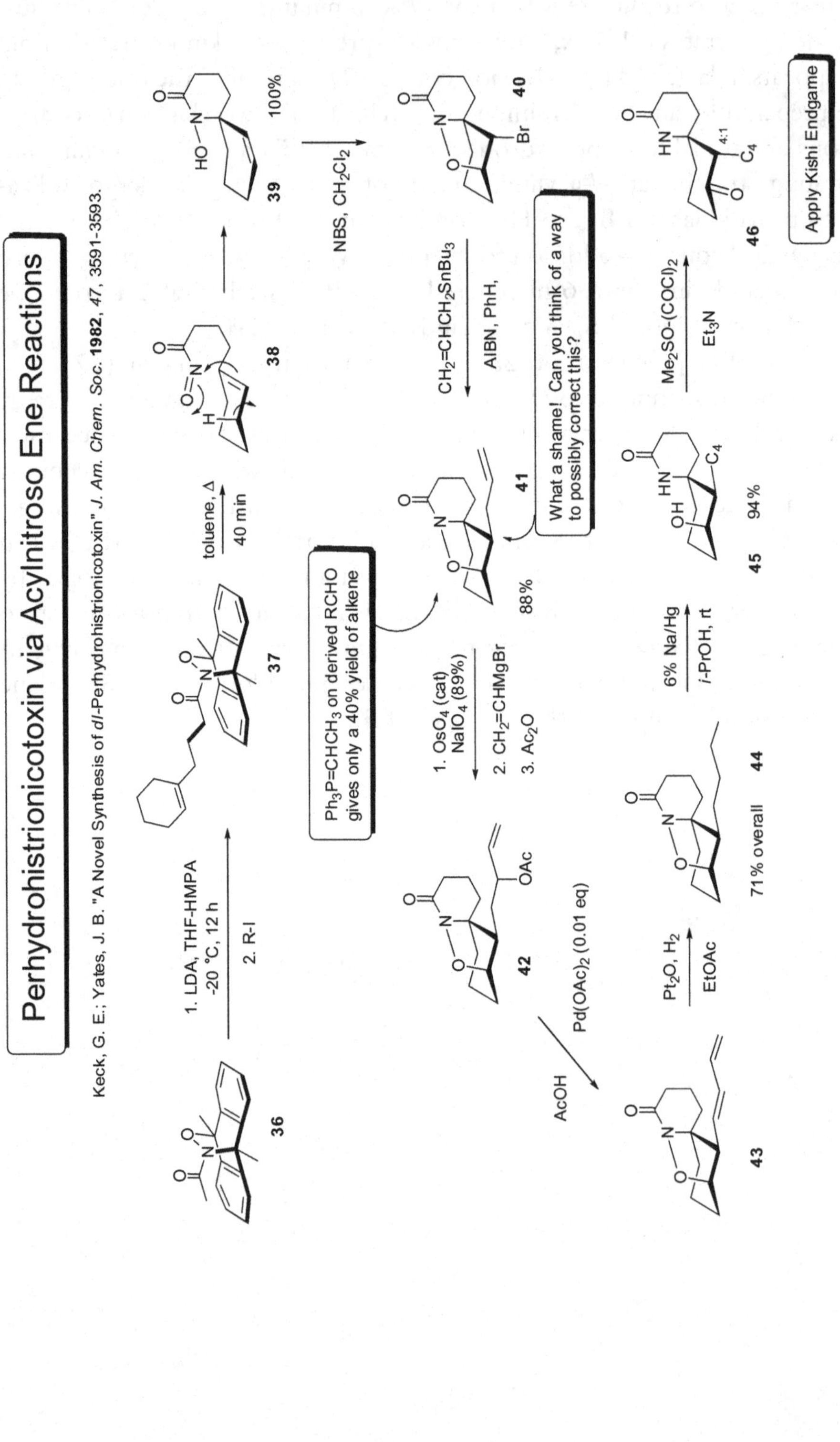

Histrionicotoxin-5

Histrionicotoxin-5

The next perhydrohistrionicotoxin synthesis we will examine was reported by the Keck group (Utah). This synthesis featured chemistry of *N*-acylnitroso compounds, just as did the Keck synthesis of pyrrolizidines (Chapter 4). Rather than approaching this in a "retrosynthetic manner", suffice it to say that this work intersects with the Kishi synthesis at the point of keto-lactam **46**. Now we will jump right into the synthesis.

The key reaction was an intramolecular ene reaction of acylnitroso compound **38**. Recall that acylnitroso compounds are very reactive and thus, the plan (as in the Keck synthesis of pyrrolizidines) was to generate this species by a *retro*-Diels-Alder reaction of **37**, in turn prepared by alkylation of the enolate derived from **36**. The ene reaction worked well to provide **39**. An electrophile initiated cyclization of **39** provided bromide **40**. The next task was introduction of the 4-carbon chain. The plan was to accomplish this using another reaction developed largely in the Keck laboratories, the allylation of free radicals using allyl tri-*n*-butylstannane.[6] This reaction also proceeded in good yield, but the sole product was **41**. The stereochemistry of the pendant chain was opposite to that required by the target. Addition of the fourth carbon took considerable effort, but was accomplished in good yield as shown, to ultimately provide **44**. Reduction of the N—O bond gave **45**. Swern oxidation of the alcohol followed by epimerization provided **46** (along with 20% of the diastereomeric ketone), and the synthesis was completed using the Kishi endgame.

This synthesis featured some innovative chemistry, but suffered in terms of stereochemistry. It is a shame that the allylation did not proceed with the desired stereochemistry because this would have rendered the synthesis extremely efficient.

First Synthesis of Racemic Histrionicotoxin

Carey, S. C.; Aratani, M.; Kishi, Y. "A Total Synthesis of *dl*-Histrionicotoxin" *Tetrahedron Lett.* **1985**, *26*, 5887-5890.

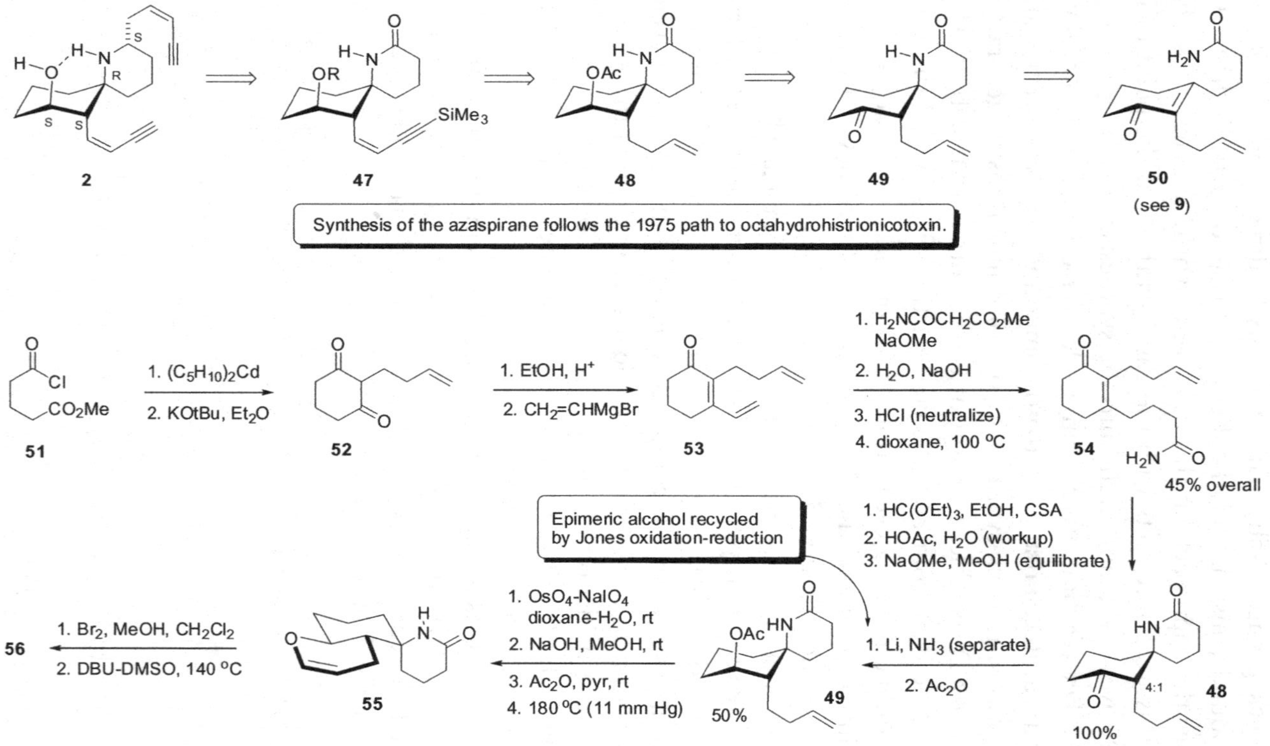

Histrionicotoxin-6

Histrionicotoxin-6

Many of the dart-poison frog alkaloids carry conjugated eneyne groups. This is part of the challenge for laboratory syntheses of these alkaloids. Thus, histrionicotoxin (**2**) is more of a challenge than perhydrohistrionicotoxin (**5**). As a result, fewer syntheses have been reported. We will look at three very different approaches to **2**, beginning with the Kishi synthesis of racemic **2**. The beginning of this synthesis resembles the Kishi approach to **5**. The plan was to roughly follow a path from **50** → **49** → **48**. The 3-buten-1-yl group in **48** was to serve as a handle for introducing the 4-carbon eneyne, and the lactam was to be a handle for introduction of the 5-carbon enyne, not necessarily in that order.

The preparation of **54** (same as **50** and similar to **9**) passed through intermediates with only odd-difunctional relationships and used "normal" carbonyl chemistry (1,2-carbonyl addition, aldol-dehydration, conjugate addition, hydrolysis-decarboxylation). Conditions were also found to coax **54** to undergo the intramolecular conjugate addition to provide **48** (and its epimer α to the ketone). An unselective reduction of the ketone provided **49** after esterification of the intermediate alcohol.

The strategy for controlling stereochemistry of the 4-carbon enyne was to establish the *cis*-olefin in a ring, and then open the ring. Thus, the butenyl sidechain was degraded to an aldehyde and then converted to cyclic enol ether **55**. Haloetherification of the olefin, followed by dehydrohalogenation of the secondary bromide, gave unsaturated acetal **56** (see Histrionicotoxin-7). Note that this transformation is formally just a halogenation-dehydrohalogenation of an aldehyde.

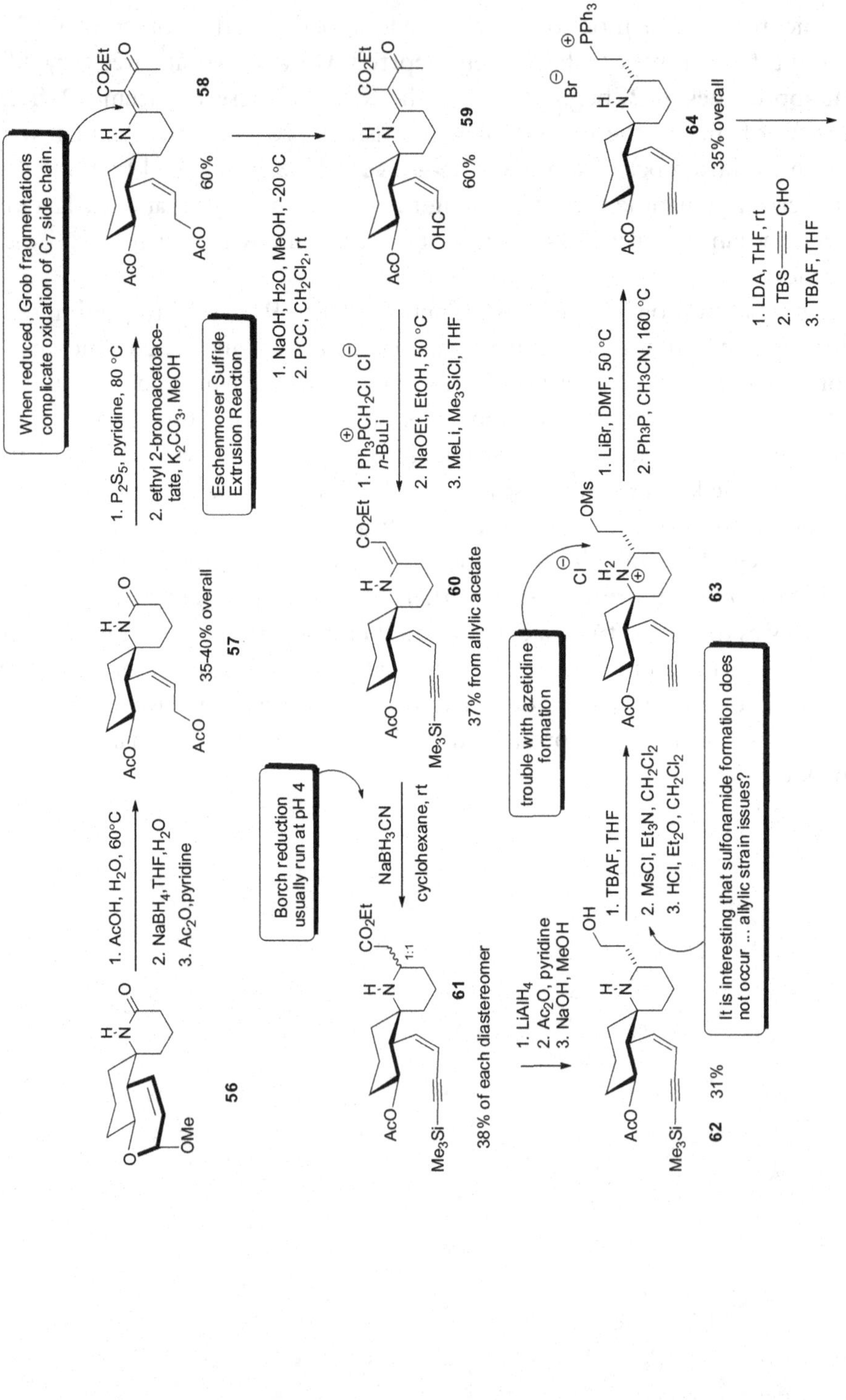

Histrionicotoxin-7

Histrionicotoxin-7

With the double bond in place, the ring was "opened" to provide **57**. Completion of the 4-carbon eneyne would eventually require homologation of the 3-carbon sidechain to a terminal alkyne via carbonyl addition chemistry to an aldehyde. It is notable that the aldehyde was "stored" at the alcohol oxidation state, presumably to avoid problems of olefin isomerization that might have been encountered in the aldehyde.

Next, work began on the 5-carbon eneyne. The plan for this portion of histrionicotoxin was to introduce a two-carbon unit (β-hydroxyethyl group) that could ultimately be reacted with an appropriate propargaldehyde (3-carbon unit) via a Wittig reaction. The first steps were toward introduction of the 2-carbon unit. Lactam **57** was converted to corresponding thiolactam, and an Eschenmoser sulfide contraction was used to prepare **58**.[7] Returning to the 4-carbon eneyne, the primary acetate was hydrolyzed. The resulting alcohol was oxidized to **59**. Wittig olefination of **59** followed by sequential removal of the acetyl group and conversion of the intermediate vinyl chloride to a TMS-protected terminal acetylene, provided **60** (Corey-Fuchs reaction).[8]

Installation of the rest of the 5-carbon eneyne came next. Sodium cyanoborohydride reduction of **60** gave a 1:1 mixture of diastereomeric β-aminoesters **61**. Lithium aluminum hydride reduction of **61** gave a diol which was converted to **62** using an esterification-hydrolysis sequence. It is interesting that the esterification reaction does not acylate the piperdine nitrogen. Perhaps allylic strain in the *N*-acyl derivative (not observed) is responsible for this unusual selectivity.[9] Removal of the TMS group and conversion of the primary alcohol to a mesylate gave **63** (as the amine hydrochloride). This amine salt was then converted to phosphonium salt **64** in the standard manner. Generation of the phosphorus ylid derived from **64**, Wittig olefination using TMS-propargaldehyde, and removal of the TMS group, gave **65** with excellent selectivity for the *Z*-geometrical isomer. The acetate was then hydrolyzed to give histrionicotoxin (**2**) (see Histrionicotoxin-8).

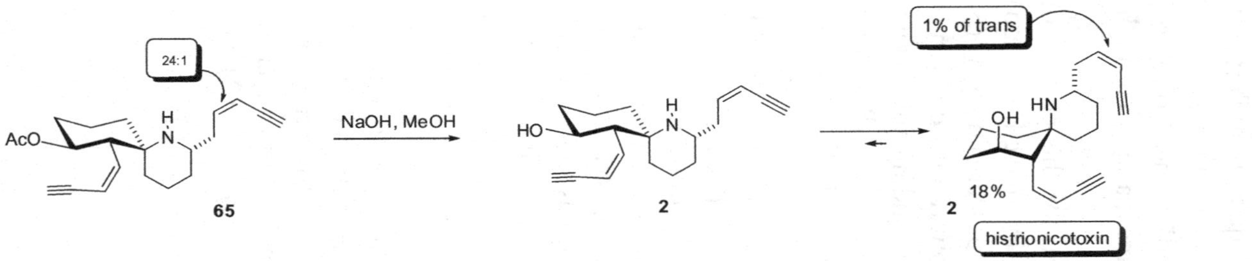

An Enantioselective Synthesis of Histrionicotoxin

Stork, G.; Zhao, K.; "Total Syntheses of (-)-Histrionicotoxin and (-)-Histrionicotoxin 235H" *J. Am. Chem. Soc.* **1990**, *112*, 5875.

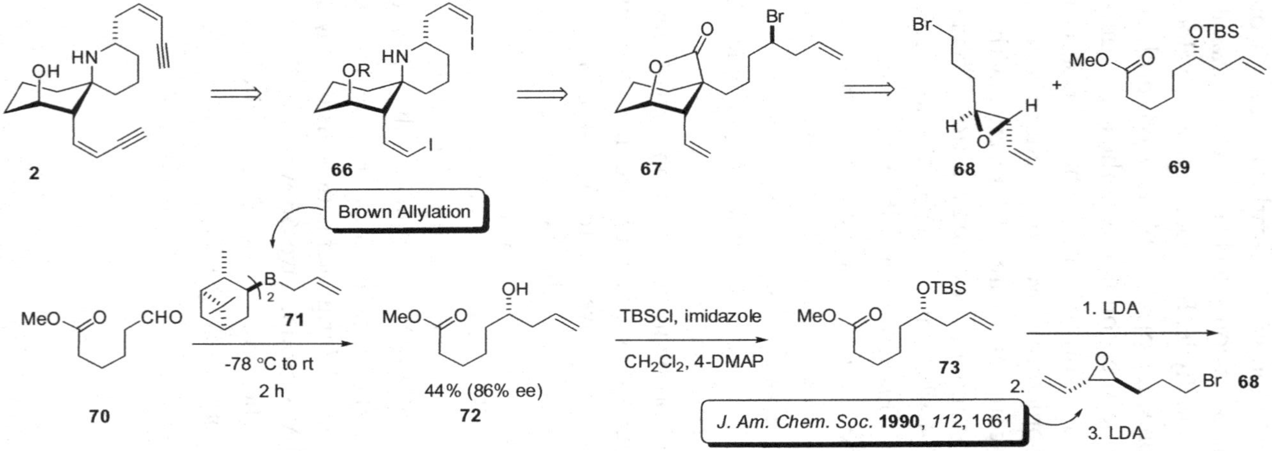

Histrionicotoxin-8

Histrionicotoxin-8

We will now examine an enantioselective route developed by Stork and Zhao. The plan was to use a "double Sonogashira" coupling to move from *bis*-vinyl iodide **66** to histrionicotoxin (**2**).[10] The *bis*-vinyl iodide was to be installed by a "double Wittig" reaction of a dialdehyde. The two olefins in **67** were to serve as "stable" precursors of this dialdehyde. The piperidine ring of **66** was to be installed by an intramolecular S_N2 reaction of a primary amine, wherein the bromide in **67** was to serve as the alkylating agent, and the lactone carbonyl group was to serve as a precursor to the amine. Lactone **67** was to be assembled by a "double alkylation" of ester **69** with *bis*-electrophile **68**. The hope was that the allylic position of vinyloxirane would be more reactive than the primary bromide, thus dictating the sequence of events associated with the double alkylation. This strategy offers an opportunity for enantioselectivity and, in fact, ***demands*** that both **68** and **69** be prepared as single enantiomers to avoid problems with diastereomer formation in the double alkylation. Also note that the regioselectivity projected for the oxirane opening is related to selectivity we saw in the opening of a geminally activated vinylcyclopropane in the Abraham approach to prostaglandins (Prostaglandins-27). So this part of the plan was not so far-fetched.

Aldehyde **73** was prepared from aldehyde **70** using a "Brown Allylation" to control absolute stereochemistry in the preparation of **72**.[11] Bromide **68** was prepared using a Sharpless epoxidation to control absolute stereochemistry.[12] Conversion of **73** to the corresponding enolate, alkylation with **68**, and addition of more LDA to generate a new enolate (**74**) gave a reasonable yield of **75** (see Histrionicotoxin-8/9).

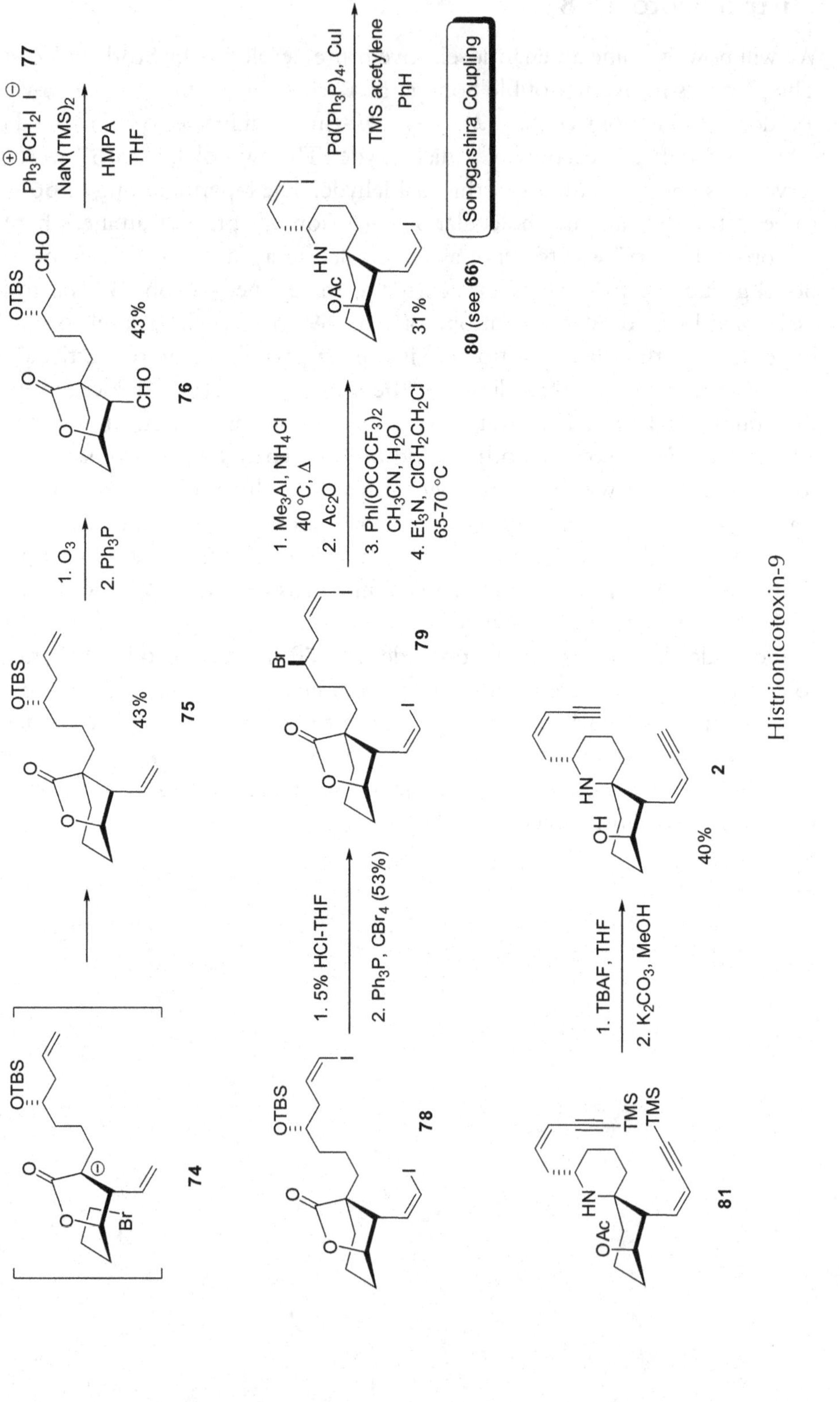
74
OTBS
Br
43%
75
1. O3
2. Ph3P
76
CHO
Ph3PCH2I I 77
NaN(TMS)2
HMPA
THF
1. 5% HCl-THF
2. Ph3P, CBr4 (53%)
78
79
1. Me3Al, NH4Cl
40 °C, Δ
2. Ac2O
3. PhI(OCOCF3)2
CH3CN, H2O
4. Et3N, ClCH2CH2Cl
65-70 °C
OAc
HN
31%
80 (see 66)
Pd(Ph3P)4, CuI
TMS acetylene
PhH
Sonogashira Coupling
81
TMS
1. TBAF, THF
2. K2CO3, MeOH
OH
40%
2
Histrionicotoxin-9

Histrionicotoxin-9

The design of the double alkylation is interesting. It was projected that the first alkylation would afford a lactone (enolate). The stereochemistry of the three stereogenic centers in **74** is derived from the stereochemistry of the starting materials. The stereochemistry of the reaction leading from **74** to **75** is dictated by the lactone stereogenic center carrying the 3-bromopropyl group. If lactonization preceeded the second alkylation, the stereochemical course of this transformation was guaranteed.

Continuing with the synthesis, ozonolysis of the *bis*-olefin gave dialdehyde **76**. *Bis*-Wittig olefination of **76** using the ylid derived from phosphonium salt **77** provided **78** with good control of *Z*-olefin geometry, a reaction developed largely in the Stork laboratories. The TBS protecting group was removed and the resulting alcohol was converted (with inversion of configuration) to bromide **79** (same as **67**). The lactone was opened to provide a primary amide. The secondary alcohol was protected as an acetate, a Curtius-type rearrangement was used to introduce the primary amine, and an intramolecular *N*-alkylation gave piperidine **80** (see projected intermediate **66**). The synthesis was completed by introducing the two *cis*-eneynes (Sonogoshira coupling), and removing the alcohol and acetylene protecting groups.

Nitrone Cycloaddition Route to Histrionicotoxin

Davison, E. C.; Fox, M. E.; Holmes, A. B.; Roughley, S. D.; Smith, C. J.; Williams, G. M.; Davies, J. E.; Raithby, P. R.; Adams, J. P.; Forbes, I. T. Press, N. J.; Thompson, M. J. "Nitrone dipolar cycloaddition routes to piperidines and indolizidines. Part 9. Formal Ssynthesis of (-)-pinidine and total synthesis of (-)-histrionicotoxin, (+)-histrionicotoxin and (-)-histrionicotoxin 235A" *J. Chem. Soc., Perkin 1* **2002**, 1494-1514. See also Smith, C. J.; Holmes, A. B.; Press, N. J. "The total synthesis of alkaloids (-)-histrionicotoxin 259A, 285C and 285E" *J. Chem. Soc., Chem. Commun.* **2002**, 1214-1215. For a related approach see Karatholuvhu, M. S.; Sinclair, A.; Newton, A. F.; Alcaraz, M-L.; Stockman, R. A.; Fuchs, P. L. "A Concise Total Synthesis of DL-Histrionicotoxin" *J. Am. Chem. Soc.* **2006**, *128*, 12656-12657.

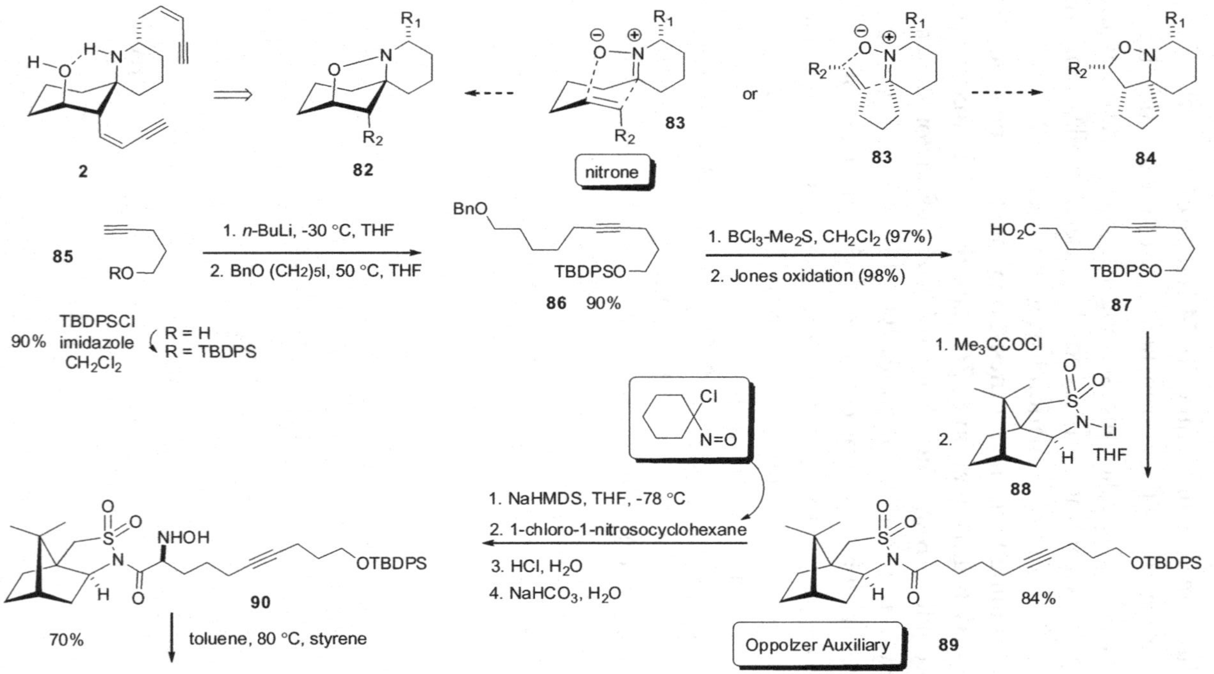

Histrionicotoxin-10

Histrionicotoxin-10

The final synthesis we will examine was reported by Holmes (Cambridge, UK) and a closely related synthesis has been reported by Fuchs (Purdue). We will examine the Holmes synthesis. This synthesis keys off the 1,3-*N,O* relationship in histrionicotoxin (**2**). Recall that this should trigger "nitrone cycloaddition" as a possible route to consider to this target. This suggests **82** as a reasonable target that could be moved forward to histrionicotoxin. The issue here was one of regiochemistry in the key cycloaddition. Would nitrone **83** undergo cycloaddition to provide **82** or **84**?

The key here turned out to be the choice of R_2 and a fine balance of steric and electronic effects. Whereas I refer you to the primary literature for the gory details, it turns out that when R_2 = C≡CTMS, the cycloaddition gave **84**, while when R_2 = C≡N (a more electron withdrawing group) the desired path was followed to provide a compound of type **82**. The synthesis is a nice example of how persistence, attention to detail, and a willingness to explore options resulted in achieving a synthesis via a plan that was initially discouraging in the lab. Let's go through some of the details.

The nitrone that led to completion of the synthesis is structure **97**, shown on Histrionicotoxin-11. This was prepared as a single enantiomer using asymmetric hydroxylamination chemistry developed by Oppolzer.[12] Thus alcohol **85** was converted to *N*-acylsulfonamide **89** via a reaction sequence that we will not bother to discuss. Generation of the enolate of **89**, followed by hydroxylamination using 1-nitrosochlorocyclohexane as a key reagent, gave hydroxylamine **90** in good yield and with a high diastereomeric excess.

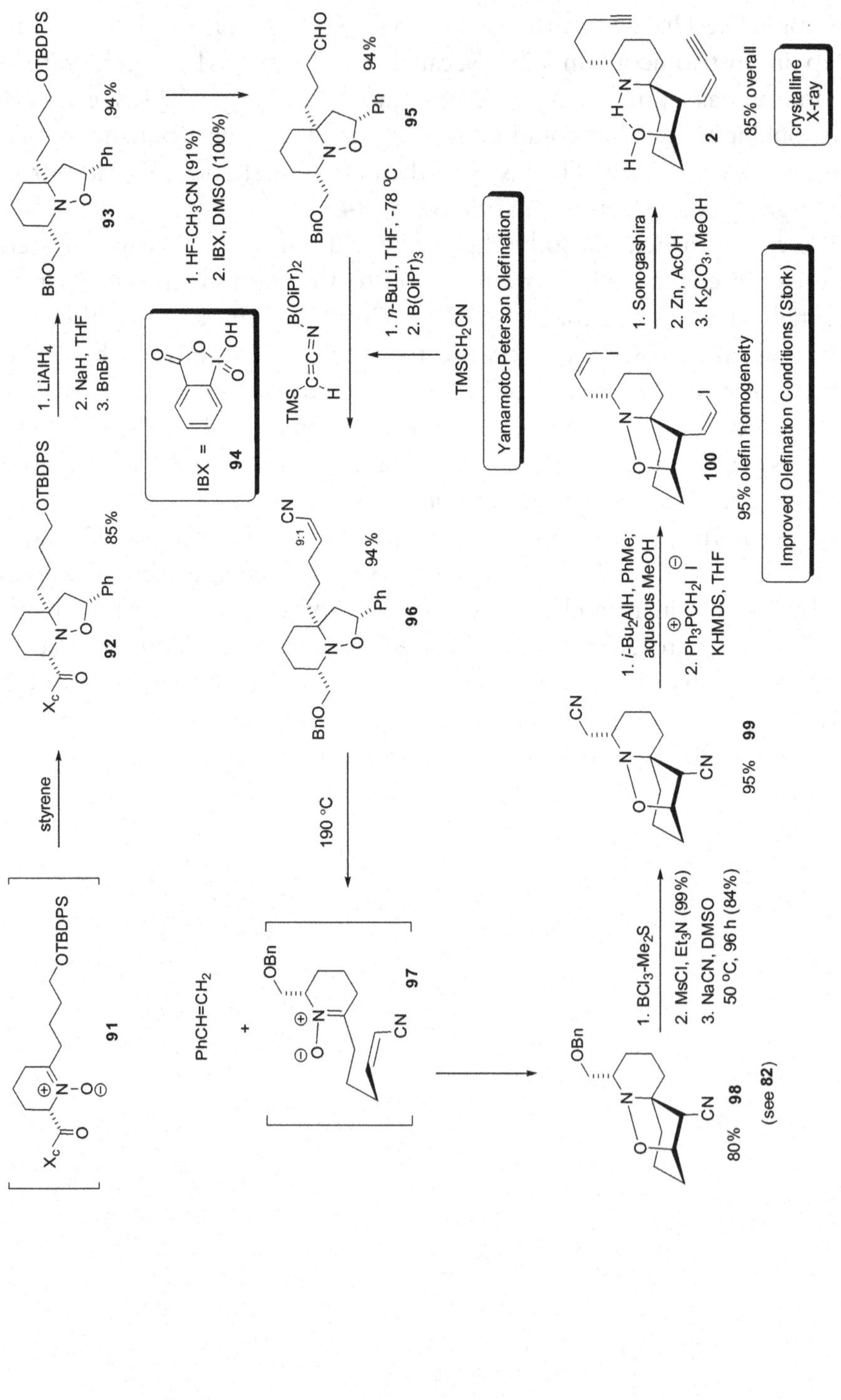

Histrionicotoxin-11

Histrionicotoxin-11

When **90** was heated in toluene, the hydroxylamine added to the alkyne to afford nitrone **91** (after tautomerization of a presumed intermediate *N*-hydroxyenamine) which was trapped by styrene in an intermolecular 1,3-dipolar cycloaddition to provide **92**. An oxidation state adjustment (with removal of the chiral auxiliary) gave **93**. The TBDPS protecting group was removed and the resulting primary alcohol was oxidized with IBX (**94**) (related to the Dess-Martin periodinane) to give aldehyde **95**. Application of the Yamamoto variation of the Peterson olefination gave **96** with decent control over olefin geometry.[13]

Heating of **96** gave the desired nitrone cycloaddition product (**98**) in excellent yield. This reaction involved a *retro*-1,3-dipolar cycloaddition, followed by the key intramolecular cycloaddition, another nice example of the use of a *retro*-cycloaddition to generate a reactive intermediate for use in a pericyclic reaction (recall the generation of acylnitroso compounds via a retro-Diels-Alder reaction).

The synthesis was completed by removal of the benzyl protecting group, homologation of the resulting primary alcohol, conversion of *bis*-nitrile **99** to a *bis*-aldehyde, Wittig olefination to give *bis*-vinyl iodide **100**, a double Sonogoshira reaction to install the eneynes, and reduction of the N—O bond. It is notable that this strategy was also used to prepare a number of other histrionicotoxins that differed by virtue of oxidation state along the 4- and 5-carbon sidechains.

This completes our look at histrionicotoxin. Let's move on to pumiliotoxin-C (**3**) a simpler target that has also received lots of attention.

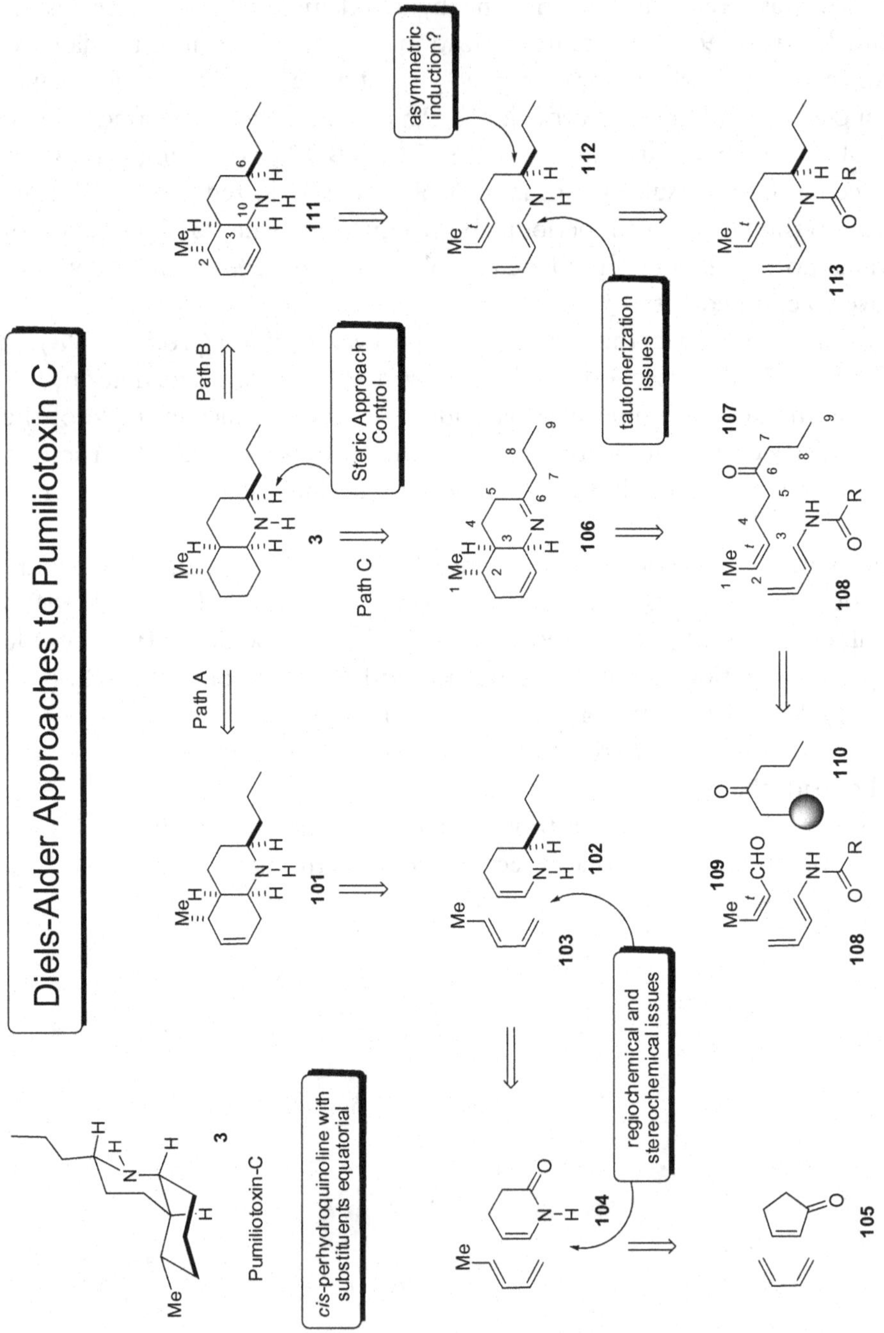

Pumiliotoxin-1

Pumiliotoxin-1

Pumiliotoxin-C (**3**) has been an extremely popular target for synthesis. This alkaloid is a simple *cis*-decahydroquinoline with four stereogenic centers. The most stable conformation of **3** places the two ring substituents (propyl and methyl groups) on equatorial sites. One of the rings is a cyclohexane and thus, it is tempting to develop strategies that revolve around Diels-Alder reactions. We examine several syntheses that adopted this approach, and then glance at a few syntheses that use alternative approaches.

Working backwards from **3**, there are six places one could "retrosynthetically" place a double bond in the cyclohexane substructure. Two are shown here. Path A leads back to cyclohexene **101**. Disconnection of **101** leads back to dienophile **102** and diene **103** (piperylene). This is not a good diene-dienophile pair. Enamine **102** is electron-rich, whereas **103** is also electron-rich. Furthermore, enamine **102** would exist in its tautomeric imine form. Even if **102** were to undergo cycloaddition with **103**, reaction would likely occur opposite the *n*-propyl group and afford an unusable diastereomer. One way to deal with both the tautomerization and stereochemical problems would be to use enamide **104** as the dienophile. This would require installation of the propyl sidechain at a later stage of the synthesis. Enamide **104**, however, is still not an electron-deficient dieneophile and regiochemical problems would be likely to plague any cycloaddition that might occur. This problem can be fixed by moving to **105** as the dienophile. The nitrogen could be installed with retention of stereochemistry using a Beckman rearrangement, but cycloadditions between **105** and piperylene would surely give the wrong regiochemistry. The regiochemistry problem can be solved by changing the diene to 1,3-butadiene and introducing the methyl group at a later stage of the synthesis. Overall this plan requires a lot of "fixing" to enable construction of the ring system using a Diels-Alder approach.

Path B works back through cyclohexene **111** and an intramolecular Diels-Alder reaction of dienamine **112**. Once again, enamine tautomerization is likely to present big problems, but once again this can be overcome by using dienamide **113**. This plan provides an opportunity to control absolute stereochemistry if **113** could be prepared in enantiopure form. The dienophile olefin geometry would afford the proper relative stereochemistry at C_2 and C_3, but stereochemistry relative to C_6 and C_{10} is not guaranteed.

Path C is an intermolecular variant of Path B. Thus, reduction of imine **106** would be expected to occur from the convex face of the perhydroquinoline to give the proper stereochemistry at C_6. Imine **106** might be expected to result from a *endo*-cycloaddition between dienamide **108** and

First Total Synthesis of Pumiliotoxin C

Oppolzer, W.; Frostl, W.; Weber, H. P. "The Total Synthesis of *dl*-Pumiliotoxin-C" *Helv. Chim. Acta* **1975**, *58*, 593. Oppolzer, W.; Flaskamp, E. "An Enantioselective Synthesis and the Absolute Configuration of Natural Pumiliotoxin-C" *Helv. Chim. Acta* **1977**, *60*, 204. Oppolzer, W.; Flaskamp. E.; Bieber. L. W. "Efficient Asymmetric Synthesis of Pumiliotoxin C via Intramolecluar [4+2] Cycloaddition" *Helv. Chim. Acta* **2001**, *84*, 141-145. For a related approach to pumiliotoxin-C and other Dendrobatid alkaloids see Banner, E. J.; Stevens, E. D.; Trudell, M. L.; *Tetrahedron Lett.* **2004**, *45*, 4411-4414.

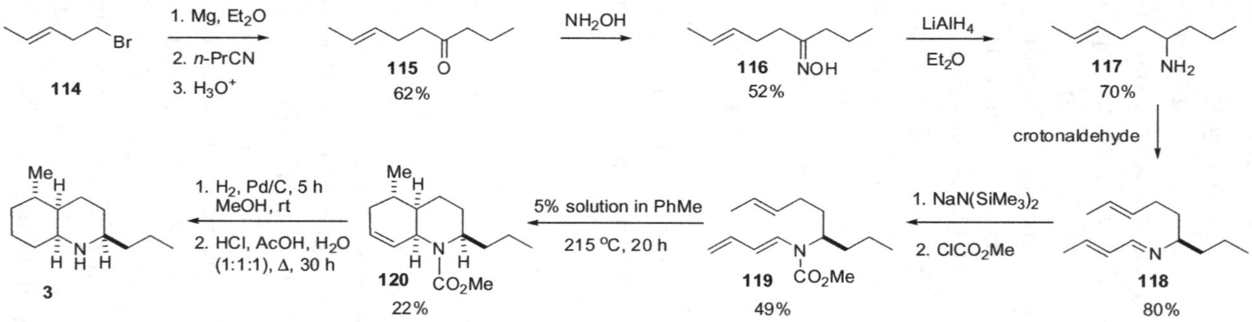

Chiral Pool Synthesis

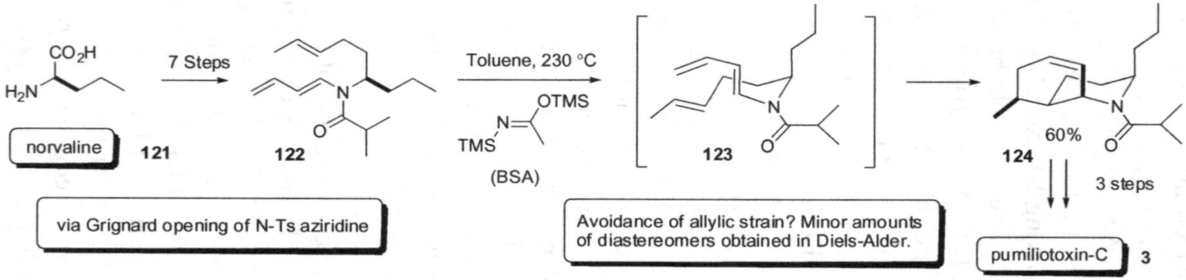

Pumiliotoxin-2

trans-olefin **107**, followed by deprotection of the nitrogen to liberate an amine. As shown here, the Diels-Alder would be plagued by regiochemical problems and poor dienophile electronic properties. But this could be fixed by using crotonaldehyde (**109**) as the dienophile followed by introduction of C_5-C_9 using carbonyl addition chemistry. In fact this looks like the most "secure" route when it comes to ensuring relative stereochemistry throughout the synthesis. Let's begin with the first synthesis of pumiliotoxin-C, reported by the Oppolzer group in 1975.

Pumiliotoxin-2

The initial target in the Oppolzer synthesis was dienamide **119**. This was prepared starting with homoallylic bromide **114**. Formation of the Grignard reagent and reaction with propionitrile gave ketone **115**. Preparation of oxime **116** was followed by reduction to provide **117**. Imine formation with crotonaldehyde was followed by deconjugative *N*-acylation to provide **119**. The critical Diels-Alder reaction proceeded to give **120** in 22% yield at 215 °C. The other major product (37%) was derived from "formal" hydrolysis of the dienamide. The synthesis was completed by catalytic hydrogenation of the olefin and removal of the carbamate protecting group under acidic conditions.

The synthesis was modified, to provide an improved yield in the cycloaddition and introduce stereoselectivity at the impending C_6 stereogenic center, by using norvaline (**121**) as the starting material and dienamide **122** as the cycloaddition substrate. Using BSA as a water scavenger improved the yield of the cycloaddition product (**124**) to 60%. The only isolated side products were minor amounts of diastereomeric cycloadducts. One might imagine that the major product is derived from the "approach of dienophile to diene" shown in structure **123**. Allylic strain with the amide is minimal in this cyclization transition state. The synthesis was completed in much the same manner as from **120**.

Intermolecular Diels-Alder Route to Pumiliotoxin-C

Overman, L. E.; Jessup. P. J. "Synthetic Applications of *N*-Acylamino-1,3-dienes. An Efficient Stereospecific Total Synthesis of *dl*-Pumiliotoxin C, and a General Entry to *cis*-Decahydroquinoline Alkaloids" *J. Am. Chem. Soc.* **1978**, *100*, 5179.

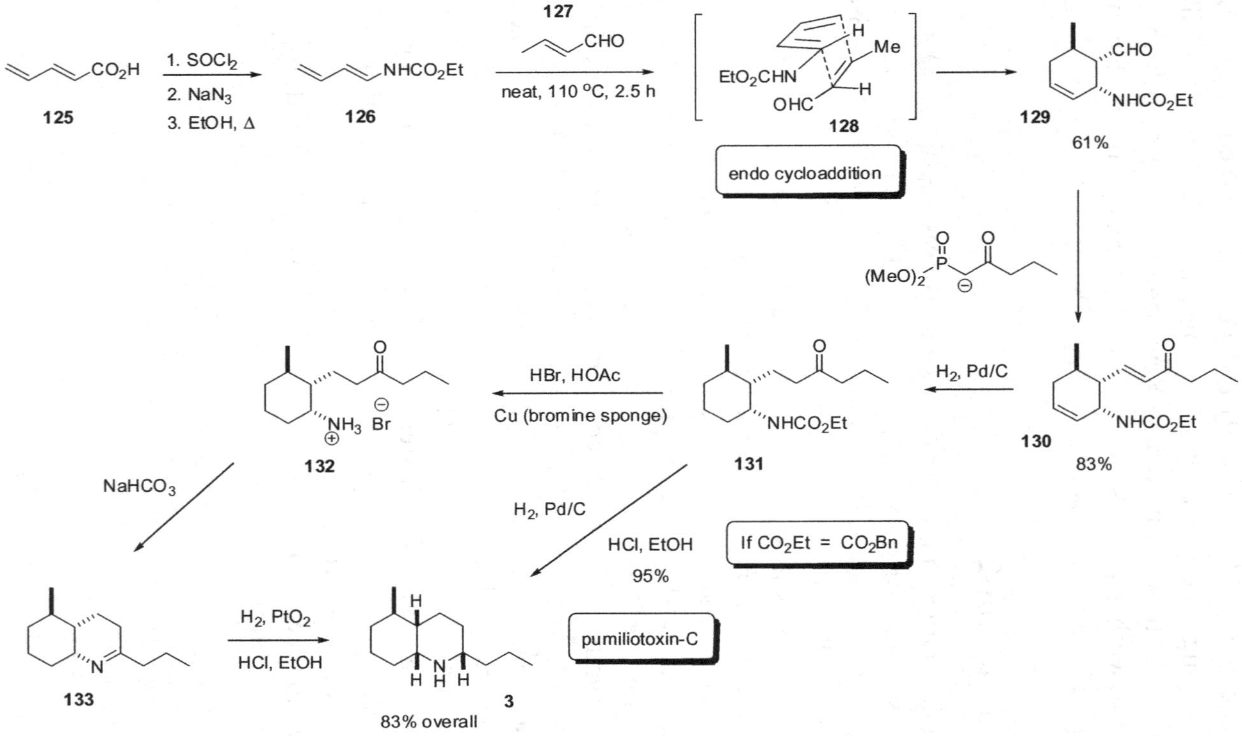

Pumiliotoxin-3

Pumiliotoxin-3

The Overman group reported a synthesis that proceeded along the disconnection shown in Path C of Pumiliotoxin-1. The key cycloaddition between dienamide **126** (from **125**) and crotonaldehyde (**127**) gave **129** with excellent control over regiochemistry and stereochemistry. The remaining carbons were introduced via a Horner-Wadworth-Emmons reaction. Catalytic hydrogenation of the olefins converted cycloadduct **130** to keto-carbamate **131**. Acid promoted "hydrolysis" of the carbamate gave ammonium salt **132**. Conversion of **132** to its free base resulted in formation of imine **133**. Catalytic hydrogenation of the imine provided pumiliotoxin-C (**3**). When the reaction sequence was performed using the benzyl carbamate corresponding to **126**, conversion of **130** ($CO_2Et \rightarrow CO_2Bn$) to **3** was accomplished in a single step! This synthesis provides a nice example of an intermolecular Diels-Alder reaction having some advantages over its intramolecular counterpart.

Additional Diels-Alder Approaches to Pumiliotoxin-C

Ibuka, T.; Inubushi, T.; Saji, I.; Tanaka, K.; Masaki, N. "Total Synthesis of *dl*-Pumiliotoxin C Hydrochloride and its Crystal Structure" *Tetrahedron Lett.* **1975**, 323-326. See also Inubushi, Y.; Ibuka, T. "Synthesis of Pumiliotoxin C, A Toxic Alkaloid from Central American Arrow Poison Frog, Dentrobates Pumilio and D. Auratus" *Heterocycles* **1977**, *8*, 633-660 (this is a brief review).

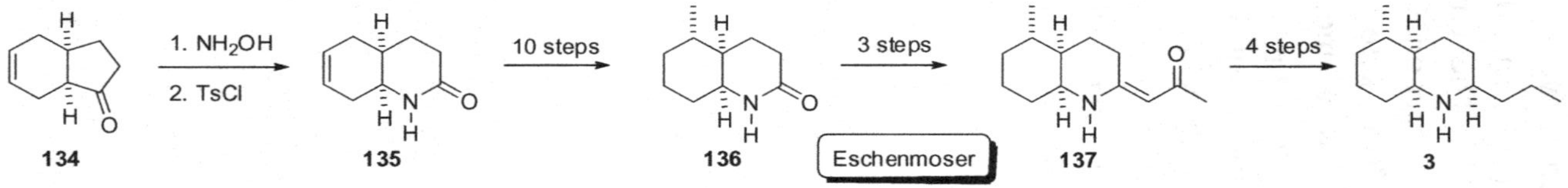

Ibuka, T.; Mori, Y.; Inubushi, Y. "A New Stereoselective Synthesis of *dl*-Pumiliotoxin C Using Novel 1,3-Bis(trimethylsilyloxy)-1,3-dienes" *Tetrahedron Lett.* **1976**, 3169-3172.

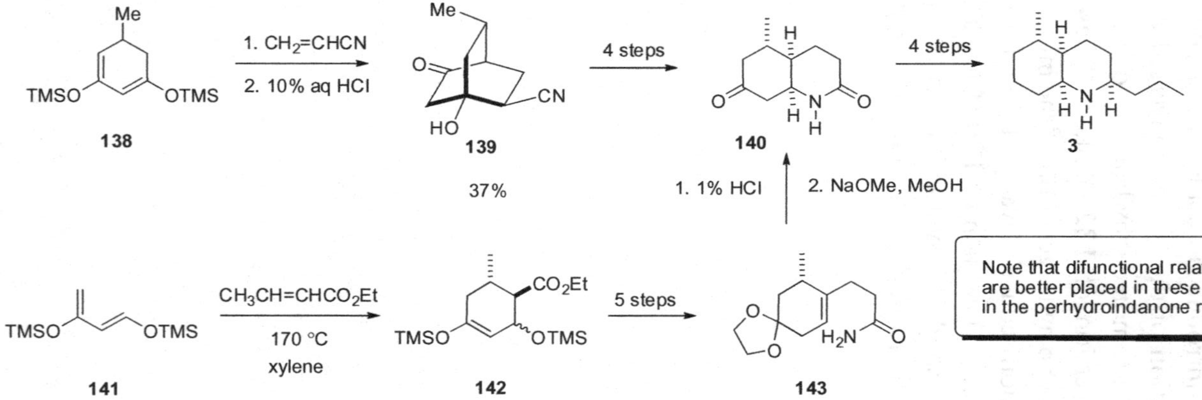

Note that difunctional relationships are better placed in these routes than in the perhydroindanone route

Pumiliotoxin-4

Pumiliotoxin-4

The synthesis shown on this panel follows Path A as delineated in Pumiliotoxin-1. The initial cycloadduct (from cyclopentenone and 1,3-butadiene) was nicely converted to lactam **135** by formation of the oxime and subsequent reaction with *p*-toluenesulfonyl chloride. The conversion of **135** to **136** required 10 steps. Not being able to carry the methyl group in the diene (see Pumiliotoxin-1) was certainly costly. Another 7 steps were needed to introduce the propyl side chain.

A somewhat related approach that makes good use of odd-difunctional relationships in its design involved the Diels-Alder reaction between **138** and acrylonitrile to provide **139** after hydrolysis of an intermediate silyl enol ether. Bromination of **139**, reduction of the α-bromoketone to an alcohol, treatment of the resulting bromohydrin with zinc in acetic acid, and treatment of the resulting olefin with 15% $HClO_4$ in acetic acid gave lactam **140**. The synthesis of **3** was completed in a manner similar to the synthesis from **136**. A related synthesis that proceeded through **140** (derived from **141** and ethyl crotonate) was also reported by the Inubushi group.

A Biomimetic Approach to Pumiliotoxin-C

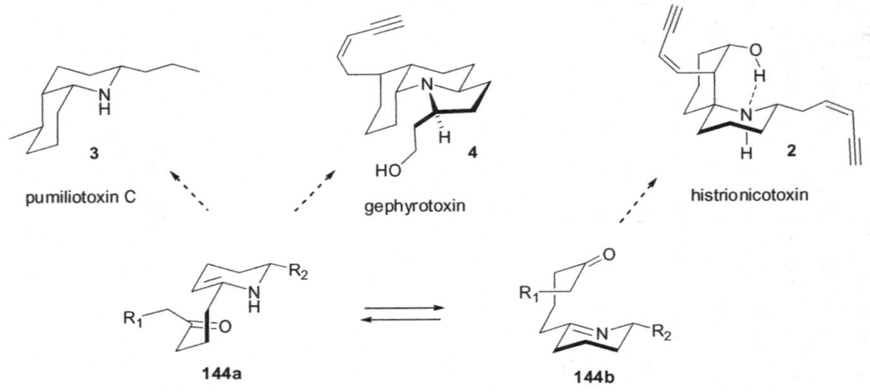

Bonin, M.; Royer, J.; Grierson, D. S.; Husson, H.-P. "Asymmetric Synthesis VIII: Biogenetically Patterned Approach to the Chiral Total Synthesis of (-)-Pumiliotoxin-C" *Tetrahedron Lett.* **1986**, *27*, 1569-1572. For a review discussing this "general approach" to alkaloids see Husson, H.-P. "A New Approach to the Asymmetric Synthesis of Alkaloids" *J. Nat. Prod.* **1985**, *6*, 894-906.

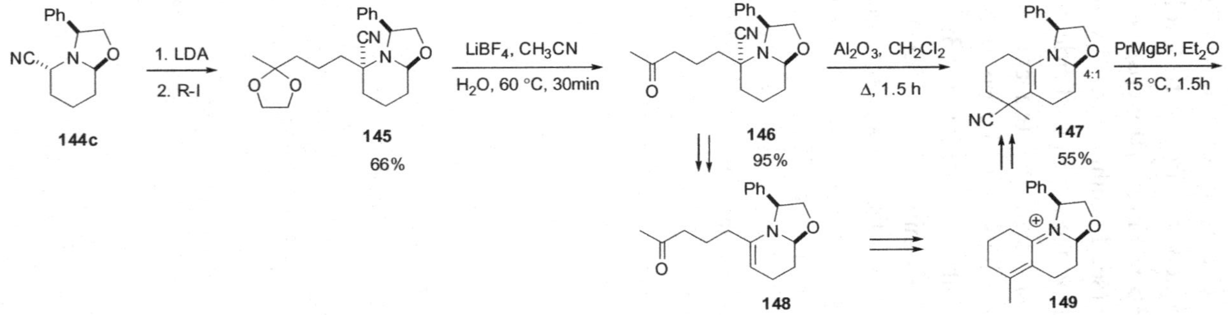

Pumiliotoxin-5

Pumiliotoxin-5

We have already seen that speculation (or hard information) about the biosynthesis of natural products can suggest strategies for laboratory syntheses. It has been speculated that the dart-poison alkaloids under consideration here arise from acyclic compounds (most likely polyketide-derived). For example enamine **144a** (perhaps derived from an acyclic aminoketone) might undergo an intramolecular aldol-type reaction to afford the decahydroquinoline skeleton of pumiliotoxin C (**3**) or gephyrotoxin (**4**). Depending on the nature of R_1 and R_2, one or the other of these alkaloids could result "downstream". On the other hand, the imine tautomer of **144a** (**144b**) might undergo an intramolecular Mannich reaction to afford the azaspiro[5.5]undecane core of histrionicotoxin (**2**). Let's look at an enantioselective approach to pumiliotoxin-C that mimics this biosynthetic proposal.

The synthesis begins with **144c**, prepared from phenylglycinol and glutaraldehyde. Deprotonation of **144c** and alkylation of the resulting anion provided **145**. Hydrolysis of the acetal gave **146** which was converted to perhydroquinoline **147** (as a mixture of stereoisomers) upon treatment with alumina. The transformation of **146** to **147** follows a mechanistic pathway that mimics the biosynthetic proposal. Ionization of the nitrile (recall the Stork approach to reserpine) and loss of a proton might give enamine **148**. An intramolecular "aldol-dehydration" might give iminium ion **149**. 1,4-Addition of cyanide to the iminium ion would provide **147**.

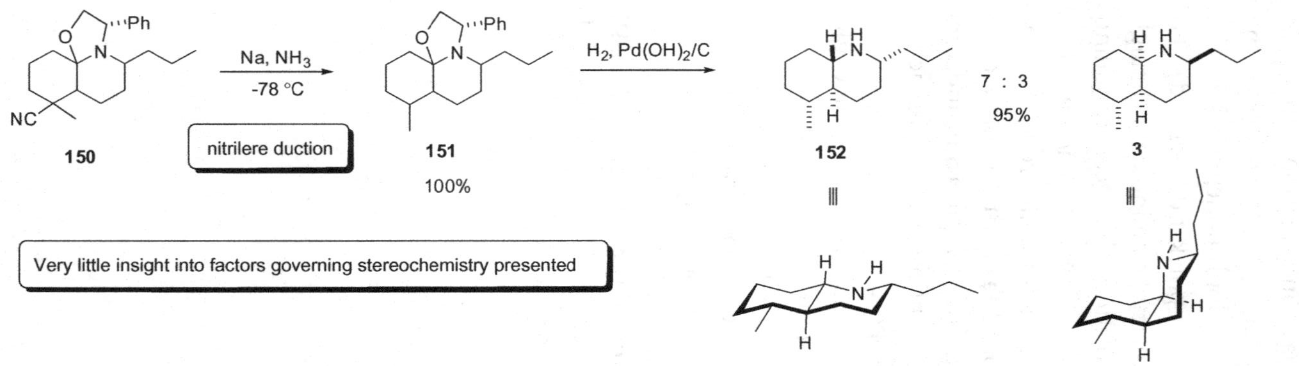

separated by chromatography

Acylnitroso Diels-Alder Approach

Naruse, M.; Aoyagi, S.; Kibayashi, C. "Total Synthesis of (-)-Pumiliotoxin C by Aqueous Intramolecular Acylnitroso Diels-Alder Approach" *Tetrahedron Lett.* **1994**, *35*, 9213-9216.

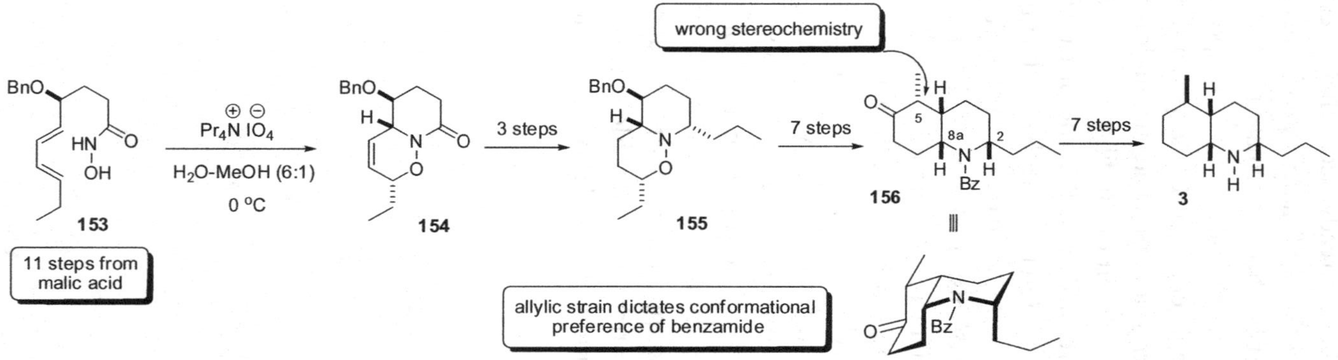

Pumiliotoxin-6

Pumiliotoxin-6

Treatment of **147** with propylmagnesium bromide gave **150**, and sodium-ammonia removal of the nitrile provided **151**. Hydrogenolysis of the benzylic C–N bond, ionization of the *N,O*-acetal to an iminium ion (and possibly tautomerization to an enamine) followed by hydrogenation, gave **152** and pumiliotoxin-C (**3**). The yields through this sequence were very good, but the route suffered in terms of diastereoselectivity. Nonetheless, the final product was produced as a single enantiomer. Notice that this is a "chiral auxiliary" approach to the problem of enantioselectivity, but the chiral auxiliary is sacrificed in the process.

We will spend the rest of our time focusing on selected steps of a few syntheses, all of which were designed to provide single enantiomers of pumiliotoxin-C. The first one uses malic acid (a common dicarboxylic acid found in apples) as the source of enantioselectivity, and features the now familiar nitroso-Diels-Alder chemistry. Thus, hydroxamic acid **153** was prepared and oxidized to the key intermediate. Cycloaddition occurred to provide **154** which was then converted to pumiliotoxin via **155** and **156** in a rather laborious sequence of reactions. One of the more interesting problems encountered in this sequence was the conversion of keto-benzamide **156** to **3**. Notice that **156** has the "wrong" stereochemistry for the pendant methyl group. This is presumably because **156** prefers a conformation in which C_2 and C_{8a} are axially disposed, and thus when the methyl group is equatorially disposed (thermodynamically most stable site) it is cis (not trans) to the propyl group. Can you suggest how the stereochemistry at C_5 was "fixed"?

Some Enantioselective Approaches to Pumiliotoxin-C

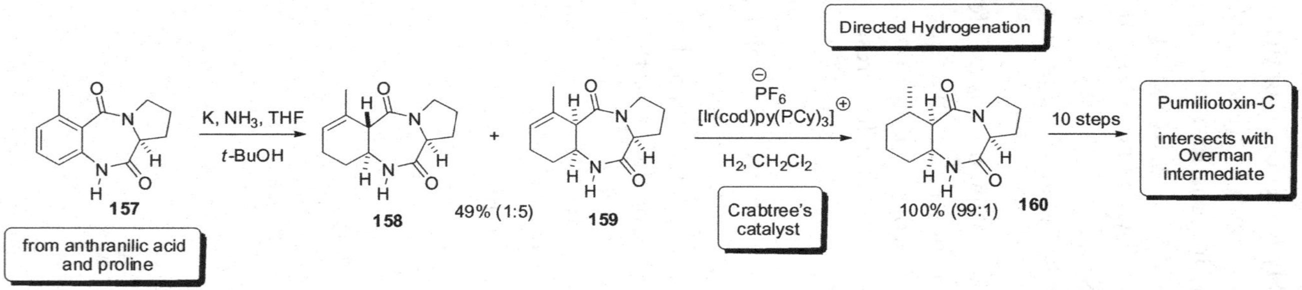

Schultz, A. G.; McCloskey, P. J.; Court, J. J. "Enantioselective Conversion of Anthranilic Acid Derivatives to Chiral Cyclohexanes. Total Synthesis of (+)-Pumiliotoxin-C" *J. Am. Chem. Soc.* **1987**, 109, 6493. Also see Schultz, A. G.; McCloskey, P. J. "Carboxamide and Carbalkoxy Group Directed Stereoselective Iridium-Catalyzed Homogeneous Olefin Hydrogenations" *J. Org. Chem.* **1985**, *50*, 5905.

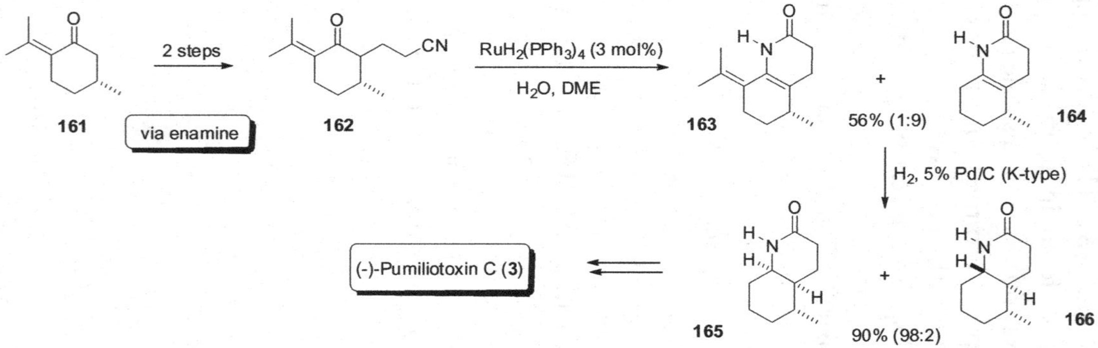

Murahashi, S.-I.; Sasao, S.; Saito, E.; Naota, T. "Ruthenium-Catalyzed Hydration of Nitriles and Transformation of δ-Ketonitriles to Ene-Lactams: Total Synthesis of (-)-Pumiliotoxin C" *Tetrahedron*, **1993**, *49*, 8805.

Pumiliotoxin-7

The next synthesis uses a chiral auxiliary approach in which the auxiliary is not sacrificed. The synthesis begins with anthranilic acid derivative **157**. A potassium-ammonia reduction provided a mixture of diastereomers **158** and **159**. A directed hydrogenation of the major diastereomer (**159**), using Crabtree's cationic iridium catalyst, provided **160**.[14] This material was moved forward to pumiliotoxin-C through intermediates we encountered in the Overman approach (compare the substitution pattern and relative stereochemistry of **160** with **131** on Pumiliotoxin-3).

Murahashi reported a synthesis that used the monoterpene pulegone (**161**) as a source of chirality and featured novel ruthenium catalyzed chemistry, retro-aldol-dehydration and nitrile hydrolysis reactions. Thus **161** was converted to **162** via reaction of its enamine with acrylonitrile. Treatment of **162** with an appropriate ruthenium catalyst, in aqueous dimethoxyethane, gave enamides **163** (minor) and **164** (major). This transformation featured the aforementioned nitrile hydrolysis accompanied by cyclization of the intermediate keto-amide to the enamide. In the case of **164**, this was accompanied by the aforementioned retro-aldol-dehydration. Catalytic hydrogenation of the mixture of enamides gave **166** (minor) and **165** (major). The latter compound was transformed to pumiliotoxin-C (**3**) as we have already seen in previous syntheses. This synthesis was clearly performed to highlight methodology, and it does so quite effectively. Notice that the 1,5-difunctional relationship present in **162** was constructed using "normal" conjugate addition chemistry.

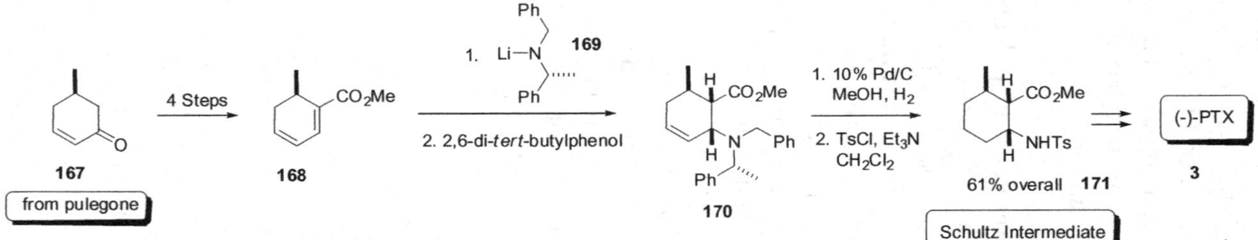

Davies, S. G.; Bhalay, G. *Tetrahedron: Asymmetry* **1996**, *7*, 1595-1596

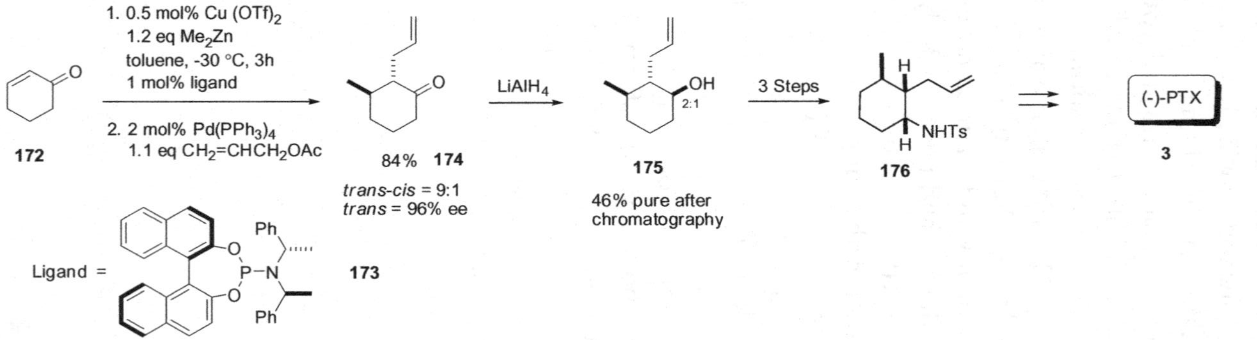

Dijk, E. W.; Panella, A.; Pinho, P.; Naasz, R.; Meetsma, A.; Minnaard, A. J.; Feringa, B. L. "The Asymmetric Synthesis of (-)-Pumiliotoxin C using Tandem Catalysis" *Tetrahedron* **2004**, *60*, 9687-9693.

For a review of 7 recent approaches to pumiliotoxin-C [Habermehl (1998), Bach (1998), Comins (1993), Kunz (1999), Mori (2001), Stille (1993), Padwa (2000)] see Sklenicka, H. M.; Hsung, R. P. "Recent Approaches to *cis*-Azadecalins: Synthesis of Dendrobatid Alkaloid Pumiliotoxin C" *Chemtracts-Organic Chemistry* **2002**, *15*, 391-401.

Pumiliotoxin-8

Davies has developed methodology for the preparation of β-aminoesters. One application of this methodology was to the synthesis of **171**, related to the Schultz and Overman intermediates, carried on to pumiliotoxin-C in the usual manner. Pulegone once again served as the source of chirality for the methyl-bearing carbon. This was converted to **168**, which then reacted with chiral lithium amide **169** to provide **170** after protonation of the intermediate enolate with 2,6-di-*tert*-butylphenol. Hydrogenolysis of the benzylic C–N bonds and the olefin, followed by formation of the sulfonamide, gave **171**.

The final synthesis we will consider uses asymmetric catalysis to establish absolute stereochemistry. Thus, treatment of cyclohexenone (**172**) with dimethylzinc and catalytic Cu(I) in the presence of chiral ligand **173** proceeded with good asymmetric induction. Alkylation of the intermediate enolate using allyl acetate in the presence of Pd(0) provided **174** with 96% ee. This ketone was reduced to provide a mixture of alcohols which were separated and converted to **171** by degrading the allylic side chain to a carboxyl group, and displacing the alcohol with a nitrogen nucleophile.

This ends our look at pumiliotoxin-C. Other syntheses have been reported and some of the more recent approaches have been reviewed in a comparative manner, suggested reading for those interested (see reference at bottom of Pumiliotoxin-8). We will next move to the structurally related alkaloid gephyrotoxin (**4**).

First Synthesis of *dl*-Gephyrotoxin

Fujimoto, R.; Kishi, Y.; Blount, J. F. "Total Synthesis of *dl*-Gephyrotoxin" *J. Am. Chem. Soc.* **1980**, *102*, 7154-7156.

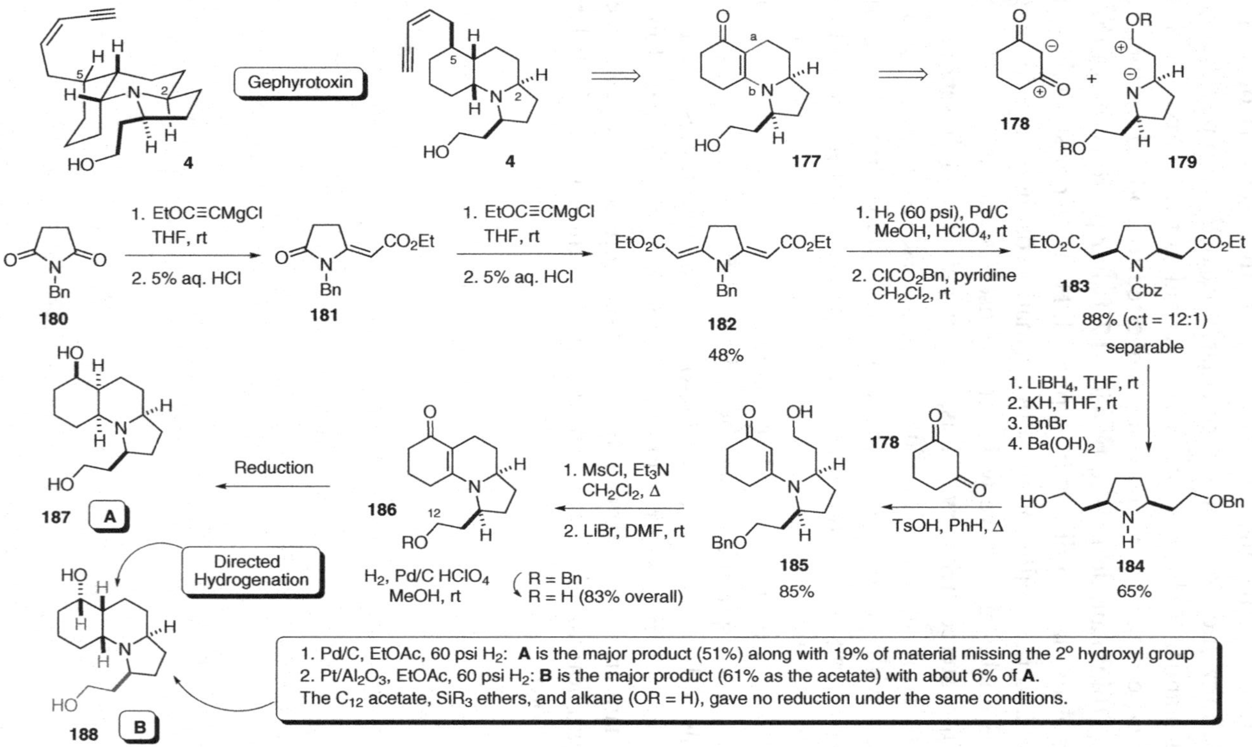

1. Pd/C, EtOAc, 60 psi H_2: **A** is the major product (51%) along with 19% of material missing the 2° hydroxyl group
2. Pt/Al_2O_3, EtOAc, 60 psi H_2: **B** is the major product (61% as the acetate) with about 6% of **A**.

The C_{12} acetate, SiR_3 ethers, and alkane (OR = H), gave no reduction under the same conditions.

Gephyrotoxin-1

Gephyrotoxin-1

I am going to indulge myself a little with this section. Gephyrotoxin (**4**) was the first alkaloid my group worked on when I began my independent career in 1978 and thus, it holds a special place for me in the world of natural products chemistry. This alkaloid resembles pumiliotoxin-C to the extent that it has a *cis*-perhydroquinoline substructure with substituents at C_2 and C_5 (quinoline numbering). The stereochemistry at the ring stereogenic centers differs at C_2 from pumiliotoxin-C. The fundamental structure and absolute stereochemistry was determined by X-ray crystallography (anomolous dispersion technique). We will look at three approaches to this alkaloid with care, and then take a quick look at the critical portions of **5** additional syntheses (not an exhaustive list).

The first synthesis (of racemic material) was reported by Kishi and Fujimoto in 1980. Tricyclic vinylgous amide **177** was projected to be a key intermediate in the Kishi synthesis. Two of the five stereogenic centers of gephyrotoxin are set in this intermediate. The carbonyl group was to serve as a handle to introduce the 5-carbon *cis*-eneyne and stereoselective reduction of the olefin was to handle stereochemistry at the *cis*-perhydroquinoline ring fusion. Local difunctional relationships in **177** are "odd" and thus, one might anticipate that there should be a "carbonyl chemistry route" to this compound if one purchases the 5-membered ring. Working backward from **177** by disconnection of the bonds "a" and "b", the plan was to couple **178** and **179** along the polar lines indicated in Gephyrotoxin-1. The symmetrical nature of **179**, and a consideration of difunctional relationships, leads back to a succinimide derivative as a starting point for the synthesis.

The synthesis began with *N*-benzylsuccinimide (**180**). Treatment of this imide with the Grignard reagent derived from ethoxyacetylene gave **181** after an acidic workup. This reaction nicely illustrates the equivalence of this acetylide with the enolate of ethyl acetate. It has practical advantages, however, over the enolate. It is stable at elevated temperatures and is less sterically hindered. On the other hand, ethoxyacetylene is a less friendly reagent to handle (and is more expensive) than ethyl acetate. Another important point is that imides are ketone-like in their reactivity. They should not be confused with amides! Treatment of **181** with the same acetylide gave **182**. Once again it is important to recognize that the carbonyl group in **181** is imide-like (rather than amide-like). In fact it is a vinylogous imide (see Functional Groups-5 and the Principle of Vinylogy). Catalytic hydrogenation of **182**, with concomitant hydrogenolysis of the *N*-benzyl group, followed by acylation of the pyrroldine nitrogen, gave **183**. The major stereoisomer had cis stereochemistry, the reduction of the first olefin presumably controlling the stereochemical course of the second olefin reduction. Modification of the two acetic acid sidechains converted **183** to **184**. Reaction of pyrrolidine **184** with dione **178** gave enamine **185**. The hydroxyl group was activated as the mesylate, transformed to the bromide, and an intramolecular alkylation of the enamine ensued to give **186** (R = benzyl). The next task was diastereoselective reduction of the olefin. It turns out that **186** (R = Bn, Ac, $OSiR_3$) gave no reduction or selectivity under typical conditions for catalytic hydrogenations. In the presence of acid, it was possible to remove the benzyl group to provide **186** (R = H; same as **177**). Hydrogenation of this substrate over Pd/C in ethyl acetate gave mainly **187** (**A**) with the wrong stereochemistry at the ring fusion. This is the result one might have expected based on steric accessibility of the enone to the surface of a heterogeneous (or for that matter a homogeneous) catalyst. When the catalyst was changed to Pt on alumina, however, the desired stereoisomer (**188** = **B**) was obtained in good yield. The notion is that the hydroxylethyl group (via an interaction of the alcohol with alumina) directed the catalyst to the desired surface of the enone. Such directing effects had been previously noted, but this is a dramatic application of this phenomenon to a complex problem in synthesis.[15] The primary alcohol was then selectively acylated to give **189** (see Gephyrotoxin-2).

Gephyrotoxin-2

The next stage of the synthesis called for "reductive homologation" of the C_5 alcohol by two carbons, to ultimately serve as a handle for introduction of the *cis*-eneyne unit. Oxidation of **189** provided the C_5 ketone. Application of the aforementioned acetylide addition chemistry, and cleavage of the acetate (MeMgCl), gave **191** as a mixture of geometrical isomers. Protection of the alcohol as either an acetate or TBDPS ether provided **192** and **193**, respectively.

As a prelude of things to come, reduction of **190** with either "big" ($LiEt_3BH$) or "little" ($NaBH_4$) hydrides gave only **189**. It appears that **190** adopts a conformation in which only the "axial" face of the ketone is exposed for nucleophilic addition. The same seems to be qualitatively true of unsaturated esters **191** and **192**. For example, catalytic hydrogenation of **191** over rhodium on alumina provided an equal mixture of **194** and **195** ($R = CO_2Et$). It is surprising that this reaction provided as much of the desired stereoisomer (**195**) as it did! Dissolving metal reduction of **192** gave mainly the product derived from protonation of the presumed carbanionic intermediate from the sterically most accessible (or perhaps stereoelectronically prefered) axial site at C_5. Heterogeneous catalytic hydrogenation of **193**, however, did provide the desired stereochemistry at C_5. It was suggested that the "huge" *tert*-butyldiphenylsilyl (TBDPS) group simply discouraged the catalyst from approaching the "top" face of the olefin, relative to the alternative face. This remote "blocking" effect is interesting, and is related to the strategy used by Corey to control stereochemistry at C_{15} in his approach to the prostaglandins (Prostaglandins-11).

Continuing with the synthesis, reduction of **195** ($R = CO_2Et$) gave the corresponding primary alcohol ($R = CH_2OH$). Oxidation of the alcohol gave the aldehyde (**195** where $R = CHO$). A Wittig reaction between the interesting ylid derived from **196**, and careful acid hydrolysis of the intermediate allene, gave unstable *cis*-enal **197**. Application of the Corey-Fuchs terminal acetylene synthesis completed the synthesis of gephyrotoxin (**4**).

The synthesis described above produced racemic gephyrotoxin. An enantioselective synthesis was performed using pyrroglutamic acid as the point of departure. The starting material clearly delivered the absolute stereochemistry that had been assigned to the natural product based on crystallography. The specific rotation, however, of this synthetic material was opposite in sign (same in magnitude) to that reported for the natural material. So this produced a dilema. Something is wrong somewhere, or perhaps the frogs used to isolate the material used for optical measurements differed from those used to isolate material used in the crystallographic studies, and produced enantiomeric alkaloids. The production of one enantiomer of a natural product by one plant source, and the

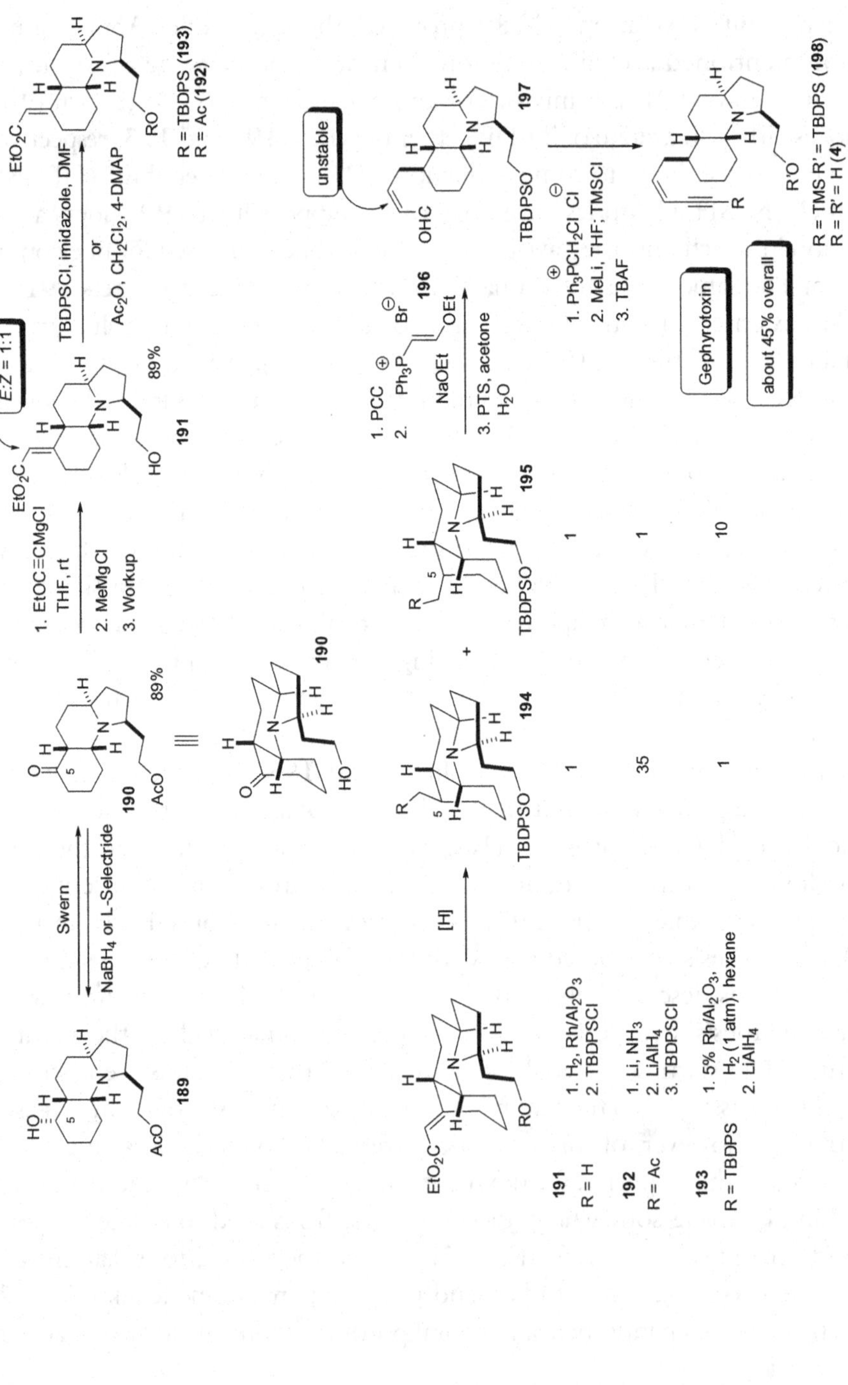

Gephyrotoxin-2

other by another plant source, is known. Thus this suggestion is not without some merit. Still, this "stereochemical problem" remains unresolved.[16]

Gephyrotoxin-3

The next synthesis features stereochemical aspects of *N*-acyliminium ion cyclizations and extends the Kishi-Fujimoto work on remote control of hydrogenation stereochemistry. The idea was that an intermediate of type **199** could serve as a precursor to gephyrotoxin (**4**). The amide would serve as a handle for introducing the hydroxyethyl side chain, while an appropriate C_5 substitutuent would serve as a precursor to the *cis*-eneyne unit. It was felt that **199** might be obtained by cyclization of *N*-acyliminium ion **200**. We have already seen that such cyclizations can be used to prepare indolizidinones, the azabicyclo[4.3.0]nonane substructure present in **199** (see Histrionicotoxin-3). The cyclization was to set stereochemistry at C_2 relative to three stereogenic centers already set in the cyclization substrate. The critical question was whether an *N*-acyliminium ion of type **200** would cyclize from conformation **201M**, which would provide the needed stereochemistry at C_2, or from conformation **201m**, which would provide stereochemistry opposite to that required at C_2. Whereas **201M** has two axial substituents on the cyclohexane, it was anticipated that allylic strain in conformation **201m** would render cyclization from that conformation higher in energy. A vinyl group was selected as the choice of "R" because, as will be seen, it introduced a useful element of pseudo-symmetry into the synthetic plan.

The synthesis of the projected *N*-acyliminium ion precursor (**208**) began with a cycloaddition between cyclohexenone (**202**) and 1,3-butadiene. This acid-promoted cycloaddition initially gave the *cis*-cycloadduct, but the acid also promoted epimerization of the ring juncture to a thermodynamic 9:1 mixture of products from which **203** was isolated in reasonable yield. Reduction of **203** occurred with the expected "axial delivery" of hydride to provide largely **204**. A Mitsunobu reaction gave succinimide **205**.

The next task was to degrade the cyclohexene to a pair of vinyl groups. This was accomplished by ozonolysis of the double bond and a reductive workup to provide diol **206**. Chemistry developed by Grieco and Sharpless was then used to formally dehydrate **206** to **207**, and a DIBAL reduction gave **208** and set the stage for the key cyclization.[17] Of course the story has a good ending. The *N*-acyliminium ion cyclization gave tricyclic formate **209** in good yield with no other stereoisomers being detected.

So does allylic strain control the stereochemistry in the cyclization? If it is a factor, it is certainly not the only factor. It turns out that NMR analysis of

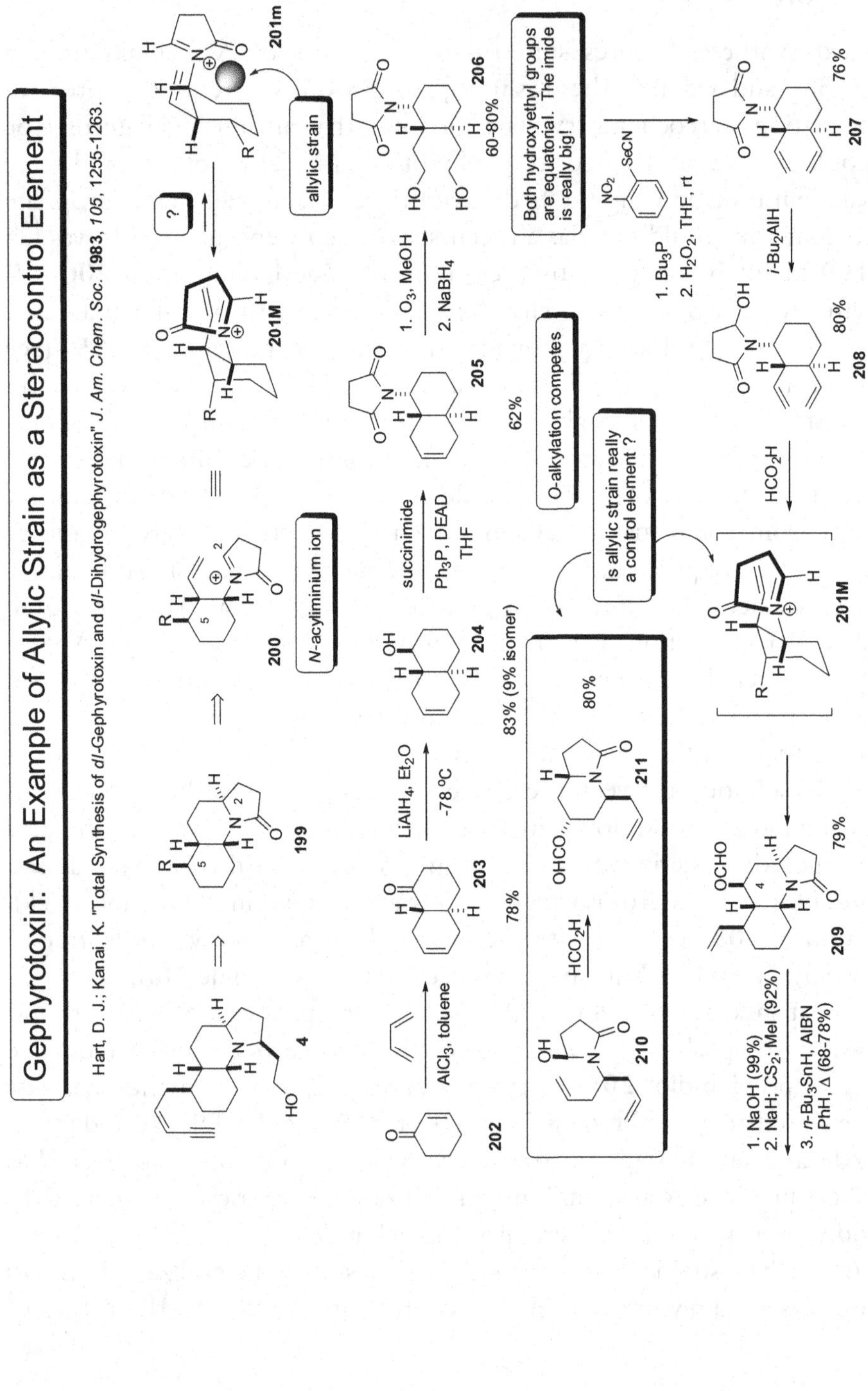

Gephyrotoxin-3

206 showed that both hydroxyethyl groups were axially disposed (in $CDCl_3$). Thus, the imide is huge. If one extrapolates to iminium ion **200**, this factor alone would surely favor conformation **201M** over conformation **201m** in the cyclization. Is allylic strain part of the picture? Most certainly yes. Compound **210** has no intrinsic bias influencing conformational preferences other than allylic strain, and it is cleanly converted to **211**.[18]

The acyliminium ion cyclization resulted in unwanted oxidation at C_4. Thus, this was removed by hydrolysis to the secondary alcohol, followed by a Barton-McCombie deoxygenation to provide **212** (Gephyrotoxin-4).[19]

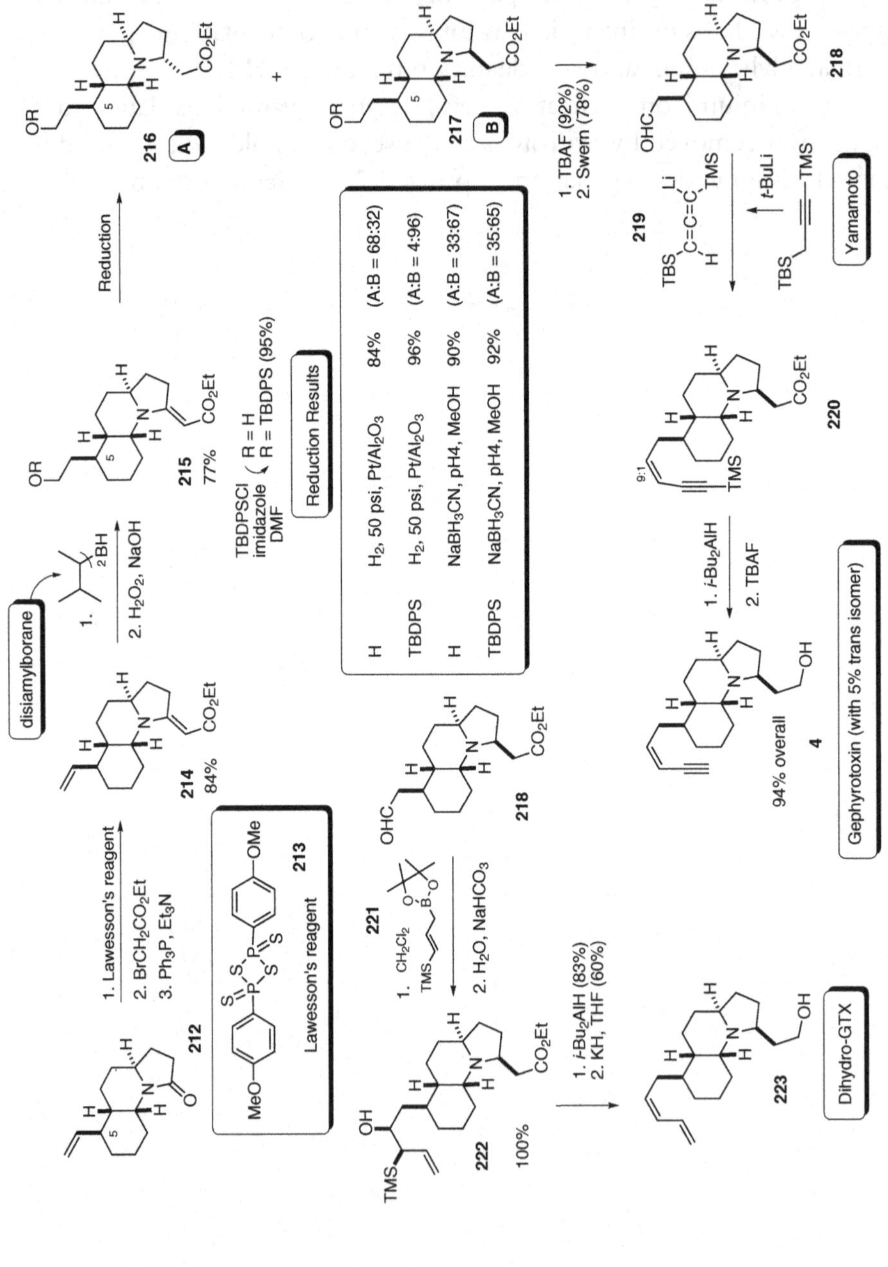

Gephyrotoxin-4

Gephyrotoxin-4

Lactam **212** was converted to the corresponding thiolactam, and an Eschenmoser sulfide contraction was used to prepare vinylogous urethane **214**. Hydroboration-oxidation gave **215** (R = H) and **215** (R = TBDPS) after silylation of the primary alcohol. The fifth stereogenic center was then introduced by reduction of the vinylogous urethane. Sodium cyanoborohydride reduction at pH4 (Borch conditions) gave a nasty mixture of **216** and **217**, regardless of the nature of the C_5 substituent. Catalytic hydrogenation of **215** (R = H) was equally ineffective in terms of stereocontrol. The Kishi-Fujimoto synthesis, which was published at this time, provided inspiration for completion of the synthesis. When **215** (R = TBDPS) was subjected to catalytic hydrogenation, excellent diastereoselectivity was observed to afford **217** as the major product. Presumably this is another example of long-range control of stereochemistry (by blocking one face of the olefin) in a heterogenous hydrogenation. With all of the stereochemistry established, the C_5 sidechain was converted to aldehyde **218**. A Yamamoto-Peterson olefination using reagent **219** provided **220** with good control over olefin geometry. An oxidation state adjustment provided gephyrotoxin (**4**). When the olefination was conducted using vinylboronate **221**, β-hydroxysilane **222** was obtained. Reduction of the ester and a *syn*-elimination completed a synthesis of another dart-poison alkaloid, dihydrogephyrotoxin (**223**).[20]

Overman's Approach to Gephyrotoxin

Overman, L. E.; Lesuisse, D.; Hashimoto, M. "Importance of Allylic Interactions and Stereoelectronic Effects in Dictating the Steric Course of the reaction of Iminium Ions with Nucleophiles. An Efficient Total Synthesis of *dl*-Gephyrotoxin" *J. Am. Chem. Soc.* **1983**, *105*, 5373-5379.

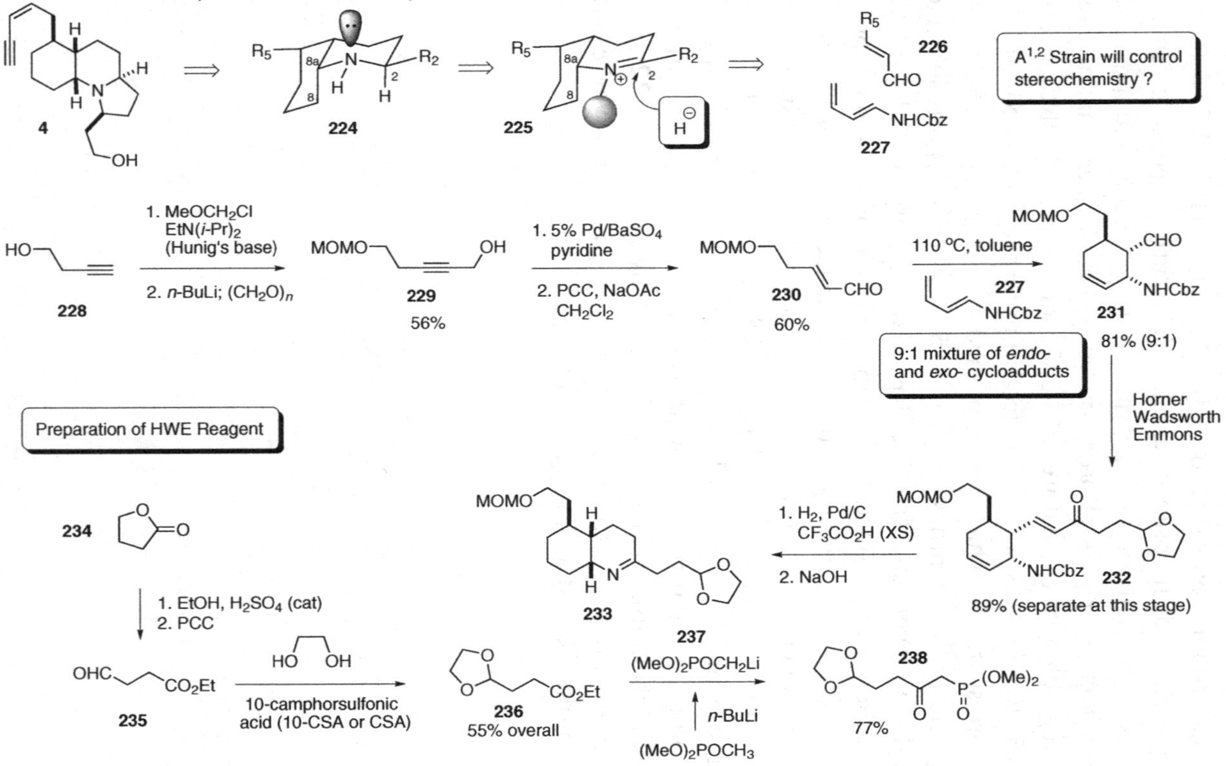

Gephyrotoxin-5

Gephyrotoxin-5

The next approach we will examine eminates from the Overman group (UC Irvine). Given the similarity in structure between pumiliotoxin-C (**3**) and gephyrotoxin (**4**), it is not surprising that their approach to these alkaloids are similar. The major difference between **3** and **4** is in stereochemistry at C_2 of the perhydroquinoline. In the Overman synthesis of pumiliotoxin-C, a C_2 imine (or iminium ion) was reduced by catalytic hydrogenation with the bowl-shaped nature of the substrate directing reduction to the convex face of the ring system. Gephyrotoxin called for delivery of hydride from the concave face of the ring system. It was hoped that an iminium ion of type **225** might predominate in solution. Why? To minimize $A^{1,2}$-strain between the nitrogen substituent and C_8-C_{8a} bond (see Problem 40). If this was the case, product formation with a *trans*-diaxial relationship between the *N*-lone pair and incoming hydride would give the required stereochemistry at C_2 (see structure **224** and recall the Stevens' stereoelectronic analysis of imine addition reactions). The rest of the strategy is the same as that used for pumiliotoxin C, except C_5 and C_2 were changed to accommodate differences in the targets.

In the forward direction, aldehyde **230** was prepared from alcohol **228** in a straightforward manner. The Diels-Alder reaction between **230** and dienamide **227** proceeded to give a mixture of endo and exo cycloadducts with **231** as the major diastereomer. Phosphonate **238** was also prepared in a straightforward manner from butyrolactone **234**. A Horner-Wadsworth-Emmons reaction between **231** and the anion derived from **238** gave **232** at which stage the mixture of diastereomers was separated. Hydrogenation of the olefins and hydrogenolysis of the Cbz group under acidic conditions gave imine **233** after basification.

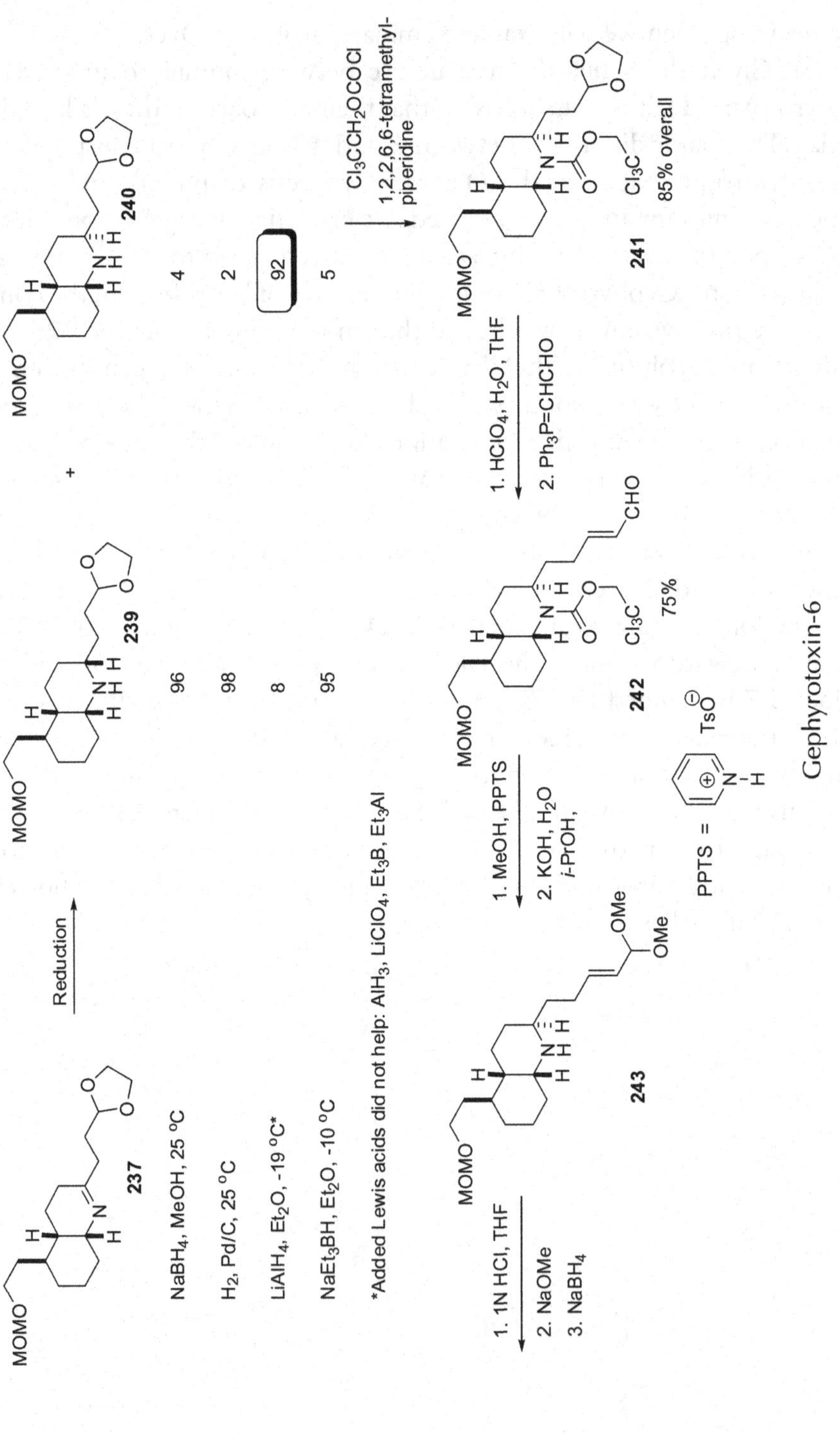
MOMO
237
Reduction
239
+
240
NaBH4, MeOH, 25 °C
96
4
H2, Pd/C, 25 °C
98
2
LiAlH4, Et2O, -19 °C*
8
92
NaEt3BH, Et2O, -10 °C
95
5
*Added Lewis acids did not help: AlH3, LiClO4, Et3B, Et3Al
Cl3CCH2OCOCl
1,2,2,6,6-tetramethyl-piperidine
241
85% overall
1. HClO4, H2O, THF
2. Ph3P=CHCHO
242
75%
CHO
1. MeOH, PPTS
2. KOH, H2O, i-PrOH,
243
OMe
OMe
1. 1N HCl, THF
2. NaOMe
3. NaBH4
PPTS =
TsO
Gephyrotoxin-6

Gephyrotoxin-6

Next came the search part of research. A sampling of reducing agents eventually revealed that the desired transformation could be accomplished by lithium aluminum hydride in ether at low temperature. One can only speculate that other reagent systems did not have the presumed stereoelectronic requirements of a hydride reduction (H_2, Pd/C for example) or did not meet the conformational requirements needed to generate **240** in preference to **239**. For examplc, if the imine nitrogen is not tightly complexed to a metal ($NaBH_4$) or if the hydride delivery agent is too large ($NaEt_3BH$), reactions from other conformations (or via boat-like conformations) might begin to compete with **225** (Gephyrotoxin-6). Regardless of the reason for success, the mechanistic analysis probably provided the hope to stimulate exploration of this route.

With the perhydroisoquinoline stereochemistry established, efforts moved to modifying the C_2 side chain to facilitate formation of the 5-membered ring. The nitrogen of **240** was protected with a Troc-group. Hydrolysis of the acetal and a Wittig reaction converted **241** to **242**. Protection of the aldehyde as an acetal and basic hydrolysis of the Troc-group provided **243**. Acetal hydrolysis and basification of the reaction appended the 5-membered ring via an intramolecular conjugate addition. Sodium borohydride reduction of the resulting aldehyde gave alcohol **244** (Gephyrotoxin-7). It is possible that reversibility of the conjugate addition allowed thermodynamics to control stereochemistry in the cyclization, but this point is not really clear.

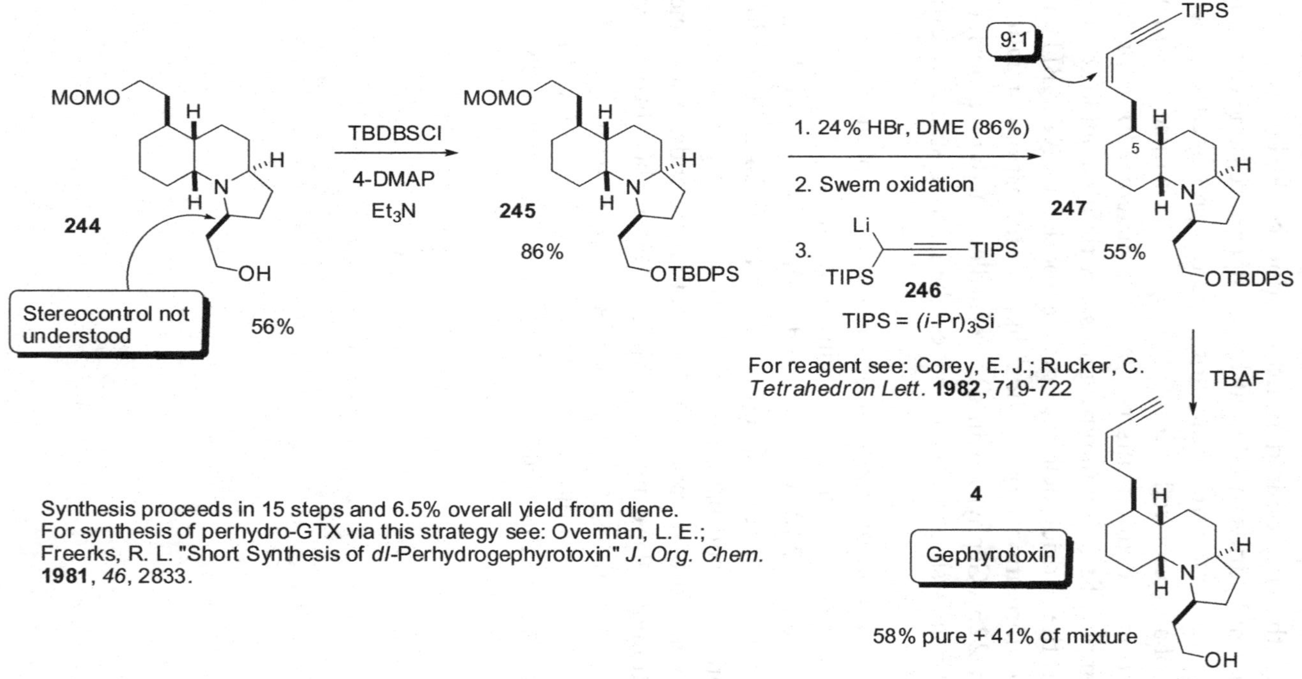

Synthesis proceeds in 15 steps and 6.5% overall yield from diene.
For synthesis of perhydro-GTX via this strategy see: Overman, L. E.; Freerks, R. L. "Short Synthesis of *dl*-Perhydrogephyrotoxin" *J. Org. Chem.* **1981**, *46*, 2833.

Gephyrotoxin-7

Gephyrotoxin-7

The hydroxyethyl group was protected as TBDPS ether **245** before moving to completion of the C_5-eneyne. The robust MOM group was removed using HBr in DME (notice that this protecting group survived several acidic conditions upon conversion of **232** to **243**). Swern oxidation provided an aldehyde. Reaction of the aldehyde with Corey-Rucker reagent **246** gave Peterson olefination product **247** with 9:1 olefin geometry favoring the Z-isomer. Fluoride was used to remove the acetylene and alcohol protecting groups to provide gephyrotoxin (**4**). Due to the efficiency with which the perhydroisoquinoline core was prepared, I think this synthesis is the most efficient and versatile route to this alkaloid.

Other Approaches to Gephyrotoxin

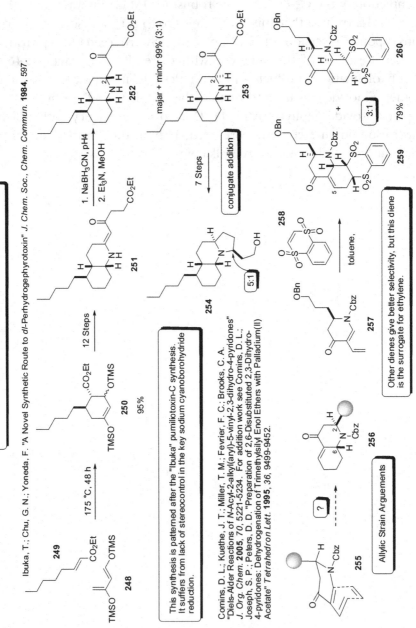

Ibuka, T.; Chu, G. N.; Yoneda, F. "A Novel Synthetic Route to *dl*-Perhydrogephyrotoxin" *J. Chem. Soc., Chem. Commun.* **1984**, 597.

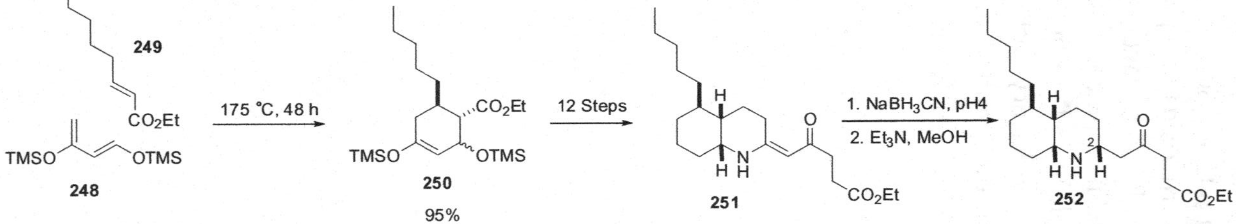

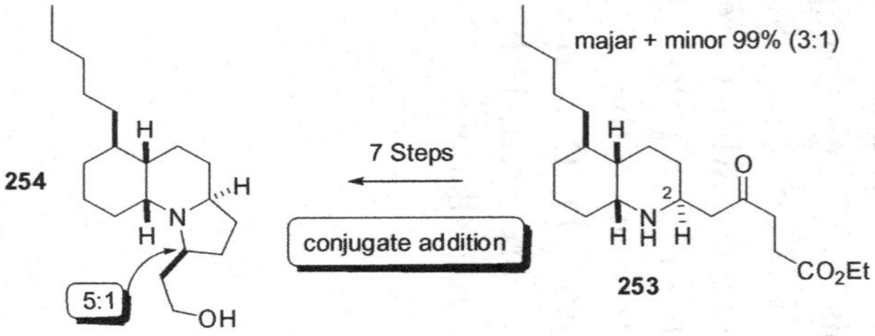

This synthesis is patterned after the "Ibuka" pumiliotoxin-C synthesis. It suffers from lack of stereocontrol in the key sodium cyanoborohydride reduction.

Comins, D. L.; Kuethe, J. T.; Miller, T. M.; Fevrier, F. C.; Brooks, C. A. "Diels-Alder Reactions of *N*-Acyl-2-alkyl(aryl)-5-vinyl-2,3-dihydro-4-pyridones" *J. Org. Chem.* **2005**, *70*, 5221-5234. For addition work see Comins, D. L.; Joseph, S. P.; Peters, D. D. "Preparation of 2,6-Disubstituted 2,3-Dihydro-4-pyridones: Dehydrogenation of Trimethylsilyl Enol Ethers with Palladium(II) Acetate" *Tetrahedron Lett.* **1995**, *36*, 9499-9452.

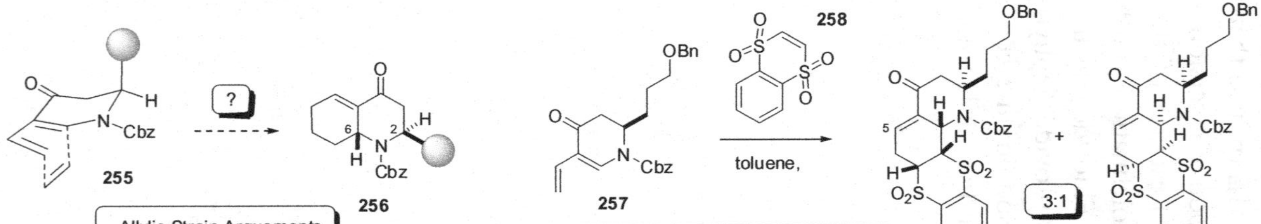

Other dienes give better selectivity, but this diene is the surrogate for ethylene.

Gephyrotoxin-8

We will now take a brief look at several additional approaches to gephyrotoxin. The Ibuka group followed a sequence (to perhydrogephyrotoxin) that resembled their approach to pumiliotoxin-C (Pumiliotoxin-4). This resulted in the synthesis of perhydroisoquinoline **251** from **248** and **249**. Reduction of vinylogous amide **251** suffered from lack of stereocontrol at C_2, providing largely **252** with the required **253** as a minor product. This result is in accord with the Overman studies. Ketoester **253** was eventually converted to perhydrogephyrotoxin (**254**).

The Commins group (North Carolina State) took an approach based on the prediction that **255** would prefer the conformation shown here to avoid allylic strain between the "gray ball" and the *N*-Cbz group. If this were true it was felt that a cycloaddition between **255** and ethylene (or more likely an ethylene equivalent) would give the required "trans" stereochemistry across the 2- and 6-positions of the product piperidone (**256**). In practice, when **257** was reacted with *bis*-sulfonylethylene **258**, this result was realized with about 3:1 facial selectivity, with a mixture of **259** (desired stereochemistry) and **260** (undesired) being obtained. The idea for proceeding from here was to reduce the C—S bonds to establish the equivalency between **258** and ethylene, and to use the enone to introduce the C_5 sidechain via a conjugate addition.

Pearson, W. H.; Fang, W. "Synthesis of Benzo-Fused 1-Azabicyclo[*m.n.*0]alkanes via the Schmidt Reaction: A Formal Synthesis of Gephyrotoxin" *J. Org. Chem.* **2000**, *65*, 7165-7174.

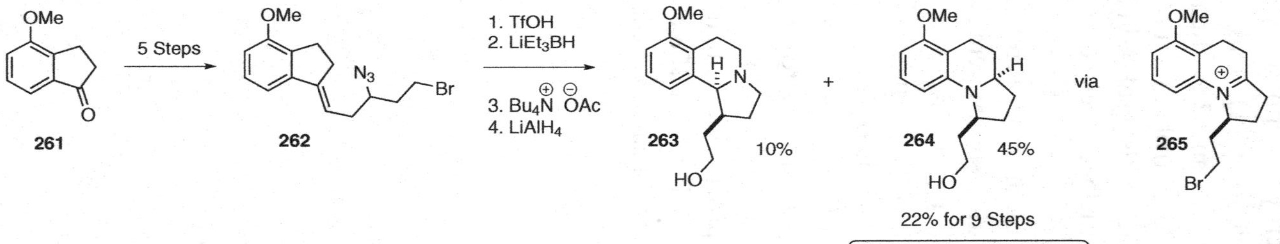

Ito, Y.; Nakajo, E.; Nakatsuka, M.; Saegusa, T. "A New Approach to Gephyrotoxin" *Tetrahedron Lett.* **1983**, *24*, 2881-2884.

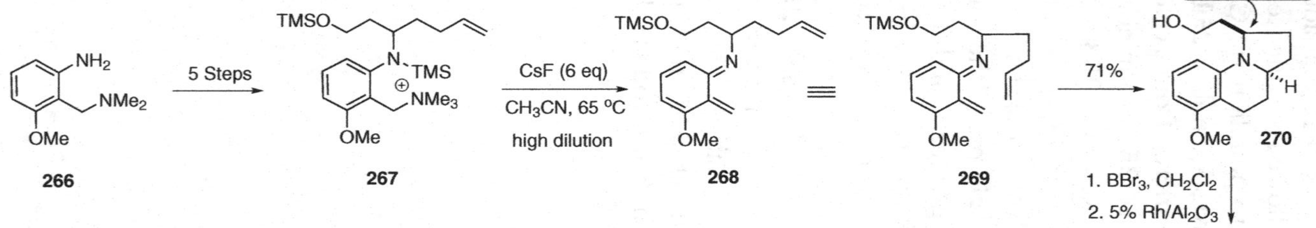

Wei, L.-L.; Hsung, R. P.; Skelnicka, H. M.; Gerasyuto, A. I. "A Novel and Highly Stereoselective Intramolecular Formal [3+3] Cycloaddition Reaction of Vinylogous Amides Tethered with α,β-Unsaturated Aldehydes: A Formal Synthesis of (+)-Gephyrotoxin" *Angew. Chem. Int. Ed.* **2001**, *40*, 1516.

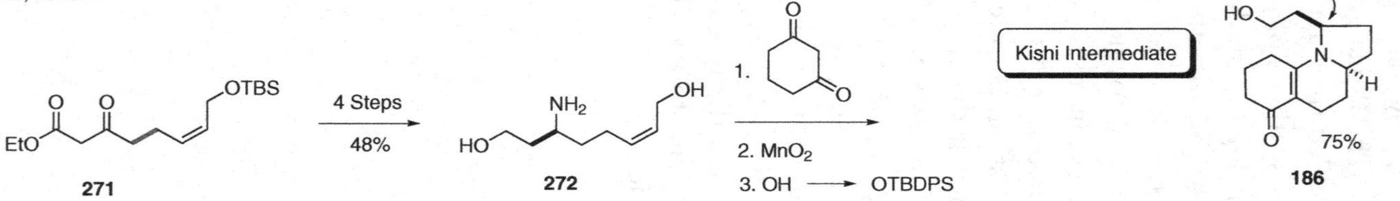

Gephyrotoxin-9

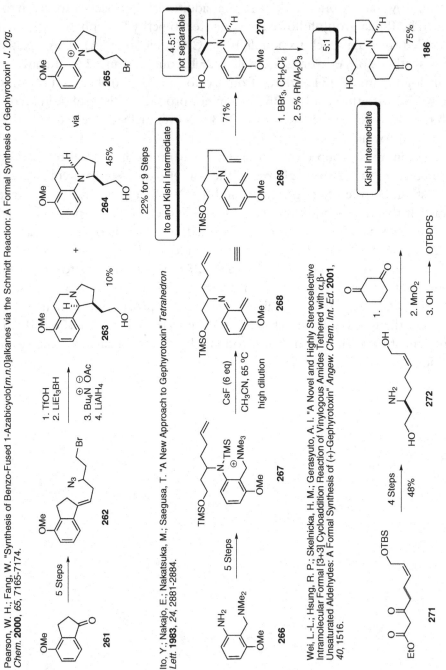

Gephyrotoxin-9

The Pearson group (University of Michigan) took an approach that relied on an *N*-insertion to prepare the tricyclic nucleus of gephyrotoxin. Thus, indanone **261** was converted to unsaturated azide **262**. Treatment of **262** with triflic acid gave a mixture of iminium ions (including **265**) that were reduced with lithium triethylborohydride, followed by conversion of the primary bromide to a primary alcohol. This sequence gave **264** as the major product (via **265**). The relationship between **264** and intermediates in the Kishi approach should be clear.

A different approach to the same tricyclic intermediate was described by the Ito group. The key step in this approach was an intramolecular cycloaddition of **268/269**, generated by a novel elimination reaction of ammonium salt **267**. Conversion to the Kishi intermediate **186** completed this partial synthesis of gephyrotoxin.

The Hsung group (Wisconsin) reported an enantioselective synthesis of Kishi-like intermediate **186** that began with the preparation of **272**. Formation of the enamine with cyclohexan-1,3-dione, followed by oxidation of the allylic alcohol and protection of the remaining primary alcohol, gave **273** (Gephyrotoxin-10).

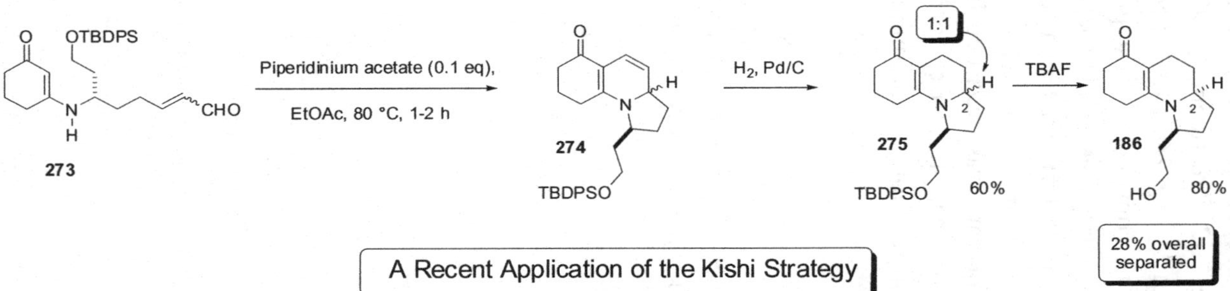

A Recent Application of the Kishi Strategy

Pu, X.; Ma, D. "Facile Entry to Decahydroquinoline Alkaloids. Total Syntheses of Lepadins A-E and H" *J. Org. Chem.* **2006**, *71*, 6562-6572.

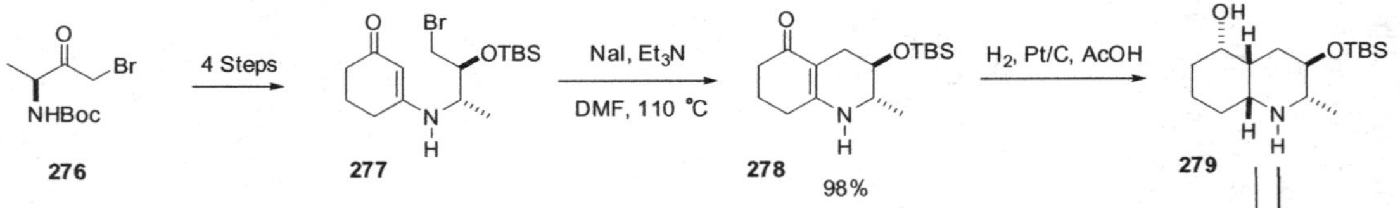

Note the relationship of this synthesis to the Kishi approach to gephyrotoxin.

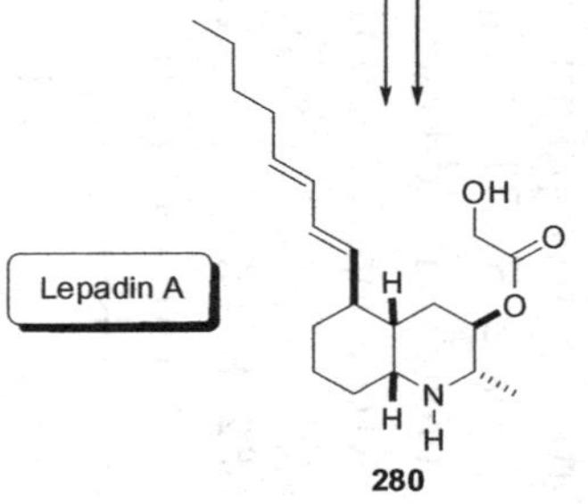

Gephyrotoxin-10

Gephyrotoxin-10

An intramolecular conjugate addition, followed by an enamine "aldol-dehydration" reaction provided **274** as a mixture of stereoisomers at C_2. Catalytic hydrogenation of the isolated olefin gave **275** and TBAF removal of the silicon protecting group provided **186** after separation from the C_2 diastereomer.

By now I imagine you recognize that the key bond constructions in the Huang synthesis resemble those in the Kishi approach. This appears again in a synthesis of lepadin A (**280**) reported in 2006 by the Ma group, which I show here without comment. Why does this approach appear again and again? One cannot say for sure, but I suggest it is that the natural reactivity pattern of the carbonyl group is what attracts researchers to this bond disconnection.

References

1. Daly, J. W.; Spande, T. F.; Garraffo, H. M. "Alkaloids from Amphibian Skin: A Tabulation of Over Eight-Hundred Compounds" *J. Nat. Prod.* **2005**, *68*, 1556–1575. Daly, J. W. "Ernest Guenther Award in Chemistry of Natural Products. Amphibian Skin: A Remarkable Source of Biologically Active Arthropod Alkaloids" *J. Med. Chem.* **2003**, *46*, 445–452.
2. Majetich, G.; Wheless, K. "Remote Intramolecular Free Radical Functionalizations: An Update" *Tetrahedron* **1995**, *51*, 7095–7129. Barton, D. H. R.; Beaton, J. M.; Geller, L. E.; Pechet, M. M "A New Photochemical Reaction" *J. Am. Chem. Soc.* **1960**, *82*, 2640–2641.
3. Speckamp, W. N.; Moolenaar, M. J. "New Developments in the Chemistry of N-Acyliminium Ions and Related Intermediates" *Tetrahedron* **2000**, *56*, 3817–3856. Speckamp, W. N.; Hiemstra, H. "Intramolecular Reactions of *N*-Acyliminium Intermediates" *Tetrahedron* **1985**, *41*, 4367–4416.
4. Maryanoff, B. E.; Zhang, H-C.; Cohen, J. H.; Turchi, I. J.; Maryanoff, C. A. "Cyclizations of *N*-Acyliminium Ions" *Chem. Rev.* **2004**, *104*, 1431–1628.
5. Stork, G. "Polycyclic Natural Products: Total Synthesis of Lycopodine" *Pure Applied Chem.* **1968**, *17*, 383–401. Stork, G.; Kretchmer, R. A.; Schlessinger, R. H. "The Stereospecific Total Synthesis of *dl*-Lycopodine" *J. Am. Chem. Soc.* **1968**, *90*, 1647–1648.
6. Keck, G. E.; Yates, J. B. "Carbon-Carbon Bond Formation via the Reaction of Trialkylallystannanes with Organic Halides" *J. Am. Chem. Soc.* **1982**, *104*, 5829–5831.
7. Roth, M.; Dubs, P.; Goetschi, E.; Eschenmoser, A. "Synthetic Methods. 1. Sulfide Contraction via Alkylative Coupling: Method for Preparation of *β*-Dicarbonyl Derivatives" *Helv. Chim. Acta* **1971**, *54*, 710–734.
8. Corey, E.; J.; Fuchs, P. L. "Synthetic Method for Conversion of Formyl Groups into Ethynyl Groups" *Tetrahedron Lett.* **1972**, *12*, 3769–3772.
9. Hoffmann, R. W. "Allylic 1,3-Strain as a Controlling Factor in Stereoselective Transformations" *Chem. Rev.* **1989**, *89*, 1841–1860. Johnson, F. "Allylic Strain in Six-Membered Rings" *Chem. Rev.* **1968**, *68*, 375–413.
10. Sonogashira, K.; Tohda, Y.; Hagihara, N. "Convenient Synthesis of Acetylenes. Catalytic Substitutions of Acetylenic Hydrogen with Bromo Alkenes, Iodo Arenes and Bromopyridines" *Tetrahedron Lett.* **1975**, *15*, 4467–4470.
11. Brown, H. C.; Bhat, K. S.; Randad, R. S. "Chiral Synthesis via Organoboranes. 21. Allyl- and Crotylboration of *α*-Chiral Aldehydes with Diisopinocampheylboron as the Chiral Auxiliary" *J. Org. Chem.* **1989**, *54*, 1570–1576. Roush, W. R.; Grover, P. T. "*N,N'*-Bis(2,2,2-trifluoroethyl)-*N,N'*ethylenetartramide: An

Improved Chiral Auxiliary for the Asymmetric Allyboration Reaction" *J. Org. Chem.* **1995**, *60*, 3806–3813.

12. Oppolzer, W.; Cintas-Moreno, P.; Tamura, O.; Cardinaux, F. "Enantioselective Synthesis of α-*N*-Alkylamino Acids via Sultam-Directed Enolate Hydroxy-amination" *Helv. Chim. Acta* **1993**, *76*, 187–196. Oppolzer, W.; Merifield, E. "Synthesis of (-)-Pinidine via Asymmetric, Electrophilic Enolate Hydroxy-amination/Nitrone Reduction" *Helv. Chim. Acta* **1993**, *76*, 957–962.
13. Yamakado, Y.; Ishiguro, M.; Ikeda, N.; Yamamoto, H. "Stereoselective Carbonyl-Olefination via Organosilicon Compounds" *J. Am. Chem. Soc.* **1981**, *103*, 5568–5570.
14. Crabtree, R. H.; Davis, M. W. "Directing Effects in Homogeneous Hydrogenation with $[Ir(cod)(PCy_3)(py)]PF_6$" *J. Org. Chem.* **1986**, *51*, 2655–2661. Stork, G.; Kahne, D. E. "Stereocontrol in Homogeneous Catalytic Hydrogenation via Hydroxyl Group Coordination" *J. Am. Chem. Soc.* **1983**, *105*, 1072–1073. Crabtree, R. H.; Felkin H.; Fillebeen-Khan, T.; Morris, G. E. "Dihydridoirridium Diolefin Complexes as Intermediates in Homogeneous Hydrogenation" *J. Organometallic Chem.* **1979**, *168*, 183–195.
15. McMurry, J. E. "Total Synthesis of Copacamphene" *Tetrahedron Lett.* **1970**, *11*, 3731–3734. Thompson, H. W. "Stereochemical Control of Reductions. The Directive of Carbomethoxy vs. Hydroxymethyl Groups in Catalytic Hydrogenation" *J. Org. Chem.* **1971**, *36*, 2577–2581.
16. Fujimoto, R.; Kishi, Y. "On the Absolute Configuration of Gephyrotoxin" *Tetrahedron Lett.* **1981**, *22*, 4197–4198.
17. Grieco, P. A.; Gilman, S.; Nishizawa, M. "Organoselenium Chemistry. A Facile One-Step Synthesis of Alkyl Aryl Selenides from Alcohols" *J. Org. Chem.* **1976**, *41*, 1485–1486. Sharpless, K. B.; Young, M. W. "Olefin Synthesis. Rate Enhancement of the Elimination of Alkyl Aryl Selenoxides by Electron-Withdrawing Substituents" *J. Org. Chem.* **1975**, *40*, 947–949.
18. Hart, D. J. "Effect of $A^{(1,3)}$ Strain on the Stereochemical Course of *N*-Acyliminium Ion Cyclizations" *J. Am. Chem. Soc.* **1980**, *102*, 397–398. Hart, D. J.; Tsai, Y-M. "*N*-Acyliminium Ions: Detection of a Hidden 2-Aza-Cope Rearrangement" *Tetrahedron Lett.* **1981**, *22*, 1567–1570.
19. Barton, D. H. R.; McCombie, S. W. "New Method for the Deoxygenation of Secondary Alcohols" *J. Chem. Soc., Perkin 1* **1975**, 1574–1585.
20. Tsai, D. J. S.; Matteson, D. S. "A Stereocontrolled Synthesis of Z and E Terminal dienes from Pinacol E-1-Trimethylsilyl-1-propene-3-boronate" *Tetrahedron Lett.* **1981**, *22*, 2751–2752. Also see reference 13 above.

Problems

1. What are the structures of products that would result from addition of *N*-acyliminium ion **24** to the "other end" of the olefin (formation of a 5-membered ring). (Histrionicotoxin-4)
2. Given that perhydrohistionicotoxin has a 1,3-*N,O* relationship, propose a Mannich reaction that might be used to contruct the spirocyclic core of **5**. What stereochemical issues would have to be addressed in your proposed Mannich reaction? (Histrionicotoxin-4)
3. Here is a practical problem. Perhydrohistrionicotoxin adopts largely conformation **5a** in $CDCl_3$. How can one distinguish between **5a** and **5e** by ^1H NMR spectroscopy (in $CDCl_3$)? [Hint: How would you distinguish between *cis*- and *trans*-4-*tert*-butylcyclohexanols by ^1H NMR?] (Histrionicotoxin-4)
4. Consider using 2-[2-(1-butyl)cyclohexenyl]ethyl iodide in the intial alkylation of **36**. Develop a plan that might convert the resulting product to the epimer of **44** at the position of *n*-butyl substitution, and not call for the stereochemical corrections seen in Histrionicotoxin-5. Discuss the "big question marks" associated with your plan. (Histrionicotoxin-5)
5. Provide the structure of all isolable intermediates proceeding from **49** → **56**.
6. Provide a mechanism for the second step (Eschenmoser sulfide contraction) of the transformation of **57** → **58**. (Histrionicotoxin-7)
7. The unsaturated ketone/ester unit of **58** is amide/urethane-like in its behavior. Thus it is not easily hydrolyzed by methoxide. Explain this "unreactive" behavior. (Histrionicotoxin-7)
8. Provide structures of intermediates and a mechanistic explanation for the conversion of aldehyde **59** to TMS-acetylene **60**. (Histrionicotoxin-7)
9. Provide the structure of isolable intermediates *en route* from **61** to **62**. Elaborate on the comments in the text regarding allylic strain relative to the "lack of reactivity" of the piperidine in this transformation, and the conversion of **62** → **63**. (Histrionicotoxin-7)
10. Suppose the enolate of **73** had reacted with the bromide faster that the vinyloxirane. How might you change tactics to overcome this problem? (Histrionicotoxin-8)
11. Show the structure of the expected product after each step going from **78** → **80**. (Histrionicotoxin-9)
12. Supply the structures of intermediates after each step in the conversion of **89** → **90**. Provide a "model" that explains the stereochemical course of the hydroxylamination reaction. (Histrionicotoxin-10)

13. Provide a mechanistic explanation for the *Z*-olefin selectivity observed in the conversion of **95** to **96**. (Histrionicotoxin-11)
14. Given that a trialkylsilylacetylene was used in the Sonogoshira reaction, provide the structure of intermediates *en route* from **98** to histrionicotoxin (**2**). (Histrionicotoxin-11)
15. Provide an estimate of the energy difference between the two chair-chair conformations of pumiliotoxin-C. (Pumiliotoxin-1)
16. Develop a Diels-Alder approach to pumiliotoxin-C that involves construction of the bonds labeled "a" (below) in a Diels-Alder reaction. Discuss problems and potential solutions (where available). (Pumiliotoxin-1)

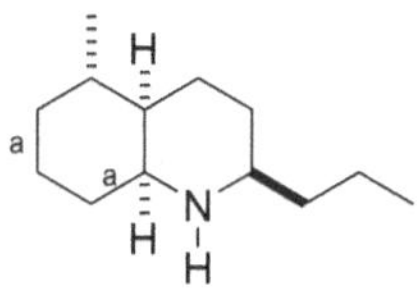

17. Describe tactics for the conversion of **121** to **122**. (Pumiliotoxin-2)
18. Explain why the most stable conformation of *cis*-*N*-benzoyl-2,6-methylpiperidine has the two methyl groups in axial sites. (Pumiliotoxin-2)

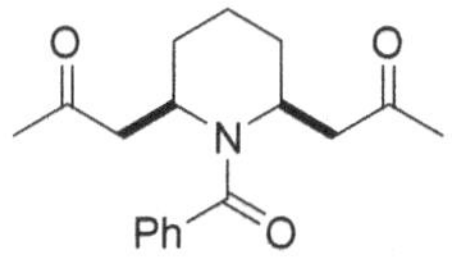

Quick, J.; Mondello, C.; Humora, M.; Brennan, T. "Synthesis of 2,6-Diacetonylpiperidine. X-Ray Difraction Analysis of its N-Benzoyl Derivative" *J. Org. Chem.* **1978**, 43, 2705–2708.

19. Explain the regiochemical course of nitrogen insertion in the conversion of **134** to **135** via a Beckman rearrangement. (Pumiliotoxin-4)
20. Provide the products expected after each step of the following reaction sequence. (Pumiliotoxin-4)

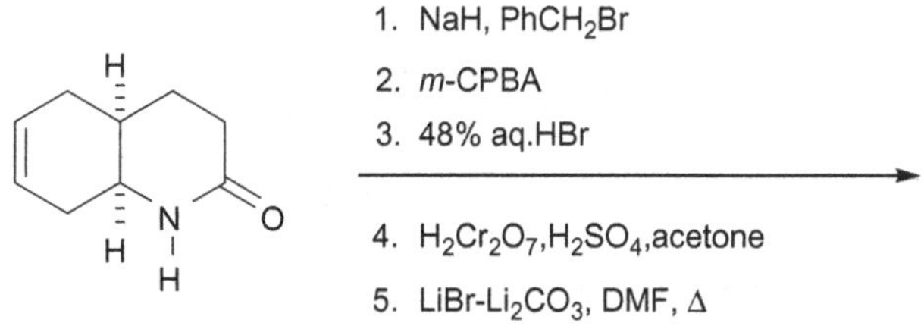

21. Suggest tactics for accomplishing the following transformation. (Pumiliotoxin-4)

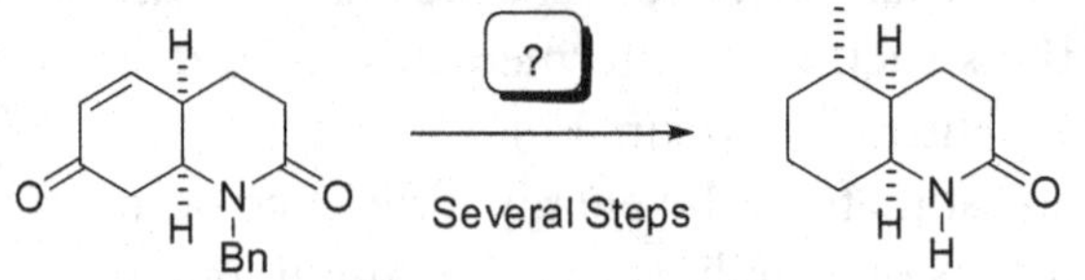

22. Suggest tactics for converting **136** → **137** → **3**. (Pumiliotoxin-4)
23. Provide intermediates *en route* from **139** → **140**. (Pumiliotoxin-4)
24. Outline syntheses of dienes **138** and **141**. (Pumiliotoxin-4)
25. Provide a mechanism for the following interesting reaction used in the conversion of **140** to **3**. (Pumiliotoxin-4)

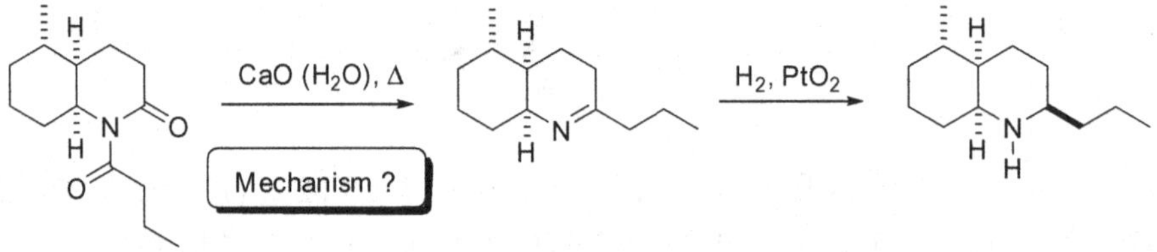

26. Provide a mechanism for the following transformation. Propose the most stable conformation of **144c** and the subsequent alkylation product **145**. (Pumiliotoxin-5)

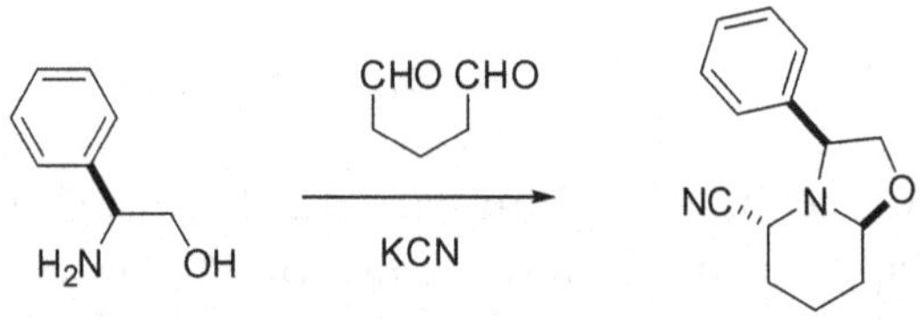

144c

27. Provide a mechanism for the conversion of **147** → **150**, and also for the conversion of **150** → **151** (you can ignore stereochemistry). (Pumiliotoxin-6)
28. Propose a synthesis of **153** from malic acid or any other commercially available starting material of your choice. (Pumiliotoxin-6)

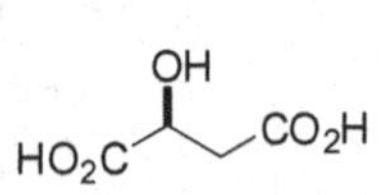

(*S*)-malic acid

29. Provide tactics for the conversion of **167** → **168**. (Pumiliotoxin-8)

30. Provide a rationale for the stereochemical course of the enolate protonation in the conversion of **168** → **170**. (Pumiliotoxin-8)
31. Suggest tactics that would convert **175** → **176**. (Pumiliotoxin-8)
32. Explain the following stereoselectivity trend. (Gephyrotoxin-1)

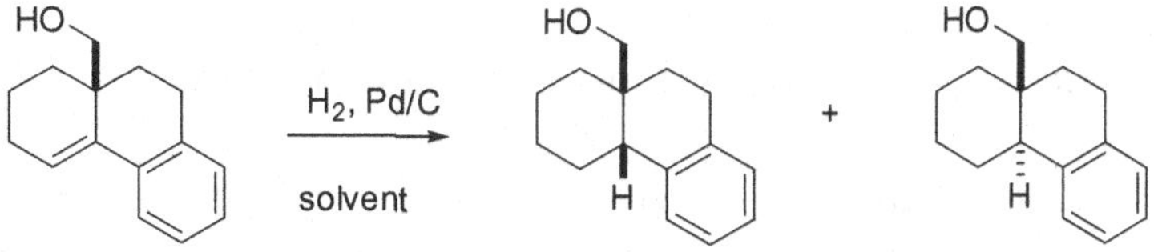

Solvent		
hexane	61	39
ethanol	6	94

Thompson, H. W.; McPherson, E.; Lences, B. L. "Stereochemical Control of Reductions. 5. Effects of Electron Density and Solvent on Group Haptophilicity" *J. Org. Chem.* **1976**, *41*, 2903–2906.

33. Provide a detailed mechanism for the Li-NH_3 reduction of **192**. Explain the stereochemical course of the reaction. (Gephyrotoxin-2)
34. Provide a mechanism, including a stereochemical explanation, for the conversion of **195** (R = CHO) → **197**. (Gephyrotoxin-2)
35. The reaction of **204** with succinimide, triphenylphosphine, and diethyl azodicarboxylate ($EtO_2CN{=}NCO_2Et$) gave (after chromatography) a 20% yield of the diastereomer of **204** with inverted alcohol stereochemistry. Provide a mechanistic explanation for this observation. (Gephyrotoxin-3)
36. Provide a stereochemical analysis for the conversion of **210** → **211**. Predict the stereochemical course of the following reaction. (Gephyrotoxin-3)

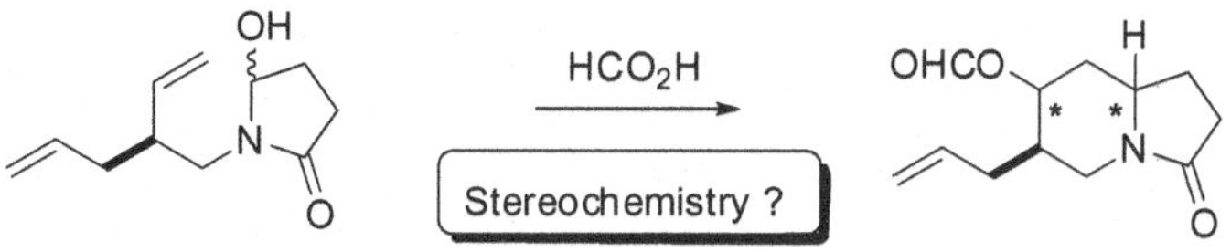

37. Suggest how NMR was used to determine the conformational preference of **206**. (Gephyrotoxin-3)
38. Provide the structure of intermediates *en route* from **206** → **207**. (Gephyrotoxin-3)
39. Provide mechanistic explanations for the observed stereocontrol in the conversions of **218** → **220** + **223**. (Gephyrotoxin-4)

40. Provide the structure of the other half-chair conformation available to *cis*-perhydroquinolinium ion **225**. Predict the product that would be obtained by a "stereoelectronically controlled reduction" from this conformation. (Gephyrotoxin-5)
41. What is the role of the trifluoroacetic acid in the conversion of **232** → **233**? (Gephyrotoxin-5)
42. Provide a reaction sequence that might convert **250** → **251**. (Gephyrotoxin-8)
43. Provide a mechanism for the conversion of **262** → **265**. Provide the structure of the minor iminium ion obtained from this reaction. Rationalize the stereochemical course of the reduction of **265** to ultimately provide **264**. (Gephyrotoxin-9)
44. Outline a synthesis of **272** beginning with preparation of **271** by construction of the bond indicated in red. (Gephyrotoxin-9)

CHAPTER 10

Morphine and Oxidative Phenolic Coupling

Morphine

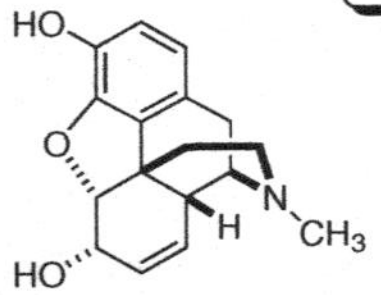

Morphine

1

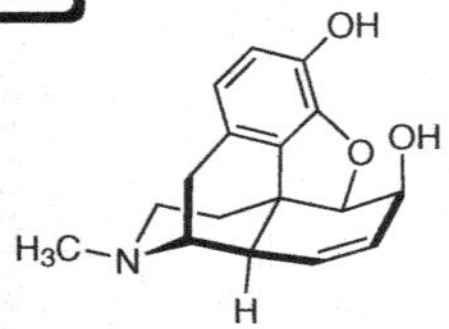

Morphine (named after Morpheus, the Greek god of dreams) is the major alkaloid present in opium. Opium is the dried sap of the unripe seed capsule of the poppy *Papaver somniferum*. High grade opium contains 9-14% of this alkaloid. Its medical properties have been known since ancient times. Morphine was not isolated in pure form until 1805. Its correct structure was not established until 1925, and it was not synthesized in the laboratory until 1952.

Pain is a major problem in medicine and relief of pain has long been a medical goal. Morphine is an analgesic, a substance that relieves pain without causing unconsciousness. It was used for large-scale relief of pain from battle wounds during the American Civil War (largely as a consequence of the invention, at about that time, of the hypodermic syringe). But morphine has serious side effects. It is addictive and also can cause nausea, a decrease in blood pressure, and a depressed breathing rate that can be fatal to the very young or the severely debilitated.

The first attempts to find a substance with morphine's benefits but without its side effects involved minor modification of its structure. Acetylation with acetic anhydride gave its diacetyl derivative (heroin), which is a good analgesic with less of a respiratory depressant effect than morphine. But heroin is severely addictive, and its abuse is a serious problem. Methylation of the phenol of morphine gave codeine, which is useful as an anticough (antitussive) agent. Unfortunately, it is less that one-tenth as effective as morphine as an analgesic.

Many compounds similar to various parts of the morphine structure have been synthesized and tested for their analgesic properties. For example, demerol is an effective analgesic with a relatively simple structure compared to that of morphine. Notice that it still retains the piperidine ring present in morphine. Methadone, which retains the nitrogen of morphine but is no longer heterocyclic, was synthesized and used as an analgesic by Germans during World War II, when natural sources of morphine were in short supply. Later it was used in substitution therapy for heroin addiction, but it, too, is addictive. The search for a perfect analgesic continues. With this background, it should be no surprise that morphine has been a popular target for synthesis, with a new synthesis appearing as late as the 50th anniversary of the 1st synthesis. We will spend the next few pages comparing and contrasting several of these syntheses.

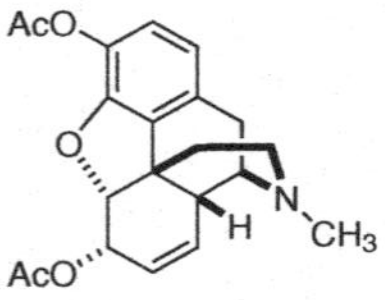

2 Heroin

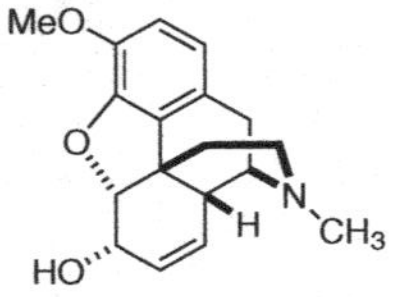

3 Codeine

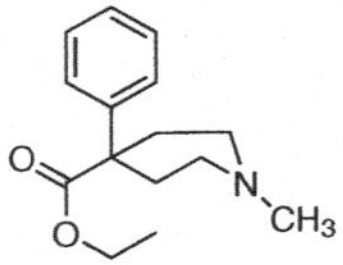

4 Demerol

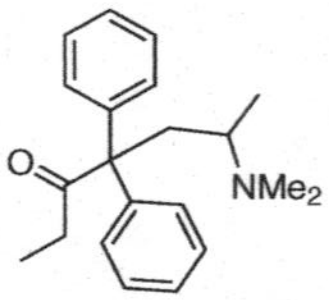

5 Methadone

For a thorough review of approaches to morphine see: Zezula, J.; Hudlicky, T. "Recent Progress in the Synthesis of Morphine Alkaloids" *Synlett* **2005**, 388-405. Novak, B. H.; Hudlicky, T.; Reed, J. W.; Mulzer, J.; Trauner, D. "Morphine Synthesis and Biosynthesis - An Update" *Current Organic Chemistry* **2000**, *4*, 343-362. Also see Magnus, P.; Sane, N.; Fauber, B. P.; Lynch, V. *J. Am. Chem. Soc.* **2009**, *131*, 16045-16047.

Morphine-1

Morphine (**1**) (named after Morpheus, the Greek god of dreams) is the major alkaloid present in opium. Opium is the dried sap of the unripe seed capsule of the poppy *Papaver somniferum*. High grade opium contains 9–14% of this alkaloid. Its medical properties have been known since ancient times. Morphine was not isolated in pure form until 1805. Its correct structure was not established until 1925, and it was not synthesized in the laboratory until 1952.

Pain is a major problem in medicine and relief of pain has long been a medical goal. Morphine is an analgesic, a substance that relieves pain without causing unconsciousness. It was used for large-scale relief of pain from battle wounds during the American Civil War (largely as a consequence of the invention, at about that time, of the hypodermic syringe). But morphine has serious side effects. It is addictive and also can cause nausea, a decrease in blood pressure, and a depressed breathing rate that can be fatal to the very young or the severely debilitated.

The first attempts to find a substance with morphine's benefits but without its side effects involved minor modification of its structure. Acetylation with acetic anydride gave its diacetyl derivative [heroin (**2**)], which is a good analgesic with less of a respiratory depressant effect than morphine. But heroin is severely addictive, and its abuse is a serious problem. Methylation of the phenol of morphine gave codeine (**3**), which is useful as an anticough (antitussive) agent. Unfortunately, it is less that one-tenth as effective as morphine as an analgesic.

Many compounds similar to various parts of the morphine structure have been synthesized and tested for their analgesic properties. For example, demerol (**4**) is an effective analgesic with a relatively simple structure compared to that of morphine. Notice that it still retains the piperidine ring present in morphine. Methadone (**5**), which retains the nitrogen of morphine but is no longer heterocyclic, was synthesized and used as an analgesic by Germans during World War II, when natural sources of morphine were in short supply. Later it was used in substitution therapy for heroin addiction, but it, too, is addictive. I remember a line from the Woody Allen film "Take the Money and Run" where one of the characters says "I used to be a heroin addict. Now I'm a methadone addict".

First Synthesis of Morphine

Gates, M.; Newhall, W. F. "Synthesis of Ring Systems Related to Morphine. I. 9,10-Dioxo-13-cyanomethyl-5,8,9,10,13,14-hexahydrophenanthrene" *J. Am. Chem. Soc.* **1948**, *70*, 2261. Gates, M. "Synthesis of Ring Systems Related to Morphine. III. 4,5-Dimethoxy-4-cyanomethyl-1,2-naphthoquinone and its Condensation with Dienes" *J. Am. Chem. Soc.* **1950**, *72*, 228. Gates, M.; Tschudi, G. "The Synthesis of Morphine" *J. Am. Chem. Soc.* **1955**, *78*, 1380.

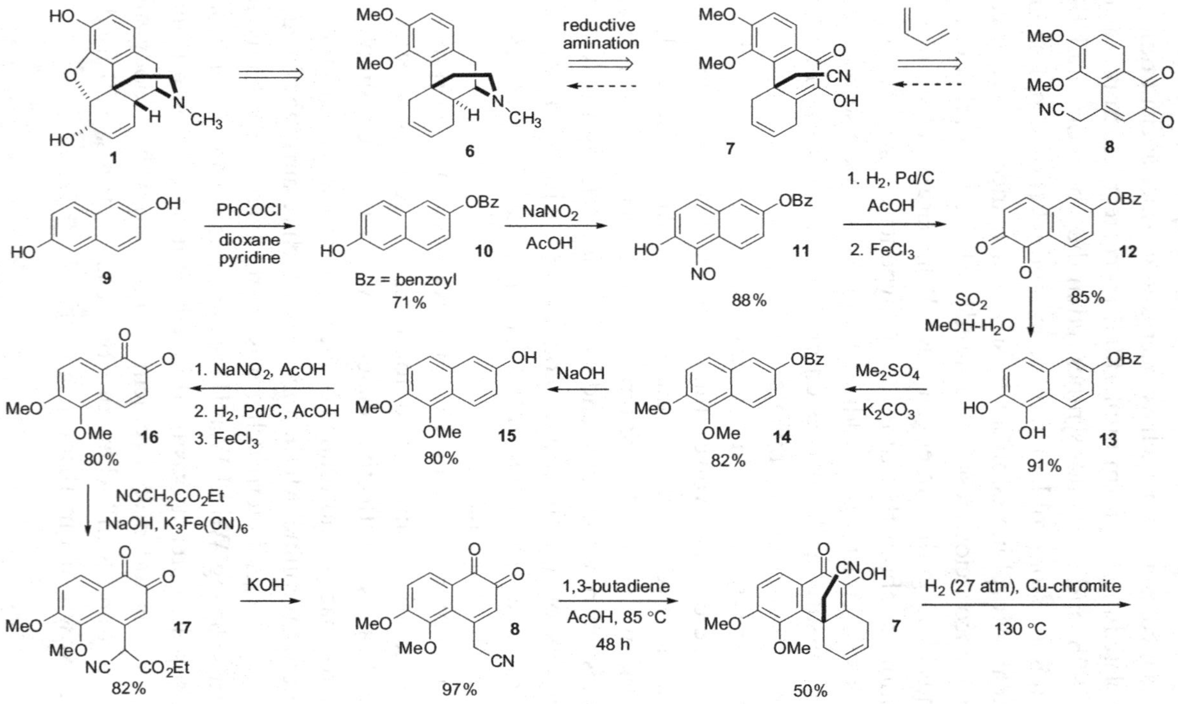

Morphine-2

Morphine-2

The search for a perfect analgesic continues. With this background, it should be no surprise that morphine has been a popular target for synthesis, with a new synthesis appearing as late as the 50th anniversary of the first synthesis. We will spend the next few pages comparing and contrasting several of these syntheses.

The first synthesis of morphine was reported by Marshall Gates (University of Rochester). It was a remarkable achievement. The plan is outlined briefly here. The idea was that if compound **6** could be obtained, it would be possible to move forward to morphine. In parallel studies it was shown that morphine could be degraded to **6**, thus providing a "relay point" for any synthesis effort, since morphine was in abundant supply and synthetic **6** would no doubt be difficult to obtain in large quantities. Structure **6** could clearly be obtained through a cycloaddition of an appropriate dienophile with 1,3-butadiene. The question is which dienophile? The choice ulitimately became *o*-naphthoquinone **8**. A cycloaddition between this electron-deficient dienophile and 1,3-butadiene would provide **7**, which one might convert to **6** via a reductive amination and additional oxidation state adjustments.

The dienophile synthesis began with 2,6-dihydroxynaphthalene (**9**). Protection of one of the phenolic groups as a benzoate was followed by *o*-nitrosation to provide **11**. Reduction of the nitroso group to a primary amine was followed by oxidation of the *o*-aminophenol to *o*-naphthoquinone **12**. The quinone was reduced to the hydroquinone (**13**) and the hydroxyl groups were protected. The benzoate was hydrolyzed and the resulting phenol (**15**) was subjected to the same oxidation sequence to provide *o*-naphthoquinone **16**. Treatment of this "enone" with ethyl cyanoacetate and base under oxidative conditions gave **17**. Ester hydrolysis followed by decarboxylation gave the desired dienophile **8**. The key cycloaddition was accomplished in the presence of acetic acid to give the desired cycloadduct **7**.

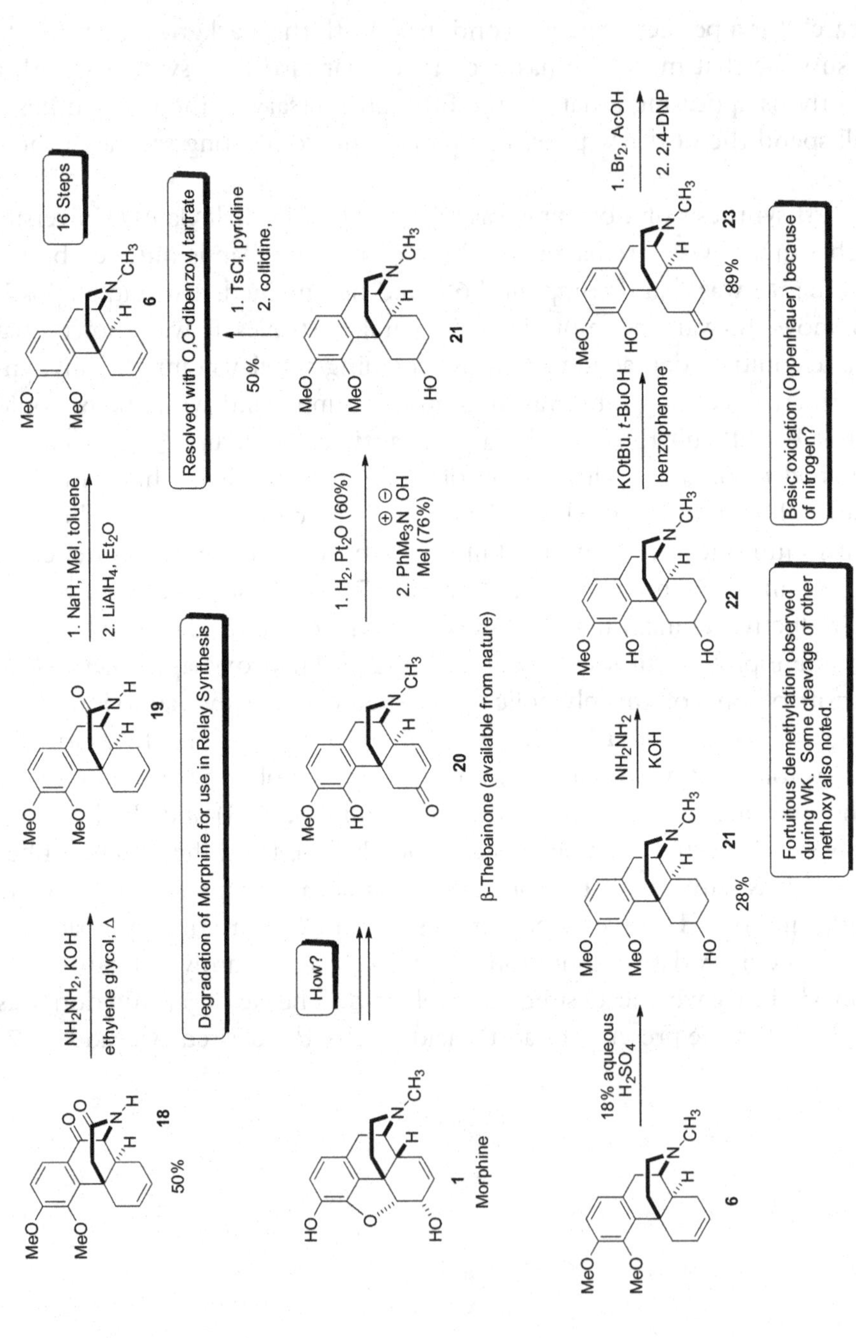
MeO
MeO
O
O
N
H
H
18
50%
NH_2NH_2, KOH
ethylene glycol, Δ
19
1. NaH, MeI, toluene
2. $LiAlH_4$, Et_2O
6
CH_3
16 Steps
Degradation of Morphine for use in Relay Synthesis
Resolved with O,O-dibenzoyl tartrate
How?
HO
Morphine
1
20
β-Thebainone (available from nature)
1. H_2, Pt_2O (60%)
2. $PhMe_3N$ OH
MeI (76%)
21
50%
1. TsCl, pyridine
2. collidine,
18% aqueous
H_2SO_4
28%
NH_2NH_2
KOH
22
KOtBu, t-BuOH
benzophenone
23
89%
1. Br_2, AcOH
2. 2,4-DNP
Fortuitous demethylation observed during WK. Some cleavage of other methoxy also noted.
Basic oxidation (Oppenhauer) because of nitrogen?

Morphine-3

Morphine-3

The conversion of 7 to keto-amide **18** required that the cyano group be hydrolyzed to a primary amide, and that reductive ring closure occur from nitrogen to the ketone generated upon protonation of the enol group present in 7. This was accomplished using copper chromite as a hydrogenation catalyst with ethanol as the solvent. This transformation is remarkable for its selectivity. Notice that the olefin is not disturbed. The intimate details of the mechanism of this process are not clear and I welcome your speculation.*

Continuing with the synthesis, a Wolf-Kishner reduction provided amide **19**. The *N*-methyl group was installed, followed by reduction of the lactam to provide key intermediate **6**. This material was available from morphine (**1**) and was used as a relay point for this synthesis. Hydration of **6** using aqueous sulfuric acid gave a low yield of **21**. This is an example of sophomore organic chemistry being applied to a complex molecule. Indeed Gates used what was available in the day, but this is clearly a weak point in the synthesis. An unexpected demethylation, observed as a minor reaction during the conversion of **18** → **19**, was applied to **21** to selectively remove one phenolic methyl group and provide **22**. Oxidation of **22** under basic conditions gave ketone **23**. Bromination of **23**, followed by 2,4-DNP promoted dehydrohalogenation, gave hydrazone **24** (see Morphine-4). Epimerization accompanied this transformation to establish the stereochemistry required by morphine γ to the enone.

*I recall a story from graduate school days (told by Gordon Rivers): The postdoc presents a remarkable result at group meeting. A graduate student in the audience asks "how does that work?". The postdoc replies "very well!". Reminds me of one of my favorite radio shows: "Whattaya Know? Not Much!".

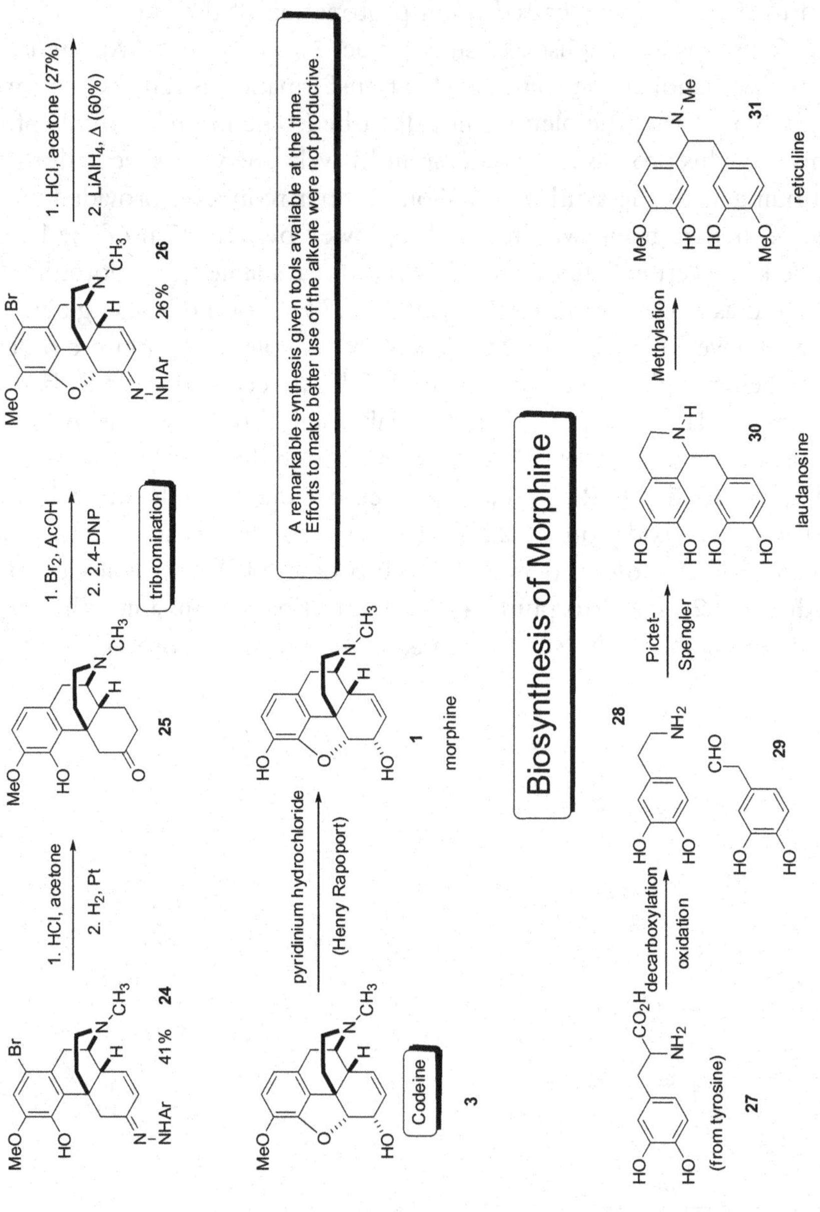
1. HCl, acetone
2. H2, Pt
1. Br2, AcOH
2. 2,4-DNP
tribromination
1. HCl, acetone (27%)
2. LiAlH4, Δ (60%)
41% 24
25
26% 26
pyridinium hydrochloride
(Henry Rapoport)
Codeine
3
1
morphine
A remarkable synthesis given tools available at the time.
Efforts to make better use of the alkene were not productive.
Biosynthesis of Morphine
(from tyrosine)
27
decarboxylation
oxidation
28
29
Pictet-
Spengler
30
laudanosine
Methylation
31
reticuline

Morphine-4

Morphine-4

Hydrolysis of the hydrazone followed by catalytic hydrogenation of the olefin and hydrogenolysis of the aryl bromide gave keto-phenol **25**. Treatment of **25** with bromine in acetic acid, followed by reaction of the intermediate tribromide with 2,4-dinitrophenylhydrazine, gave **26**. Hydrolysis of the hydrazone was followed by reduction of the enone to provide codeine (**3**). Demethylation of codeine provided morphine (**1**). This is a remarkable synthesis given tools that were available at the time.

It has been established that the biosynthesis of morphine follows a path from tyrosine derivative **27** to laudanosine (**30**) via a Pictet-Spengler type cyclocondensation of phenethylamine **28** and aldehyde **29**. Dimethylation of **30** then provides reticuline (**31**).

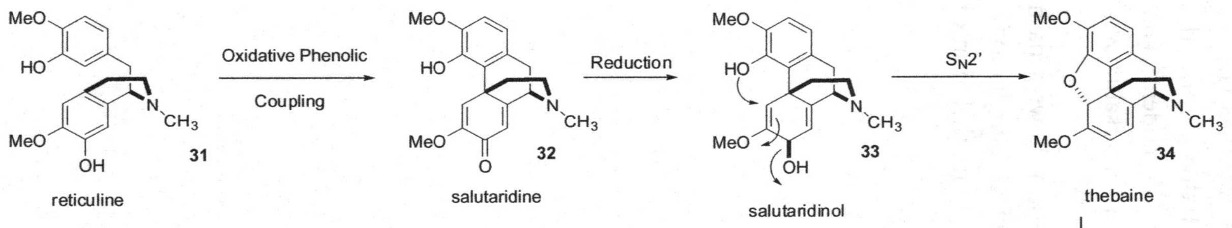

Nature's route to the morphinane alkaloids is very efficient.There have been many attempts to mimic this in the laboratory. We will look at a few and look at the problems associated with translating nature's plan to reality in the hands of man.

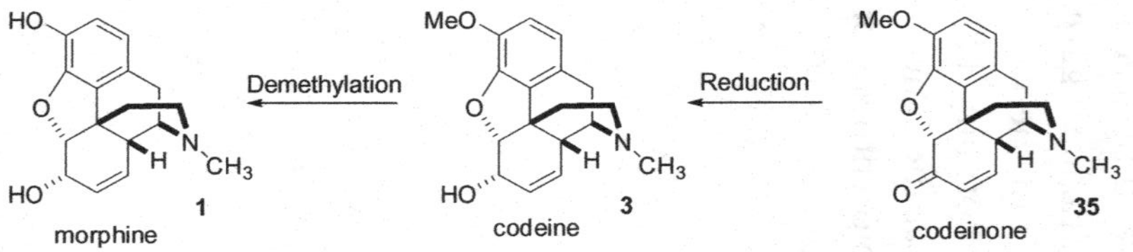

The First Biomimetic Synthesis of Morphine: Isotope Dilution Method

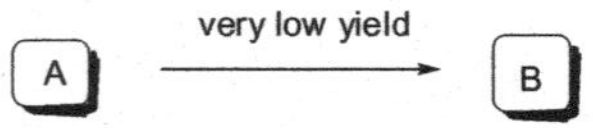

Suppose the yield is so low there is no chance of isolating B and determining yield.

Isotope Dilution Method

1. Prepare radiolabeled A
2. Run reaction
3. Add cold B
4. Isolate B
5. Purify to constant radioactivity
6. Determine % radiolabel in B
7. Calculate yield of B

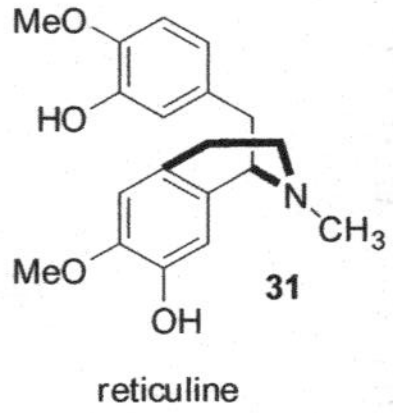

Tritiate aromatic ring with acid and tritiated water. Check for position of substitution using heavy water (all positions of aromatic ring deuterated).

Barton, D. H. R.; Bhakuni, D. S.; James, R.; Kirby, G. W. "Phenol Oxidation and Biosynthesis. Part XII. Stereochemical Studies Related to the Biosynthesis of Morphine" *J. Chem. Soc. (C)* **1967**, 128.

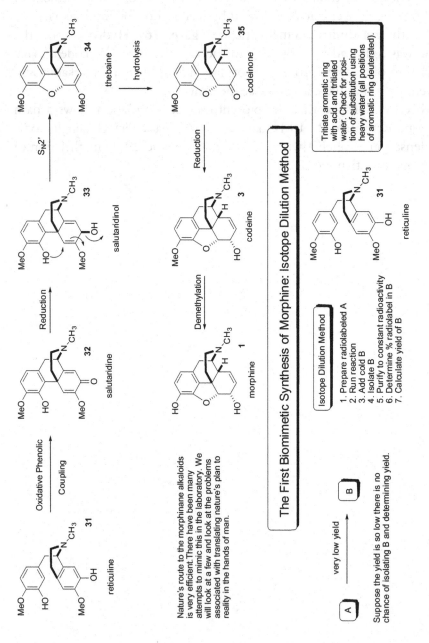

Morphine-5

The key step in the biosynthesis of morphine involves the "oxidative phenolic coupling" of reticuline (**31**) to salutaridine (**32**). This step can be viewed mechanistically as (1) oxidation of the two aromatic rings to phenoxy radicals followed by an intramolecular radical-radical coupling or (2) oxidation of one ring to a radical cation or cation, followed by an intramolecular electrophilic aromatic reaction. This process is very important in the biosynthesis of a number of natural products, and is a process that nature has used to "cross-link" peptides containing aromatic residues. The biosynthesis of morphine continues with reduction of salutaridine (**32**) to salutaridinol (**33**) followed by an intramolecular S_N2' reaction to give thebaine (**34**). Dienol ether hydrolysis to codeinone (**35**), reduction of the ketone to codeine (**3**) and *O*-demethylation completes the biosynthesis of morphine (**1**).

Nature's route to the morphinane alkaloids is very efficient. There have been many attempts to mimic this in the laboratory. We will look at a few and look at problems associated with translating nature's plan to reality in the hands of chemists.

The first biomimetic synthesis of morphine was described by Sir Derek Barton (Imperial College in London). Barton showed that oxidation of reticuline (**31**) with potassium ferricyanide gave salutaridine (**32**) in 0.015% yield. Since salutaridine had been converted to morphine, this constituted a synthesis of the target alkaloid. How does one isolate a product in 0.015% yield? Not easily. In this case, the "isotope dilution" method was used to establish yield. Reticuline was prepared in radioactive "hot" form (tritiation in the aromatic ring). The oxidation reaction was run and a known amount of "cold" salutaridine was added to the reaction mixture. The salutaridine was then isolated and purified to a constant level of radioactivity. The amount of "hot" salutaridine derived from reticuline was calculated based on the radioactivity that had been incorporated, and the percentage yield of salutaridine (from reticuline) was thus determined. The details are given on Morphine-6. The bottom line is that this demonstrated the feasability of this route to morphine, but it did not provide a practical route to the natural product.

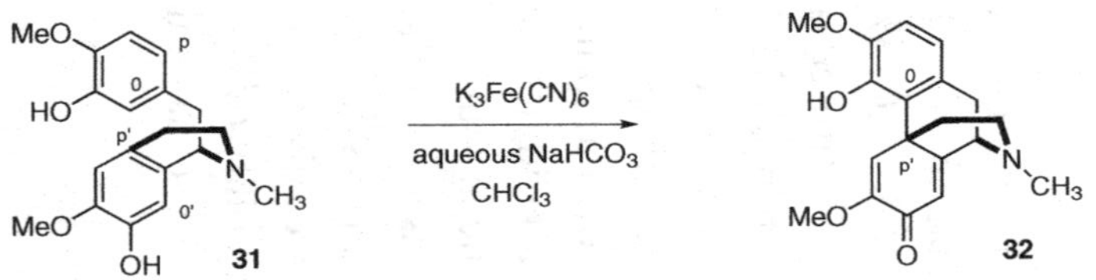

reticuline salutaridine

1. Use 52 mg of hot racemic reticuline
2. Run reaction
3. Add 50 mg of cold (+)-salutaridine
4. Isolate 45 mg of (+)-salutaridine after crystallization to constant activity
5. Yield = 0.015% (0.03% of racemate)

This paper also describes results of significance in determining stereochemical aspects of the conversion of salutaridine to morphine (stereochemical course of S_N2')

There are problems associated with conducting this reaction in the laboratory: (1) ortho-para' coupling vs para-ortho' coupling or ortho-ortho' coupling (both of which give aporphines or para-para' coupling (which gives a less hindered morphinane) (2) oxidation regiochemistry and oxidation of the amine.

morphinane

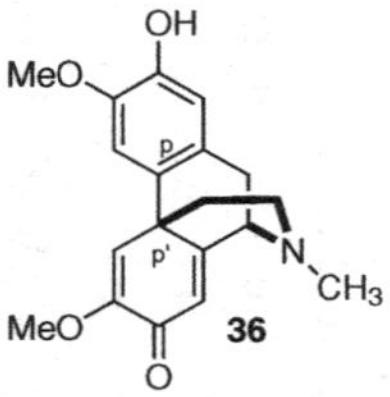

36

para-para'

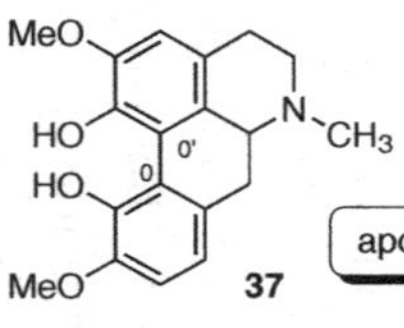

37

ortho-ortho'

aporphines

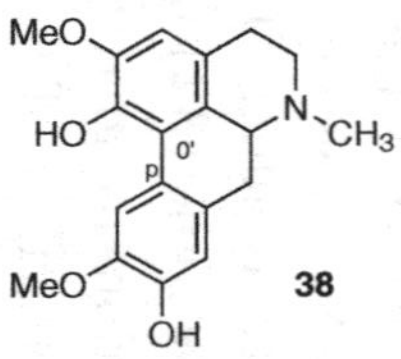

38

para-ortho'

A number of other chemical oxidizing agents have been applied to this transformation (MnO_2-silica gel, $AgCO_3$-Celite, $VOCl_3$) using either the tertiary amine or reticuline derivatives in which the *N*-methyl group is replaced with a carbamate. None of these deal adequately with the aforementioned problems. Electrochemical oxidations sometimes provide reasonable yields of morphinanes, but the para-para' isomers dominate, possibly for steric reasons. Here is one attempt to get around the regiochemical problem.

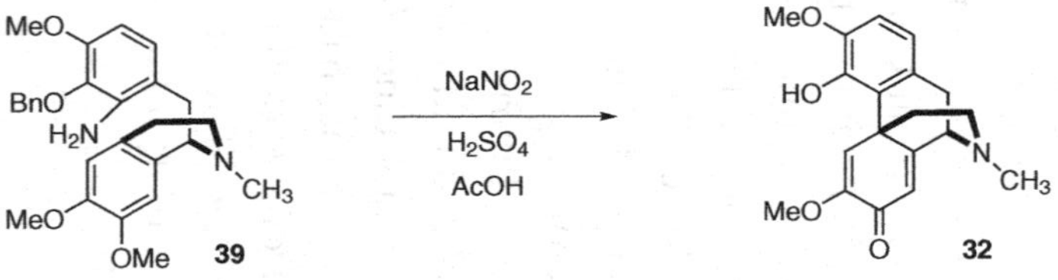

1.1%

Kametani, T.; Ihara, M.; Fukumoto, K.; Yagi, H. "Studies on the Syntheses of Heterocyclic Compounds. Part CCC. Syntheses of Salutaridine, Sinoacutine, and Thebaine. Formal Total Syntheses of Morphine and Simomenine" *J. Chem. Soc. (C)* **1969**, 2030-2033.

Morphine-6

What are the problems associated with this reaction in the laboratory? The substrate could undergo a variety of oxidative-phenolic couplings, each from a different conformation (or from several different conformations). If you examine the structure of reticuline (**31**) in the upper left-hand corner of the page, para′-ortho coupling provides salutaridine (**32**). However, para-para′ coupling would provide **36**, ortho-ortho′ coupling would provide **37** and para-ortho′ coupling would provide **38** (the latter two compounds belong to a family of alkaloids called aporphines). The site of oxidation (which aromatic ring is oxidized) can also influence coupling selectivities. Finally, oxidation of the tertiary amine can cause problems, although this is solved by *N*-acylation.

A number of chemical oxidizing agents have been applied to this transformation, using either the tertiary amine or reticuline derivatives in which the *N*-methyl group is replaced with a carbamate. None of these adequately deal with the aforementioned problems. Electrochemical oxidations sometimes provide reasonable yields of morphinanes, but the para-para′ isomers dominate, possibly for steric reasons.

One indirect solution to the site-of-oxidation problem was addressed by Kametani (Japan). His group reported the conversion of **39** to **32** in 1.1% yield (almost a 100-fold improvement over the Barton conditions), by performing the oxidation "site-selectively" using aryl diazonium chemistry, but this was still quite a low yield.

Oxidative Phenolic Coupling Route to Morphinanes

Schwartz, M. A.; Mami, I. S. "A Biogenetically Patterned Synthesis of the Morphine Alkaloids" *J. Am. Chem. Soc.* **1975**, *97*, 1239-1240.

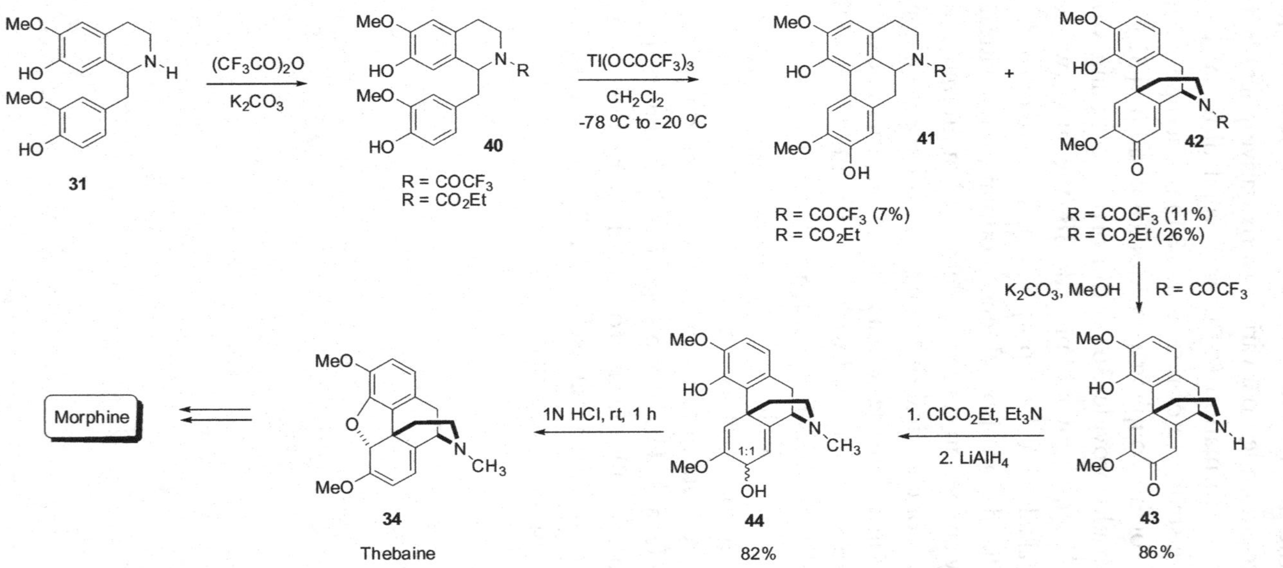

Morphine-7

Morphine-7

Here is one of the better attempts to accomplish oxidative phenolic coupling in the lab. Schwartz (Florida State University) reported that oxidation of carbamates of type **40** derived from reticuline (**31**) gave morphinanes of type **42** in 11–26% yield upon oxidation with thallium trifluoroacetate. Aporphines of type **41** were also obtained in lower yields. Whereas the carbamate forces the "benzylic group" to be pseudo-axial on the tetrahydroisoquinoline core (a requirement for morphinane formation), the reason for regioselectivity is not clear to this author. Nonetheless this provides access to morphine via **42** (R = CO_2Et). Reduction of the carbamate to a methyl group was accompanied by conversion of the enone to allylic alcohol **44**. Treatment of **44** with acid provided thebaine (**34**), and the transformation of thebaine to morphine is known (see Morphine-5).

Alternatives to Oxidative Phenolic Coupling 1: The Fuchs Approach

The following approaches revolve around construction of same key bond formed in the biomimetic approaches. They depart from the biomimetic approach, however by not "insisting" that a biaryl coupling be used to construct that bond. This "thought process" leads to a number of interesting approaches. We will examine three such approaches that involve carbanions, free radicals, and organopalladium species as key reactive intermediates for constructing this bond. Although it will not be covered here, an approach that involves cationic intermediates has also been described [*Tetrahedron Lett.* **1967**, 4055]. We will start with an approach developed by the Fuchs group that revolves around vinyl sulfone addition chemistry. (Toth, J. E.; Hamann, P. R.; Fuchs, P. L. "Studies Culminating in the Total Synthesis of (*dl*)-Morphine" *J. Org. Chem.* **1988**, *53*, 4694-4708).

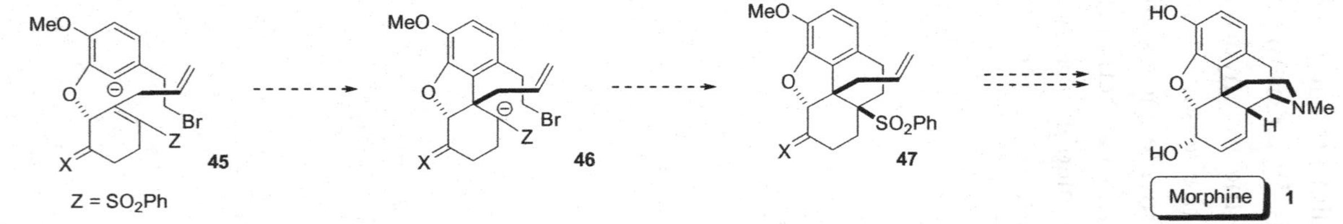

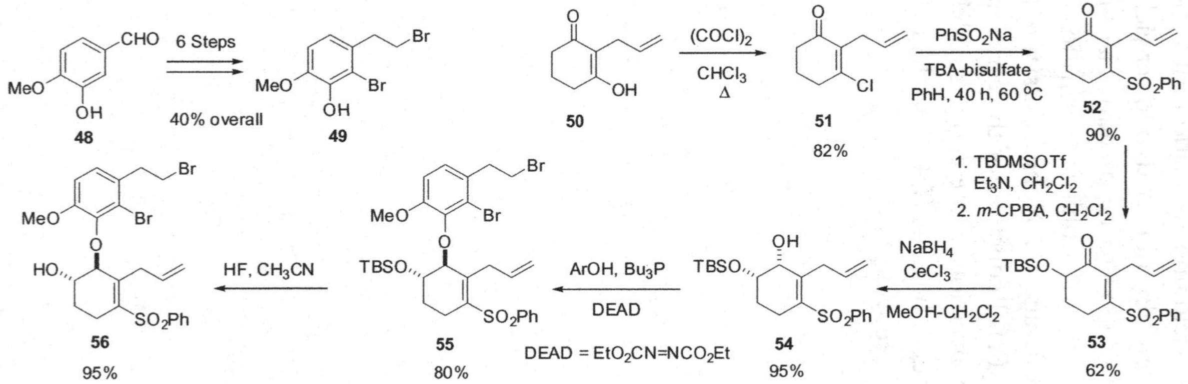

Morphine-8

Now we will turn to approaches that revolve around construction of the same key bond formed in the biomimetic approaches. They depart from the biomimetic approach, however, by not "insisting" that a biaryl coupling be used to construct that bond. This "thought process" leads to a number of interesting approaches. We will examine three such approaches that involve carbanions, free radicals, and organopalladium species as key reactive intermediates for constructing this bond. Although it will not be covered here, an approach that involves cationic intermediates has also been described.[1] We will start with an approach developed by the Fuchs group (Purdue University) that revolves around vinyl sulfone addition chemistry.

Fuchs imagined that a carbanion of type **45** might undergo an intramolecular conjugate addition to provide **46**. A subsequent intramolecular alkylation would give **47**, which then might be converted to morphine in a sequence of steps that would require degradation of the allyl sidechain and construction of the C–N bond required to complete the synthesis of morphine (**1**). This plan nicely addresses the regiochemical problems posed by the oxidative phenolic coupling approach through use of intramolecular reactions. The requirements for "Z" were that it "supports" the proposed cyclization-alkylation and that it be converted to an "H". As an application of vinyl sulfone chemistry developed over a period of years in the Fuchs group, a phenylsulfonyl group was chosen to serve as "Z". Thus, the first task was the preparation of a precursor of **57** (Morphine-9), a precursor of **45** where an aryl bromide was to be the source of the carbanion, and a secondary alcohol was to play the role of =X.

The synthesis began with the preparation of dibromide **49** from isovanillin (**48**). This bromide was to be coupled to allyic alcohol **54** using an S_N2 reaction. Alcohol **54** was prepared from cyclohexan-1,3-dione derivative **50** as follows. Treatment of this vinylogous acid with oxalyl chloride provide vinylogous acid chloride **51**. A vinylogous nucleophilic acyl substitution reaction, using sodium benzenesulfinate as the nucleophile, provided **52**. The adjacent hydroxyl group was introduced by reaction of an intermediate silyl enol ether with *m*-chloroperoxybenzoic acid. Reduction of the ketone from the least hindered face gave **54**. A Mitsunobu reaction was used to couple **54** with phenol **49** to provide **55** in excellent yield.[2] Removal of the TBS protecting group gave **56**.

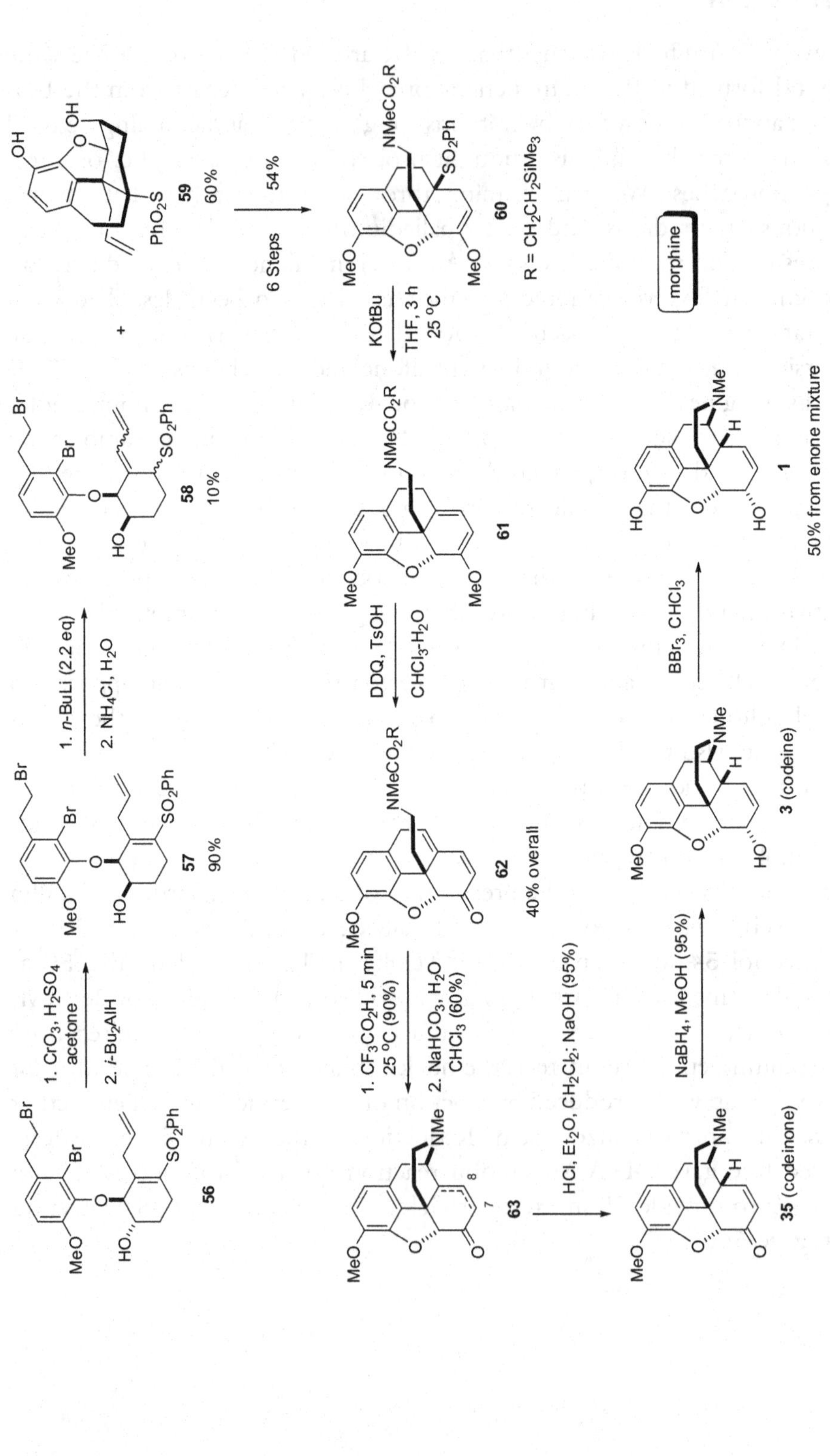

56
1. CrO_3, H_2SO_4 acetone
2. i-Bu_2AlH
57
90%
1. n-BuLi (2.2 eq)
2. NH_4Cl, H_2O
58
10%
+
59
60%
6 Steps
54%
60
R = $CH_2CH_2SiMe_3$
KOtBu
THF, 3 h
25 °C
61
DDQ, TsOH
$CHCl_3$-H_2O
62
40% overall
1. CF_3CO_2H, 5 min 25 °C (90%)
2. $NaHCO_3$, H_2O $CHCl_3$ (60%)
63
HCl, Et_2O, CH_2Cl_2; NaOH (95%)
35 (codeinone)
$NaBH_4$, MeOH (95%)
3 (codeine)
BBr_3, $CHCl_3$
1
50% from enone mixture
morphine

Morphine-9

Morphine-9

An oxidation-reduction sequence adjusted stereochemistry to provide key intermediate **57**. Treatment of **57** with an excess of *n*-BuLi gave the desired addition-alkylation product **59** in an excellent yield (45% or 60% if one accounts for recovered **57**) along with a small amount of diene **58**. This minor product presumably is the result of γ-deprotonation of the vinyl sulfone followed by α-protonation of the resulting carbanion. In reality it took a lot of substrate adjustment to finally achieve the desired annulation reaction. For example, **56** has all of the structural elements that would be needed to proceed with the plan, so what happened? Treatment of **56** with *n*-BuLi gave none of the desired cyclization product. Evidence indicated that γ- and γ'-deprotonation of **56** was one cause of problems with this substrate. In retrospect (or most likely recognized ahead of time by Fuchs), the proposed cyclization required a unique set of rates to succeed: (1) metal-halogenation exchange (ArBr to ArLi) had to be faster than sulfone deprotonation and (2) cyclization had to be faster than intramolecular and/or intermolecular deprotonation of the sulfone by the resulting dianion (ArLi and ROLi). One can imagine other complications as well. For example, carbanion **46** could undergo an intramolecular dehydrohalogenation. Regardless, the "search" part of research brought this part of the synthesis to a successful conclusion.

Moving forward from **59**, six steps were required to convert this compound to **60**. Vicinal dihydroxylation of the olefin was followed by oxidative cleavage of the intermediate diol using lead tetraacetate. Reductive amination of the resulting aldehyde with methylamine, followed by acylation of the intermediate secondary amine gave the desired carbamate. Swern oxidation of the secondary alcohol,[3] followed by enol ether formation gave **60**. Elimination of *p*-toluenesulfinic acid from **60** provided **61**. Oxidation of this dienol ether to dienone **62** was followed by release of the secondary amine, followed by a conjugate addition reaction to establish the critical C–N bond. The remainder of the synthesis followed known chemistry. The mixture of enones **63** was converted to codeinone (**35**), codeine (**3**) and then morphine (**1**).

Alternatives to Oxidative Phenolic Coupling 2: The Parker Approach

Parker, K. A.; Fokas, D. "Convergent Synthesis of *dl*-Dihydroisocodeine in 11 Steps by the Tandem Radical Cyclization Strategy. A Formal Total Synthesis of *dl*-Morphine" *J. Am. Chem. Soc.* **1992**, *114*, 9688-9689.

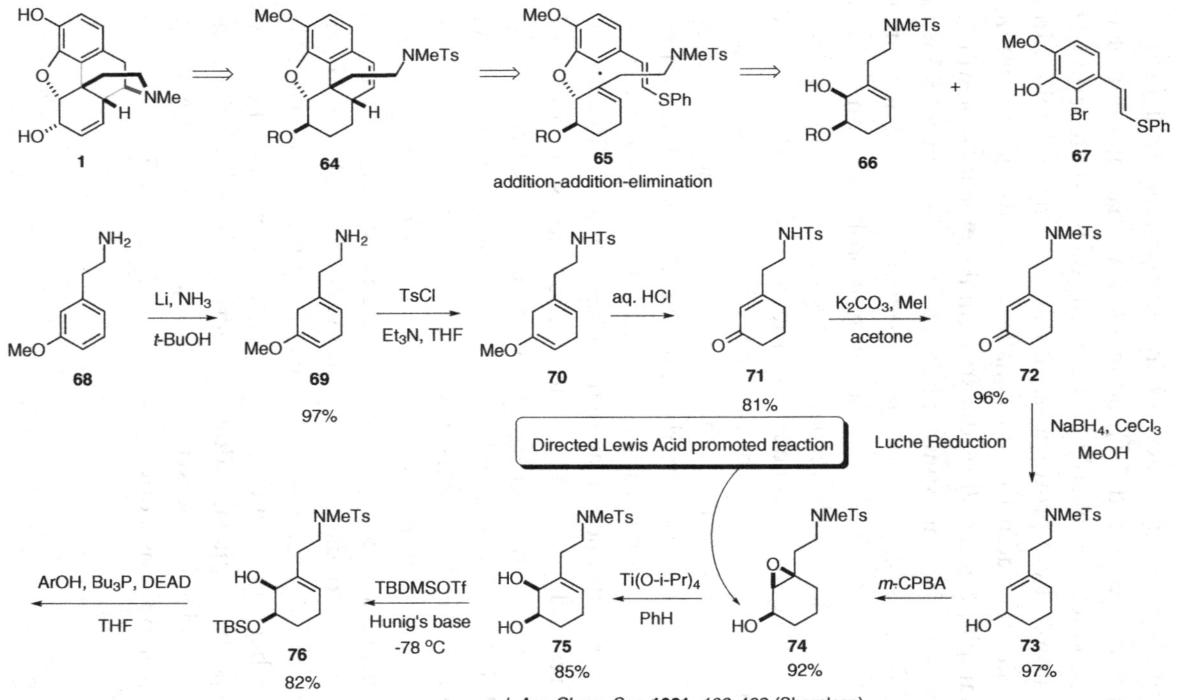

J. Am. Chem. Soc. **1981**, *103*, 462 (Sharpless)

Morphine-10

Morphine-10

Here is another alternative to oxidative phenolic coupling. Kathy Parker and her group (Brown University) described a synthesis of morphine that focuses on the same key bonds as seen in the Fuchs synthesis, but using totally different chemistry. The Parker plan was, once again, to construct the key C–N bond late in the synthesis from an intermediate of type **64**. Parker also focused on initial construction of the same C–C bonds, but this time using a free radical addition-addition-elimination sequence beginning with a radical of type **65**. The precursor of this radical (see **78** on Morphine-11) was to be an aryl bromide. This bromide was to be assembled from diol **66** and aryl bromide **67**, structures that are similar to the pieces used in the Fuchs synthesis (**49** and **54**).

The synthesis of **66** began with *m*-methoxyphenethylamine (**68**). Birch reduction of **68** gave **69**, which was converted to sulfonamide **70**. Enol ether hydrolysis was accompanied by conjugation of the olefin to provide enone **71**. Alkylation of the nitrogen gave **72** and reduction of the ketone using the Luche conditions gave allylic alcohol **73**.[4] Hydroxyl-directed epoxidation of **73** gave **74**, and a titanium-mediated opening of the epoxide, directed by the alcohol, gave diol **75**.[5] Protection of the homoallylic alcohol gave **76** (a derivative of **66**). A Mitsunobu coupling of **76** with phenol **67** (whose preparation will not be described here) gave **77** (Morphine-11).

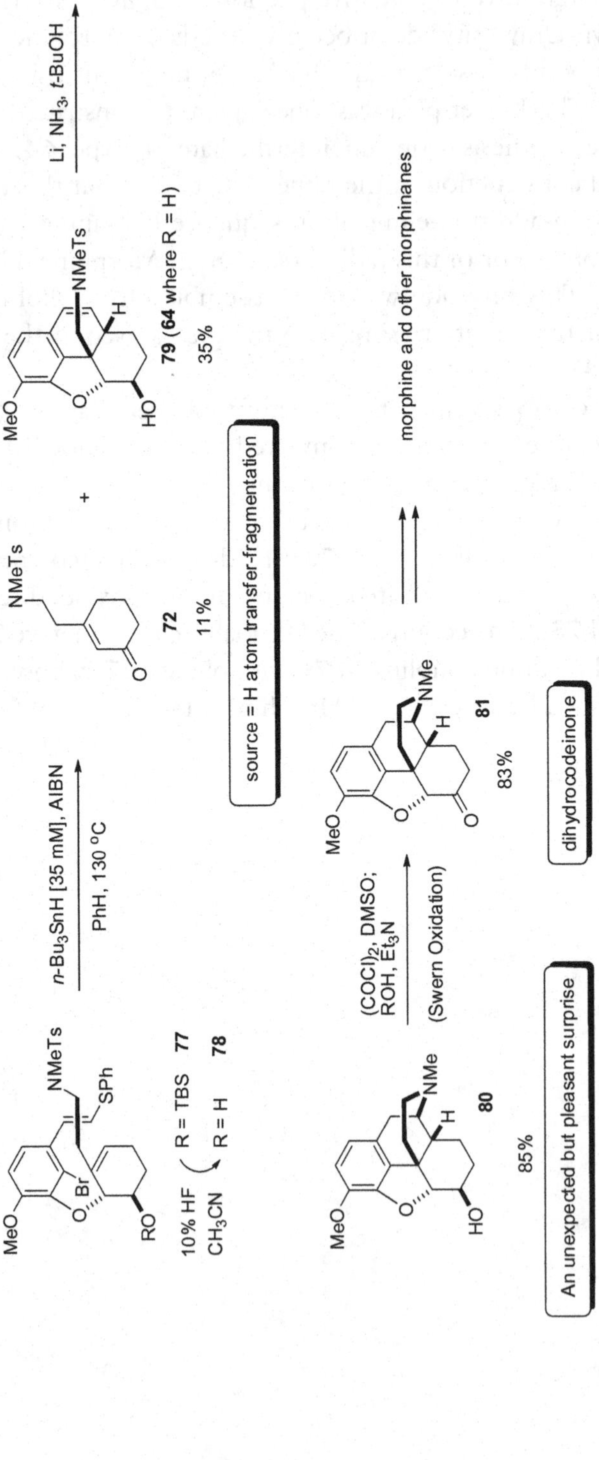
MeO
NMeTs
SPh
Br
RO
R = TBS 77
R = H 78
10% HF
CH3CN
n-Bu3SnH [35 mM], AIBN
PhH, 130 °C
72
11%
+
79 (64 where R = H)
35%
Li, NH3, t-BuOH
source = H atom transfer-fragmentation
NMe
HO
H
80
85%
(COCl)2, DMSO;
ROH, Et3N
(Swern Oxidation)
81
83%
morphine and other morphinanes
An unexpected but pleasant surprise
dihydrocodeinone

Morphine-11

Morphine-11

The TBS group was removed to provide **78**. The anticipated free radical gave **79** (**64** where R = H) as had been hoped.[6] This reaction involves sequential formation of the aryl radical, addition to the olefin to give a secondary radical (5-*exo* cyclization rather than 6-*endo* cyclization), a second radical cyclization to give a benzylic radical (6-*endo* cyclization rather than 5-*exo* cyclization), and a radical fragmentation. The expelled thiophenoxy radical was reduced by the tri-*n*-butyltin hydride to provide thiophenol and tri-*n*-butyltin radicals to continue the chain. Formation of **72** reveals that a 1,5-hydrogen atom transfer competes with the initial radical cyclization. This key reaction (as in the Fuchs synthesis) depends on a fine balance of competing reaction rates. It is notable that in the first cyclization, the 5-*exo* process wins over a possible 6-*endo* cyclization as expected, but in the second cyclization the reverse is true. The last stage of the synthesis was a pleasant surprise. Reductive remove of the *N*-tosyl group (Li, NH_3, *t*-BuOH) gave **80** in outstanding yield. Presumably this reaction generated a nitrogen-centered radical that cyclized to afford the critical C–N bond. A Swern oxidation provided dihydrocodeinone which had previously been taken on to morphine and other morphinanes.

Alternatives to Oxidative Phenolic Coupling 3: The Overman Approach

Hong, C. Y.; Overman, L. E. "Preparation of Opium Alkaloids by Palladium Catalyzed Bis-Cyclizations. Formal Total Synthesis of Morphine" *Tetrahedron Lett.* **1994**, *35*, 3453. Also see: Hong, C. Y.; Kado, N.; Overman, L. E. "Asymmetric Synthesis of Either Enantiomer of Opium Alkaloids and Morphinans. Total Synthesis of (-)- and (+)-Dihydrocodeinone and (-)- and (+)-Morphine" *J. Am. Chem. Soc.* **1993**, *115*, 11028-11029. Heerding, E. A.; Hong, C. Y.; Kado, N.; Look, G. C.; Overman, L. E. "Simple Method for Controlling Stereoselection in Mannich Cyclization Reactions of Aldehydes" *J. Org. Chem.* **1993**, *58*, 6847-6948.

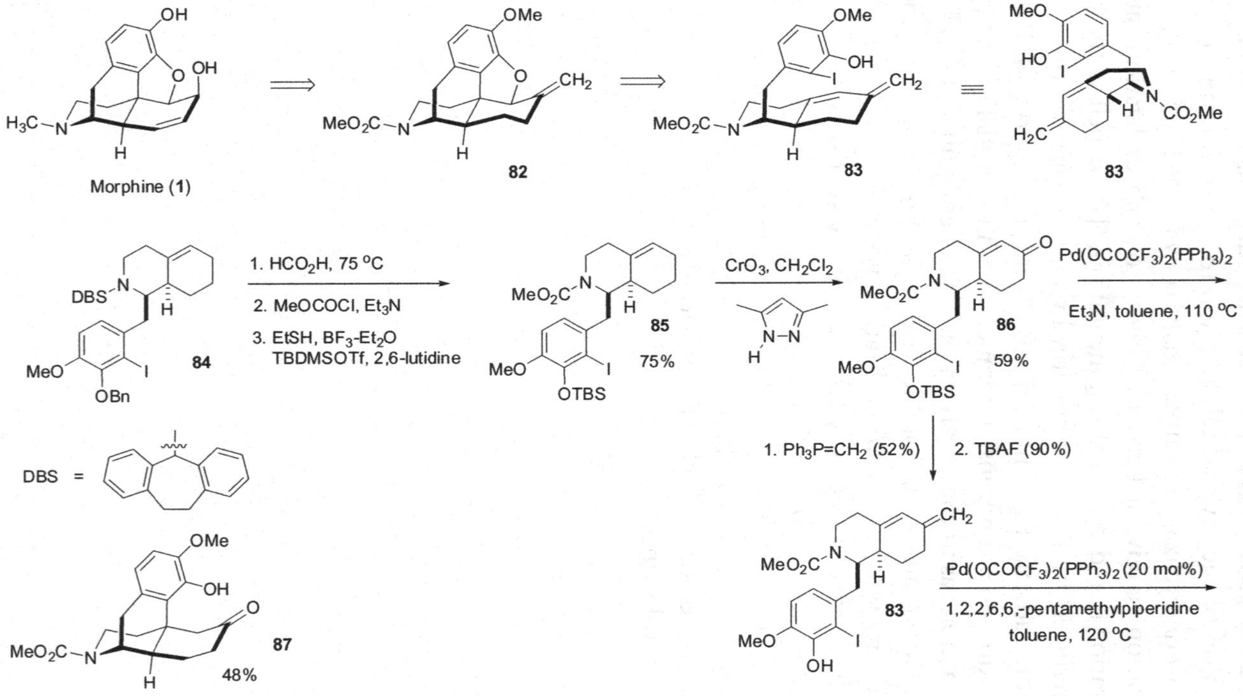

Morphine-12

Here is another approach whose biomimetic nature is more apparent than the carbanion and free radical routes we just examined. Overman (UC Irvine) anticipated that **82** could be taken to morphine through a series of standard transformations. This olefin was to be prepared from **83** via an interesting palladium-mediated reaction (a variation of the Heck arylation).[7] Substrate **83** can clearly be recognized as an isoquinoline that is more highly reduced than in reticuline (**31**), the natural substrate used in morphine biosynthesis. In essence, the oxidation states of the two aromatic rings that nature couples have been purposely differentiated.

I will not go through the synthesis of **84** (related to **83**) as this can be an exercise for you. Removal of the *N*-DBS protecting group, allylic oxidation (Corey-Fleet reagent)[8] of the alkene to enone **86**, and a Wittig methylenation, provided cyclization substrate **83**.

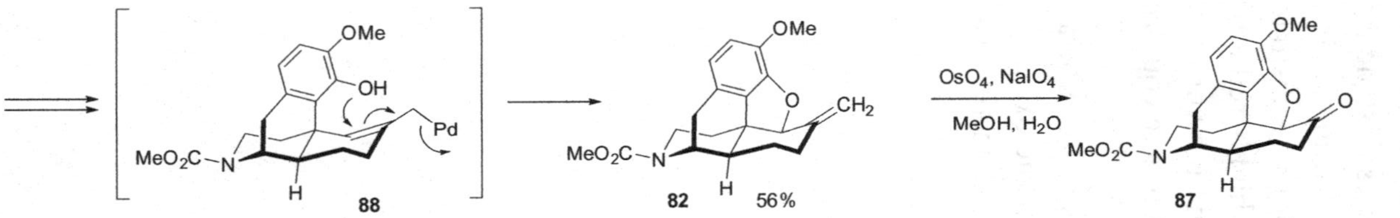

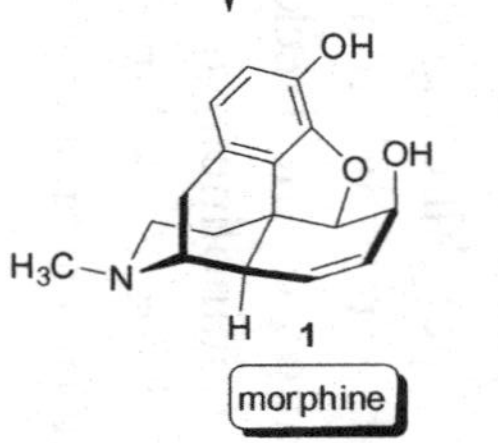

Iijima, I.; Rice, K. C.; Silverton, J. V. *Heterocycles* **1977**, *6*, 1157

Mulzer's Conjugate Addition Route to Morphine

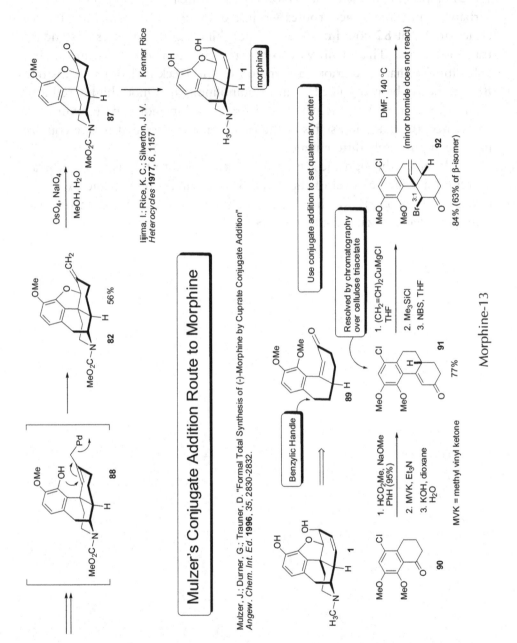

Mulzer, J.; Durner, G.; Trauner, D. "Formal Total Synthesis of (-)-Morphine by Cuprate Conjugate Addition" *Angew. Chem. Int. Ed.* **1996**, *35*, 2830-2832.

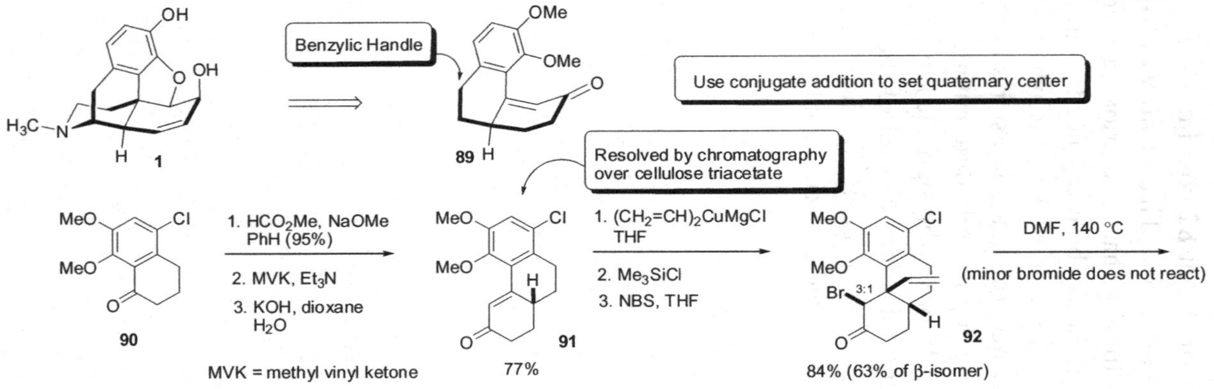

Morphine-13

Morphine-13

The projected Pd-mediated "double cyclization" was accomplished to give **82** in a good yield through a presumed intermediate of type **88**. Johnson-Lemieux oxidative cleavage of the double bond gave ketone **87** which had previously been converted to morphine by the Rice group (NIH).

We will now move away from biomimetic synthesis and examine a perhydrophenanthrene-based approach reported by Mulzer. The plan was to prepare tetrahydrophenanthrene **89**, use conjugate addition chemistry to set the quaternary center, and use the indicated benzylic position as a handle for construction of the same C–N bond central to the Gates, Fuchs and Parker syntheses. Tetralone **90** will serve as our point of departure.

Annulation chemistry similar to that we encountered in the steroid area was used to provide **91**. The enantiomers of **91** were separated by chromatography over a chiral support. A vinylic cuprate was used to generate the quaternary center. The resulting enolate was trapped as a trimethylsilyl enol ether, bromination of which provided α-bromoketone **93** as a mixture of diastereomers. Heating **92** converted the major diastereomer to dihydrobenzofuran **93**. The minor stereoisomer did not react.

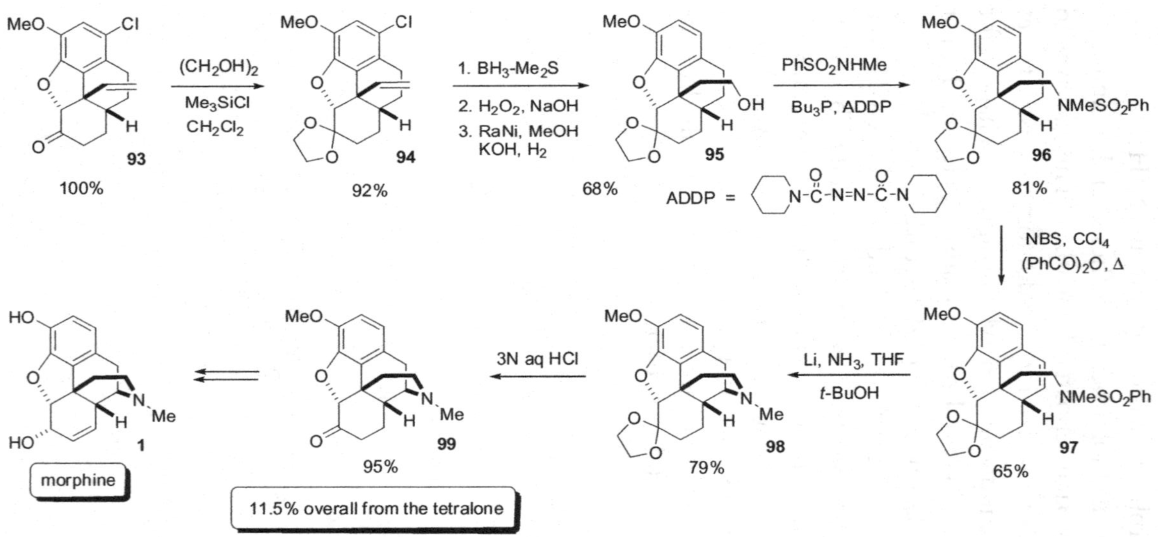

A Recent Synthesis

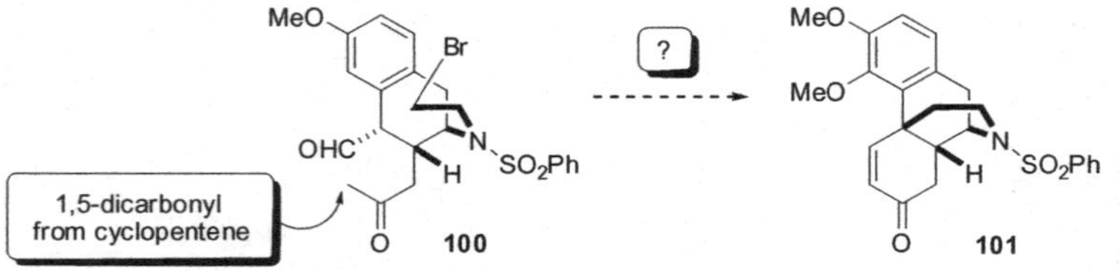

The critical bond construction is the same as in the Mulzer synthesis, but with the opposite polarity. This deposits residual functionality (after bond formation) at a different location.

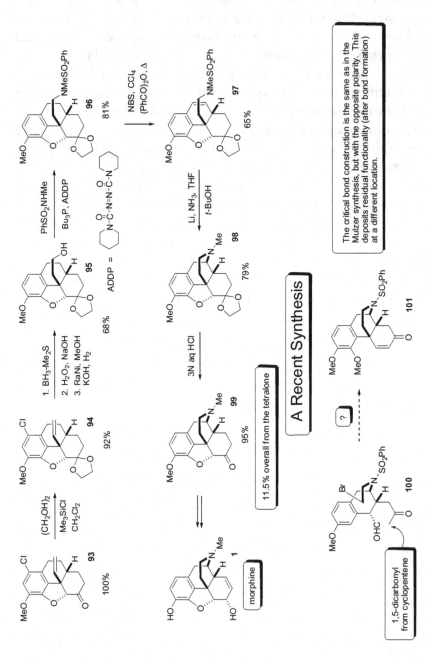

Taber, D. F.; Neubert, T. D.; Rheingold, A. L. "Synthesis of (-)-Morphine" *J. Am. Chem. Soc.* **2002**, *124*, 12416-12417. This article contains excellent references to other morphine syntheses.

Morphine-14

The ketone was protected and the resulting **94** was put through some standard paces to give first alcohol **95** and then sulfonamide **96**. You can see that this route is headed for an intersection with the Parker route. All that is missing is the olefin needed for the *N*-centered radical cyclization. This was introduced using a free radical bromination-dehydrohalogenation. The resulting olefinic sulfonamide (**97**) cyclized to **98** upon treatment with lithium in ammonia. Ketal hydrolysis provided dihyrocodeinone (**99**) which we have encountered before. This synthesis is quite efficient, handles regiochemical issues nicely, and provides either enantiomer of the natural product.

Taber (University of Delaware) described an approach that focuses on the same bond construction used by Mulzer to set the quaternary center, but uses the opposite sense of polarity in construction of this bond. The key transformation was projected to be **100** to **101** via a combination of aldol-dehydration and intramolecular alkylation reactions (not in any particular order). Compound **100** is highly functionalized. Perhaps in an attempt to reduce the level of functionality in precursors, it was to be generated in a manner we first saw in Chapter 2, by the oxidative cleavage of a cyclopentene.

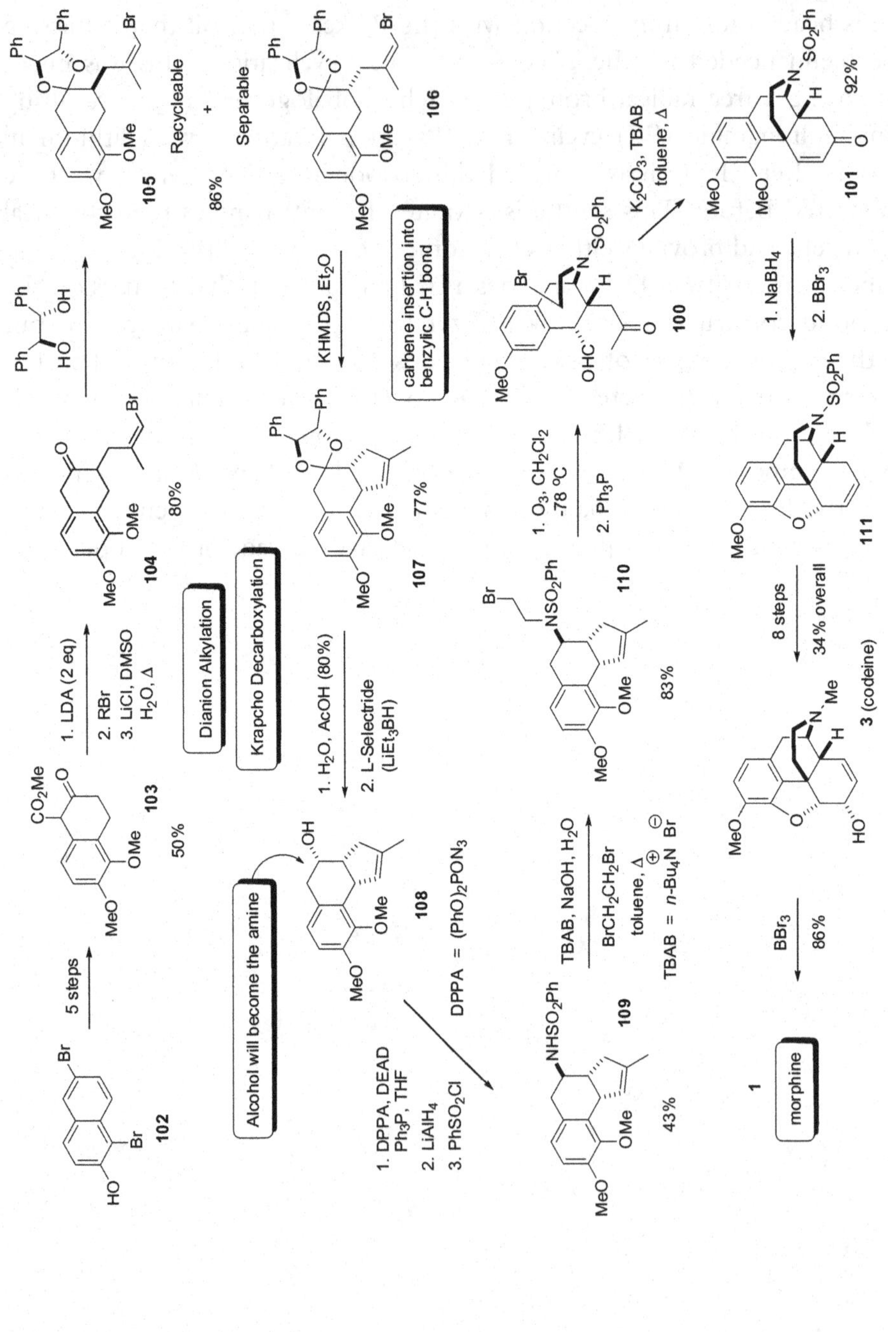
5 steps
1. LDA (2 eq)
2. RBr
3. LiCl, DMSO
H2O, Δ
Dianion Alkylation
Krapcho Decarboxylation
Alcohol will become the amine
1. H2O, AcOH (80%)
2. L-Selectride
(LiEt3BH)
KHMDS, Et2O
carbene insertion into
benzylic C-H bond
Recycleable
Separable
1. DPPA, DEAD
Ph3P, THF
2. LiAlH4
3. PhSO2Cl
DPPA = (PhO)2PON3
TBAB, NaOH, H2O
BrCH2CH2Br
toluene, Δ
TBAB = n-Bu4N Br
1. O3, CH2Cl2
-78 °C
2. Ph3P
K2CO3, TBAB
toluene, Δ
1. NaBH4
2. BBr3
8 steps
34% overall
BBr3
86%
morphine
1
3 (codeine)
102
103
104
105
106
107
108
109
110
100
101
111
50%
80%
86%
77%
43%
83%
92%

Morphine-15

Morphine-15

We will not go through every detail of the synthesis. Suffice it to say that the known 1,6-dibromo-2-naphthol (**102**) was first converted to β-ketoester **103**. The dianion derived from **103** was then alkylated with 1,3-dibromo-2-methylpropene. Krapcho decarbomethoxylation of the β-ketoester gave **104**.[9] Ketone **104** was then resolved via chromatographic separation of diastereomeric acetals **105** and **106**. The undesired acetal (**105**) could be recycled. The desired acetal could was converted to **107** via insertion of a carbene (α-elimination of HBr from **106**) into the benzylic C–H bond. This is an unusual and creative approach to this compound that illustrates the power of such insertion reactions.[10] Hydrolysis of the acetal followed by reduction of the intermediate ketone from the sterically most accessible face gave **108**. The alcohol was converted to the corresponding azide (with inversion of configuration), which was reduced to the amine and converted to sulfonamide **109**. Alkylation of **109** under phase transfer catalysis conditions provided **110**, the penultimate intermediate in the projected synthesis of **101**. Ozonolysis of the olefin and a reductive workup with triphenylphosphine provided **100**. Treatment of **100** with base provided **101** as planned. It was shown (by isolation of an intermediate product) that the sequence of events was intramolecular alkylation of the aldehyde enolate followed by the aldol-dehydration. Sodium borohydride reduction of **101**, followed by treatment of the allylic alcohol with boron tribromide, gave **111** with the morphinane ring system intact. A lengthy, but reasonably efficient sequence of reactions converted **111** to morphine (**1**) via codeine (**3**).

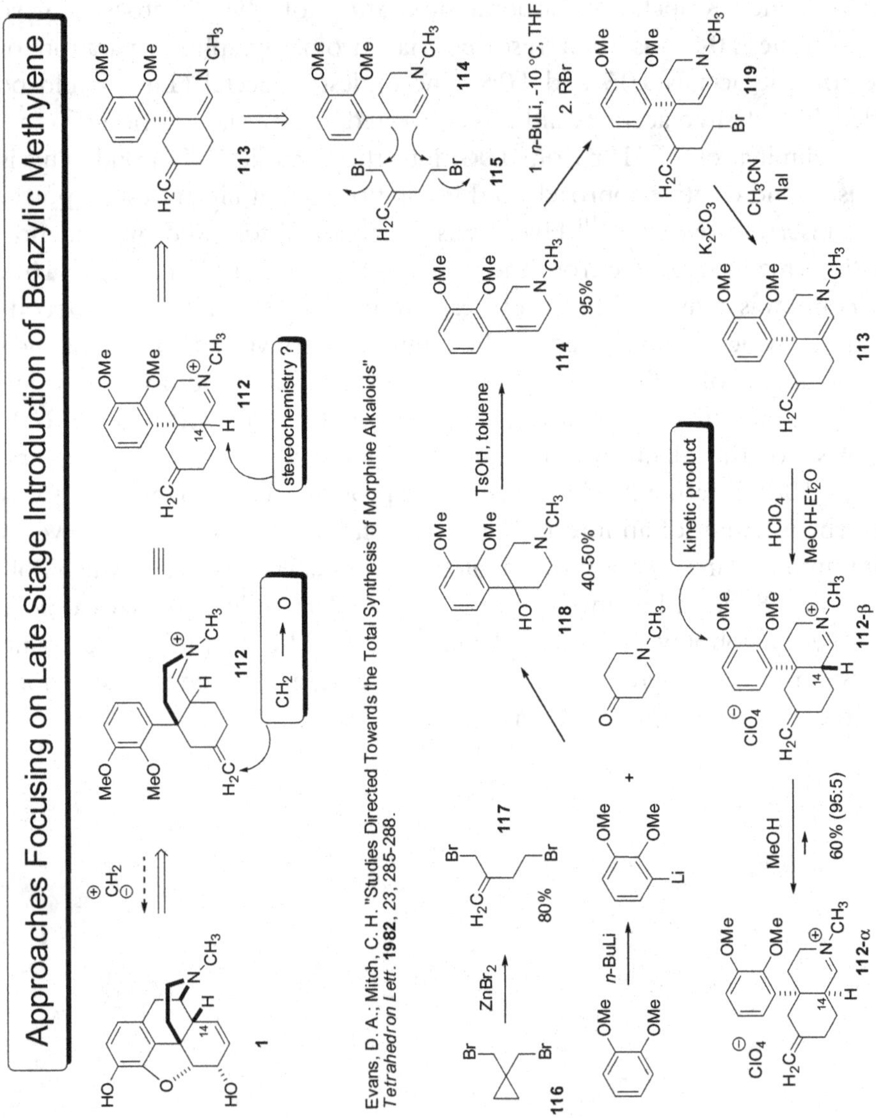
Approaches Focusing on Late Stage Introduction of Benzylic Methylene
HO
HO
N-CH3
H
14
1
CH2
112
MeO
MeO
H2C
N-CH3
H
CH2 → O
OMe
OMe
H2C
14
N-CH3
H
112
stereochemistry ?
113
114
115
Br
Br
Evans, D. A.; Mitch, C. H. "Studies Directed Towards the Total Synthesis of Morphine Alkaloids" Tetrahedron Lett. 1982, 23, 285-288.
116
ZnBr2
80%
117
n-BuLi
Li
+
N-CH3
HO
118
40-50%
TsOH, toluene
95%
1. n-BuLi, -10 °C, THF
2. RBr
119
K2CO3
CH3CN
NaI
kinetic product
HClO4
MeOH-Et2O
ClO4
112-β
MeOH
60% (95:5)
112-α

Morphine-16

Morphine-16

The final approach we will examine revolves around construction of an angularly arylated perhydroisoquinoline of type **112**, functionalized to allow introduction of the benzylic methylene group. This was to be accomplished using a zwitterionic CH_2 equivalent. It was imagined that such an equivalent would behave as a nucleophile, and add to the iminium ion, and then as an electrophile toward the aromatic ring. Iminium ion **112** was to be generated by protonation of enamine **113**, which was to be assembled by alkylation of the anion derived from **114** with *bis*-electrophile **115**. Replacement of the ketone in morphine with a methylene was expected to control regioselectivity in the *bis*-alkylation. Note that this also reduces the number of electrophilic sites in the alkylating agent. There are stereochemical issues associated with this approach. Attack of the nucleophile on iminium ion **112** would have to generate a *cis*-relationship between the entering nucleophile and the aryl group, otherwise the anticipated electrophilic aromatic substitution reaction would be impossible. In addition, protonation of **113** would set stereochemistry at what becomes C_{14} of morphine and thus, stereoselectivity here is also an issue.

The alkylating agent (**117** = **115**) was prepared from **116** via a modified Julia olefin synthesis. 4-Aryl-1-piperidinol **118** was prepared by *o*-metallation of catechol dimethyl ether followed by reaction of the organolithium reagent with *N*-methyl-4-piperidinone. Dehydration of **118** provided **114**. Metallation of **114** with *n*-BuLi, followed by reaction with **117**, gave **119**. Conversion of the primary bromide to an iodide was followed by intramolecular alkylation of the enamine to provide **113**. Kinetic protonation of **113** gave iminium perchlorate **112-*β*** (*trans*-fused ring system). Upon dissolving in methanol this equilibrated with *cis*-fused iminium perchlorate **112-*α***. This set the stage for the key reaction.

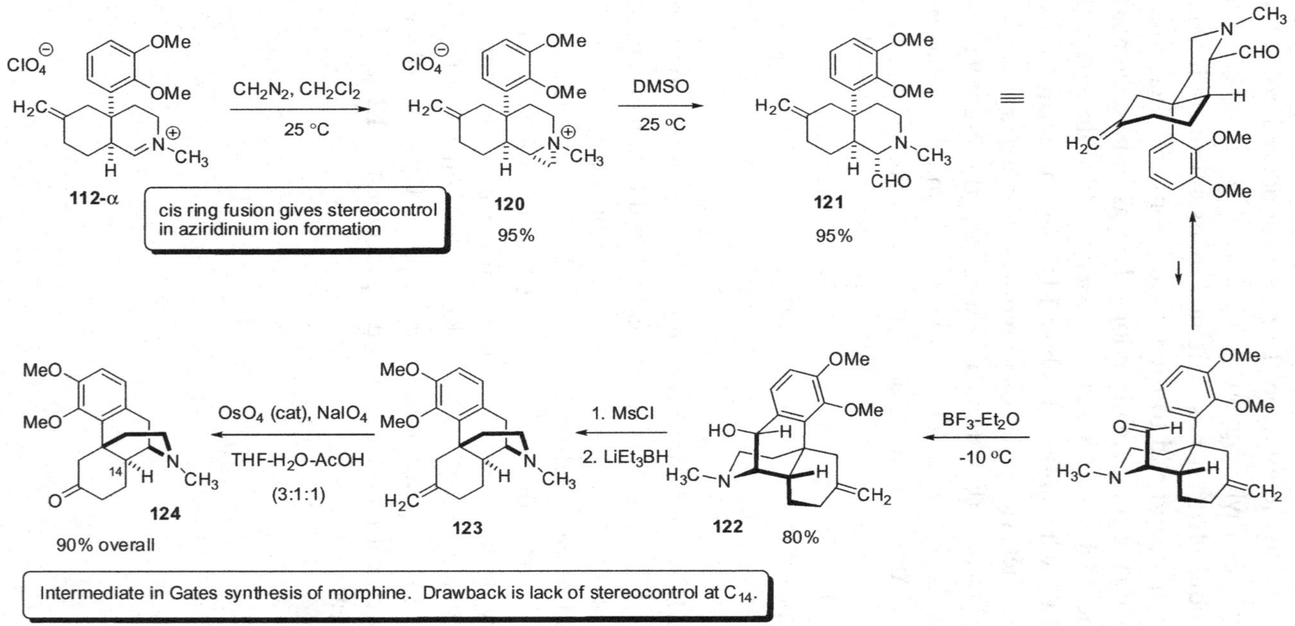

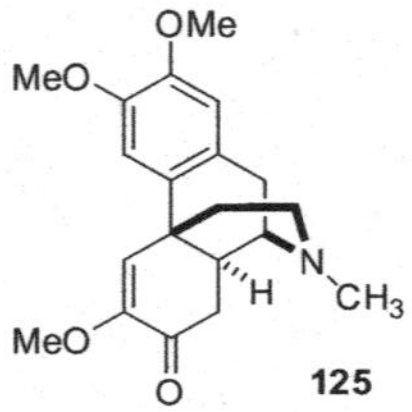

125

A strategically related synthesis of O-methylpallidinine (left) follows: McMurry, J. E.; Farina, V. "Total Synthesis of *O*-Methylpallidinine" *Tetrahedron Lett.* **1983**, *24*, 4653-4656.

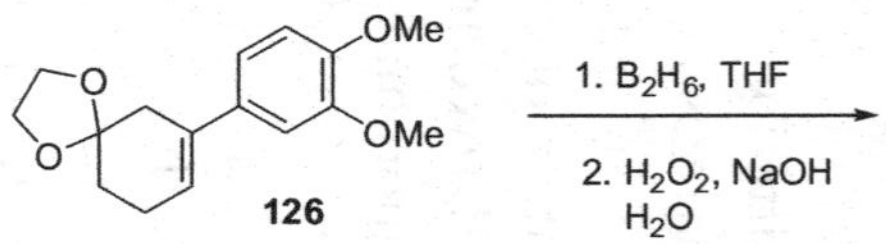

126

Synthesis?

Morphine-17

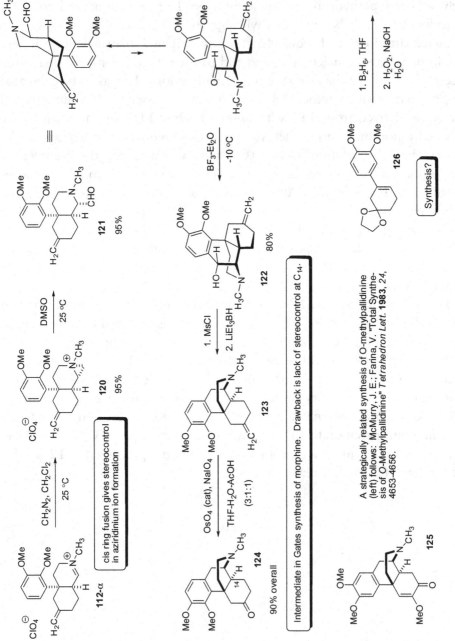

Morphine-17

Treatment of **112-α** with diazomethane gave aziridinium ion **120**. The cis ring fusion apparently directs attack of the diazomethane to its convex face. In addition, rather than the aromatic ring behaving as the "nucleophile" toward the methylene cation equivalent, the perhydroisoquiniline nitrogen intervened. This problem was corrected by oxidizing the aziridinium ion to aldehyde **121** using a Kornblum-type oxidation.[11] This substrate was suitable for the electrophilic aromatic substitution as reaction with boron trifluoride etherate provided **122**, via an intramolecular EAS reaction from the less stable chair-chair conformation available to **121**. The benzylic hydroxyl group was removed via reduction of the derived mesylate. A Johnson-Lemieux oxidation converted **123** to **124**. Ketone **124** was an intermediate in the Gates synthesis and thus, this completed a formal synthesis of morphine.

One drawback of this synthesis is that it initially provides the wrong stereochemistry at C_{14}. There are morphinanes, however, that have $C_{14\alpha}$ stereochemistry, so this is not all bad. One such morphinane is *O*-methylpallidinine (**125**), and a strategically similar approach to this alkaloid will be the last synthesis we will consider in this chapter.

John McMurry (Cornell University) described a synthesis of **125** that begins with preparation of aryl cyclohexene **126** (see Problem 21). Hydroboration-oxidation of this alkene gave ketone **127**.

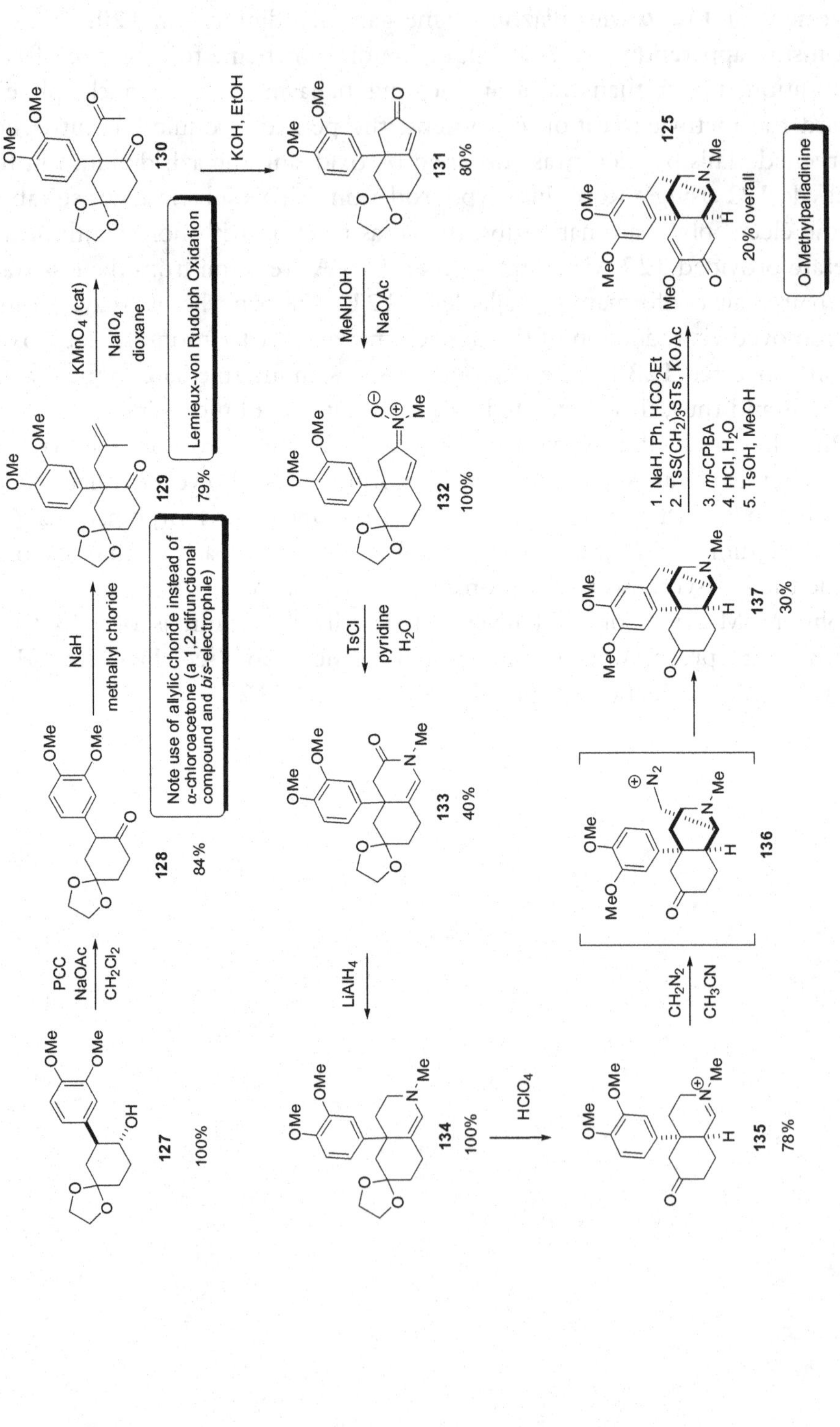
PCC
NaOAc
CH_2Cl_2
NaH
methallyl chloride
Note use of allylic choride instead of α-chloroacetone (a 1,2-difunctional compound and bis-electrophile)
$KMnO_4$ (cat)
$NaIO_4$
dioxane
Lemieux-von Rudolph Oxidation
KOH, EtOH
MeNHOH
NaOAc
TsCl
pyridine
H_2O
$LiAlH_4$
$HClO_4$
CH_2N_2
CH_3CN
1. NaH, Ph, HCO_2Et
2. $TsS(CH_2)_3STs$, KOAc
3. m-CPBA
4. HCl, H_2O
5. TsOH, MeOH
127 100%
128 84%
129 79%
130
131 80%
132 100%
133 40%
134 100%
135 78%
136
137 30%
125
20% overall
O-Methylpalladinine

Morphine-18

Morphine-18

Oxidation of **127** with PCC, buffered to discourage acetal hydrolysis, gave **128**. Alkylation of **128** with methallyl chloride, followed by oxidative cleavage of the olefin, gave **130**. The transformation of **128** to **130** involves construction of a 1,4-difunctional relationship. The use of an allylic halide in the alkylation, rather than an α-haloacetone (a 1,2-difunctional compound), avoids the use of a *bis*-electrophile that might suffer from regiochemical problems. Treatment of **130** with base, promoted an aldol-dehydration to provide **131**. The ketone was then converted to nitrone **132**, and a Beckman-type rearrangement provided **133**. Lithium aluminum hydride reduction of the lactam gave enamine **134**. This intermediate is similar to enamine **113** (Morphine-16), but there is a difference in the placement of the methoxy groups on the aromatic ring. It turned out that reaction of **134** with perchloric acid, followed by diazomethane, directly gave "methylene insertion product" **137**, via iminium ion **135** and presumably alkyldiazonium salt **136** as intermediates. The rate of capture of the intermediate **136** by the aromatic ring was apparently faster in **136** than in the comparable ion derived from **112**. The synthesis of O-methylpallidinine was completed by a series of oxidation state adjustments in the cyclohexanone ring.

This concludes our look at morphine and indeed, at alkaloids. In closing I would like to make a suggestion, first made to me by one of my former teachers, Henry Rapoport. Rap was trying to convince me (and other students in a class) that heterocyclic chemistry is actually much easier that hydrocarbon chemistry (by this I think he meant terpenoid chemistry). To paraphrase what he said, "Making carbon-heteroatom bonds is easy relative to making carbon-carbon bonds. There are electronegativity differences between the atoms, but carbon is carbon and you have to mess with it to be able to make a carbon-carbon bond. I am surprised more of you don't want to do heterocyclic chemistry". Of course he was trying to sell us on his area of research interest, but that simple statement helped me be less afraid of nitrogen (for example) as a young student of synthesis. It made a difference to me and I hope what we have looked at here may do the same for you.

References

1. Rice, K. C. "Synthetic Opium Alkaloids and Derivatives. A Short Total Synthesis of (±)-Dihydrothebainone, (±)-Dihydrododeinone, and (±)-Nordihydrocodeinone as an Approach to a Practical Synthesis of Morphine, Codeine and Congeners" *J. Org. Chem.* **1980**, *45*, 3135–3137. Morrison, G. C.; Waite, R. O.; Shavel, J. Jr. "Alternate Route in the Synthesis of Morphine" *Tetrahedron Lett.* **1967**, *9*, 4055–4056.
2. Mitsunobu, O. "The Use of Diethyl Azodicarboxylated and Triphenylphosphine in Synthesis and Transformation of Natural Products" *Synthesis* **1981**, 1–28.
3. Mancuso, A. J.; Swern, D. "Activated Dimethyl Sulfoxide: Useful Reagents for Synthesis" *Synthesis* **1981**, 165–185. Mancuso, A. J.; Brownfain, D. S.; Swern, D. "Structure of the Dimethyl Sulfoxide-Oxalyl Chloride Reaction Product. Oxidation of Heteroaromatic and Diverse Alcohols to Carbonyl Compounds" *J. Org. Chem.* **1979**, *44*, 4148–4150. Mancuso, A.; J.; Huang, S-L.; Swern, E. "Oxidation of Long-Chain and Related Alcohols to Carbonyls by Dimethyl Sulfoxide "Activated" by Oxalyl Chloride" *J. Org. Chem.* **1978**, *43*, 2480–2482. Omura, K.; Swern, D. "Oxidation of Alcohols by "Activated" Dimethyl Sulfoxide. A Preparative Steric and Mechanistic Study" *Tetrahedron* **1978**, *34*, 1651–1660.
4. Luche, J. L. "Lanthanides in Organic Chemistry. 1. Selective 1,2-Reductions of Conjugated Ketones" *J. Am. Chem. Soc.* **1978**, *100*, 2226–2227.
5. Morgans, D. J. Jr.; Sharpless, K. B.; Traynor, S. G. "Epoxy Alcohol Rearrangements: Hydroxyl-Mediated Delivery of Lewis Acid Promoters" *J. Am. Chem. Soc.* **1981**, *103*, 462–464.
6. For reviews and books see: Julia, M. "Cyclizations by Radical Reactions" *Rec. Chem. Progr.* **1964**, *25*, 3–29. Julia, M. "Free Radical Cyclizations" *Pure Appl. Chem.* **1967**, *15*, 167–183. Julia, M. "Free-Radical Cyclizations" *Acc. Chem. Res.* **1971**, *4*, 386–392. Beckwith, A. L. J. "Some Guidelines for Radical Reactions" *J. Chem. Soc., Chem. Commun.* **1980**, 482–483. Beckwith, A. L. J. "Regioselectivity and Stereoselectivity in Radical Reactions" *Tetrahedron* **1981**, *37*, 3073–3100. Hart, D. J.; "Free-Radical Carbon-Carbon Bond Formation in Organic Synthesis" *Science* **1984**, *223*, 883–887. Beckwith, A. L. J. "Mechanism and Applications of Free Radical Cyclization" *Rev. Chem. Int.* **1986**, *7*, 143–154. Giese, B. "Radicals in Organic Synthesis: Formation of Carbon-Carbon Bonds" Pergamon Press, **1986** (294 pages). Stork, G. "Radical Cyclization in the Control of Regio- and Stereochemistry" *Bull. Chem. Soc. Jap.* **1988**, *61*, 149–151. Giese, B.; Kopping, B.; Gobel, T.; Dickhout, J.; Thoma, G.; Kulicke, K. J.; Trach, F. "Radical Cyclization Reactions" *Organic Reactions* **1996**, *48*, 301–856.

7. Link, J. T. "The Intramolecular Heck Reaction" *Organic Reactions* **2002**, *60*, 157–534. Shibasaki, M.; Vogl, E. M. "Heck Reaction" Comprehensive Asymmetric Catalysis I-III **1999**, *1*, 457–487.
8. Corey, E. J.; Fleet, G. W. J. "Chromium Trioxide-3,5-Dimethylpyrazole complex as a Reagent for Oxidation of Alcohols to Carbonyl Compounds" *Tetrahedron Lett.* **1973**, *15*, 4499–4501.
9. Krapcho, A. P. "Recent Synthetic Applications of the Dealkoxycarbonylation Reaction. Part 1. Dealkoxycarbonylations of Malonate Esters. *ARKIVOC* (Gainesville, FL, USA) **2007**, 1–53. Krapcho, A. P. "Recent Synthetic Applications of the Dealkoxycarbonylation Reaction. Part 2. Dealkoxycarbonylations of β-Keto Esters, α-Cyano Esters and Related Analogues" *ARKIVOC* (Gainesville, FL, USA) **2007**, 54–120. Krapcho, A. P.; Jahngen, E. G. E. Jr.; Lovey, A. J.; Short, F. W. "Decarbalkoxylations of Geminal Diesters and β-Keto Esters in Wet Dimethyl Sulfoxide. Effect of Added Sodium Chloride on the Decarbalkoxylation Rates of Mono- and Disubstituted Malonate Esters" *Tetrahedron Lett.* **1974**, *16*, 1091–1094. Krapcho, A. P.; Lovey, A. J. "Decarbalkoxylations of Geminal Diesters, β-Keto Esters, and α-Cyano Esters Effected by Sodium Chloride in Dimethyl Sulfoxide" *Tetrahedron Lett.* **1973**, *15*, 957–960.
10. Chen, M. S.; White, M. C. "A Predictably Selective Aliphatic C-H Oxidation Reaction for Complex Molecule Synthesis" *Science* **2007**, *318*, 783–787. Davies, H. M. L. "Recent Advances in Catalytic Enantioselective Intermolecular C-H Functionalization" *Angew. Chem. Int. Ed.* **2006**, *45*, 6422–6425. Tambar, U. K.; Ebner, D. C.; Stoltz, B. M. "A Convergent and Enantioselective Synthesis of (+)-Amurensinine via Selective C-H and C-C Bond Insertion Reactions" *J. Am. Chem. Soc.* **2006**, *128*, 11752–11753. Feldman, K. S. "Alkynyliodonium Salts in Organic Synthesis" in Strategies and Tactics in Organic Synthesis, Harmata, M. Ed.; **2004**, *4*, 133–170. Doyle, M. P. "Synthetic Carbene and Nitrene Chemistry" *Reactive Intermediate Chemistry* **2004**, 561–592. Hinman, A.; Du Bois, J. "A Stereoselective of (−)-Tetrodotoxin" *J. Am. Chem. Soc.* **2003**, *125*, 11510–11511. Salzmann, T. N.; Ratcliffe, R. W.; Christensen, B. G.; Bouffard, F. A. "A Stereocontrolled Synthesis of (+)-Thienamycin" *J. Am. Chem. Soc.* **1980**, *102*, 6161–6163. Cory, R. M.; McLaren, F. R. "Bicycloannulation. Carbon Atom Insertion: An Efficient Synthesis of Ishwarane" *J. Chem. Soc., Chem Commun.* **1977**, 587–488.
11. Kornblum, N.; Powers, J. W.; Anderson, G. J.; Jones, W. J.; Larson, H. O.; Levand, O.; Weaver, W. M. "New and Selective Method of Oxidation" *J. Am. Chem. Soc.* **1957**, *79*, 6562.

Problems

1. Discuss the regioselectivity in the nitrosation of **10**. (Morphine-2)
2. Describe the sequence of events leading from **16** → **17**. (Morphine-2)
3. Suggest tactics for the conversion of morphine (**1**) to *β*-thebainone (**20**). (Morphine-3)
4. Provide mechanisms for the conversion of **21** → **22** and the oxidation of **22** → **23**. (Morphine-3)
5. Suggest a structure for the tribromide derived from **25**, and a mechanism for its transformation to **26**. (Morphine-4)
6. The 2,4-DNP chemistry in the Gates synthesis is quite interesting. Is the following transformation mechanistically related? (Morphine-4)

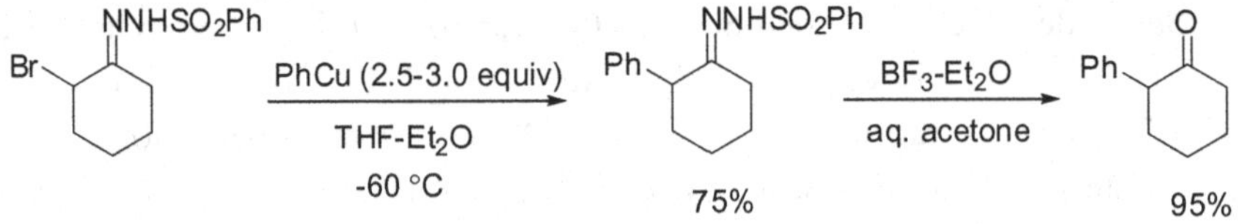

Sacks. C. E.; Fuchs, P. L. "*α*-Arylation of Carbonyl Groups. Utilization of the *p*-Toluenesulfonylazo Olefin Functional Group as an Enolonium Synthon" *J. Am. Chem. Soc.* **1975**, *97*, 7273–7274. See also Stork, G.; Ponaras, A. A. "*α*-Alkylation and Arylation of *α*,*β*-Unsaturated Ketones" *J. Org. Chem.* **1976**, *41*, 2937–2939.

7. Show the structural relationship between the following natural products and oxidative phenolic coupling precursors. (Morphine-5)

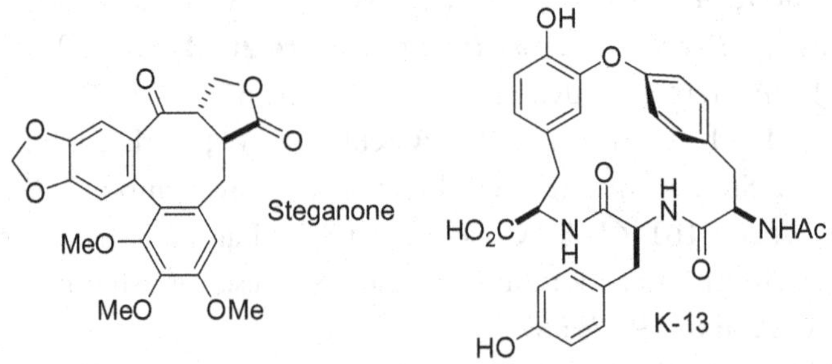

8. What is the basis for the comment that the carbamate (in **40**) forces the "benzylic group" to be pseudo-axial on the tetrahydroisoquinoline core? (Morphine-7)
9. Provide tactics for the conversion of **48** → **49**. (Morphine-8)

10. Outline several syntheses of **50**. Use acyclic starting materials. Use catechol dimethyl ether as a starting material. (Morphine-8)
11. Provide the structures of intermediates *en route* from **59** → **60**. Suggest reagents for accomplishing transformations where they are missing from the text. (Morphine-9)
12. Provide a mechanistic interpretation of the conversion of **74** → **75**. Note that no reaction is expected in the absence of the free hydroxyl group. (Morphine-10)
13. Outline a reaction sequence that will accomplish the following transformation. (Morphine-10)

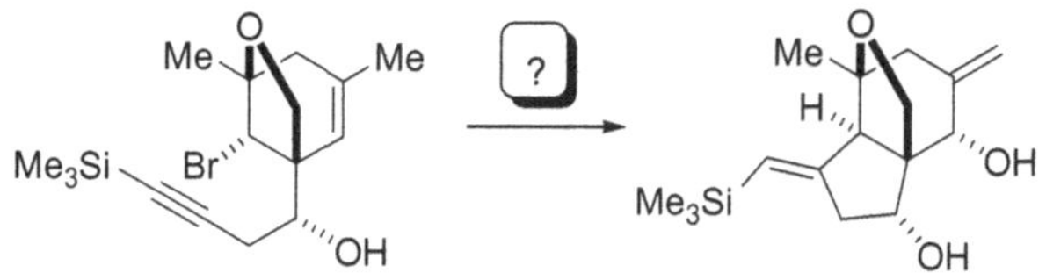

14. Speculate regarding the success of the second free radical cyclization in the conversion of **78** → **79** (why 6-*endo* over 5-*exo*). (Morphine-11)
15. Suggest a mechanism for the conversion of **92** → **93**. (Morphine-13)
16. Outline a reaction sequence that would convert **102** → **103**. (Morphine-15)
17. Mesembrine is an angularly arylated perhydroindole alkaloid that is structurally similar to morphine. Design several syntheses of mesembrine that use some of the strategies described in Chapters 8–10. Compare your approaches to those reported in the literature. (Morphine-16)

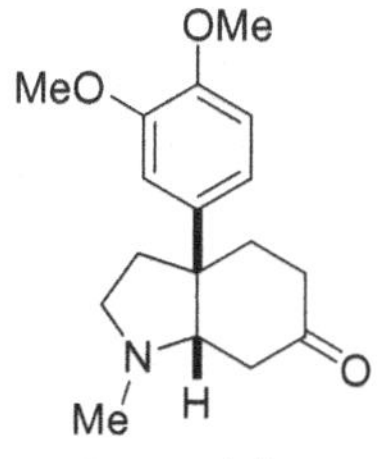

mesembrine

18. Draw structures of **112-*β*** and **112-*α*** that show their respective conformations. (Morphine-16)
19. Provide mechanisms for the conversion of **112-*α*** → **120** and of **120** → **121**. (Morphine-17)

20. Outline a synthesis of **126** from commercially available materials. (Morphine-17)
21. Provide a mechanism for the conversion of **132** → **133**. (Morphine-18)
22. Provide the structure of intermediates after each step in the conversion of **137** → **125**. (Morphine-18)

CHAPTER 11

Olefin Synthesis and Cecropia Juvenile Hormone

Cecropia Juvenile Hormone

Trost, B. M. "The Juvenile Hormone of *Hyalophora Cecropia*" *Acc. Chem. Res.* **1970**, *3*, 120.

One field to which organic synthesis has contributed is the identification and production of insect chemical signaling agents (growth hormones, sex hormones). These are of interest (in part) as a means of contolling insect populations. One such hormone is the *Cecropia* juvenile hormone. This compound is involved in the molting process of the moth. The developmental sequence of many insects involves three stages: larval, pupal and adult. In some insects a "molt" occurs that results in the insect proceeding directly from the larval to the adult stage. These molts are controlled by hormones. In *Hyalophora cecropia*, the "juvenile hormone" results in retention of juvenile characteristics as each molt occurs. The more "juvenile hormone" is present, the more juvenile characteristics are retained after the molt. The Cecropia juvenile hormone was isolated in 1965 by the Roeller group in the Department of Zoology at the University of Wisconsin from the abdomen of the male adult (0.5 μg/abdomen).Only 200 μg of material was available at any point in time. Based on NMR and mass spectral analyses, a structure (without stereochemical detail) was proposed. Due to the stereochemical ambiguities and lack of material, synthesis was used to ultimately determine its structure as shown below. The first synthesis was stereorandom (no doubt on purpose) and provided a number of isomers. Bioassays were used to determine which isomer was the true hormone.

Once the structure was known, many syntheses were reported and all of a sudden the literature was loaded with stereoselective methods for olefin synthesis. Although it is only a guess, it is probable that the discovery of Cecropia juvenile hormone (CJH) was responsible (in part) for this burst of activity. Another reason would be the interest in carbocycle synthesis via polyolefin cyclizations (the development of steroid and terpene chemistry) which slightly preceeded the discovery of CJH. We will start by looking at the Trost group synthesis (the first synthesis) of CJH and then examine several other syntheses wherein the focus is stereoselective olefin synthesis.

Dahm, K. H.; Trost, B. M.; Roller, H. "The Juvenile Hormone. V. Synthesis of the Racemic Juvenile Hormone" *J. Am. Chem. Soc.* **1967**, *89*, 5292-5294.

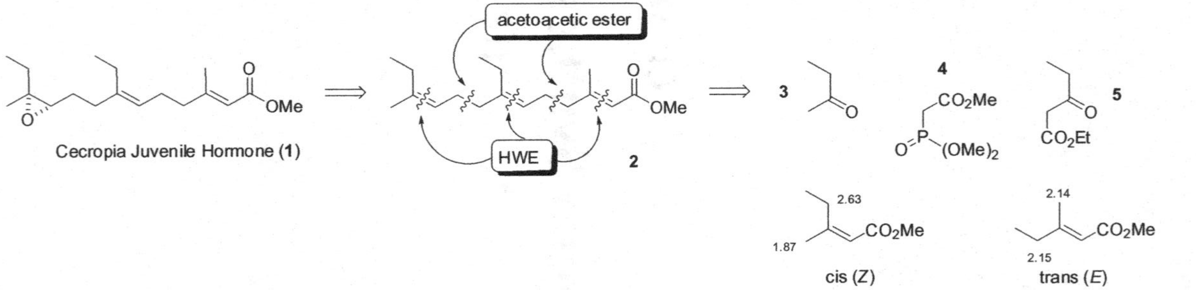

In β–substituted α, β–unsaturated esters, the β–group cis to the CO_2R is more deshielded than the β–group trans to the CO_2R.

Cecropia Juvenile Hormone-1

Cecropia Juvenile Hormone-1

This chapter focuses on selected aspects of stereoselective olefin synthesis. This is an important topic for a number of reasons: (1) When olefins are targets it is simply necessary to have a litany of methods available for their synthesis. Olefins are ubiquitous in nature as we have already seen (for examples see the prostaglandin and poison-dart alkaloid sidechains). (2) Olefin stereochemistry can be used to control non-olefinic stereochemical relationships that must be established during the course of a target-oriented synthesis. For examples, see the syntheses of pyrrolizidine alkaloids where olefin addition reactions were used to control vicinal stereochemistry (also see Prostaglandins-23), or the Johnson approach to steroids where olefin stereochemistry is translated into vicinal stereochemistry (see Steroids-14 and 17), or the importance of olefin geometry in transfer-of-chirality reactions (see Prostaglandins-24 and 25). We will explore the topic of stereoselective olefin synthesis within the context of Cecropia Juvenile Hormone (**1**) in this chapter and explore additional methods in Chapter 12. So what is Cecropia juvenile hormone and why was it of interest as a target for synthesis? For a discussion see page 446 (CJH-1).[1]

The synthetic plan was to prepare **1** by regioselective epoxidation of triene **2**. This is a reasonable idea because one of the olefins is electron-deficient, and of the other two, the "internal" olefin might be expected to be sterically more conjested if **2** coils into a ball-like structure. The plan was to build "from left-to-right" using an iterative (repetitive) reaction sequence. Thus, it was imagined that 2-butanone (**3**) would undergo a Horner-Wadsworth-Emmons (HWE) reaction with β-ketophosphonate **4** to provide a mixture of α,β-unsaturated ester geometrical isomers. This reaction would establish what was to become the "left-hand" olefin of **2**. Little stereoselectivity was expected in this reaction because the difference in size between methyl and ethyl groups is minimal. It was planned to convert the ester to a halide (bromide or iodide) and then perform an acetoacetic ester synthesis using β-ketoester **5** as the nucleophile.[2] This would provide a ketone that could be subjected to another HWE reaction-separation-conversion to halide-alkylation sequence to introduce the "middle" olefin of **2**. A partial repeat of this sequence (stopping after the HWE reaction) was to complete the synthesis of **2**. Let's move to the synthesis.

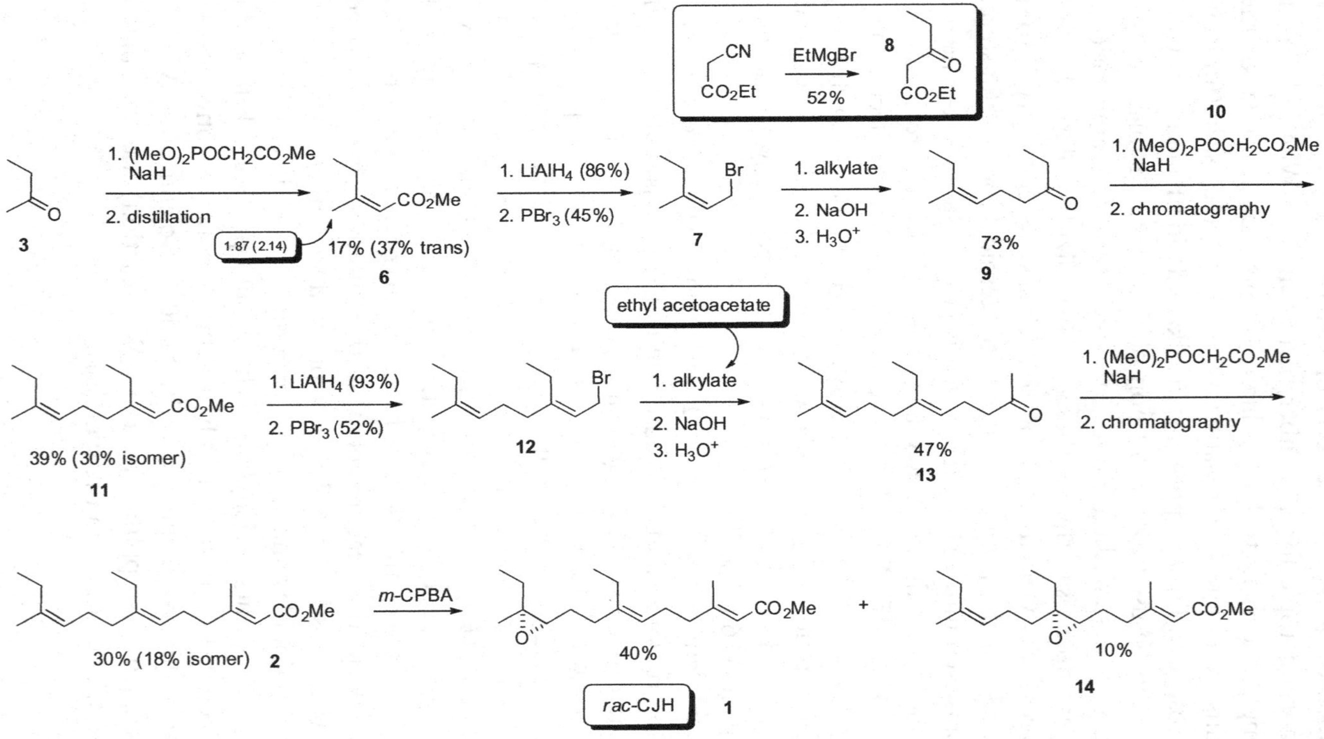

Although this synthesis was stereorandom, it was necessary because the goal was structure determination. This is the kind of synthesis that is sometimes pursued in the early stages of a medicinal chemistry effort where the goal is activity and structure is secondary. In otherwords, the structure of the specific target follows from biological activity.

Cecropia Juvenile Hormone-2

Cecropia Juvenile Hormone-2

The initial HWE reaction did provide a mixture of stereoisomeric olefins, from which **6** was isolated by a spinning band distillation. Reduction to the allylic alcohol was followed by conversion to bromide **7**. *β*-Ketoester **8** was prepared from ethyl cyanoacetate and used to convert **7** to ketone **9** using a classical "acetoacetic ester synthesis".[2] Note that the ester, which is eventually sacrificed, controls the regiochemical course of the alkylation reaction. A repeat of the HWE reaction gave **11**, which was converted to ketone **13** via bromide **12** in the standard manner. A final HWE reaction provided **2**. Epoxidation of **2** proceeded with modest regioselectivity to give **1** along with a small amount of the regioisomeric epoxide **14**.

Although this synthesis was stereorandom, it was necessary because the goal of the synthesis was structure determination. This is the kind of synthesis that is sometimes pursued in the early stages of a medicinal chemistry effort where the goal is activity and structure is secondary. In other words, the structure of the specific target follows from biological activity. We will now move to some stereoselective syntheses.

Claisen Rearrangements: The Faulkner Approach

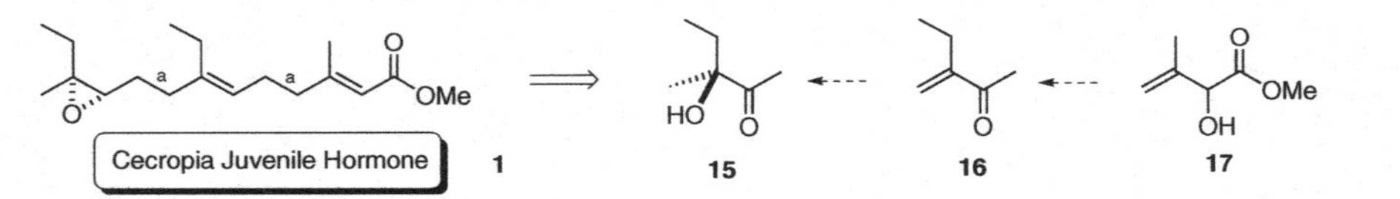

1. S_N2'
2. Reduce
3. S_N2'
4. Reduce
5. diol ⟶ epoxide

Claisen Rearrangement as an "enforced" S_N2' reaction.

Faulkner, D. J.; Petersen, M. R. "Application of the Claisen Rearrangement to the Synthesis of Trans Trisubstituted Olefinic Bonds. Synthesis of Squalene and Insect Juvenile Hormone" *J. Am. Chem. Soc.* **1973**, *95*, 553.

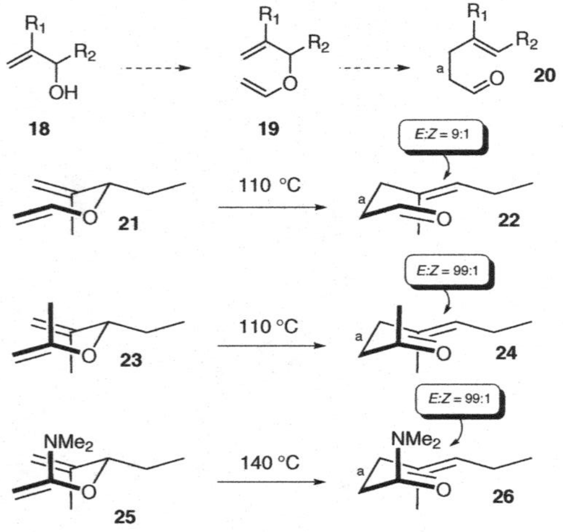

Results are consistent with Perrin-Faulkner analysis: Perrin, C. L.; Faulkner, D. J.; *Tetrahedron Lett.* **1969**, 2873.

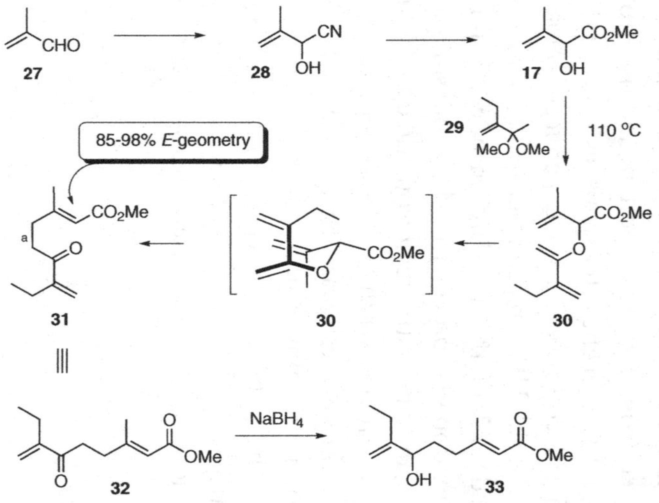

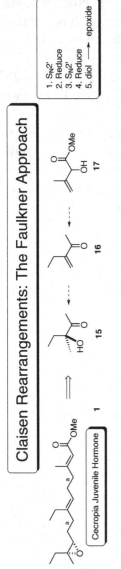

Cecropia Juvenile Hormone-3

Cecropia Juvenile Hormone-3

Johnson and Faulkner reported a synthesis of **1** that revolved around stereoselective synthesis of trisubstituted olefins via the Claisen rearrangement. The plan was to build CJH (**1**) from **15–17** (or derivatives thereof) starting with allylic alcohol **17**. How might one disconnect **1** into these pieces? Imagine the enol or enolate of **16** reacting with an appropriate derivative of **17** via an S_N2' reaction. This would establish the right-hand "a"-bond shown in structure **1** (the olefin conjugated to the carbomethoxy group). As we will see, the Claisen rearrangement is a reaction that accomplishes this transformation with good control of olefin geometry.[3] Reduction of the resulting ketone to the corresponding allylic alcohol, and repetition of this process using the enolate of **15**, would provide an α-hydroxyketone that could be carried forward to CJH via reduction of the ketone, and conversion of the resulting vicinal diol to the epoxide.

Let's first examine some stereochemical aspects of the Claisen rearrangement. This transformation involves conversion of an allylic alcohol (**18**) to an allyl vinyl ether (**19**), followed by thermal rearrangement to a γ,δ-unsaturated carbonyl compound (**20**). Notice that this reaction looks like a reaction between an enol of a carbonyl compound, acetaldehyde for the specific reaction shown, with an allylic alcohol derivative with S_N2' regiochemistry. There is ample evidence that this rearrangement takes place via a transition state in which the allyl vinyl ether is chair-like. The terminal olefin substituents occupy axial or equatorial sites, the internal olefin substituents occupy axial sites, and substituents at the original carbinol center occupy equatorial or axial sites that minimize 1,3-diaxial interactions in the rearrangement transition state. Thus, if one of the carbinol subsitutents is hydrogen, it occupies an axial site. This transition state analysis predicts (in accord with fact) that **19** will provide the trisubstituted olefin geometry indicated in structure **20**. Specific examples are the conversion of **21**, **23** and **25** → **22**, **24** and **26**, respectively. Note that the rearrangement of **21** is less stereoselective than the rearrangements of **23** or **25**. This is because one of the internal olefin substitutents in **21** is hydrogen and neither of the internal olefin substituents in **23** and **25** are hydrogens. You should work the problem at the end of this chapter to see if you understand, in depth, the reason for this difference.

Application of this olefin synthesis to the CJH problem began with conversion of methacrolein (**27**) to allylic alcohol **17** via cyanohydrin **28**. Reaction of **17** with ketal **29** provided **31**/**32** with good olefin stereoselectivity, presumably via allyl vinyl ether intermediate **30**. Reduction of **32** with sodium borohydride gave allylic alcohol **33** and set the stage for the second "forced" S_N2' reaction.

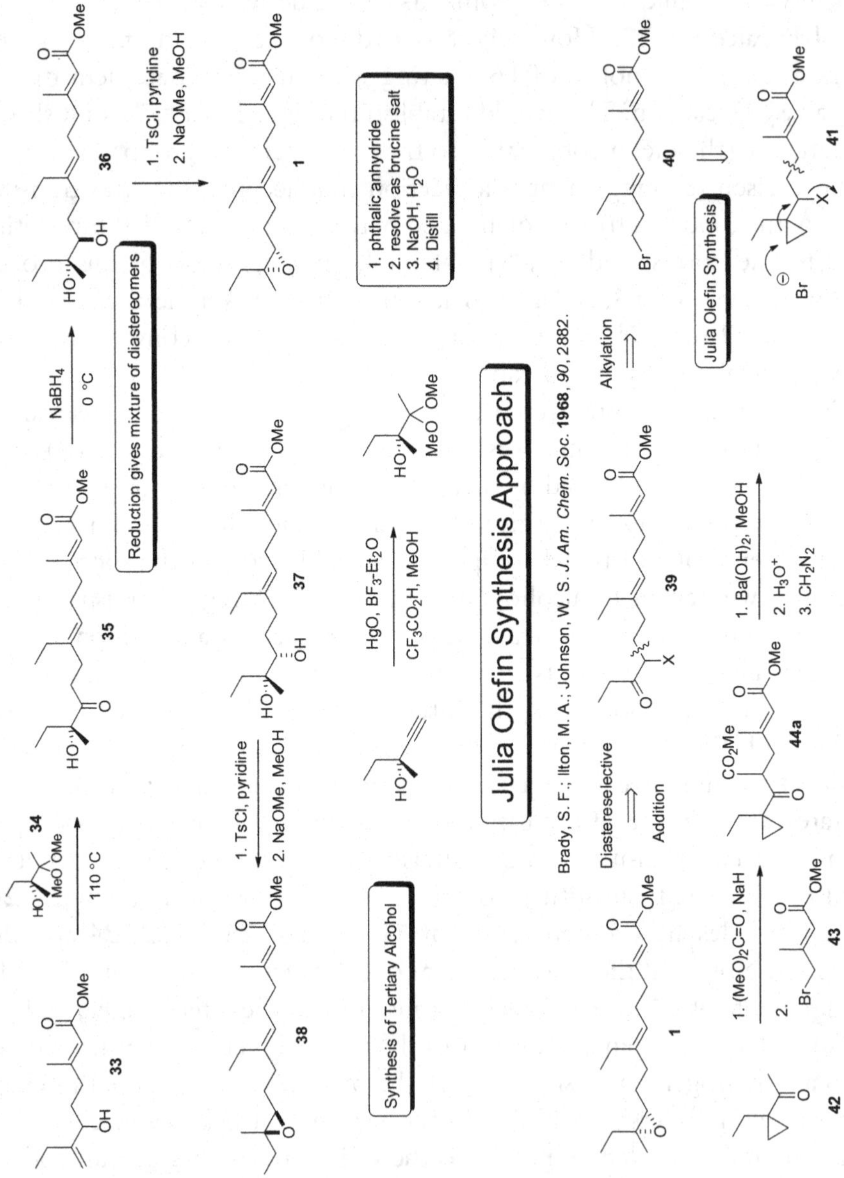

Cecropia Juvenile Hormone-4

Cecropia Juvenile Hormone-4

Reaction of **33** with **34** gave **35** with excellent olefin stereoselectivity. When racemic **33** was used, racemic **35** was obtained. When a single enantiomer of **34** was used (synthesis outlined without comment), a single enantiomer of **35** was obtained. Sodium borohydride reduction of **35** gave a mixture of diols **36** and **37**. Conversion of the secondary alcohol of **36** to a tosylate, followed by treatment with base, gave CJH (**1**) via a Williamson ether synthesis. Diastereomeric diol **37** was converted to CJH diastereomer **38** in a similar manner. This synthesis not only demonstrated the importance of the Claisen rearrangment as methodology for stereoselective olefin synthesis, but established the absolute configuration of Cecropia Juvenile Hormone.

The Johnson group also examined a synthesis that relied on the Julia olefin synthesis for construction of the central trisubstituted olefin.[4] In this plan, **1** was to be prepared from an α-haloketone of type **39** via diastereoselective addition of a methyl group to the ketone, followed by a Williamson ether synthesis. Ketone **39** was to be prepared from **40** using an acetoacetic ester synthesis. Compound **40** was to be prepared from **41** using the Julia olefin synthesis. A make-or-break aspect of this plan was the stereochemical course of the Julia synthesis. Of course it was anticipated that the proper stereochemistry would result as will be seen shortly.

Julia substrate **41** was prepared as follows. Cyclopropylketone **42** was converted to the corresponding β-ketoester using sodium hydride and dimethyl carbonate. The anion derived from the β-ketoester was alkylated with allylic bromide **43** to provide **44a**. Ester hydrolysis, decarboxylation of the intermediate β-ketoacid, and esterification of the terminal carboxylic acid using diazomethane, gave ketone **44b** (CJH-5).

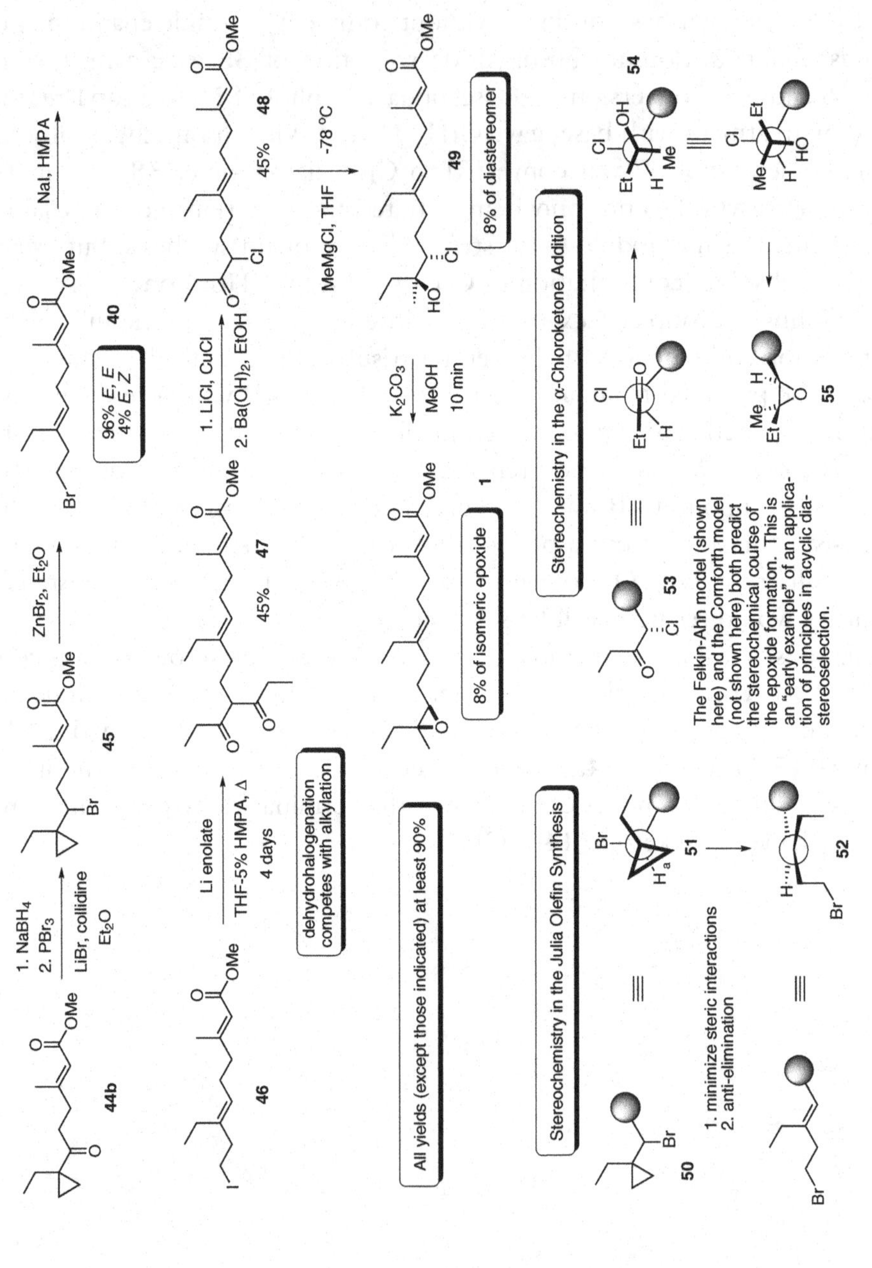

Cecropia Juvenile Hormone-5

Cecropia Juvenile Hormone-5

Reduction of **44b** gave the corresponding alcohol, which was converted to bromide **45**. Treatment of **45** with $ZnBr_2$ provided the desired olefin (**40**) with excellent control of stereochemistry. The observed stereoselectivity was rationalized as follows. It was reasoned that **45** (see structure **50** for a truncated version of this substrate) would react from conformation **51** in which (1) steric interactions between the carbon chain and cyclopropane are minimized (notice that the C–H bond, rather than the C–C bond, bisects the cyclopropane "a" bond) and (2) the C–Br and cyclopropane C–C bonds are disposed such that the reaction occurs via an *anti*-elimination (this is an important point that we will revisit shortly).

Continuing with the synthesis, **40** was converted to iodide **46** using a Finkelstein reaction. Alkylation of heptan-3,5-dione with **46** provided **47**. Chlorination of **47** followed by deacylation of the resulting non-enolizable 1,3-diketone gave **48**. Treatment of **48** with methylmagnesium chloride gave **49** contaminated with a small amount of the diastereomeric chlorohydrin. A Williamson ether synthesis gave **1** as a racemic mixture. The stereochemical course of the Grignard addition reaction could be rationalized using either the Cornforth or Felkin-Ahn models for asymmetric induction.[5,6] The Felkin-Ahn model is shown here using a truncated version of **48** (see conversion of **53** → **54**). This is an "early application" of principles in acyclic diastereoselection to a target-oriented synthesis problem.

Julia Olefin Synthesis Substituent Effects

Disubstituted Olefin Synthesis

Julia, M. *Bull. Soc. Chim. Fr.* **1960**, 1072; **1961**, 1849.

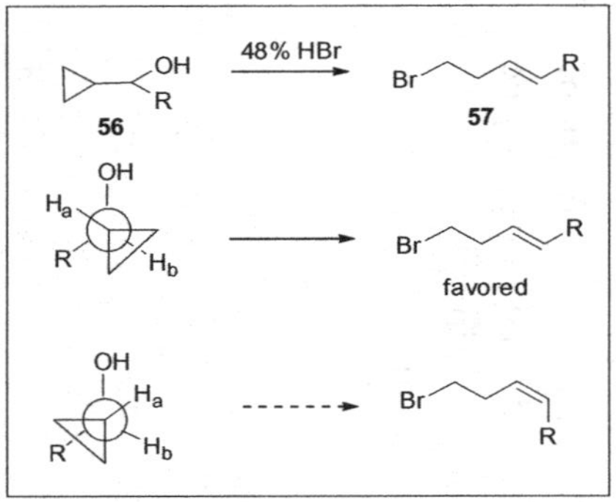

Nonselective Trisubstituted Olefin Synthesis

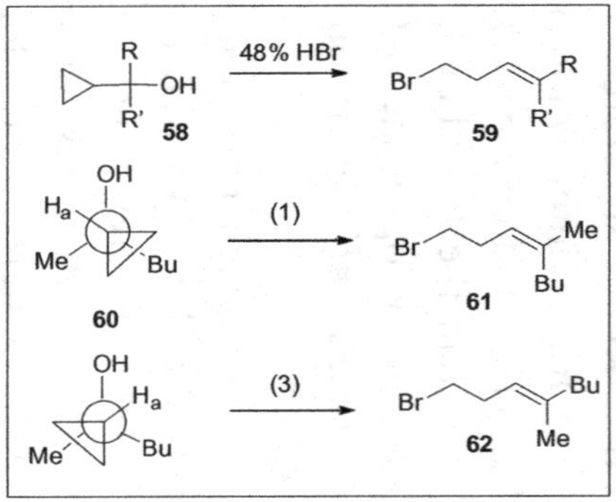

Brady, S. F.; Ilton, M. A.; Johnson, W. S. *J. Am. Chem. Soc.* **1968**, *90*, 2882.

Stereoselective Trisubstituted Olefin Synthesis

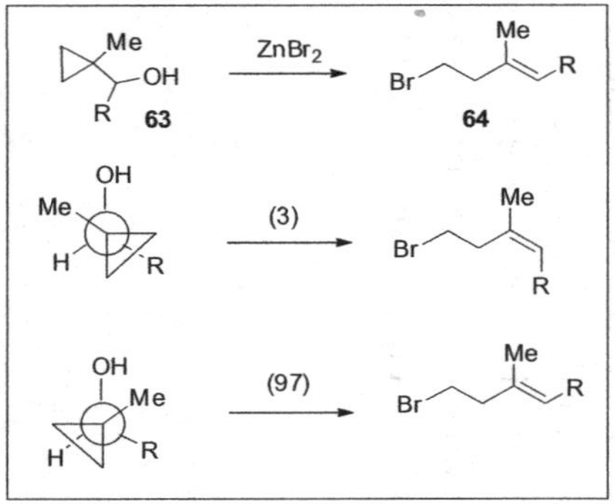

Stereoselective Trisubstituted Olefin Synthesis

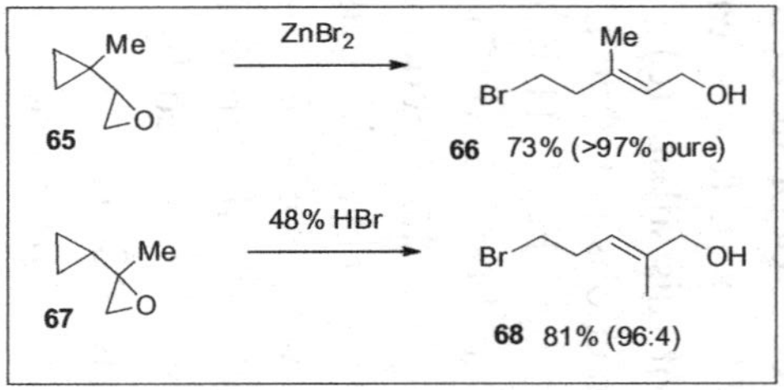

For another CJH synthesis see: Mori, K. "Synthesis of Compounds with Juvenile Hormone Activity - XII. A Stereoselective Synthesis of 6-Ethyl-10-methyldodeca-5-*trans*,9-*cis*-dien-2-one, A Key Intermediate in the Synthesis of C_{18}-*Cecropia* Juvenile Hormone" *Tetrahedron*, **1972**, *28*, 3747.

Nakamura, H.; Yamamoto, H.; Nozaki, H. *Tetrahedron Lett.* **1973**, 111

Cecropia Juvenile Hormone-6

Cecropia Juvenile Hormone-6

A number of variations of the Julia olefin synthesis are summarized in CJH-6. Substrates of type **56** provide *E*-olefins (**57**) with high selectivity. The rationale is the same as that used to explain selectivity in the Johnson synthesis of Cecropia Juvenile Hormone (H_b bisects cyclopropane rather than R). We have already seen that substrates of type **63** give trisubstituted olefins of type **64** with excellent stereoselectivity. Substrates of type **58**, however, do not give good stereoselectivity because there is little size differential between R and R′. Thus, substrate **60** gives a 1:3 mixture of **61** and **62**, respectively. Finally, epoxides (rather than bromides or alcohols) can serve as the initiating group in this process to provide interesting trifunctional products (halide, olefin, alcohol) with good control of olefin stereochemistry. For example, **65** and **67** can be converted to **66** and **68**, respectively, under typical Julia olefin synthesis conditions.

The Syntex-Zoecon Route to CJH

Zurfluh, R.; Wall, E. N.; Sidall, J. B.; Edwards, J. A. "Synthetic Studies on Insect Hormones. VII. An Approach to Stereospecific Synthesis of Juvenile Hormones" *J. Am. Chem. Soc.* **1968**, *90*, 6224. For full papers see: Henrick, C. A.; Schaub, F.; Siddall, J. B. "Stereoselective Synthesis of the C-18 Cecropia Juvenile Hormone" *J. Am. Chem. Soc.* **1972**, *94*, 5374-5378. Anderson, R. J.; Henrick, C. A.; Sidall, J. B.; Zurfluh, R. "Stereoselective Synthesis of the Racemic C-17 Juvenile Hormone of Cecropia" *J. Am. Chem. Soc.* **1972**, *94*, 5379-5386.

The Syntex approach relies on a series of fragmentation reactions (Grob fragmentations) that unravel a bicyclic ring system (perhydroindan) into the target acyclic structure. The "retrosynthetic analysis" leading to this route requires recognition that fragmentation of 2,2-disubstituted cycloalkane-1,3-diols can afford ketonic trisubstituted olefins with control of olefin geometry, a substructure of use as an intermediate in any projected synthesis of CJH. The Syntex synthesis is follows an ingenious design.

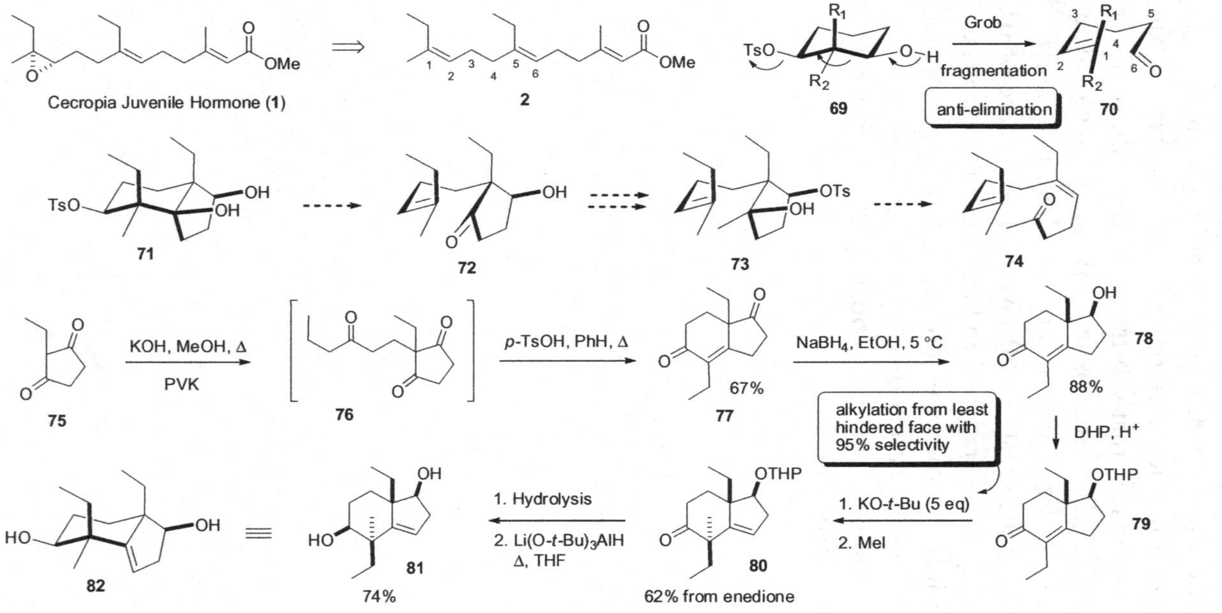

Cecropia Juvenile Hormone-7

Cecropia Juvenile Hormone-7

When the mechanisms by which insects develop and communicate (via pheremones) were first being elucidated, the economic potential of controlling insect populations through manipulation of these processes was recognized. In part for these reasons, Zoecon (the equivalent of a start-up company nowadays) and Syntex (a non-profit research institute at the time) developed a synthesis of CJH (**1**). This synthesis relied on a series of fragmentation reactions (Grob fragmentations)[7] that unravelled a bicyclic ring system (perhydroindan) into the target acyclic structure. The "retrosynthetic analysis" leading to this route required recognition that fragmentation of 2,2-disubstituted cycloalkane-1,3-diols would afford ketonic trisubstituted olefins with control of olefin geometry. For example, tosylate **69** would be expected to provide **70**. This ingenious design led to development of the famous Syntex-Zoecon route to CJH. The plan for CJH was to prepare fragmentation substrate **71**. It was expected that this would fragment to provide **72**, which was to be manipulated to provide **73**. Fragmentation of **73** was to then provide **74**, which would be converted to **1** following the path developed by Trost (CJH-2).

Before we look at the details of the synthesis, let's relate this fragmentation to something simple. Consider the first olefin synthesis you are likely to have learned, dehydrohalogenation of an alkyl halide. Dehydrohalogenations are *anti*-eliminations (with rare exceptions). The stereochemistry of the resulting olefin is a function of the vicinal stereochemistry of the dehydrohalogenation substrate, and the conformation(s) in which the leaving group and β-hydrogen can adopt an *anti*-relationship. Work some of the problems presented below to test your understanding of this reaction. The fragmentation of **69** to **70** can also be regarded as an *anti*-elimination. The leaving group is a tosylate and the β-substitutent is a carbon atom. But this carbon atom has electronic properties that are not unlike the properties of the β-hydrogen in a dehydrohalogenation (it becomes electron deficient in the elimination). Carrying this analogy further, the Julia olefin synthesis can be viewed as a "fragmentation" reaction ... the C–X bond ionizes, the adjacent σ-bond is cleaved with formation of an olefin, and the remaining cation (or partially charged carbon) is captured by a nucleophile. It is mechanism that unifies these approaches to stereoselective olefin synthesis.

The synthesis began with cyclopentan-1,3-dione **75**. A Robinson annelation gave **77** via intermediate trione **76**. Reduction of the more electrophilic of the two ketones gave **78**, which was protected as tetrahydropyranyl ether **79**. Deconjugative alkylation of **79** gave **80**. Hydrolysis of the THP protecting group and reduction of the ketone (pseudo-axial delivery of hydride) provided **81**/**82**.

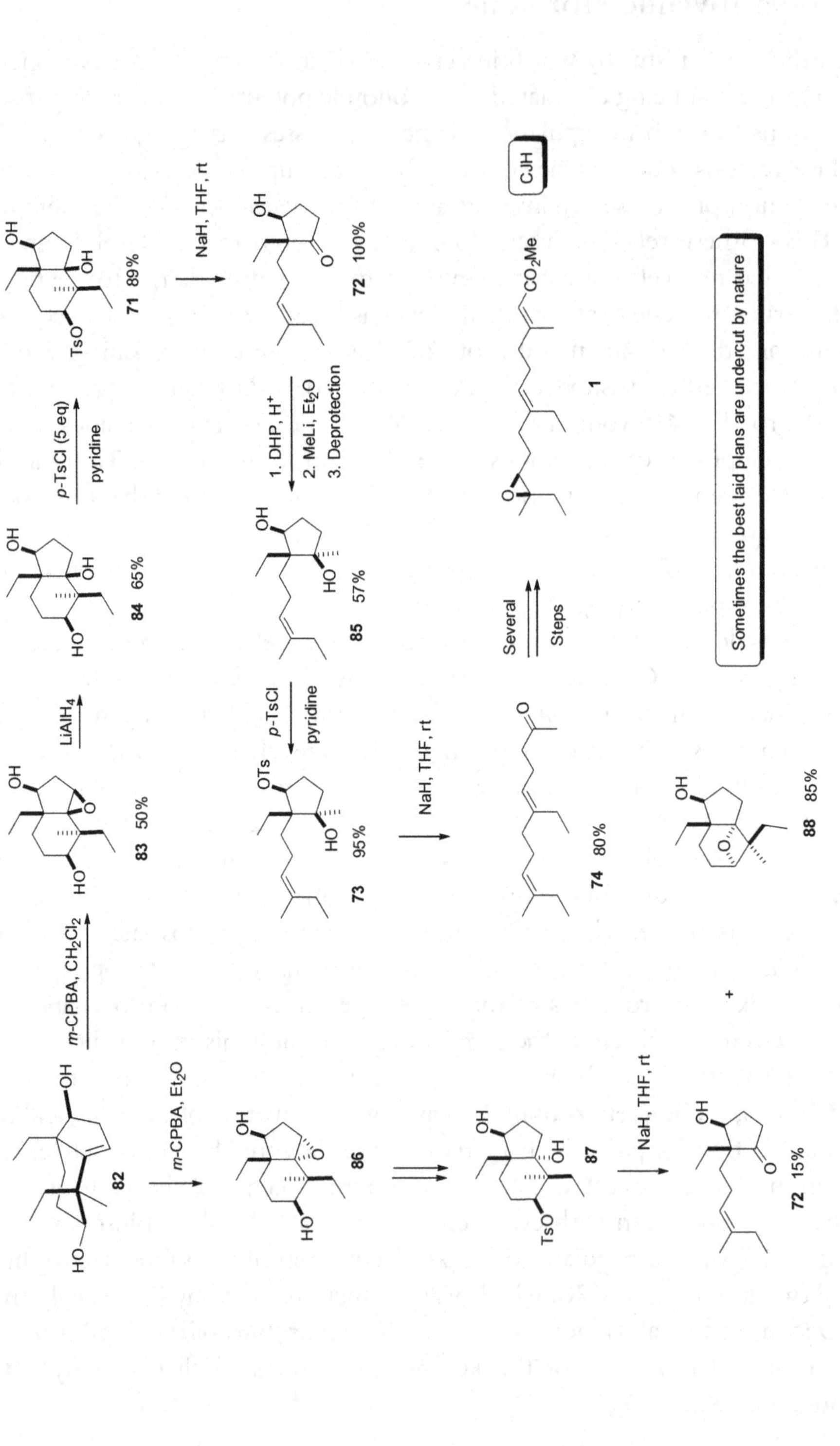

Cecropia Juvenile Hormone-8

Cecropia Juvenile Hormone-8

Oxidation of **82** with *m*-CPBA in dichloromethane provided **83**. Reduction of the epoxide with lithium aluminium hydride gave **84**, which was converted to fragmentation substrate **71**. The fragmentation occurred in high yield to provide the desired geometrical isomer **72**. Protection of the secondary alcohol, addition of methyllithium to the ketone, and removal of the THP protecting group gave diol **85**. Conversion of the secondary alcohol to tosylate **73** was followed by the second fragmentation reaction to give **74**, an intermediate in the Trost synthesis of CJH (**1**).

It is notable that tosylate **87**, derived from isomeric epoxide **86**, available in turn by epoxidation of **82** in diethyl ether rather than dichloromethane, is also disposed to undergo fragmentation to **72**. Unfortunately, when **87** was subjected to sodium hydride, a Williamson ether synthesis (**87** → **88**) became competitive with the desired fragmentation. Sometimes the best laid plans are undercut by nature. Regardless, the Syntex-Zoecon synthesis remains one of the least obvious, yet creative and successful, routes to Cecropia Juvenile Hormone.

Synthesis-Driven Methodology Development

Corey, E. J.; Katzenellenbogen, J. A.; Gilman, N. W.; Roman, S. A.; Erickson, B. W. "Stereospecific Total Synthesis of the *dl*-C_{18} Cecropia Juvenile Hormone" *J. Am. Chem. Soc.* **1968**, *90*, 5618. The purpose of this work was to prepare CJH, to develop a new iterative approach to stereoselective trisubstituted olefins, and to illustrate the use of other new synthetic methods. We will skip the "analysis" and simply proceed through the synthesis.

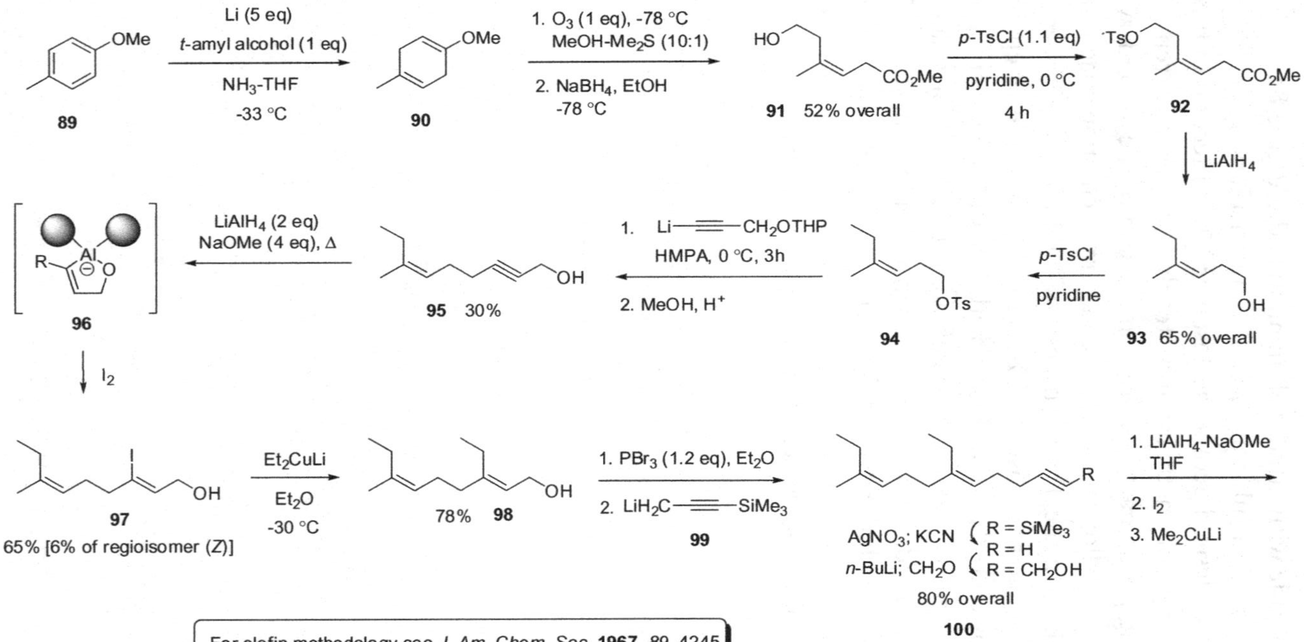

For olefin methodology see *J. Am. Chem. Soc.* **1967**, *89*, 4245

Cecropia Juvenile Hormone-9

Cecropia Juvenile Hormone-9

The purpose of the next synthesis was to develop a new, iterative, and stereoselective approach to trisubstituted olefins, and also to illustrate the use of other new synthetic methods. We will skip the "analysis" and go directly to the synthesis, which runs from "left-to-right". The first stage of the synthesis called for preparation of alcohol **93**. The strategy used to prepare **93** involved starting with all carbons intact in the form of *p*-methoxytoluene (**89**). Birch reduction of **89** provided **90**. Regioselective ozonolysis of the more electron rich olefin, followed by a reductive work-up, gave **91**. Appropriate adjustment of oxidation states at the terminal carbons provided **93**. This portion of the synthesis uses a ring to set trisubstituted olefin geometry. This approach to controlling olefin stereochemistry is less common nowadays, but is still occasionally used.

Conversion of **93** to tosylate **94**, followed by an acetylide displacement to give **95**, set the stage for application of the new trisubstituted olefin synthesis. It had been known for some time that reduction of propargylic alcohols with lithium aluminum hydride, followed by hydrolysis of the reaction mixture, gave *E*-allylic alcohols with excellent stereoselectivity. Corey and Katzenellenbogen found that, under appropriate conditions, a presumed intermediate vinylalanate (**96**) could be captured by iodine to give trisubstituted iodoalkene **97** with high stereoselectivity. Furthermore, they were able to couple iodoalkenes with lithium dialkylcuprates with retention of stereochemistry to provide trisubstituted alkenes. In the CJH synthesis, **97** was converted to **98** using Et_2CuLi. We will see variations of this coupling strategy again in the next chapter. Chain extension of **97** to **100** ($R{=}CH_2OH$), using organolithium **99** as a key reagent, set the stage for repetition of this reaction sequence. Repetition of the reduction-iodination-coupling sequence provided allylic alcohol **101** (see CJH-10).

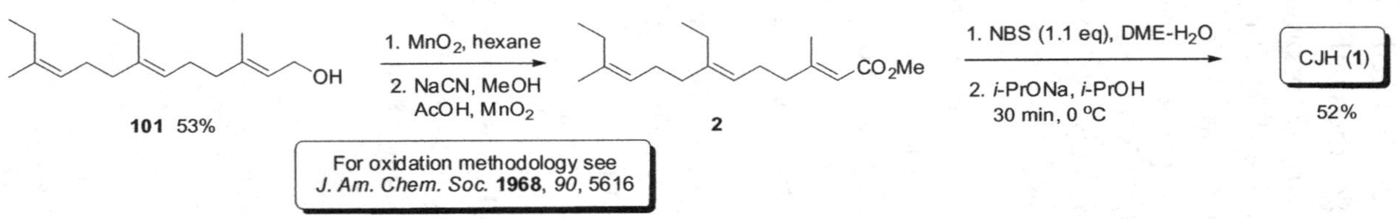

Modification of the Schlosser Modification of the Wittig Reaction

Corey, D. J.; Yamamoto, H.; "Simple Stereospecific Synthese of C_{17}- and C_{18}-Cecropia Juvenile Hormones (Racemic) from a Common Intermediate" *J. Am. Chem. Soc.* **1970**, *92*, 6636.

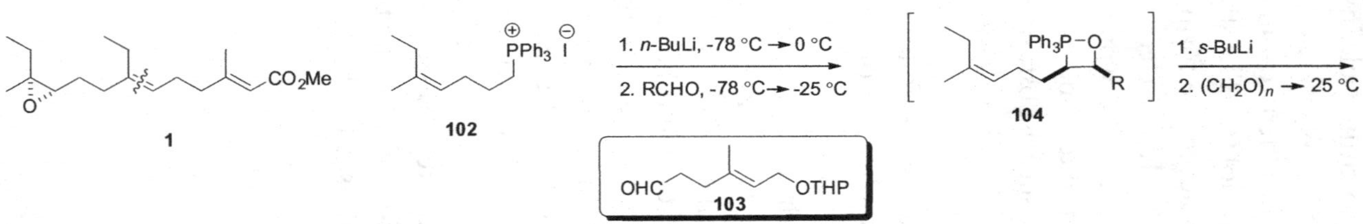

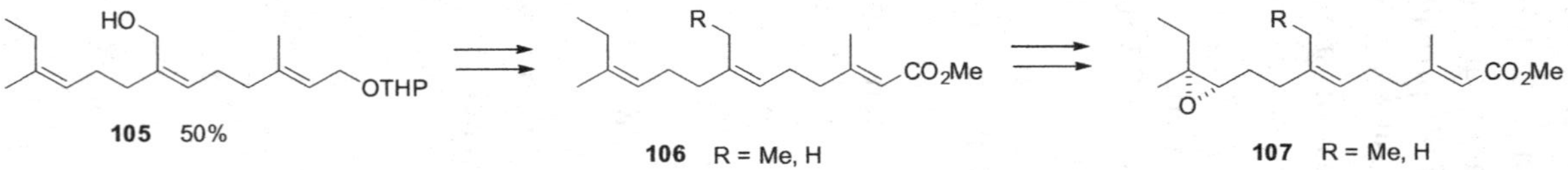

This is a variation of the Schlosser modification of the Wittig reaction [Schlosser, M.; Coffinet, D. *Synthesis* **1971**, 380 and *Synthesis* **1972**, 575]. Vedejs has published interesting work that provides a rational for the stereochemical and regiochemical course of the key reaction: Vedejs, E.; Snoble, K. A. J. "Direct Observation of Oxaphosphatanes from Typical Wittig Reactions" *J. Am. Chem. Soc.* **1973**, *95*, 5778.

Cecropia Juvenile Hormone-10

Cecropia Juvenile Hormone-10

Oxidation of **101** to ester **2** and epoxidation of the sterically most accessible olefin, completed the synthesis of CJH. The conversion of **101** to **2** is notable, and relies on the selective oxidation of allylic alcohols to aldehydes by manganese dioxide. The sequence of events is: (1) oxidation of **101** to the corresponding aldehyde, (2) cyanohydrin formation, (3) oxidation of the cyanohydrin to an acyl cyanide and (4) methanolysis of the acyl cyanide to provide the ester.

Another synthesis of CJH eminating from the Corey labs involved a modification of the Schlosser modification of the Wittig Reaction. Whereas Wittig reactions between non-stablilized phosphoranes and aldehydes usually provide Z-olefins, Schlosser reported a variation of this reaction that provided *E*-olefins. His group showed that when non-stabilized phosphoranes are reacted with aldehydes in the presence of lithium halides (as opposed to the salt-free conditions that favor *Z*-olefins), followed by deprotonation of the intermediate adduct, and reprotonation of the resulting *β*-oxido ylid, *E*-olefins are obtained. Corey and Yamamoto reported a variation of this reaction in which the intermediate *β*-oxido ylid was treated with formaldehyde, resulting in formation of trisubstituted allylic alcohols. For example, **102** reacted sequentially with aldehyde **103** and formaldehyde to provide **105**. This material was then carried on to CJH intermediate **106**, and the CJH derivative where R=H.

Still Synthesis of CJH

Still, W. C.; McDonald, J. H. III; Collum, D. B.; Mitra, A. "A Highly Stereoselective Synthesis of the C_{18} Cecropia Juvenile Hormone" *Tetrahedron Lett.* **1979**, 593-594.

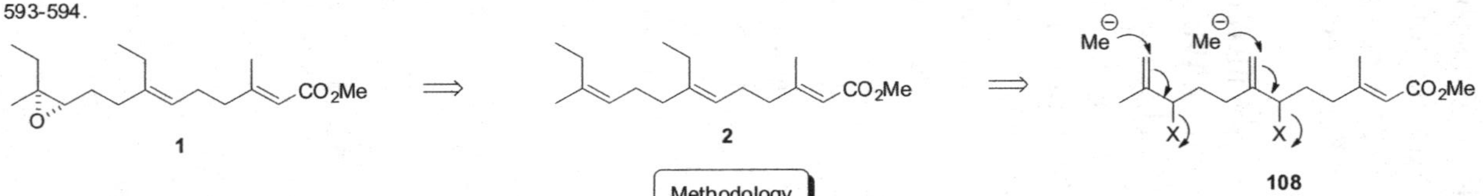

Methodology

Still, W. C.; Mitra, A. "A Highly Stereoselective Synthesis of *Z*-Trisubstituted Olefins via [2,3]-Sigmatropic Rearrangement. Preference for a Pseudoaxially Substituted Transition State" *J. Am. Chem. Soc.* **1978**, *100*, 1927. For additional reading see: Still, W. C.; Sreekumar, C." α-Alkoxyorganolithium Reagents. A New Class of Configurationally Stable Carbanions for Organic Synthesis" *J. Am. Chem. Soc.* **1980**, *102*, 1201-1202.

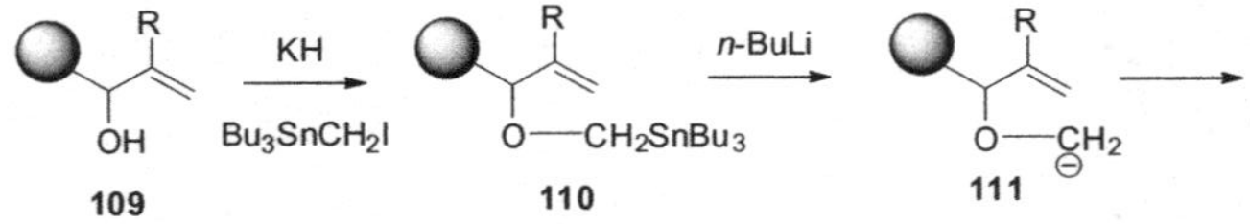

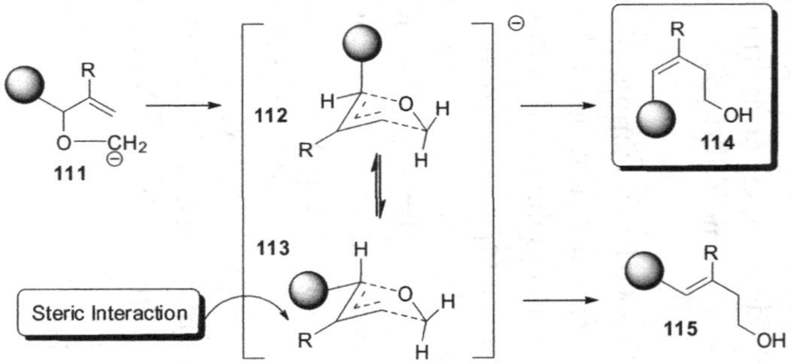

When the ball is an *n*-butyl group and R = Me, the stereoselectivity is greater than 95%. When R = H the selectivity nearly vanishes (*Z*:*E* = 60:40). Similar reduction in selectivity is observed when the Me group is moved to the olefin terminus. An early transition state is proposed.

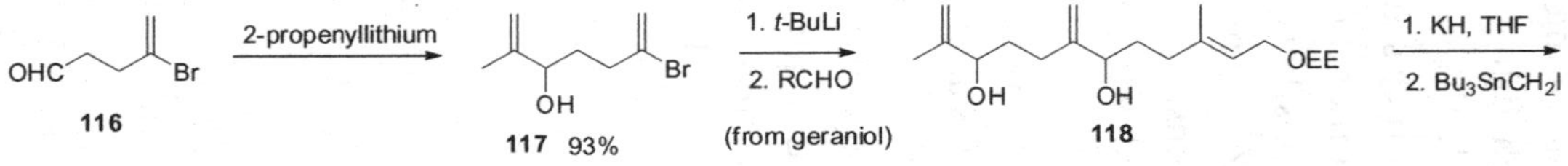

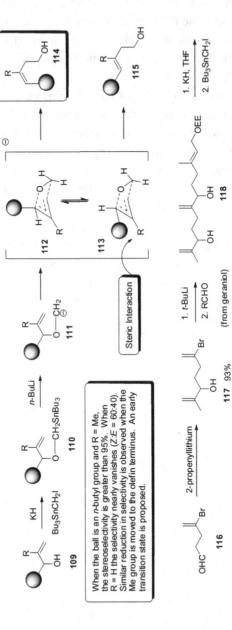

Cecropia Juvenile Hormone-11

Cecropia Juvenile Hormone-11

A conceptually interesting approach to the synthesis of CJH involves the reaction of allylic substrates of type **108** with methyl carbanion equivalents in an S_N2' manner. Several syntheses have followed this approach. The Still-Collum-Mitra synthesis used enforced S_N2' methodology, a 2,3-sigmatropic rearrangement of allylic alcohols (**108** where X=OH). The "big questions" in this approach were (1) what would be used as a "CH_3^-" equivalent and (2) would the desired olefin stereocontrol be achieved?

The basic methodology is illustrated with structures **109–115**. The notion was that an allylic alcohol of type **109** would be converted to **110**, a precursor of carbanion **111**. Carbanion **111** would undergo 2,3-sigmatropic rearrangement from either conformation **112** or **113** to provide homoallylic alcohol **114** or **115**, respectively. The hydroxyl group would later be converted to a hydrogen to establish the necessary equivalency. The plan worked remarkably well with substrates of type **109**. When R and the "ball" were methyl and *n*-butyl groups respectively, the ratio of **114**:**115** was greater than 95:5. It was suggested that the rearrangment occurred via an early transition state in which steric interactions between the methyl and butyl substitutents were minimized (**112** lower energy than **113**). In accord with this suggestion, substrates where R=H gave nearly equal mixtures of *Z*- and *E*-olefins.

In the application to CJH, aldehyde **116** (prepared as described on CJH-12) was converted to **117**. Metallation of **117** followed by addition of the vinyllithium reagent to the appropriate aldehyde (prepared as described on CJH-12) gave *bis*-allylic alcohol **118**. Derivatization of **118** gave *bis*-sigmatropic rearrangement substrate **119**.

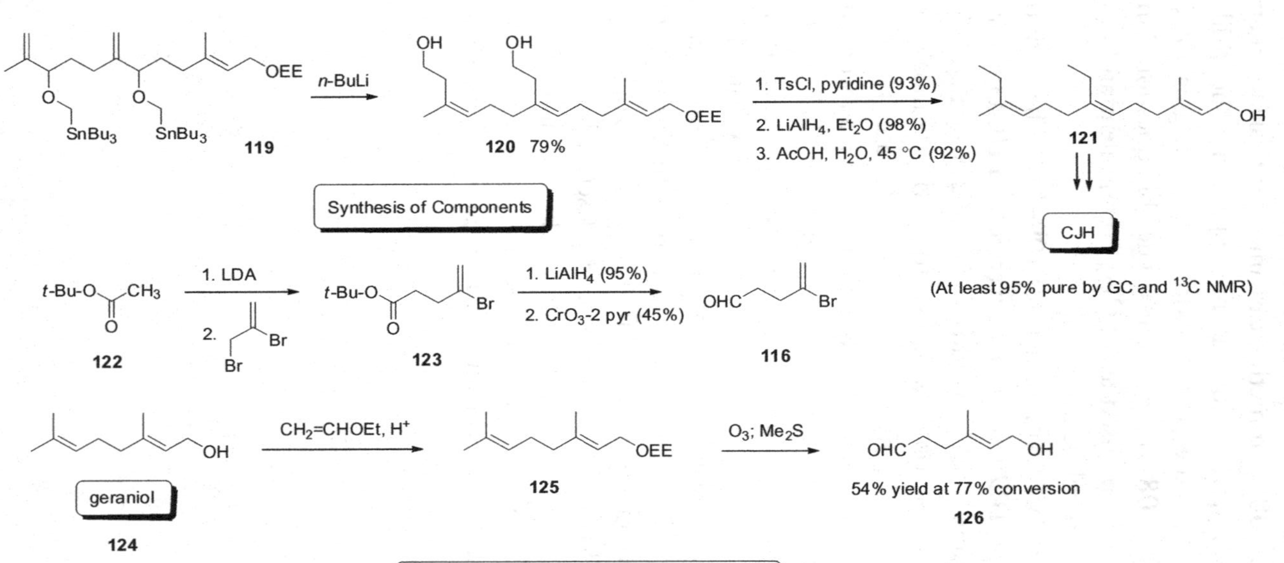

Strategically Related Syntheses

van Tamelen, E. E.; McCormick, J. P. "Synthesis of Cecropia Juvenile Hormone from *trans, trans*-Farnesol" *J. Am. Chem. Soc.* **1970**, *92*, 737-738.

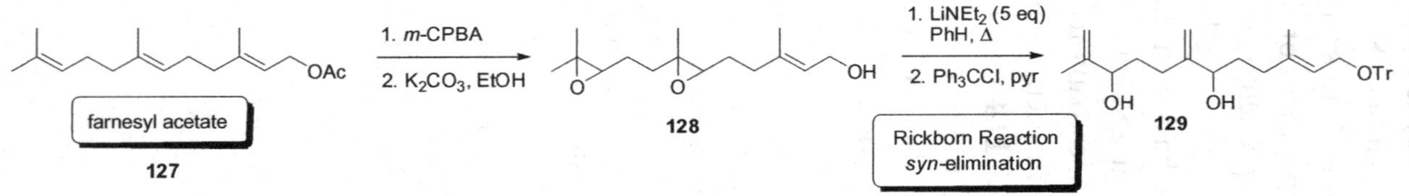

Cecropia Juvenile Hormone-12

Cecropia Juvenile Hormone-12

Metallation of **119** followed by the double 2,3-sigmatropic rearrangement provided *bis*-homoallylic alcohol **120**. The aforementioned "CH_3^-" equivalency was established by converting the hydroxymethyl groups to methyl groups (via reduction of the derived tosylates). Hydrolysis of the ethoxyethyl protecting group gave **121** which was converted to CJH following established procedures.

A strategically related, although tactically quite different, synthesis of CJH had been reported by the VanTamelen group (Stanford) some years earlier. In this approach farnesyl acetate (**127**), a commercially available sesquiterpene, was converted to *bis*-epoxide **128**. The inductive electron-withdrawing effect of the acetoxy group presumably helped with regioselectivity in this transformation. The *bis*-epoxide was then converted to *bis*-allylic alcohol **129** (compare with **118** on CJH-11) using a reaction developed by Rickborn, followed by selective protection of the primary alcohol as a trityl ether.[8]

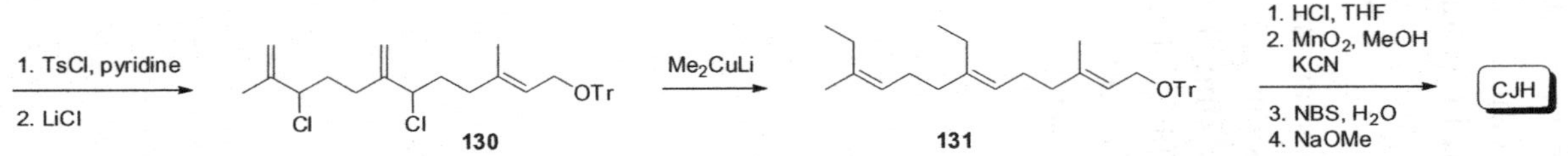

The sequence produces a mixture of four olefin geometrical isomers in nearly equal amounts (one of which is CJH). Thus, the cuprate chemistry is not stereoselective.

Tanaka, S.; Yamamoto, H.; Nozaki, H.; Sharpless, K. B.; Michaelson, R. C.; Cutting J. D. "Stereoselective Epoxidations of Acyclic Allylic Alcohols by Transition Metal-Hydroperoxide Reagents. Synthesis of *dl*-C_{18} Cecropia Juvenile Hormone from Farnesol" *J. Am. Chem. Soc.* **1974**, *96*, 5254.

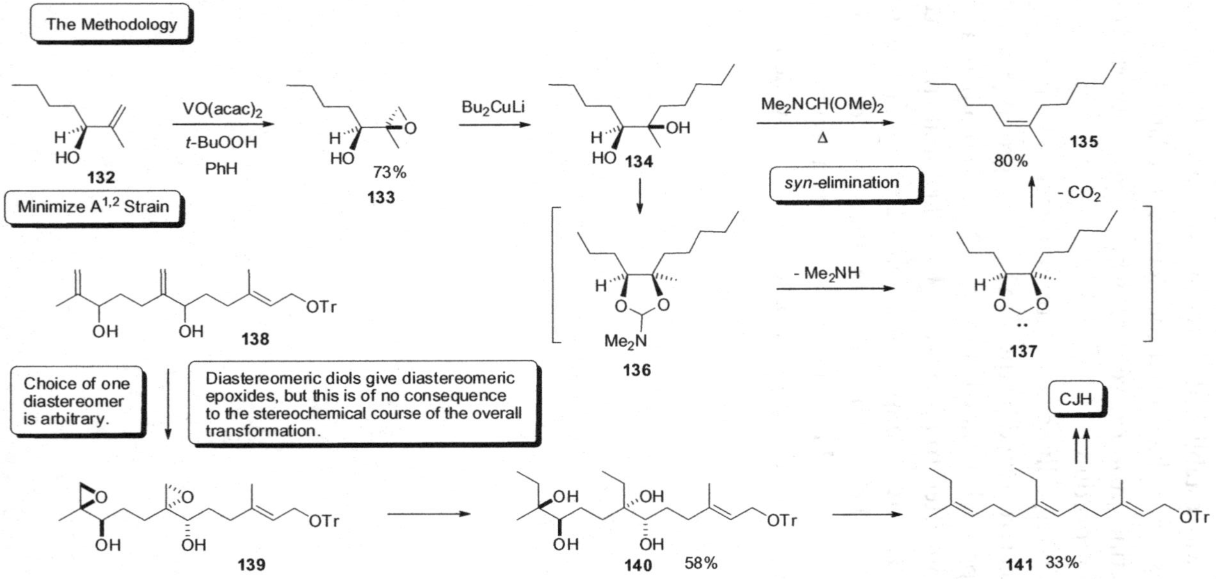

Cecropia Juvenile Hormone-13

Cecropia Juvenile Hormone-13

Diol **129** was converted to the corresponding *bis*-chloride (**130**) via an intermediate *bis*-tosylate. Treatment of **130** with lithium dimethylcuprate gave **131** via a double S_N2' reaction. This reaction, however, gave a nearly statistical mixture of the four possible geometrical isomers and thus, upon completion of the synthesis, a mixture of products was obtained that included some CJH. Therefore, the tactics used by the Still group for accomplishing this double S_N2' strategy were better than those employed by the VanTamelen group.

Another methodology-driven synthesis of CJH that is strategically related to the Still and VanTamelen syntheses was reported by the Sharpless group. This method relies on a trisubstitued olefin synthesis that revolves around a diastereoselective epoxidation reaction developed in the Sharpless laboratories. Sharpless had shown that allylic alcohol **132** underwent a highly diastereoselective epoxidation to give **133** upon exposure to *tert*-butyl hydroperoxide in the presence of an appropriate vanadium catalyst. This reaction presumably results from the conformation of **132** that minimizes 1,2-allylic strain (between the methyl and butyl groups for example). Covalent binding of both the alcohol and *tert*-butyl hydroperoxide to the vanadium result in directed delivery of "oxygen" to the expected face of the olefin. Treatment of **133** with lithium dibutylcuprate gave **134** (epoxide opening at primary carbon, rather than tertiary carbon). Treatment of **134** with *N,N*-dimethylformamide dimethyl acetal gave **135**. This overall *syn*-elimination of vicinal hydroxyl groups is imagined to occur by conversion of **134** to amide acetal **136**, α-elimination of dimethylamine to provide carbene **137**, and chelatropic extrusion of carbon dioxide to give **135**. This method was applied to a CJH synthesis by first preparing **139** (from **129**). Reaction of this *bis*-epoxide with lithium dimethylcuprate gave **140** and a *bis*-elimination reaction gave **141** (same as **131**), an intermediate in the VanTamelen synthesis of CJH.

Set Olefin Stereochemistry in a Ring

Kondo, K.; Negishi, A.; Matsui, K.; Tunemoto, D.; Masamune, S. "A New Approach to the Stereospecific Total Synthesis of Racemic Cecropia Juvenile Hormone" *J. Chem. Soc., Chem. Commun.* **1972**, 1311-1312. For a closely related approach see: Demonte, J.-P.; Hainaut, D.; Toromanoff, E. "Sur une nouvelle synthese, stereospecifique, de l'hormone juvenile en C_{18} Hyalophora cecropia" *Comptes Rendus Acad. Sci. Paris* **1973**, *277*, 49-51.

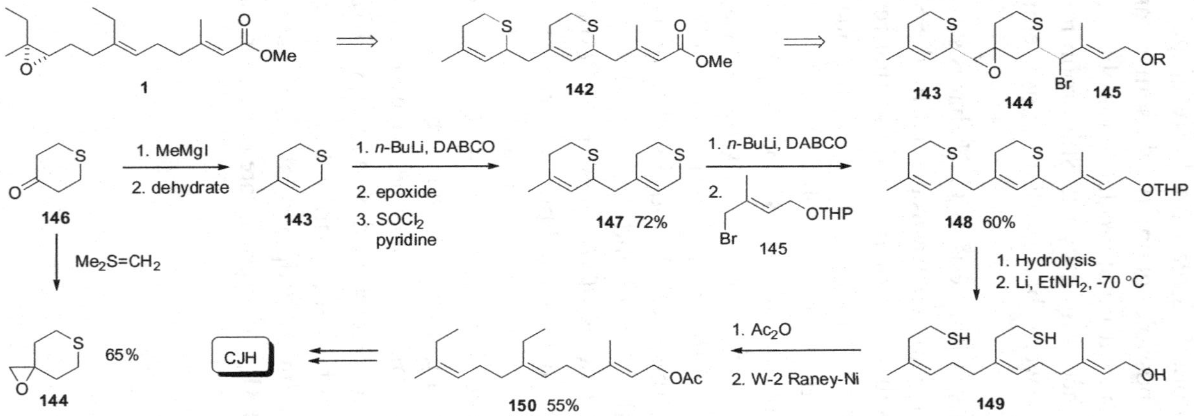

Another Iterative Approach

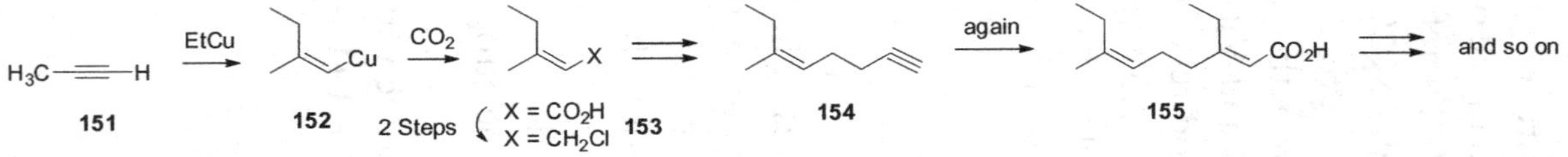

Chuit, C.; Cahiez, G.; Normant, J.; Villieras, J. "Synthese Stereospecifique Recurrente de Structures Apparentees a Celle de l'Hormone Juvenile de Hyalphora Cecropia" *Tetrahedron* **1976**, *32*, 1675-1680.

Cecropia Juvenile Hormone-14

Cecropia Juvenile Hormone-14

Two more syntheses of CJH are presented here. The first synthesis relies on setting olefin stereochemistry by incorporating it into a ring (recall the first Corey synthesis of CJH shown on CJH-9). The idea was to use a sulfur-containing heterocycle to establish double bond geometry and then reduce the two carbon-sulfur bonds to liberate the acyclic olefin. Thus, it was hoped that **142** would serve as a precursor of **1** that could be stitched together from **143–145**.

In the forward direction, tetrahydrothiapyrone **146** served as the precursor of both **143** and **144**. Metallation of **143** (an allylic sulfide) using *n*-butyllithium, followed by a reaction of the derived carbanion with **144**, gave **147** after dehydration of the intermediate tertiary alcohol. Metallation of **147** followed by alkylation of the resulting anion with bromide **145**, provided **148**. Hydrolysis of the THP ether followed by reduction of the allylic C–S bonds with lithium in diethylamine, gave *bis*-sulfide **149**. This reaction would have been expected to proceed via allyic radical and/or anion intermediates and thus, loss of olefin regiochemistry might have been anticipated, but the reaction seems to have proceeded without a major problem. Esterification of the alcohol (and most likely the thiols) and reduction of the homoallyic C–S bonds with Raney-Ni, gave **150**. The synthesis of CJH (**1**) was completed in the usual manner.

The final CJH synthesis we will examine, albeit briefly, is outlined using structures **151** to **155**. This is another iterative approach (from the Normant group in France) that uses *syn*-additions of organometallics to alkynes to establish olefin stereochemistry. This is an approach to olefin synthesis that has gained popularity in recent years, with a variety of reagents having been developed to "carbometallate" alkynes.[9] The Normant synthesis began with a reaction between propyne (**151**) and ethyl copper to provide vinylic organometallic **152**. Carboxylation of **152** took place with retention of olefin geometry to provide carboxylic acid **153** (X=CO_2H). The acid was converted to the corresponding allylic chloride, which was converted to alkyne **154** using methodology that resembles that used by Corey in his CJH synthesis (CJH-9). Repetition of the sequence provided **155** and so on to CJH.

In closing this chapter I point out that whereas CJH (**1**) has been used as the context for our discussion of olefin synthesis, the lasting value of the research presented here is surely the methodology developed in pursuit of CJH (or illustrated by application to CJH syntheses). Furthermore I have not been anywhere near exhaustive in presenting olefin syntheses. We will see more in the next chapter. I also hope you recognize that the olefin syntheses we examined in this chapter are greater in number than they are in terms of strategy. To summarize, we have seen methods that set stereochemistry in β-elimination reactions (Julia synthesis, Sharpless synthesis, use of Wittig reactions), by converting cyclic olefins to acyclic olefins (ring-opening reactions such as those used by Corey and Masamune), by stereocontrolled S_N2' reactions (Johnson-Faulkner, Still), and by stereocontrolled addition reactions of alkynes (Corey and Normant). As you learn more about olefin synthesis, you will surely come across other strategies, or other tactics, that accomplish olefin synthesis using the broad strategies presented in this chapter.

References

1. Eisner, T.; Meinwald, J. "The Chemistry of Sexual Selection" *Proc. Nat. Acad. Sci. (USA)* **1995**, *92*, 50–55. Karlson, P. "The Chemistry of Insect Hormones and Insect Pheromones" *Pure Appl. Chem.* **1967**, *14*, 75–87. Karlson, P. "Chemie und Biochemie Der Insektenhormone" *Angew. Chem.* **1963**, *75*, 257–265.
2. Houser, C. R.; Hudson, B. E. Jr. "The Acetoacetic Ester Condensation and Certain Related Reactions" *Organic Reactions* **1944**, *1*, 266–302.
3. Castro, A. M. M. "Claisen Rearrangement over the Past Nine Decades" *Chem. Rev.* **2004**, *104*, 2939–3002. Rhoads, S. J.; Raulins, N. R. "Claisen and Cope Rearrangements" *Organic Reactions* **1975**, *22*, 1–252.
4. Julia, M.; Julia, S.; Yu, T. S. "Stéréochemie des Alcool β,γ-Éthyléniques Issus de la Tranposition Homoalliques" *Bull. Soc. Chim. Fr.* **1961**, 1849–1853. Julia, M.; Julia, S.; Guégan, R. "Préparation de Composés Terpéniques et Apparentés, à Partir de Méthyl Cyclopropyl Cétone" *Bull. Soc. Chim. Fr.* **1960**, 1072–1079.
5. Cornforth, J. W.; Cornforth, R. H.; Mathew, K. K. "A General Stereoselective Synthesis of Olefins" *J. Chem. Soc.* **1959**, 112–127. Evans, D. A.; Siska, S. J.; Cee, V. J. "Resurrecting the Cornforth Model for Carbonyl Addition: Studies on the Origin of 1,2-Asymmetric Induction in Enolate Additions to Heteroatom-Substituted Aldehydes" *Angew. Chem. Int. Ed.* **2003**, *42*, 1761–1765.
6. Ahn, N. T.; Eisenstein, O. "Theoretical Interpretation of 1,2-Asymmetric Induction. The Importance of Antiperiplanarity" *Nouv. J. Chim.* **1977**, *1*, 61–70. Cherest, M.; Felkin, H.; Prudent, N. "Tortional Strain Involving Partial Bonds. The Stereochemistry of the Lithium Aluminum Hydride Reduction of Some Simple Open-Chain Ketones" *Tetrahedron Lett.* **1968**, *10*, 2199–2204.
7. Grob, C. A.; Kiefer, H. R.; Lutz, H.; Wilkens, H. "The Stereochemistry of Synchronous Fragmentation" *Tetrahedron Lett.* **1964**, *6*, 2901–2904.
8. Rickborn, B.; Thummel, R. P. "Stereoselectivity of the Base-Induced Conversion of Epoxides to Allylic Alcohols" *J. Org. Chem.* **1969**, *34*, 3583–3586.
9. For two representative reviews see: Negishi, E. "Bimetallic Catalytic Systems Containing Titanium, Zirconium, Nickel and Palladium. Their Applications to Selective Organic Syntheses" *Pure Appl. Chem.* **1981**, *53*, 2333–2356. Fallis, A. G.; Forgione, P. "Metal Mediated Carbometallation of Alkynes and Alkenes Containing Adjacent Heteroatoms" *Tetrahedron* **2001**, *57*, 5899–5913.

Problems

1. Propose a mechanism for the Horner-Wadsworth-Emmons reaction. The reaction of aldehydes with **10** provides 1,2-disubstituted olefins with high *E*-selectivity. Provide (or look up) a rationale for this observation. Then explain why the reaction of **3** with **10** shows little stereoselectivity. (CJH-2)
2. Rearrangement from the other chair-like conformation available to **21**, **23** and **25** (not shown in CJH-3) results in formation of the minor products (*Z*-geometrical isomers). Provide 3-dimensional structures of these transition geometries (as in CJH-3) and indicate the steric interaction that is responsible for the increased stereoselectivity in the reactions of **23** and **25** relative to **21**. (CJH-3)
3. Provide a mechanism for the conversion of **17** + **29** → **30**. (CJH-3)
4. Provide the transition state structure of the allyl vinyl ether that results in the conversion of **33** + **34** → **35**. (CJH-4)
5. Propose a practical synthesis of **42** and **43**. (CJH-4)
6. Provide the structure of the minor epoxide obtained from the conversion of **48** → **1**. (CJH-5)
7. Discuss the factors behind selection of the "reactive conformation" in the Felkin-Ahn model for conversion of **53** to **54**. How would Cornforth have rationalized this stereochemical result? [For an interesting system where the Cornforth and Felkin-Ahn models predict different stereochemical results see Evans, D. A.; Siska, S. J.; Cee, V. J. "Resurrecting the Cornforth Model for Carbonyl Addition: Studies on the Origin of 1,2-Asymmetric Induction in Enolate Additions to Heteroatom-Substituted Aldehydes" *Angew. Chem. Int. Ed.* **2003**, *42*, 1761–1765] (CJH-5)
8. Provide an explanation for the regiochemical course of the following reactions (indicate transition state structures). (CJH-7)

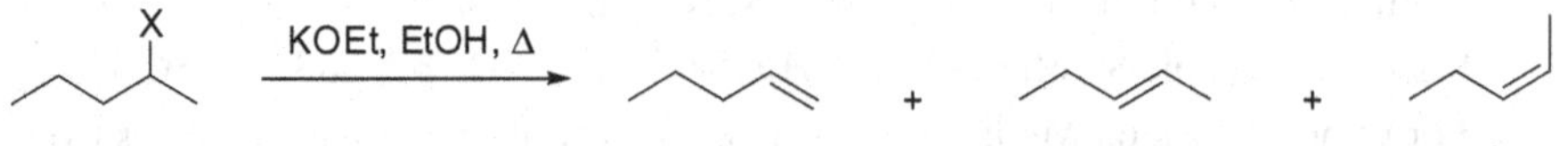

1 2t 2c

X	Yield of Olefins	Ratio of Olefins (1 : 2t : 2c)
Br	73%	31 : 51 : 18
OTs	50%	48 : 34 : 18
$^{\oplus}SMe_2$	60%	87 : 8 : 5
$^{\oplus}NMe_3$	71%	98 : 1 : 1

Brown, H. C.; Wheeler, O. H. "Steric Effects in Elimination Reactions. IX. The Effect of the Steric Requirements of the Leaving Group on the Direction of Bimolecular Elimination in 2-Pentyl Derivatives" *J. Am. Chem. Soc.* **1956**, *78*, 2199–2202. Also see Bartsch, R. A.; Bunnett, J. F. "Orientation of Olefin-Forming Elimination in Reactions of 2-Substituted Hexanes with Potassium *tert*-Butoxide-*tert*-Butyl Alcohol and Sodium Methoxide-Methanol" *J. Am. Chem. Soc.* **1969**, *91*, 1376–1382.

9. Provide the products expected from the following olefin-forming reactions. (CJH-7)

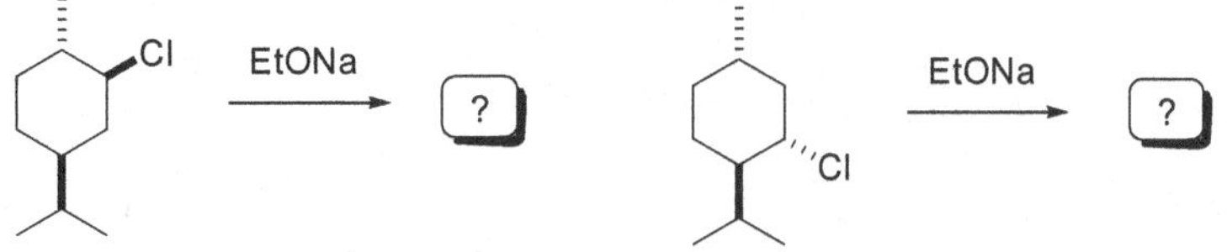

Huckel, W.; Tappe, W.; Legutke, G. "Elimination Reactions and Their Steric Course" *Ann. Chem.* **1940**, *543*, 191–230.

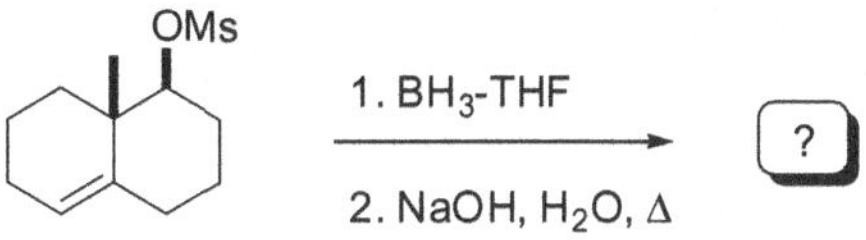

Marshall, J. A.; Bundy, G. L. "A New Fragmentation Reaction. The Synthesis of 1-Methyl-trans,trans-1,6-cyclodecadiene" *J. Am. Chem. Soc.* **1966**, *88*, 4291–4292.

10. Draw the conformation of **87** that is disposed to fragment to **72**. (CJH-8)
11. Outline a synthesis of **93** using the Corey-Katzenellenbogen synthesis to set stereochemistry. How might **93** be prepared using a Julia olefin synthesis? Comment on stereochemical aspects of the Julia route to **93**. (CJH-9)
12. Provide the structures of intermediates *en route* from **101** to **2**. Explain the sequence of events that leads from **I** to **II**. (CJH-10)

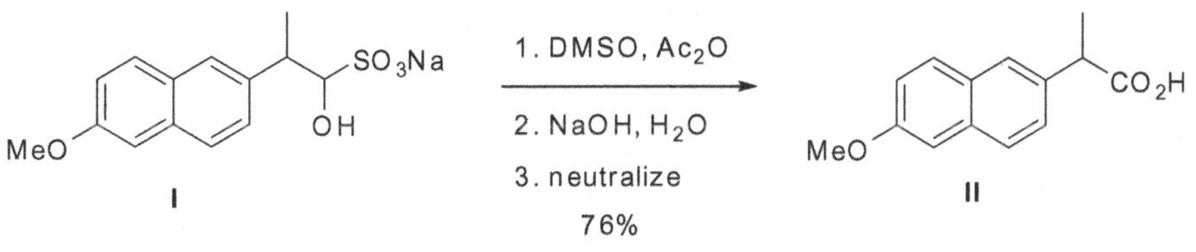

Wuts, P. G. M.; Bergh, C. L. "The Oxidation of Aldehyde Bisulfite Adducts to Carboxylic Acids and their Derivatives with Dimethyl

Sulfoxide and Acetic Anhydride" *Tetrahedron Lett.* **1986**, *27*, 3995–3998.

13. Provide the structures of the four products expected from the reaction of **130** with lithium dimethylcuprate. (CJH-13)
14. An interesting olefin synthesis is presented below. To which of the aforementioned strategies does this belong? (CJH-14)

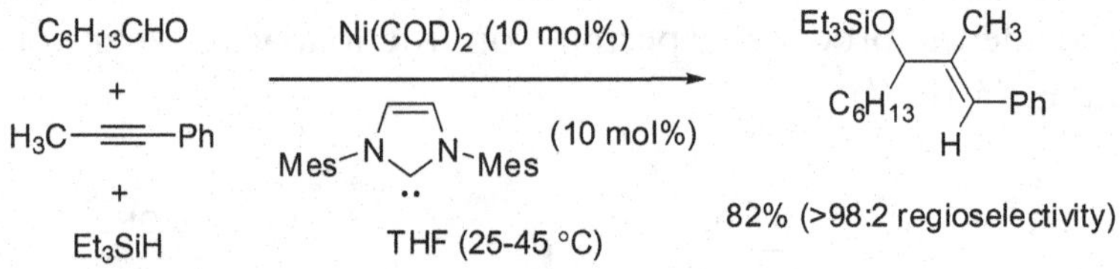

Mahandru, G. M.; Liu, G.; Montgomery, J. "Ligand-Dependent Scope and Divergent Mechanistic Behavior in Nickel-Catalyzed Reductive Couplings of Aldehydes and Alkynes" *J. Am. Chem. Soc.* **2004**, *126*, 3698–3699.

CHAPTER 12

A Recent Example of Structure Determination Through Total Synthesis and Convergent Syntheses: Lasonolide-A

Total Synthesis and Structure Determination: A Modern Example

Lasonolide A provides a modern example of synthesis playing a role in natural product structure determination. There are other examples for sure, but I have selected this one because of a personal interest in the problem, and because it also is a polyolefin that will let us continue our discussion of diastereoselective olefin synthesis. Lasonolide A is a marine natural product produced by a sponge (or microorganisms that live on the sponge ... these situations can be difficult to differentiate). The structure was assigned as shown below [Horton, P. A.; Koehn, F. E.; Longley, R. E.; McConnell, O. J. *J. Am. Chem. Soc.* **1994**, *116*, 6015-6016] based largely on NMR studies. The stereochemistry at C_{28} was not determined, nor was its absolute configuration. The natural product showed activity against several tumor cell lines with activity (IC_{50} values) in the ng/mL region. It also inhibited cell adhesion in several cell lines. Thus, lasonolide A promptly became an attractive target for synthesis: (1) to establish the structure and (2) to provide more material for biological evaluation. Since the isolation of lasonolide A, the lasonolides have grown into a small family of natural products [Wright, A. E.; Chen, Y.; Winder, P. L.; Pitts, T. P.; Pomponi, S. A.; Longley, R. E. "Lasonolides C-G, Five New Lasonolide Compounds from the Sponge *Forcepia* sp." *J. Nat. Prod.* **2004**, *67*, 1351-1355. The newer lasonolides differ in the C_{23} sidechain and esterification of the C_{22} hydroxymethyl group (laurate esters)

It turns out that there were some mistakes in the original structure assignment. The actual structure is shown below. Thus, synthesis determined the stereochemistry at C_{28} (relative to other centers) and corrected the assignments of olefin geometry at $C_{17\text{-}18}$ and $C_{25\text{-}26}$. It is interesting that the natural product was reported to be dextrorotatory (+), but by synthesis it was shown that biological activity actually resides in the levorotatory (-) enantiomer. We will not go through all of the syntheses of the many stereoisomers that were prepared *en route* to the first synthesis of lasonolide A, but will jump right into the first total synthesis performed by the Lee group in Korea: Lee, E.; Song, H. Y.; Kang, J. W.; Kim, E.-S.; Jung, C.-K.; Joo, J. M. "Lasonolide A: Structural Revision and Synthesis of the Unnatural (-)-Enantiomer" *J. Am. Chem. Soc.* **2002**, *124*, 384-385. For a full paper see: Song, H. Y.; Joo, J. M.; Kang. J. W.; Kim, D.-S.; Jung, C.-K.;Kwak, H. S.; Park, J. H.; Lee, E.; Hong, C. Y.; Jeong, S. W.; Jeon, K.; Park, J. H. "Lasonolide A: Structural Revision and Total Synthesis" *J. Org. Chem.* **2003**, *68*, 8080-8087. For work that indicates biological activity to reside in the (-)-enantiomer see: Lee, E.; Song, H. Y.; Joo, J. M.; Kang, J. W.; Kim, D. S.; Jung, C. K.; Hong, C. Y.; Jeong, S. W.; Jeon, K. "Synthesis of (+)-Lasonolide A: (-)-Lasonolide A is the Biologically Active Enantiomer" *Bioorg. Med. Chem. Lett.* **2002**, *12*, 3519-3520.

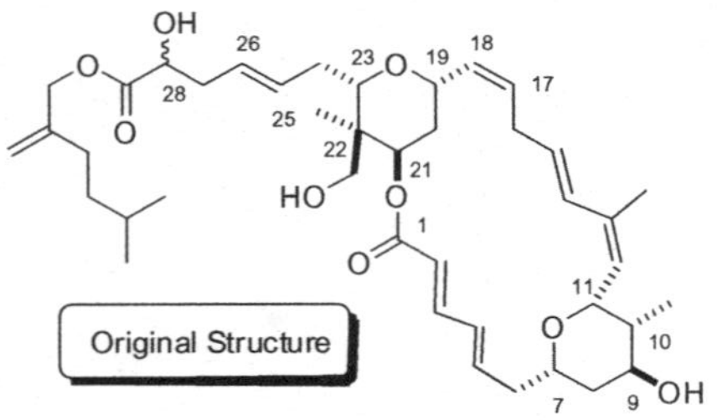

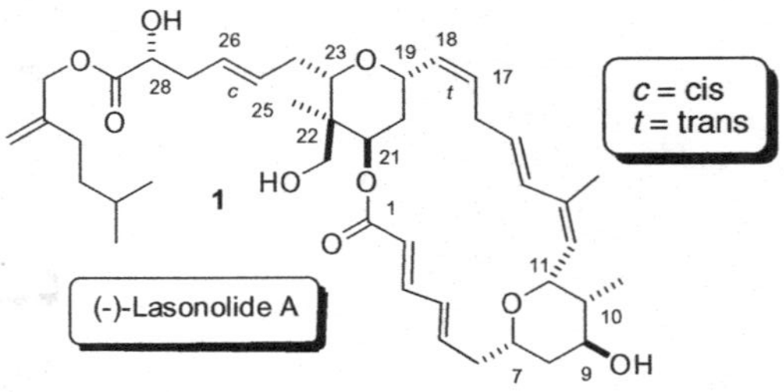

The enantiomer is 10^3 times less active in assays against several tumor cell lines.

Lasonolide-1

Lasonolide-1

The next molecule we will consider is lasonolide A (**1**). Lasonolide A provides a modern example of synthesis playing a role in natural product structure determination. There are other examples for sure, but I have selected this one because of a personal interest in the problem, and because it also is a polyolefin that will let us continue our discussion of diastereoselective olefin synthesis.

Lasonolide A (**1**) is a marine natural product produced by a sponge (or microorganisms that live on the sponge ... these situations can be difficult to differentiate). The structure was assigned based largely on NMR studies. The stereochemistry at C_{28} was not determined, nor was the absolute configuration of lasonolide A determined. The natural product showed activity against several tumor cell lines with activity (IC_{50} values) in the ng/mL region. It also inhibited cell adhesion in several cell lines. Thus, lasonolide A promptly became an attractive target for synthesis: (1) to establish the structure and (2) to provide more material for biological evaluation. Since the isolation of lasonolide A, the "lasonolides" have grown into a small *family* of natural products. The newer lasonolides differ in the C_{23} sidechain, and by virtue of esterification of the C_{22} hydroxymethyl group (laurate esters).

It turns out that there were some mistakes in the original structure assignment. The actual structure is shown in Lasonolide-1 along with the original structure. Thus, synthesis determined the stereochemistry at C_{28} (relative to other centers) and corrected the assignments of olefin geometry at C_{17}-C_{18} and C_{25}-C_{26}. It is interesting that the natural product was reported to be dextrorotatory (+), but by synthesis it was shown that biological activity actually resides in the levorotatory (−) enantiomer. Thus there is still some confusion regarding the absolute stereochemical details of the natural product (see Gephyrotoxin-2 in Chapter 9 for another example of this problem). We will not go through all of the syntheses of the many stereoisomers that were prepared *en route* to the first synthesis of lasonolide A, but will jump right into the first total synthesis performed by the Lee group at Seoul National University in Korea.

First Synthesis of (-)-Lasonolide A

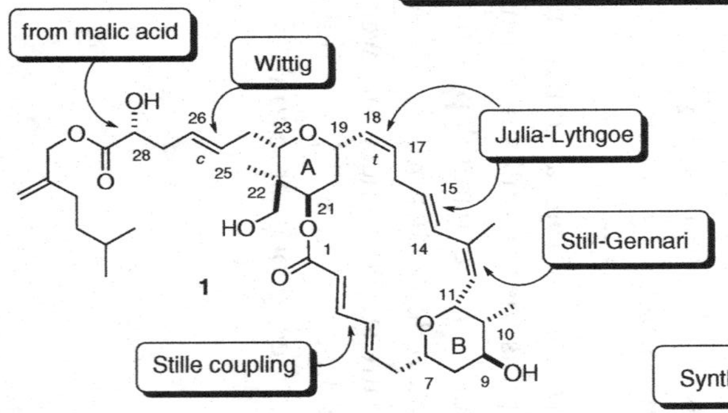

Outline of Plan

1. Convergent approach calling for three pieces: the A-ring THP, the B-ring THP, and the hydroxyester side chain.

2. Connect the two THP pieces by (a) Julia-Lythgoe olefin synthesis followed by (b) intramolecular Stille coupling.

3. Add the hydroxyester side chain by a Wittig reaction.

Synthesis of Upper THP

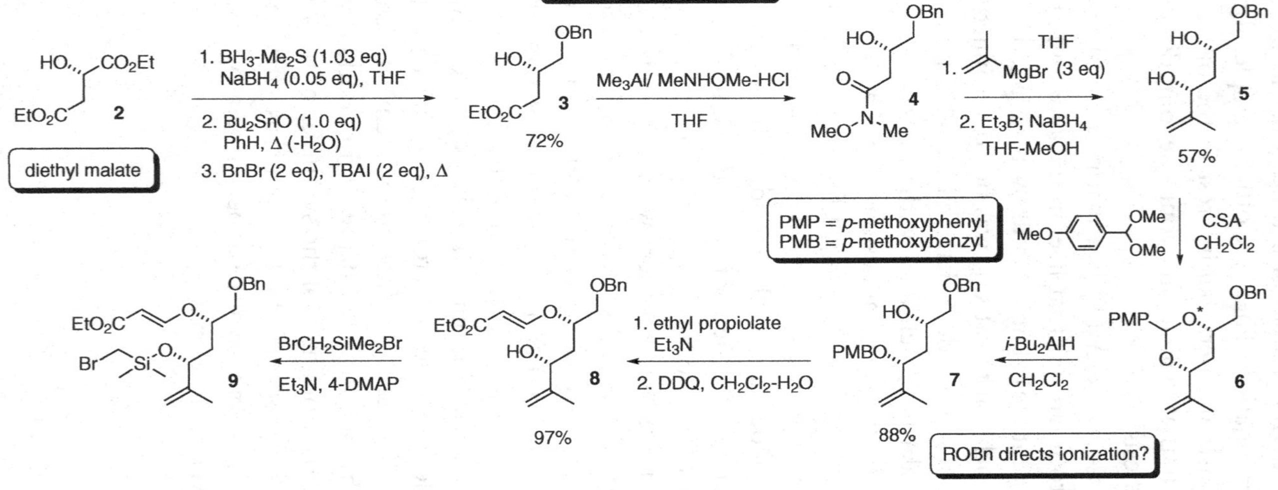

Lasonolide-2

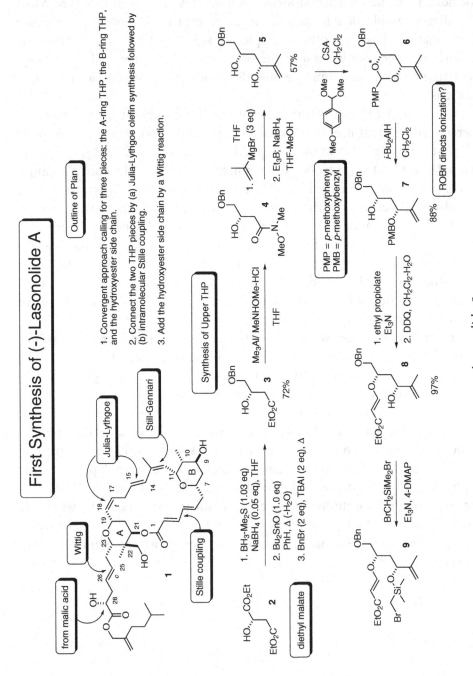

Lasonolide-2

Lasonolide A (**1**) has 9 stereogenic centers. Eight of these appear in two "clusters", the tetrahydropyran (THP) rings labelled A and B. The ninth stereogenic center is isolated from the THP rings and appears at C_{28}. Thus, lasonolide A demands a convergent strategy in which the absolute stereochemistry is set in pieces, and then the pieces are assembled. The three pieces selected by Lee appear on the next few pages. They include the A-ring (**18**), the B-ring (**31**) and the C_{28}-containing side chain (**37**). You will notice that these pieces lack C_1, C_2 and C_3. The selection of these pieces dictated the manner in which they were to be assembled. A Julia-Lythgoe-Kocienski olefin synthesis was to be used to connect the A- and B-ring pieces (construction of the C_{14}-C_{15} olefin), and an intramolecular Stille coupling was to be used to close the macrocyclic lactone, after introduction of C_1-C_3.[1,2] Finally, the hydroxyester side chain was to be introduced using a Wittig reaction to establish the C_{25}-C_{26} olefin.

A few other points are of interest before looking at the synthesis. Lasonolide A is an example of a family of natural products known as macrolides (macrocyclic lactones). We will see another example of such a natural product in Chapter 14. Whereas macrolides can often be assembled using a macrolactonization (intramolecular esterification) to close the ring, this approach was not followed by the Lee group, perhaps because this portion of the target looks sterically congested. What is the biosynthetic origin of lasonolide A? One can speculate that it is derived largely from "acetate". There are a large number of natural products that are "polyacetates" in origin. They are derived from iterative condensation reactions (Claisen condensations) between acetic acid derivatives (for example acetyl coenzyme-A units) followed by adjustment of oxidation states along the chain. Fatty acids are biosynthesized in this manner. A "signature" of natural products that arise from this biosynthetic pathway is the appearance of odd (1,3 or 1,5 or 1,7) difunctional relationships within the molecule (usually between oxygens). In the case of lasonolide A, oxygens appear at C_1, C_7, C_9, C_{11}, C_{19}, C_{21} and C_{23} (with a few extras added). We already have seen that such relationships can be constructed using carbonyl addition chemistry (Claisen and aldol condensations) (Chapter 6) and we will see that this bond construction plays a minor role in Lee's synthesis of the B-ring. The A- and B-rings of lasonolide A have similar substitution patterns (2,3,4,6-tetrasubstituted THPs). If you remove the C_{10} substitutent from the B-ring and the C_{22} substitutents from the A-ring, they have the same substitution pattern, including relative stereochemistry. Therefore, it is not surprising that the same strategy was adopted to prepare both THP rings. Finally, the strategy used to prepare the THP rings involved free radical cyclization chemistry (look ahead to **9** → **11** and **25** to **26**) originally

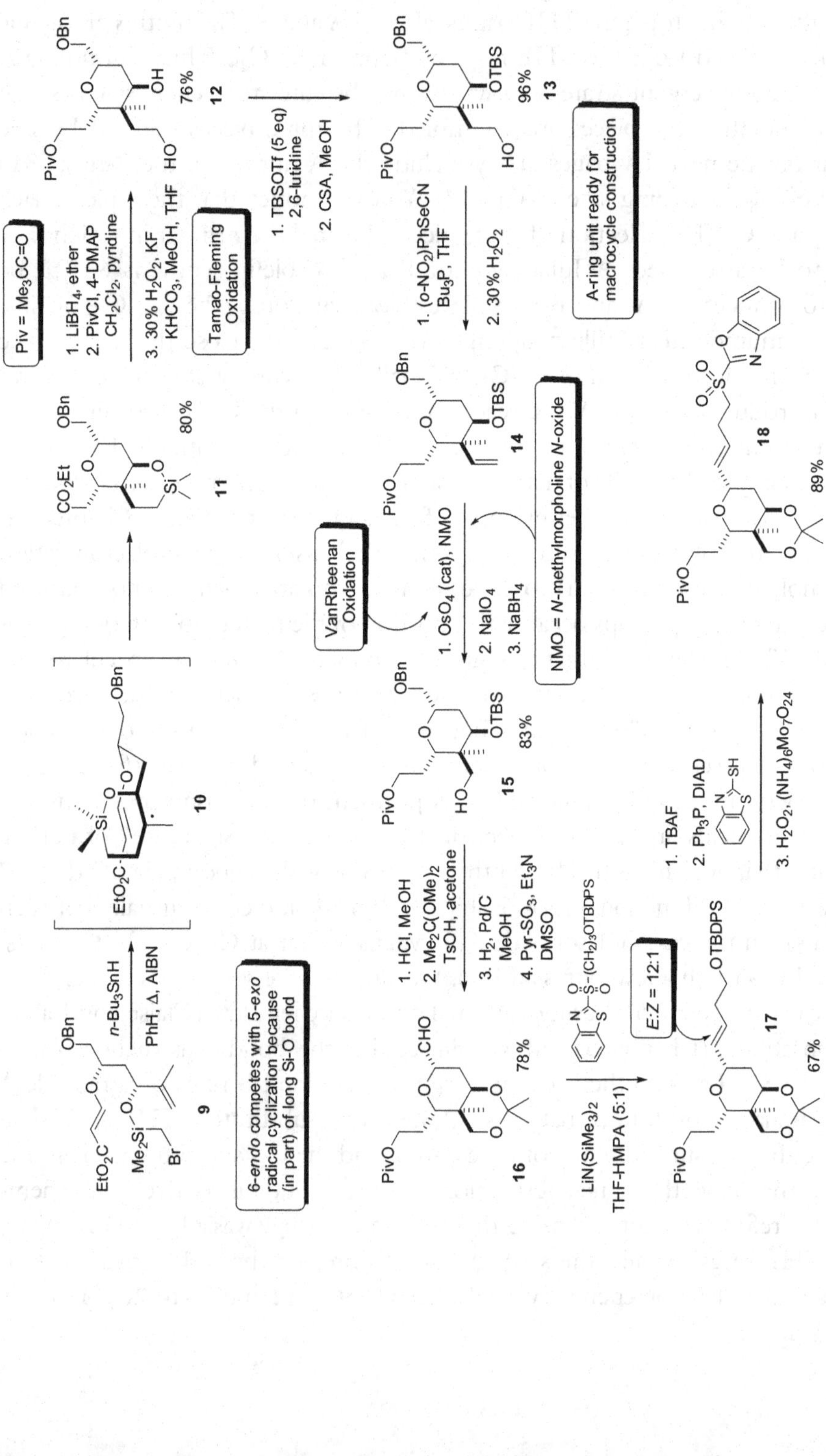

Lasonolide-3

developed for other purposes by the Lee group. Thus, this approach also was dictated by the practitioner's interest in developing and testing methodology. Let's move to the synthesis of the A-ring (**18** via **9**).

Recall that this strategy (and molecule) demands an enantioselective approach. The "chiral pool" approach was taken and the synthesis began with diethyl malate (**2**) (malic acid is an inexpensive dicarboxylic acid that can be isolated from apples, among other sources). Reduction of the ester proximal to the hydroxyl group gave a diol. Conversion of the diol to the corresponding *bis*-tin alkoxide, and alkylation of the less hindered primary alkoxide, gave **3**. Treatment of **3** with the dimethylaluminum amide, derived from *N,O*-bis(dimethyl) hydroxylamine, gave **4**, which reacted with isopropenylmagnesium bromide to afford a *β*-hydroxy ketone. These two reactions constitute a nice way of converting an ester to a ketone. Both transformations were developed by the Weinreb group (now at Pennsylvania State University).[3,4] Chelation controlled reduction of the *β*-hydroxy ketone provided diol **5** with good stereoselectivity.[5] Acetal formation, followed by reductive cleavage of **6** yielded **7**. The regioselectivity of this reaction might be controlled by direction, by the benzyloxy group, of the Lewis acidic aluminum to the oxygen of the acetal marked with an asterisk. Treatment of **7** with ethyl propiolate gave vinylogous carbonate **8**, and derivatization of the allylic alcohol provided **9**. This set the stage for the featured free radical cyclization.

Lasonolide-3

Treatment of **9** with tri-*n*-butyltin hydride gave THP **11**, presumably via sequential free radical cyclizations. The first cyclization follows a 6-*endo* course to provide **10**, which undergoes a 6-*exo* cyclization via a chair-like conformation to ultimately provide **11** with the proper stereochemistry at the two newly formed stereogenic centers. Reduction of the ester, protection of the resulting primary alcohol, and a Tamao-Fleming oxidation, converted **11** to **12**.[6,7] The hydroxyethyl group was then degraded to a hydroxymethyl. This was accomplished using the Grieco-Sharpless procedure for formal dehydration of alcohols (**13** → **14**) followed by oxidative degradation of the vinyl group to an aldehyde and subsequent reduction to alcohol **15**. Removal of the TBS protecting group was followed by formation of an acetonide, hydrogenolysis of the benzyl group, and oxidation of the resulting primary alcohol to provide aldehyde **16**. Application of a variation of the Julia-Lythgoe-Kocienski olefin synthesis gave **17** with good control of olefin geometry.[8] The synthesis of **18** was completed by removal of the silicon protecting group, displacement of the primary alcohol, and oxidation of the resulting sulfide to the sulfone oxidation state.

Synthesis of B-Ring

The substitution pattern of the B-ring is similar to that of the A-ring and thus, it is not surprising that the syntheses have a common flavor.

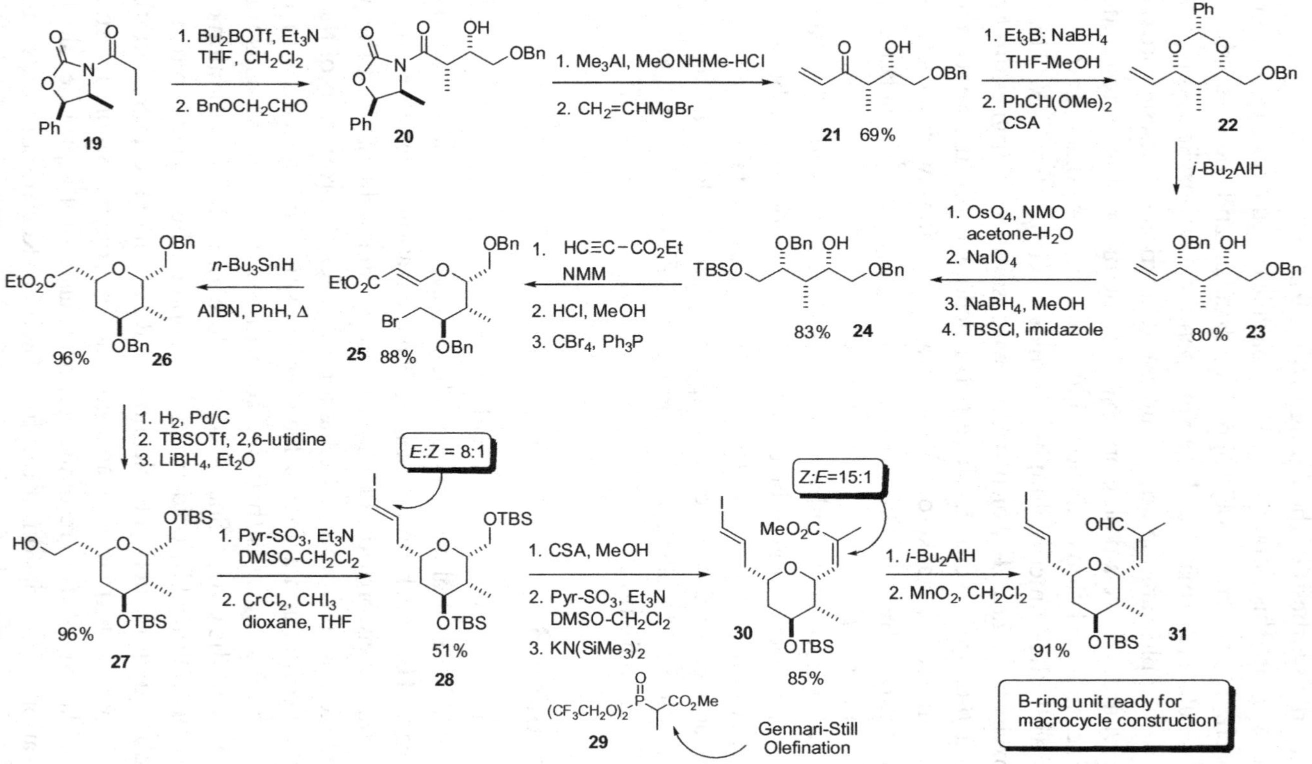

Lasonolide-4

Lasonolide-4

The substitution pattern of the B-ring is similar to that of the A-ring and thus, it is not surprising that the syntheses have a common flavor. The synthesis of the B-ring via a free radical cyclization called for the preparation of vinylogous carbonate **25**. The starting point was a diastereoselective aldol-type condensation reaction between the boron enolate of imide **19** and 2-benzyloxyacetaldehyde. The conversion of **20** to **24** followed tactics closely related to the conversion of **3** to **7** (Lasonolide-2). Alcohol **24** was transformed to the corresponding vinylogous carbonate. Hydrolysis of the TBS protecting group and treatment of the resulting alcohol with triphenylphosphine-bromine gave primary bromide **25**. The key free radical cyclization proceeded smoothly to give **26** with surprisingly good stereoselectivity.

The benzyl protecting group was removed by hydrogenolysis and replaced with a TBS group. Lithium borohydride was then used to reduce the ester to primary alcohol **27**. Oxidation of the alcohol gave an aldehyde, which reacted with chromyl chloride and iodoform to provide vinyl iodide **28** along with a small amount of the *Z*-isomer. The TBS protecting group was removed with acid, and another Moffatt-type oxidation gave the corresponding aldehyde.[9] The next stage of the synthesis called for introduction of the C_{12}-C_{13} olefin. This was accomplished using the Gennari-Still modification of the Horner-Wadworth-Emmons (HWE) reaction.[10] The normal HWE reaction of phosphonoacetates with aldehydes employs dimethyl or diethyl phosphonate anions.[11] The resulting olefins have largely *E*-geometry. This is presumably a consequence of the greater thermodynamic stability of such olefins relative to their *Z*-counterparts (regardless of whether the olefin is di- or tri-substituted. The Gennari-Still modification employs 2,2,2-trifluoroethoxy groups (for example **29**) and provides largely *Z*-olefins (for example **30**). Although the reasons for this change in stereoselectivity are not well established, it is possible that the rate determining-step of this process is changed such that olefin geometry is set in the C–C bond-forming step. In other words, the carbonyl addition may normally be reversible and elimination rate-determining, while in the Gennari-Still method, the addition may be rate-determining due to the electron-withdrawing nature of the trifluoroethoxy groups. Regardless, this HWE modification is a widely used method for the synthesis of Z-α,β-unsaturated esters. Moving forward, a reduction-oxidation sequence converted **30** to the desired B-ring unit **31**.

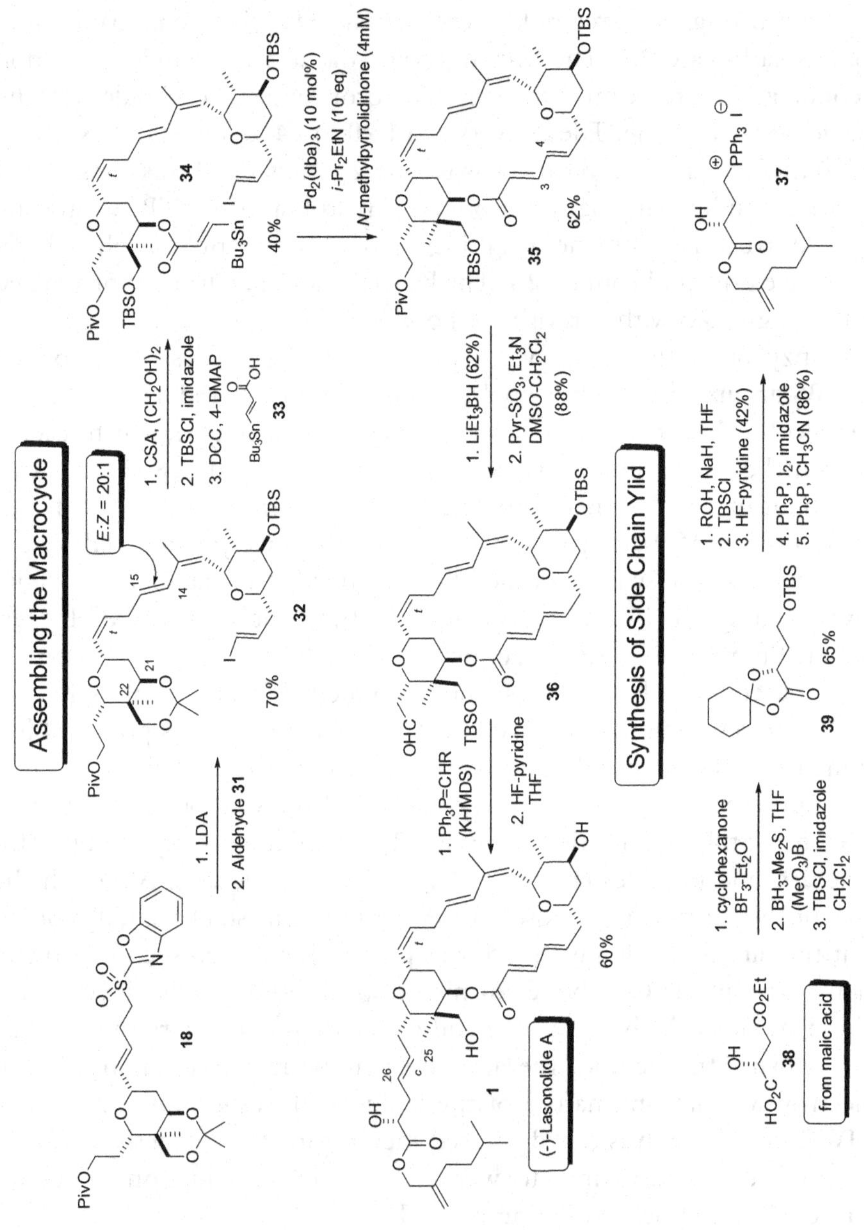

Lasonolide-5

Lasonolide-5

Sulfone **18** and aldehyde **31** were coupled using a Julia-Lythgoe-Kocienski reaction to provide **32** with good stereoselectivity. The acetal was removed in an exchange reaction with ethylene glycol, the primary C_{22} hydroxymethyl group was protected, and the secondary C_{21} alcohol was acylated using acrylic acid derivative **33**. An intramolecular Stille reaction of the resulting iodostannane (**34**) provided 20-membered ring lactone **35**.[2]

The pivalate was reductively removed at C_{25} using lithium triethylborohydride. It is notable that the lactone was not disturbed, perhaps an indication of the aforementioned steric congestion. Oxidation of the primary alcohol then gave aldehyde **36**. A Wittig reaction with the phosphorane derived from **37** gave the *Z*-olefin and removal of the C_9 protecting group completed the synthesis of lasonolide-A (**1**).

The starting material for the synthesis of **37** was **38**, prepared in turn from malic acid, both enantiomers of which are commercially available. The choice of this starting material facilitated determination of C_{28} stereochemistry as both enantiomers could be used in this coupling reaction. Note that this Wittig reaction provided largely the *Z*-olefin geometry as expected (recall introduction of the "upper" prostaglandin side chain via a similar reaction).

Second Synthesis of Lasonolide A

Kang, S. H.; Kang. S. Y.; Kim, C. M.; Choi, H.-W.; Jun, H.-S.; Lee, B. M.; Park, C. M.; Jeong, J. W. "Total Synthesis of Natural (+)-Lasonolide A" *Angew. Chem. Int. Ed.* **2003**, *42*, 4779. This synthesis follows the same bond disconnections (for assembling the macrocycle) with the exception that the final ring closure was to be accomplished by an intramolecular HWE reaction for construction of the $C_{2\text{-}3}$ bond. The details for construction of the A-ring and B-ring also differ and the synthesis is shorter. For the full paper see: Kang, S. H.; Kang. S. Y.; Choi, H.-W.; Kim, C. M.; Jun, H.-S.; Youn, J.-H.; "Stereoselective Total Synthesis of the Natural (+)-Lasonolide A" *Synthesis* **2004**, 1102-1114.

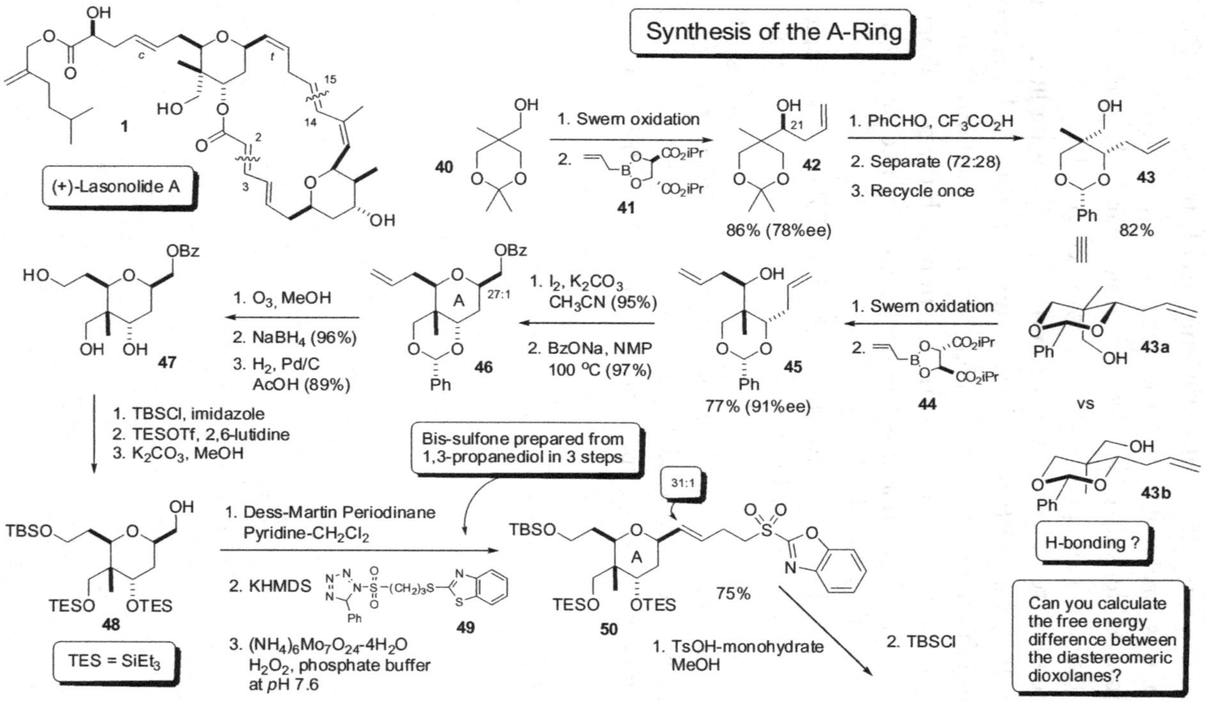

Lasonolide-6

We will examine two more syntheses of lasonolide A (**1**). The Kang group (KAIST in Korea) decided to assemble the macrocycle by a route that used pieces **51** (A-ring), **59** (B-ring) and the C_{28} triethylsilyl ether of phosphonium salt **37**. The C_{14}-C_{15} bond was constructed using Julia-Lythgoe chemistry and the macrocycle was closed using an intramolecular HWE reaction to set the C_2-C_3 olefin. The synthesis uses an effective reaction sequence for introducing β-hydroxyaldehyde units on three occasions: (1) addition of an allyl group to an aldehyde followed by (2) oxidative cleavage of the terminal olefin to reveal the aldehyde carbonyl group. The construction of such units is important in the broad field of polyacetate-derived natural products synthesis (and a modification is important to the related field of polypropionate synthesis).

Synthesis of the A-ring THP (**51**) began with acetal **40**. The primary alcohol was oxidized using the Swern conditions (another Moffat-type oxidation)[12] and the resulting aldehyde was reacted with chiral allylic boronate **41**. This type of reagent was first developed in the laboratories of Bill Roush for the asymmetric allylation of aldehydes.[13] In the current case, the yield was excellent and an 89:11 mixture of diastereomers (at the incipient C_{21}) was obtained. Treatment of **42** with benzaldehyde in the presence of acid gave **43a** along with diastereomeric acetal **43b**. Separation of the mixture, and recycle of the minor acetal (**43b**), provided an 82% overall yield of **43**. It is possible that intramolecular hydrogen bonding renders **43a** more stable than **43b**. Oxidation of the primary hydroxyl group of **43a**, followed by another aldehyde allylation, gave **45**. The allylation reagent (**44**) is the enantiomer of the reagent used in the conversion of **40** to **42**. The two reagents (**41** and **44**) provide opposite absolute stereochemistry at the newly formed stereogenic center, a nice example of reagent-controlled asymmetric synthesis (see Prostaglandins-12). Treatment of **45** with iodine (iodoetherification) was followed by a nucleophilic substitution reaction to provide benzoate **46**.[14] The double bond was cleaved, the resulting aldehyde reduced, and the acetal was removed by hydrogenolysis to provide **47**. The least hindered primary alcohol was protected as a TBS ether, the remaining alcohols were protected as TES ethers, and the benzoate was hydrolyzed to give **48**. Oxidation of the alcohol was followed by a Julia-Lythgoe-Kocienski type reaction using sulfone **49**, and the 2-thiobenzimidaxole was "armed" for the next Julia-Lythgoe-Kocienski by oxidation to the corresponding sulfone **50**. Removal of all of the protecting groups and reprotection of the two primary alcohols gave A-ring substrate **51** (Lasonolide-7).

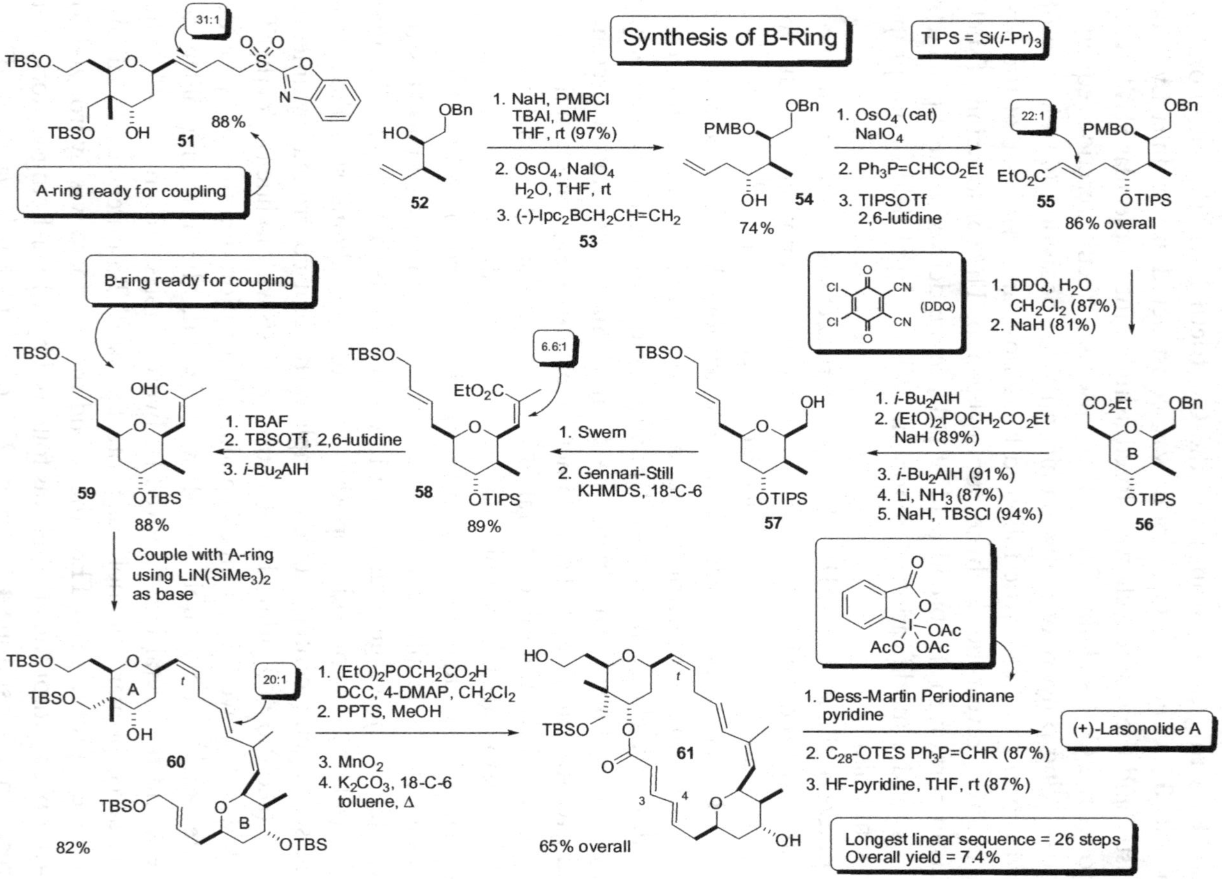

Lasonolide-7

Lasonolide-7

The B-ring THP (**59**) was prepared starting with known alcohol **52**. Protection of the secondary alcohol, cleavage of the olefin, and allylation of the resulting aldehyde with a reagent developed by the Brown group at Purdue (**53**) gave **54**.[15] Ozonolysis of the double bond was followed by a Wittig reaction and protection of the secondary alcohol to provide **55**. Oxidative removal of the *p*-methoxybenzyl protecting group, and an intramolecular conjugate addition formed B-ring THP **56**, which was converted to **59** using a standard series of reactions.

The A-ring and B-ring THPs were coupled to provide **60**, using a Julia-Lythgoe-Kocienski reaction. The C_{21} alcohol was esterfied, and three of the four TBS groups (presumably the least hindered groups) were removed. Oxidation of the allylic alcohol using activated manganese dioxide was followed by the intramolecular HWE reaction to give macrocycle **61**. Oxidation of the primary alcohol using the Dess-Martin periodinane gave the expected aldehyde. A Wittig reaction and deprotection of the silicon protecting groups (TBS and TES) completed the synthesis of lasonolide-A (**1**). This was quite an effective synthesis which illustrates, once again, the importance of convergence in planning a synthesis.

Third Synthesis of Lasonolide A

Yoshimura, T.; Yakushiji, F.; Kondo, S.; Wu, X.; Shindo, M.; Shishido, K. "Total Synthesis of (+)-Lasonolide A" *Org. Lett.* **2006**, *8*, 475-478.

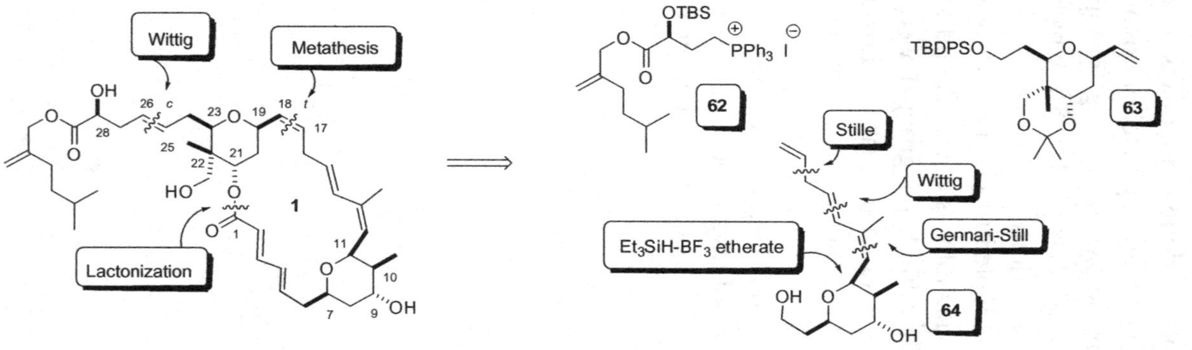

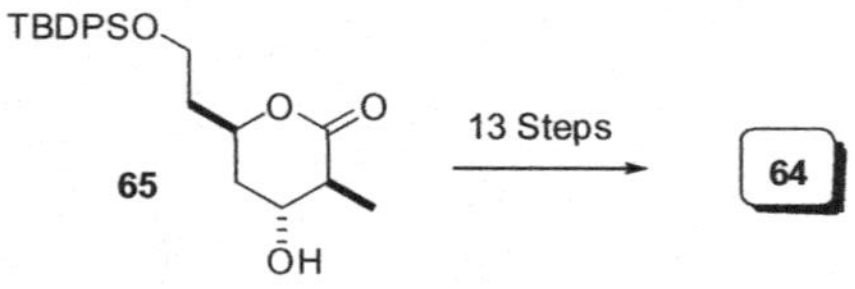

Deba, T.; Yakushiji, F.; Shindo, M.; Shishido, K. *Synlett* **2003**, 1500

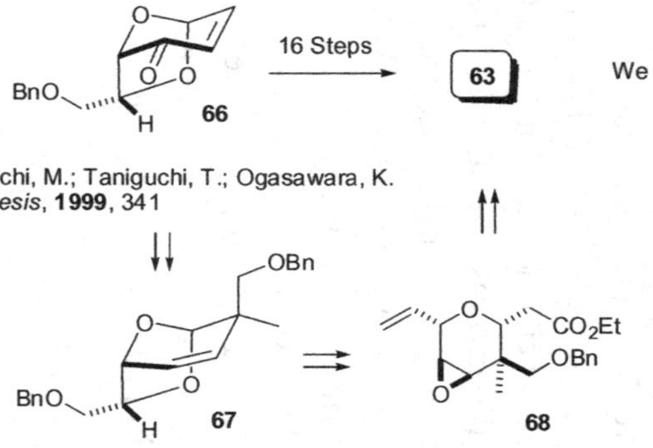

We have seen **62** before.

Takeuchi, M.; Taniguchi, T.; Ogasawara, K. *Synthesis*, **1999**, 341

There have been a number of other partial syntheses (and one formal synthesis) published. For a very interesting approach to the A-ring see: Dalgard, J. E.; Rychnovsky, S. D. "Enantioselective Synthesis of the C_{18}-C_{25} Segment of Lasonolide A by an Oxonia-Cope Prins Cascade" *Org. Lett.* **2005**, *7*, 1589-1592.

Lasonolide-8

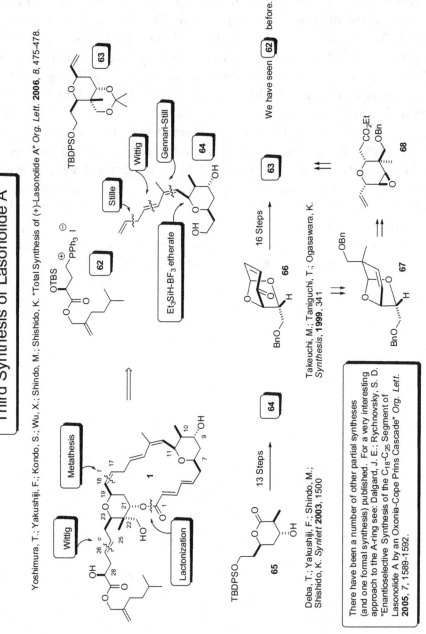

Lasonolide-8

We will look at a third synthesis of lasonolide-A in abbreviated form. The Shishido plan (University of Tokushima in Japan) was to assemble subunits **62–64**. The side chain was to be introduced via a Wittig reaction in the usual manner. The A-ring THP (**63**) and B-ring THP (**64**) were to be joined using a crossed-olefin metathesis reaction to construct the C_{17}-C_{18} double bond, and a lactonization to close the macrocycle. The olefins in **64** were to be installed sequentially using Gennari-Still, Wittig (stabilized phosphorane) and Stille coupling reactions. The C_{11} stereochemistry in **64** was to be set in an ionic reduction of a C_{11} hemiacetal.

Briefly, B-ring THP **64** was prepared from **65** (which took several steps to prepare) using a 13 reaction sequence. A-ring THP **63** was prepared in 16 steps from known compound **66** via key intermediates **67** and **68**.

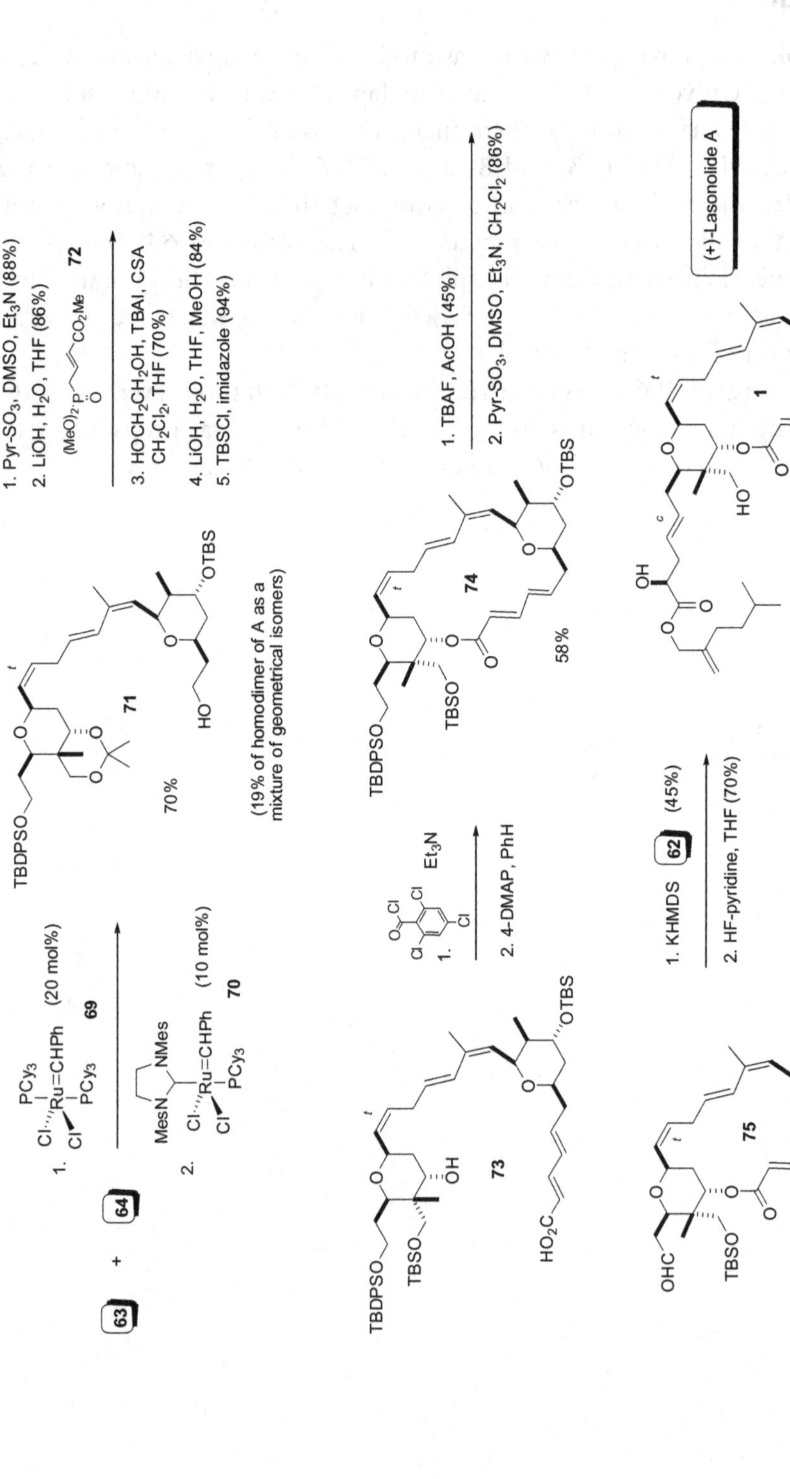

Lasonolide-9

Lasonolide-9

Sequential treatment of what is presumed to be an equimolar mixture of **63** and **64** with Grubb's olefin metathesis catalysts **69** and **70**, gave a good yield of **71** along with lesser amounts of the dimer of **64** (derived from the terminal olefin). The olefin geometry of the newly formed double bond was clearly *E* or *trans*. The primary alcohol was oxidized to the aldehyde and a vinylogous HWE reaction using phosphonate **72** introduced the C_5-C_6 double bond. An acetal exchange reaction, hydrolysis of the methyl ester, and protection of the C_{22} hydroxymethyl group gave hydroxy acid **73**. Activation of the acid by formation of a mixed anhydride, and cyclization using 4-dimethylaminopyridine as a catalyst (the Yamaguchi procedure) gave **74** in 58% overall yield from **71**.[16] Removal of the sterically most accessible silyl ether protecting group and subsequent oxidation of the primary alcohol gave aldehyde **75**. The now familiar Wittig reaction and global deprotection provided lasonolide A (**1**).

In this chapter you have been introduced to lasonolide A (**1**) as both a *polyacetate*-derived natural product and as a *macrolide*. The next chapter will focus on a biosynthetically related *polypropionate*-derived natural product, and introduce a family of natural products known as *ionophores*. The final chapter will deal with a famous *polypropionate-derived macrolide*. The targets selected to represent ionophores and polypropionate-derived macrolides have both been selected for historical reasons. Let's move on.

References

1. Julia, M.; Paris, J. M. "Syntheses a L'aide de Sulfones. V. Methode de Syntheses Generale de Doubles Liasons" *Tetrahedron Lett.* **1973**, *15*, 4833–4836. Kocienski, P. J.; Lythgoe, B.; Ruston, S. "Scope and Stereochemistry of an Olefin Synthesis from β-Hydroxy-Sulfones" *J. Chem. Soc., Perkin Trans. 1* **1978**, 829–834.
2. Farina, V.; Krishnamurthy, V.; Scott, W. J. "The Stille Reaction" *Organic Reactions* **1997**, *50*, 1–652. Pattenden, G.; Sinclair, D. J. "The Intramolecular Stille Reaction in Some Target Natural Product Syntheses" *J. Organomet. Chem.* **2002**, *653*, 261–268.
3. Lipton, M. F.; Basha, A.; Weinreb, S. M. "Conversion of Esters to Amides with Dimethylaluminum Amides: *N,N*-Dimethylcyclohexanecarboxamide" *Organic Syntheses* **1980**, *59*, 49–53. Basha, A.; Lipton, M.; Weinreb, S. M. "A Mild, General Method for Conversion of Esters to Amides" *Tetrahedron Lett.* **1977**, *18*, 4171–4174.
4. Nahm, S.; Weinreb, S. M. "*N*-Methoxy-*N*-methylamides as Effective Acylating Agents" *Tetrahedron Lett.* **1981**, *22*, 3815–3818.
5. Still, W. C.; McDonald, J. H., III "Chelation-Controlled Nucleophilic Additions. 1. A Highly Effective System for Asymmetric Induction in the Reaction of Organometallics with α-Alkoxy Ketones" *Tetrahedron Lett.* **1980**, *21*, 1031–1034. Still, W. C.; Schneider, J. A. "Chelation-Controlled Nucleophilic Additions. 2. A Highly Effective System for Asymmetric Induction in the Reaction of Organometallics with β-Alkoxy Aldehydes" *Tetrahedron Lett.* **1980**, *21*, 1035–1038.
6. Tamao, K.; Ishida, N.; Kumada, M. "(Diisopropoxymethylsilyl)methyl Grignard Reagent: A New, Practically Useful Nucleophilic Hydroxymethylating Agent" *J. Org. Chem.* **1983**, *48*, 2120–2122. Tamao, K.; Kakui, T.; Akita, M.; Iwahara, T.; Kanatani, R.; Yoshida, J.; Kumada, M. "Oxidative Cleavage of Silicon-Carbon Bonds in Organosilicon Fluorides to Alcohols" *Tetrahedron* **1983**, *39*, 983–990.
7. Fleming, I. "Silyl-to-Hydroxy Conversion in Organic Synthesis" *Chemtracts: Organic Chemistry* **1996**, *9*, 1–64.
8. Blakemore, P. R.; Cole, W. J.; Kocienski, P. J.; Morley, A. "A Stereoselective Synthesis of *trans*-1,2-Disubstituted Alkenes Based on the Condensation of Aldehydes with Metalated 1-Phenyl-1H-tetrazol-5-yl Sulfones" *Synlett* **1998**, 26–28. Blakemore, P. R. "The Modified Julia Olefination: Alkene Synthesis via the Condensation of Metallated Heteroarylalkylsulfones with Carbonyl Compounds" *J. Chem. Soc., Perkin Trans. 1* **2002**, 2563–2585.
9. Parikh, J. R.; Doering, W. v. E. "Sulfur Trioxide in the Oxidation of Alcohols by Dimethyl Sulfoxide" *J. Am. Chem. Soc.* **1967**, *89*, 5505–5507. Pfitzner, K. E.;

Moffatt, J. G. "Sulfoxide-Carbodiimide Reactions. I. A Facile Oxidation of Alcohols" *J. Am. Chem. Soc.* **1965**, *87*, 5661–5670. Pfitzner, K. E.; Moffatt, J. G.; "Sulfoxide-Carbodiimide Reactions. II. Scope of the Oxidation Reaction" *J. Am. Chem. Soc.* **1965**, *87*, 5670–5678.

10. Still, W. C.; Gennari, C. "Direct Synthesis of Z-Unsaturated Esters. A Useful Modification of the Horner-Emmons Olefination" *Tetrahedron Lett.* **1983**, *24*, 4405–4408.
11. Maercker, A. "The Wittig Reaction" *Organic Reactions* **1965**, *14*, 270–490. Maryanoff, B. E.; Reitz, A. B. "The Wittig Olefination Reaction and Modifications Involving Phosphoryl-Stabilized Carbanions. Stereochemistry, Mechanism, and Selected Synthetic Aspects" *Chemical Reviews* **1989**, *89*, 863–927. Blanchette, M. A.; Choy, W.; Davis, J. T.; Essenfeld, A. P.; Masamune, S. Roush, W. R.; Sakai, T. "Horner-Wadsworth-Emmons Reaction: Use of Lithium Chloride and an Amine for Base-Sensitive Compounds" *Tetrahedron Lett.* **1984**, *25*, 2183–2186.
12. Mancuso, A. J.; Swern, D. "Activated Dimethyl Sulfoxide: Useful Reagents for Synthesis" *Synthesis* **1981**, 165–185.
13. Roush, W. R.; Halterman, R. L. "Diisopropyl Tartrate Modified (*E*)-Crotylboronates: Highly Enantioselective Propionate (*E*)-Enolate Equivalents" *J. Am. Chem. Soc.* **1986**, *108*, 294–296. Roush, W. R.; Ando, K.; Powers, D. B.; Palkowitz, A. D.; Halterman, R. L. "Asymmetric Synthesis Using Diisopropyl Tartrate Modified (*E*)- and (*Z*)-Crotylboronates: Preparation of the Chiral Crotylboronates and Reactions with Achiral Aldehydes" *J. Am. Chem. Soc.* **1990**, *112*, 6339–6348.
14. Bartlett, P. A. "Olefin Cyclization Processes that Form Carbon-Heteroatom Bonds" in *Asymmetric Synthesis* **1984**, *3*, 411–454. Bartlett, P. A. "Olefin Cyclization Processes that Form Carbon-Carbon Bonds" in *Asymmetric Synthesis* **1984**, *3*, 341–409.
15. Brown, H. C.; Bhat, K. S.; Randad, R. S. "Chiral Synthesis via Organoboranes. 21. Allyl- and Crotylboration of α-Chiral Aldehydes with Diisopinocampheylboron as the Chiral Auxiliary" *J. Org. Chem.* **1989**, *54*, 1570–1576.
16. Inanaga, J.; Hirata, K.; Saeki, H.; Katsuki, T.; Yamaguchi, M. "A Rapid Esterification by Mixed Anhydride and its Application to Large-Ring Lactonization" *Bull. Chem. Soc. Jpn.* **1979**, *52*, 1989–1993.

Problems

1. Reactions of esters with an excess of a Grignard reagent, followed by a protic work-up, give tertiary alcohols. However, the reaction of **4** with excess Grignard, followed by protic work-up, affords a ketone. Provide a mechanistic explanation for this observation. Provide a mechanistic explanation for the following related transformation. (Lasonolide-2)

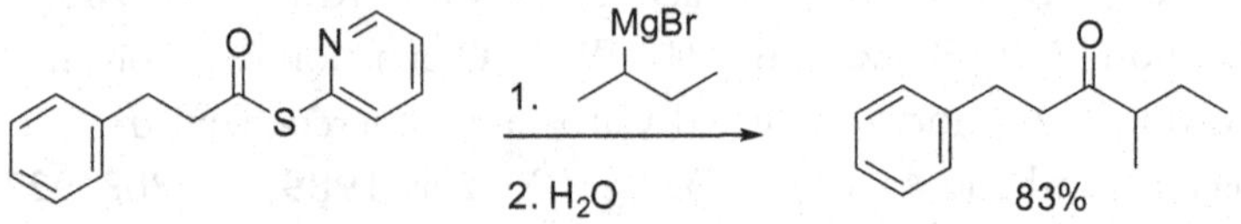

Mukaiyama, T.; Araki, M.; Takei, H. "Reactions of S-(2-Pyridyl)thioates with Grignard Reagents. Convenient method for the Preparation of Ketones" *J. Am. Chem. Soc.* **1973**, *95*, 4763–4765.

2. Cyclization of the 5-methyl-5-hexenyl radical (**1**) provides a 24:1 mixture of **II** and **III**. Explain the reversed regioselectivity in the cyclization of **9** → **10**. (Lasonolide-3)

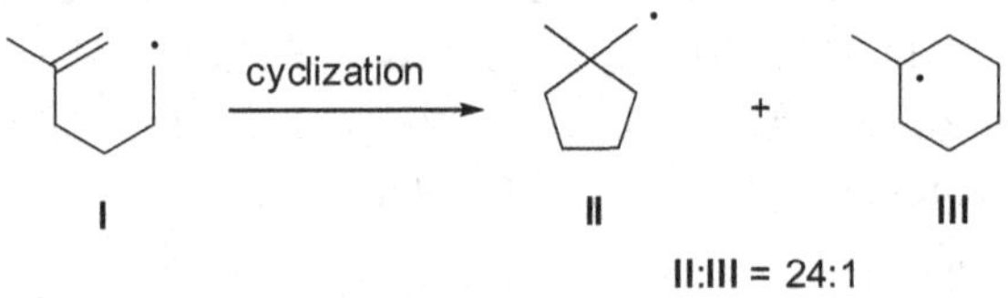

Walling, C.; Cioffari, A. "Cyclization of 5-Hexenyl Radicals" *J. Am. Chem. Soc.* **1972**, *94*, 6059–6064.

3. Provide a mechanism for the oxidation of the C-Si bond (to a C-O bond) in Tamao-Fleming reaction. (Lasonolide-3)
4. Provide the structures of the product after each step *en route* from **13** → **16**. (Lasonolide-3)
5. The conversion of **16** → **17** involves metallation of the sulfone, addition of the resulting anion to the aldehyde, intramolecular transfer of the benzothiazole to the resulting alkoxide, and an elimination reaction (or fragmentation reaction) to provide the product. Provide the mechanistic details for this transformation and speculate on the origin of stereoselectivity in the process. (Lasonolide-3)
6. Propose a transition state geometry for the cyclization of **25** → **26**. Compare this with alternative cyclization geometries and suggest why the reaction was stereoselective. (Lasonolide-4)

7. Propose a mechanism for the oxidation of **27** to the corresponding aldehyde. (Lasonolide-4)
8. Provide equations that describe the mechanistic discussion of the HWE and Gennari-Still reactions presented here. (Lasonolide-4)
9. Provide the structure of each product in the reaction sequence leading from **38** → **37**. (Lasonolide-5)
10. Assume that the partitioning between **43a** and **43b** is controlled by thermodynamics (all reactions reversible). Assume that the reactions (**42** or **43b** → **43a** + **43b**) are run at room temperature (25 °C). Estimate the free energy difference between **43a** and **43b**.
11. Provide an explanation for the diastereoselectivity (27:1) observed in the conversion of **45** → **46**. (Lasonolide-6)
12. Alcohol **52** was prepared by asymmetric "crotylation" of 2-benzyloxyacetaldehde with the "crotyl derivative" of **53**. What is the structure of this reagent? Provide a mechanistic rationale for the diastereoselectivity of the following reaction (the product is a racemic mixture). (Lasonolide-7)

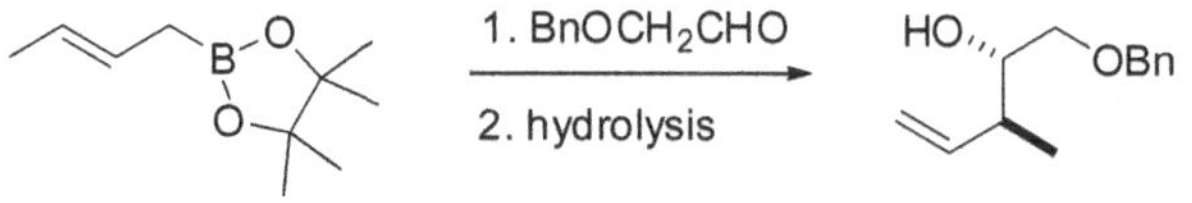

13. Discuss the stereochemical issues involved in the stereoselective closure of **55** to B-ring THP **56**. Discuss the relationship of this reaction to the conversion of **25** → **26** in the Lee synthesis (Lasonolide-7)
14. Provide the structure of the product obtained in each reaction in the conversion of **56** → **59**. (Lasonolide-7)
15. Develop a plan for the synthesis of **65** from simple starting materials. Do not worry about stereochemistry at first, but eventually discuss the stereochemical issues you will have to face given your synthetic strategy. (Lasonolide-8)
16. Outline tactics that would accomplish the following transformation. (Lasonolide-8)

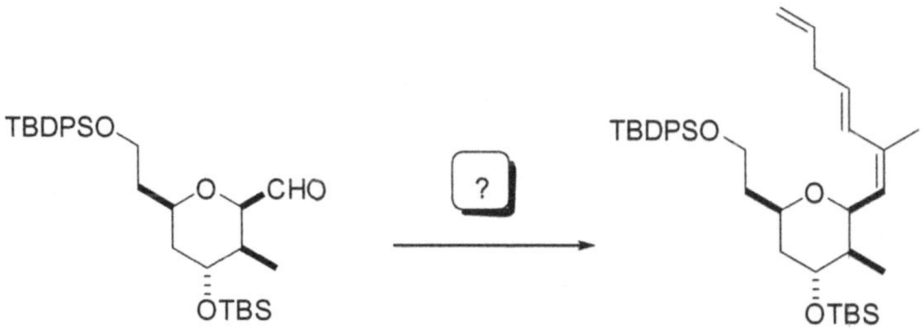

CHAPTER 13

Ionophores: Calcimycin

Ionophores: Calcimycin (A-23187)

Ionophores constitute a large family of natural products. These compounds complex ions and thus, can have interesting pharmacological properties. Three examples of ionophores are shown below. Calcimycin (originally known as A-23187) and monensin are examples of polyether ionophores. They complex metal ions much as do crown ethers. The ligating atoms in both ionophores are largely oxygen atoms. Enterobactin is a siderophore (iron complexing agent) in which the three catechols provide six oxygen ligands for metal complexation. We will focus on the synthesis of calcymicin.

Both calcimycin and monensin contain spiroacetal substructures, a common structural motif in ionophore natural products. In both cases the acetal carbon is a stereogenic center. The problem of stereocontrol at this acetal center is usually handled by thermodynamics and follows a stereochemical analysis with its foundation in the anomeric effect. The anomeric effect suggests that spiroacetals should prefer configurations that allow conformations that place both "exocyclic oxygens" on sites that are axial on the ring to which they are attached (see structure of calcimycin). The situation is similar for monensin.

Just as with lasonolide A, these targets are amenable to convergent syntheses. It is not surprising that there are features common to all of the calcimycin syntheses. There are some strategic differences, however, and these will emerge as we consider each approach. We will begin with the first synthesis of calcimycin.

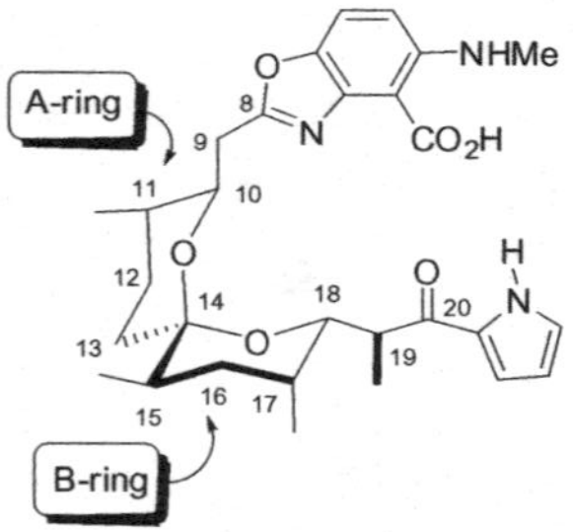

Calcimycin (A-23187) (complexes calcium) (**1**)

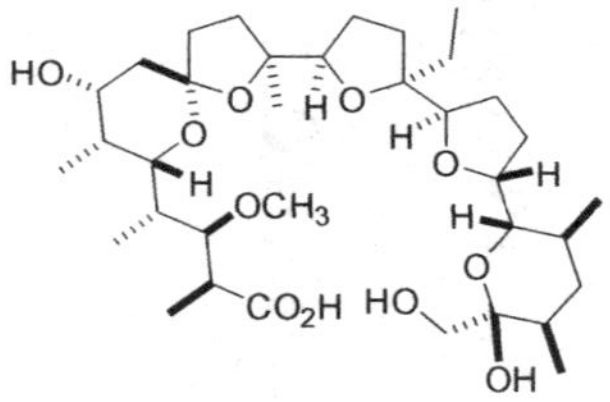

Monensin (complexes sodium) (**2**)

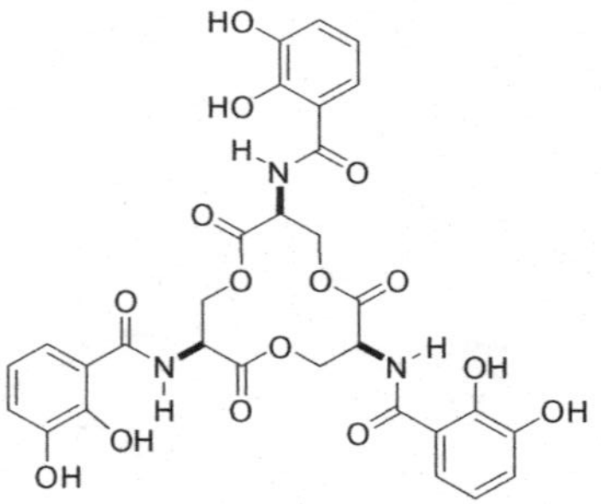

Enterobactin (complexes Fe^{+3}) (**3**)

O_a exocyclic and axial to B-ring

O_b exocyclic and axial to A-ring

The Kishi synthesis of monensin features allylic conformational analysis to predict stereochemistry of hydroboration-oxidations in acyclic systems. The Still synthesis features acyclic diastereoselection in carbonyl addition reactions (chelation control and Felkin-Ahn control).

Calcimycin-1

Calcimycin-1

This chapter will deal with calcimycin (**1**), originally known as A-23187. Calcimycin is an example of an ionophore, a large family of natural products known to complex ions in water. Other "famous" ionophores include monensin (**2**) and enterobactin (**3**), although I will not dwell on the synthesis of these natural products.[1,2] I have selected calcimycin as a target (1) for personal reasons (I wrote an NIH postdoctoral fellowship on this topic after the molecule was suggested to me by my postdoctoral mentor as an interesting target), (2) because a large number of syntheses have been completed and thus, comparison is possible, (3) because calcimycin is one of the first ionophores to attract the attention of synthetic organic chemists (many others have followed), (4) calcimycin is a polypropionate-derived natural product (at least in part) and (5) calcimycin is one of the first spiroketal-containing natural products to be prepared by total synthesis.

Any synthesis of calcimycin must deal with issues of stereochemistry along the chain labelled C_8-C_{20}. Six of the seven stereogenic centers are contained in rings and one (C_{19}) is not. The acetal carbon is particularly interesting. It is a common structural motif in many natural products (see monensin for another example). The problem of stereocontrol at this acetal center is usually handled by thermodynamics and follows from a stereochemical analysis with its foundation in the anomeric effect.[3] The anomeric effect suggests that spiroacetals should prefer configurations that allow placement of both "exocyclic oxygens" on sites that are axial on the ring to which they are attached (see structure of calcimycin).

Just as with lasonolide A, most ionophores are amenable to convergent syntheses. It is not surprising that there are features common to all of the calcimycin syntheses. There are strategic differences, however, and these will emerge as we consider each approach. We will begin with the first synthesis of calcimycin, described by the Evans group (Caltech at the time).

First Synthesis of Calcimycin

Evans, D. A.; Sacks, C. E.; Kleschick, W. A.; Taber, T. R. "Polyether Antibiotics Synthesis. Total Synthesis and Absolute Configuration of the Ionophore A-23187" *J. Am. Chem. Soc.* **1979**, *101*, 6789-6791.

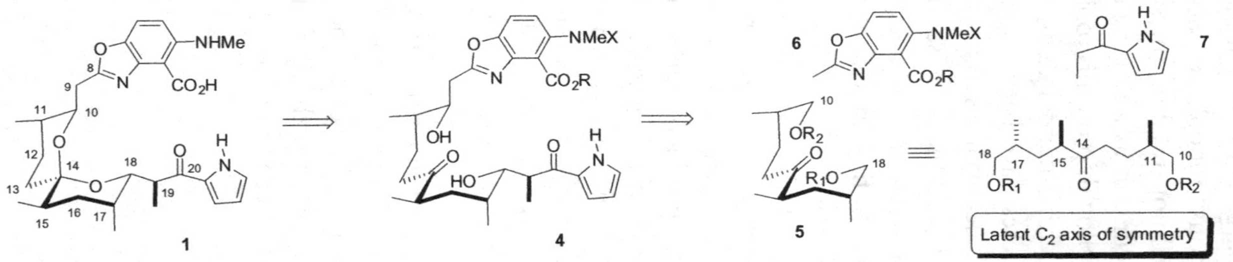

Notes on Strategy

1. Set C_{11} and C_{17} stereochemistry using latent C_2 axis of symmetry to advantage.
2. Use themodynamics to set stereochemistry at C_{15} (see note 3).
3. C_{13} and C_{15} exchange hydrogen without epimerization upon exposing calcimycin to DCl, dioxane, Δ, 18h.

Synthesis of Benzoxazole

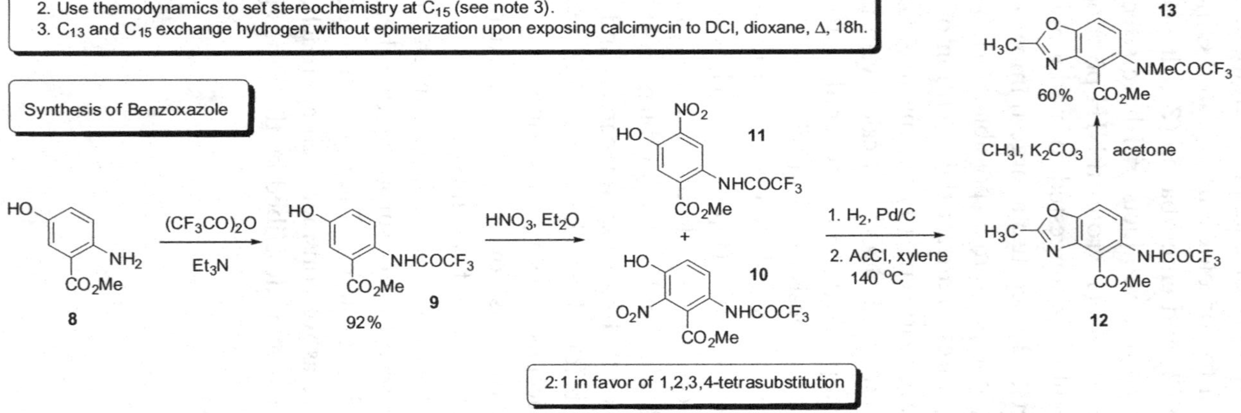

Calcimycin-2

Calcimycin-2

The plan revolved around the aforementioned notion that thermodynamics would control spiroacetal stereochemistry (C_{14}). This is a reasonable assumption given that acetal formation is a reversible process. Thus a keto-diol of type **4** was selected as a late intermediate in the projected synthesis. It was hoped that thermodynamics might also control stereochemistry at C_{15}. In principle, this center is epimerizable. The presence of a 1,3-diaxial methyl-methyl interaction in the C_{15}-epimer of **1** would be expected to render it much higher energy that **1** itself. In practice, treatment of calcimycin (**1**) with DCl in dioxane led to exchange of the C_{13} and C_{15} hydrogens without erosion of stereochemistry. This provided experimental evidence to support the plan for setting these stereogenic centers. Compound **4** was to be prepared from a protected keto-diol of type **5** (the C_{10}-C_{18} fragment of calcimycin). Addition of the anion derived from a benzimidazole of type **6** to a C_{10}-aldehyde, and an aldol reaction between 2-acylpyrrole **7** and a C_{18}-aldehyde, were to be used to assemble **4** from pieces **5–7**. There were stereochemical issues associated with both of the carbonyl additions, and also some protecting group issues associated with the "timing" of these steps, that had to be addressed during execution of the plan. It was noticed that **5** has a latent C_2-axis of symmetry (consider **5** without the C_{15} methyl group). It was hoped that this would simplify the preparation of this intermediate.

The benzimidazole was prepared from **8**. Acylation of the amino group and nitration of the resulting trifluoroacetamide (**9**) gave a mixture of regioisomers **10** and **11**. The nitro group of the **10** was reduced and the resulting *o*-aminophenol was converted to benzimidazole **12** upon reaction with acetyl chloride. *N*-Alkylation of the amide provided **13**, which ultimately played the role of **6** in the synthetic plan.

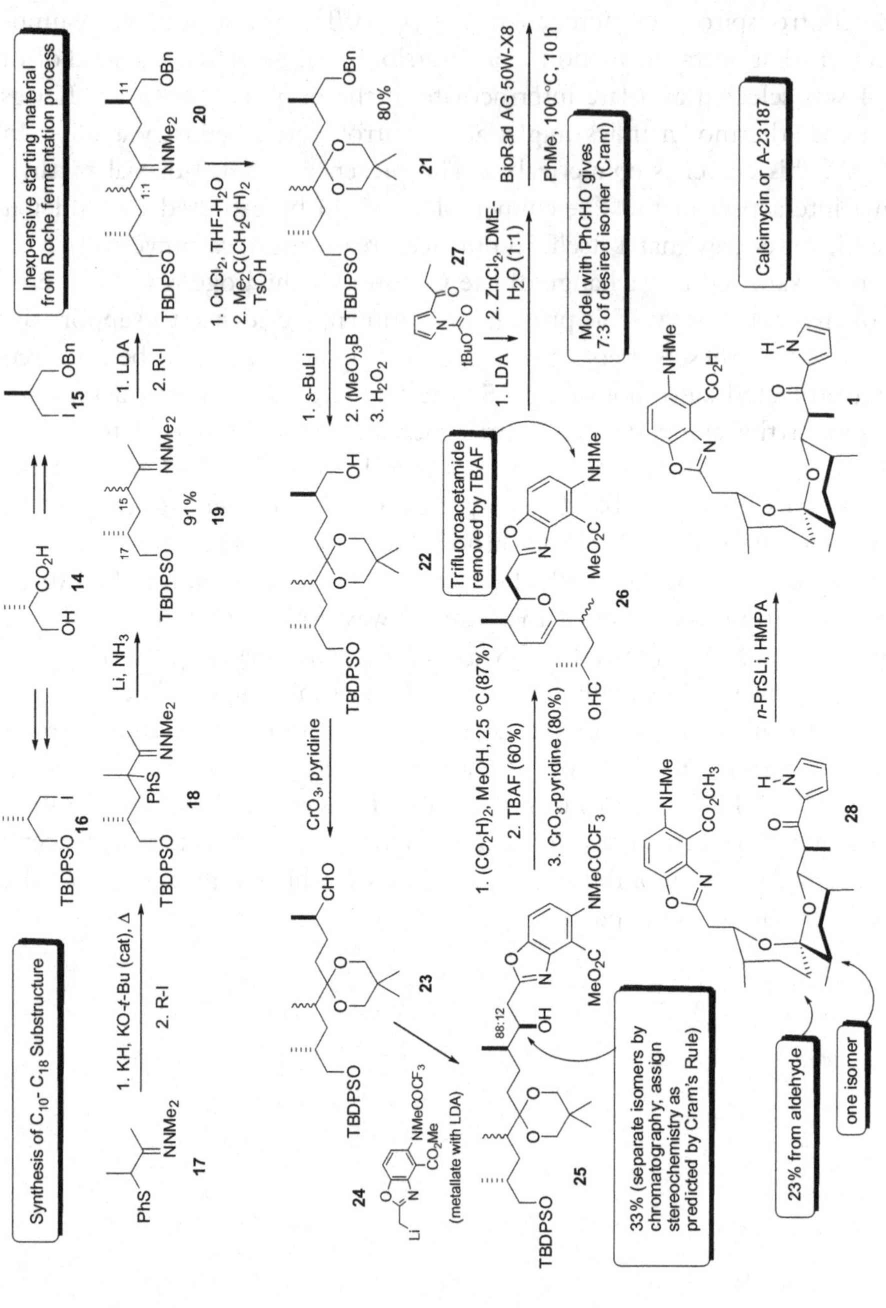

Calcimycin-3

Calcimycin-3

The C_{10}-C_{18} substructure of calcimycin was prepared using **14** as the source of absolute stereochemistry at both C_{11} and C_{17}. This commercially available starting material was converted to both **15** and **16** through a standard series of manipulations. The anion derived from hydrazone **17** was then alkylated with **16** to provide **18**. The activating group (PhS) was removed using a dissolving metal reduction. The resulting mixture of diastereomers (**19**) was again metallated at the least hindered site, and alkylation of the resulting anion with **15** gave hydrazone **20**. The hydrazone was hydrolyzed and the resulting ketone was converted to ketal **21**.

The purpose of differential protection of the C_{10} and C_{18} alcohols was to help with the aforementioned "timing" issues. It was planned to introduce the benzimidazole unit first and the pyrrole unit last. This called for initial removal of the C_{10} benzyl protecting group. This turned out to be a big problem using standard conditions such as hydrogenolysis. Eventually metallation of **21**, followed by trapping the resulting carbanion with trimethyl borate and oxidation of the resulting C–B bond, gave **22**. Collins oxidation of **22** gave aldehyde **23**. Treatment of **23** with **24** (from benzimidazole **13**) provided a separable mixture of diastereomeric alcohols. It was anticipated that **25** would be the major isomer based on expected Cram or Felkin-Ahn selectivity.[4,5] Hydrolysis of the acetal using methanolic oxalic acid provided a mixture of dihydropyrans which were separated by chromatography. Removal of the C_{18} protecting group and another Collins oxidation provided aldehyde **26**. The presumed zinc enolate of **27** reacted with aldehyde **26** to give a mixture of materials that were treated with an acidic resin to convert the aldol adduct to spiroacetal **28**. Model reactions using benzaldehyde as the electrophile suggested the aldol should occur with only modest stereoselectivity. The overall yield from **26** to **28** was low, but **28** was the only stereoisomer obtained. This suggests that the plan for controlling stereochemistry at C_{14} and C_{15} had merit, although it was possible that stereoisomers at C_{15} simply did not undergo cyclization. The synthesis was completed by deprotection of the methyl ester using S_N2 chemistry.

This synthesis of calcimycin was one of the first total syntheses of an ionophore. It was preceeded by syntheses of nonactin (Gerlach and Schmidt),[6] followed shortly by syntheses of lasalocid A (X537A) (Ireland),[7] and then by the landmark syntheses of monensin (**2**) developed by Kishi and Still.[1] The Evans synthesis illustrates the importance of symmetry in synthesis design, and documented observations that were clearly useful to those that followed. The synthesis suffered somewhat from stereocontrol, particularly

Grieco Synthesis of Calcimycin

Martinez, G. R.; Grieco, P. A.; Williams, E.; Kanai, K., Srinivasan, C. V. "Stereocontrolled Synthesis of Antibiotic A-23187 (Calcimycin)" *J. Am. Chem. Soc.* **1982**, *104*, 1436-1438.

One of the low points of the first calcimycin synthesis is introduction of the pyrrole unit via an aldol condensation. The yields are low and the stereocontrol at C_{19} is probably marginal. The Grieco synthesis disconnects calcimycin between C_{20} and the 2-position of the pyrrole. Therefore a significant difference between this approach and the Evans approach is an attempt to achieve better control of stereochemistry at C_{19}. A secondary difference is that C_{15} was to be introduced with complete control of stereochemistry rather than relying on thermodynamics for stereocontrol. It will be seen that a consequence of this plan is an increase in the length of the synthesis.

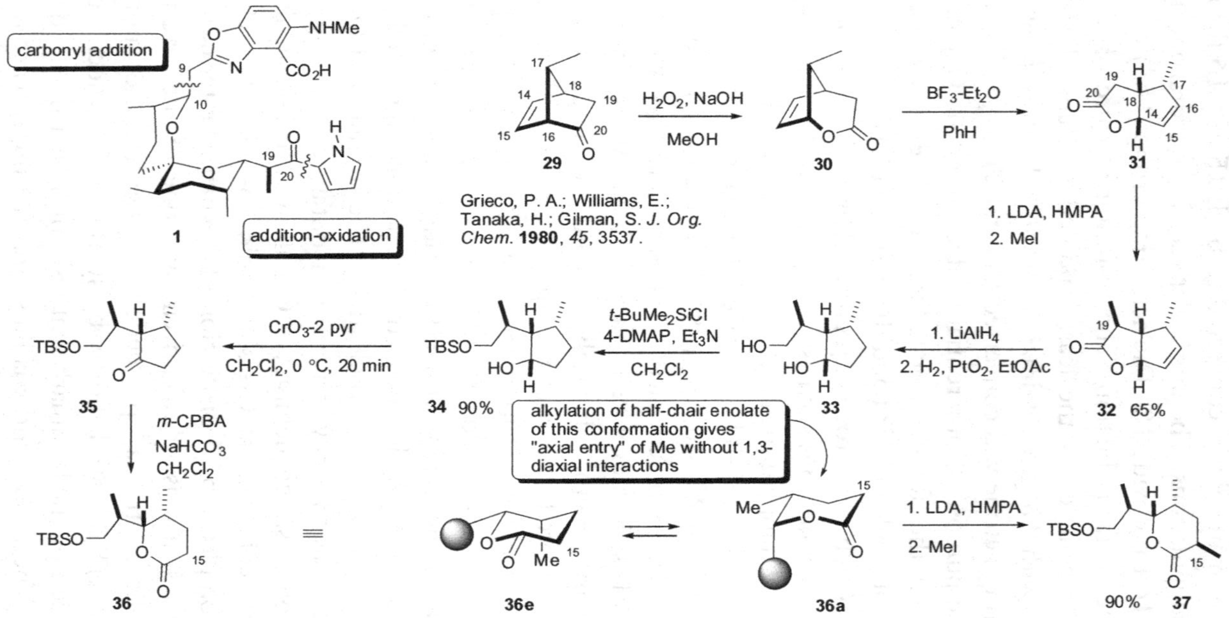

Calcimycin-4

over the C_{10} and C_{18} stereogenic centers. We will next move to a synthesis that addresses this deficiency.

Calcimycin-4

One of the low points of the Evans calcimycin synthesis was introduction of the pyrrole unit via an aldol condensation. The yields were low and the stereocontrol at C_{19} was probably marginal. The Grieco group (Indiana at the time) described a synthesis that disconnected calcimycin between C_{20} and the 2-position of the pyrrole. Therefore a significant difference between this approach and the Evans approach was an attempt to achieve better control of stereochemistry at C_{19}. A secondary difference was that C_{15} was to be introduced with complete control of stereochemistry, rather than relying on thermodynamics for stereocontrol. It will be seen that a consequence of this plan is an increase in the length of the synthesis. As a personal note, it is fun for me to discuss this synthesis because of my connection with two of the coauthors. Greg Martinez was an undergraduate student at UC Berkeley when I was a graduate student, where we worked on a project together. Kenichi Kanai was my first postdoctoral student at OSU, moving to my group from the Grieco labs at about the time this synthesis was published.

The starting point for the synthesis was norbornenone derivative **29**, prepared (in part) as described at the bottom of Calcimycin-6. This starting material contains carbons C_{14}-C_{20} (and the C_{17} methyl group) of calcimycin and thus, the initial stages of the synthesis were largely an exercise in oxidation chemistry. Given that **29** has a bicyclo[2.2.1]heptane substructure, two C–C bonds (C_{16}-C_{20} and C_{14}-C_{18}) had to be broken to arrive at an acyclic compound of type **37**. In addition, two methyl groups had to be introduced at the incipient C_{19} and C_{15} positions. Both C–C bonds were to be cleaved using Baeyer-Villiger oxidations, and the methyl groups were to be introduced by alkylation of cyclic intermediates, providing stereocontrol that was predictable.

The synthesis began with Baeyer-Villiger oxidation of **29** to provide lactone **30**. Acid was then used to rearrange **30** to lactone **31**. There is a clear relationship between these transformations and chemistry developed for the synthesis of prostaglandins. Oxabicyclo[3.3.0]octane **31** provided a rigid template for introducing the C_{19} methyl group. Thus, alkylation of the enolate derived from **31** took place from the convex face to provide **32**. A series of reductions provided **33**. Protection of the primary alcohol and oxidation of the secondary alcohol gave **35**, setting the stage for the second C–C bond oxidation. In accord with the plan, Baeyer-Villiger oxidation of **35** gave lactone **36**. Alkylation of the enolate derived from **36** gave **37** with the desired

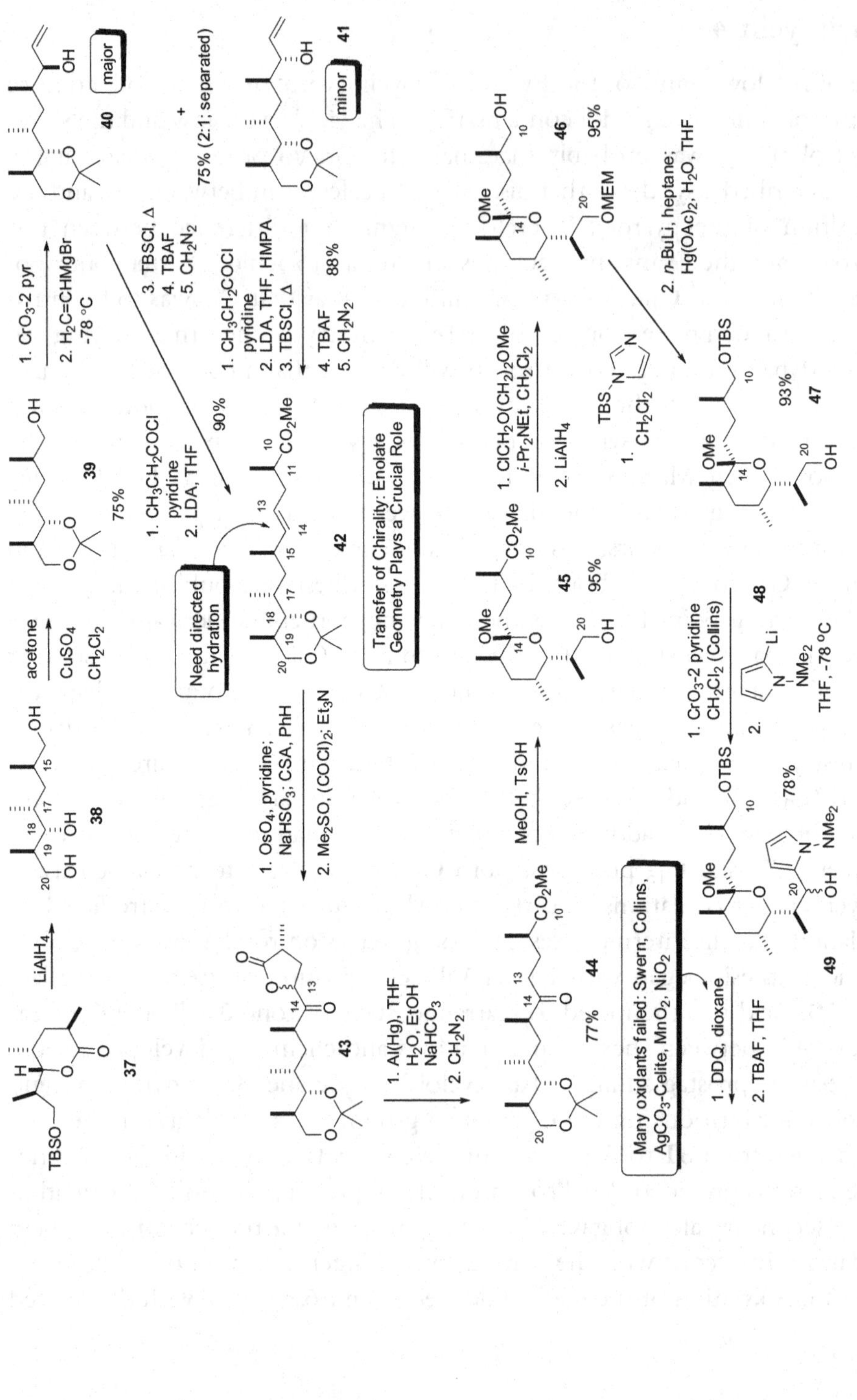

Calcimycin-5

stereochemistry at C_{15}. How can the stereoselectivity of this alkylation be explained? Alkylation of the half-chair conformation of the enolate derived from **36a**, with axial entry of the electrophile (iodomethane), would give the observed product. The same stereoelectronic course of alkylation of the enolate derived from conformation **36e**, would experience a 1,3-diaxial Me-Me interaction. This type of analysis can be useful in general in predicted the stereochemical course of many 6-membered ring enolate alkylations. It certainly works here.

Calcimycin-5

Continuing with the synthesis, reduction of lactone **37** was accompanied by renewal of the alcohol protecting group to give triol **38**. The 1,3-diol unit was converted to an acetonide (**39**). Oxidation of the residual alcohol followed by reaction of the resulting aldehyde with vinylmagnesium bromide gave a separable mixture of **40** (major) and **41** (minor). These compounds differ by virtue of relative stereochemistry at the C_{14} and C_{15} stereogenic centers. Both diastereomers were moved forward to **42** using the ploy of transfer of chirality (Prostaglandins-24). Thus, conversion of **40** to the corresponding propionate ester, enolate formation in THF, trapping the enolate to provide the silyl ketene acetal, Claisen rearrangement of this intermediate (presumably via the most stable chair conformation), and conversion of the resulting silyl ester to a methyl ester, gave **42**. When **41** was subjected to the same reaction sequence, only using THF-HMPA as solvent during enolate formation, **42** was also obtained. This sequence revolves around the "Ireland-Claisen" or "Enolate-Claisen" rearrangement.[8] It is enolate geometry (*E* from **40** and *Z* from **41**) that dictates the course of transfer of chirality in this reaction sequence. Try to work your way through the details by working the problem provided at the end of this chapter. It is notable that the reaction sequence translates 1,2-diastereoselectivity (although 2:1 is not very good) into 1,4-diastereoselectivity, much as we saw in the Taber approach to prostaglandins (Prostaglandins-25).

The next task was to convert **42** to ketone **44**. This transformation called for a regioselective hydration of the C_{13}-C_{14} olefin. This was accomplished by vicinal dihydroxylation of the olefin, which was accompanied by formation of the γ-lactone (rather that the δ-lactone). The preference for γ-lactonization (which is most often the case in such situations) left the C_{14} hydroxyl group free for oxidation to ketone **43**. Reductive cleavage of the α-C–O bond, and esterification of the resulting acid, gave **44** (see **99** to **100** in Alkaloids-12 for comparison).

Ketoester **44** contains five of the seven stereogenic centers of calcimycin. Completion of the synthesis called for introduction of the pyrrole and

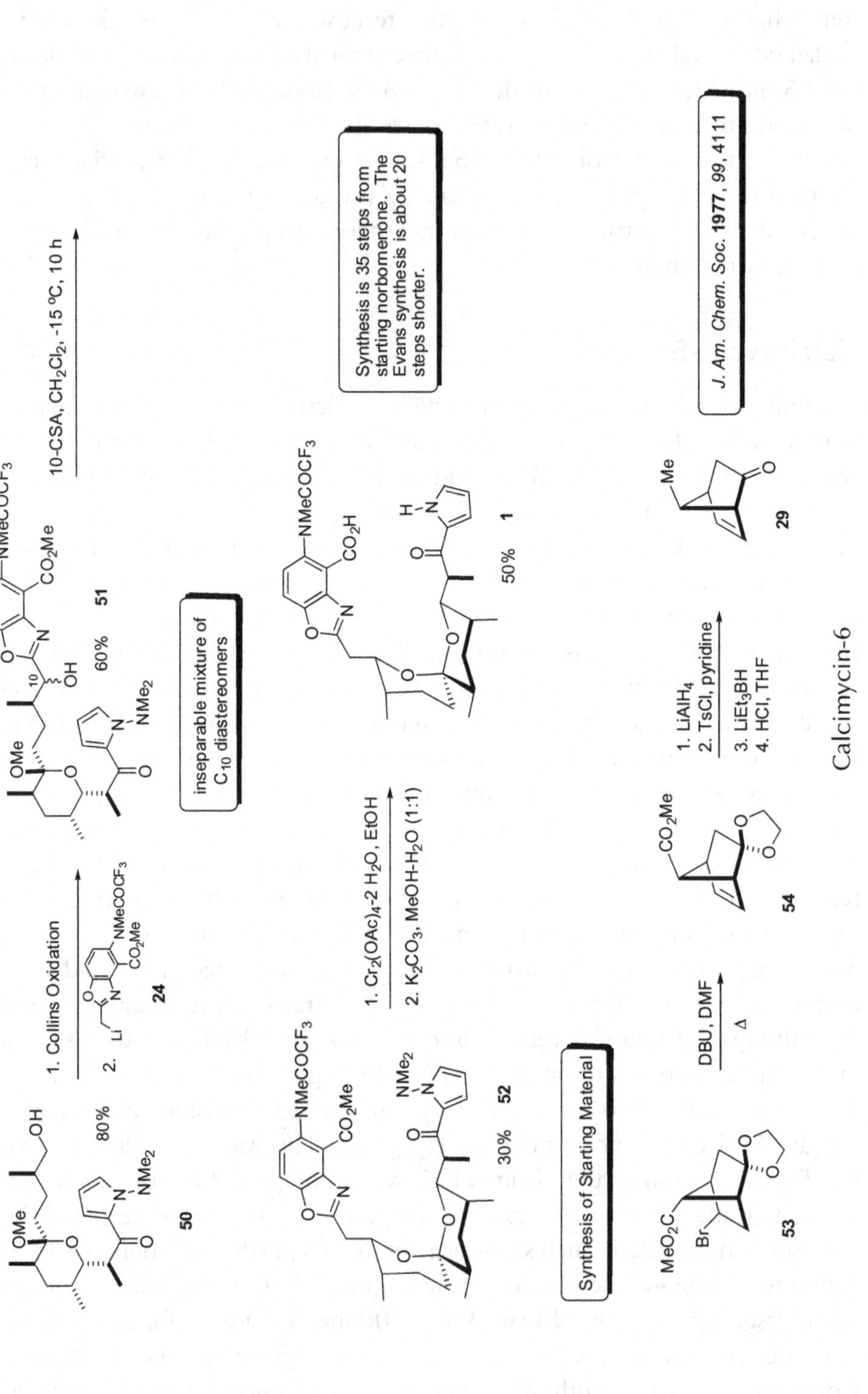

Calcimycin-6

benzimidazole fragments. Since these were to be introduced by carbonyl addition reactions, the C_{14} ketone had to be protected. This was accomplished by conversion of **44** to THP **45**. This liberated C_{20} at the alcohol oxidation state. Some protecting group manipulations provided **47** which was poised for introduction of the pyrrole group. Oxidation of the primary alcohol followed by addition of **48** to the intermediate aldehyde, gave **49** as a mixture of diastereomers. Oxidation of the pseudo-benzylic C_{20}-alcohol with DDQ (lots of the search part of research here) and removal of the C_{10} protecting group, provided **50**, ready for introduction of the benzimidazole unit (Calcimycin-6).

Calcimycin-6

The synthesis of calcimycin continued with oxidation of **50**, reaction of the aldehyde with **24** to give **51** as a mixture of C_{10}-diastereomers. Treatment of this mixture with 10-camphorsulfonic acid gave **52**. Presumably only the desired C_{10} diastereomer cyclized to the spiroacetal. The synthesis was completed by reductive cleavage of the N–N bond (pyrrole nitrogen protecting group) and hydrolysis of the trifluoroacetamide and methyl ester.

This synthesis illustrates one strategy for the preparation of acyclic molecules containing multiple stereogenic centers — use cyclic structures to control stereochemistry and then liberate the acyclic structure. The strategy is not unlike several of the Cecropia juvenile hormone syntheses we examined, where stereoselective olefin synthesis was the goal. Whereas three pieces are ultimately assembled, the synthesis of the "central fragment" is linear and there was a price to pay for this approach. It is long.

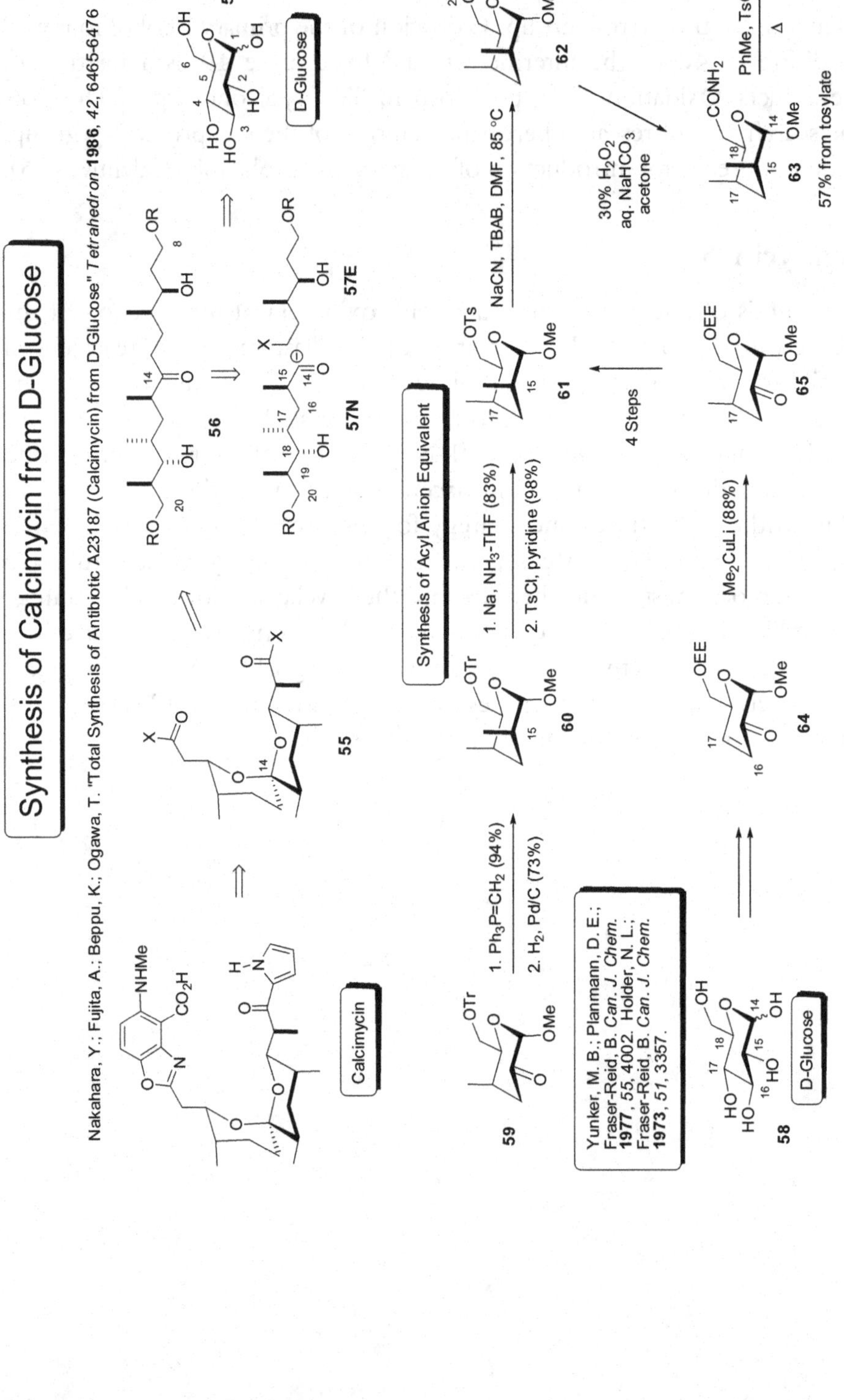
Synthesis of Calcimycin from D-Glucose
Nakahara, Y.; Fujita, A.; Beppu, K.; Ogawa, T. "Total Synthesis of Antibiotic A23187 (Calcimycin) from D-Glucose" Tetrahedron 1986, 42, 6465-6476
Calcimycin
55
56
57N
57E
58
D-Glucose
Synthesis of Acyl Anion Equivalent
1. Ph3P=CH2 (94%)
2. H2, Pd/C (73%)
59
60
1. Na, NH3-THF (83%)
2. TsCl, pyridine (98%)
61
NaCN, TBAB, DMF, 85 °C
62
30% H2O2
aq. NaHCO3
acetone
63
57% from tosylate
PhMe, TsOH
Δ
4 Steps
Me2CuLi (88%)
64
65
Yunker, M. B.; Planmann, D. E.; Fraser-Reid, B. Can. J. Chem. 1977, 55, 4002. Holder, N. L.; Fraser-Reid, B. Can. J. Chem. 1973, 51, 3357.

Calcimycin-7

Calcimycin-7

The Evans synthesis gave a single enantiomer of calcimycin, whereas the Grieco synthesis was performed with racemic starting material and thus, gave racemic calcimycin. Note that Evans' convergent strategy called for starting with enantiopure material whereas Grieco's linear strategy was amenable to a racemic synthesis. The next synthesis we will examine returns to preparation of a single enantiomer.

Carbohydrates are a rich source of chirality. They are popular members of the "chiral pool" as a starting point for enantioselective synthesis (see Prostaglandins-15).[9] The Nakahara group (RIKEN in Japan) used D-glucose as the point of departure in their synthesis of calcimycin. The plan was to prepare a compound of type **55** from a keto-diol of type **56**, and then introduce the heterocyclic end-groups. The spiroacetal precursor (**56**) was to be assembled by reaction of an acyl anion equivalent of type **57N** with an electrophile of type **57E**. Both the nucleophile and electrophile were to be prepared from D-glucose. Whereas this approach did lead to calcimycin, a glance at D-glucose reveals that a lot of work would have to be done to arrive at **57N** and **57E**. This is the down-side of the approach.

Dithiane **71** (see Calcimycin-8) was selected as the acyl anion equivalent (**57N**). The plan was to use C_1-C_6 of D-glucose as C_{14}-C_{19} of the target. This called for a one-carbon homologation at C_6 of the glucose (to install C_{20}) and a myriad of oxidation state changes and "methylations". The known ketone **59** was chosen as the point of departure. Wittig olefination of the ketone followed by catalytic hydrogenation from the sterically most accessible face gave **60**. The trityl group was then exchanged for a tosylate (**60** → **61**). An alternative route to **61** was developed from enone **64**. The C_{17} methyl group was introduced via a cuprate addition (pseudo-axial delivery of methyl), and **65** was converted to **61** in a manner resembling its preparation from **59**. It turned out the ethoxyethyl protecting group had operational advantages to the trityl protecting group upon scale-up.

Continuing with the synthesis, the incipient C_{20} was introduced as a nitrile (**61** → **62**) and the nitrile was hydrolyzed to afford amide **63**. Treatment of **63** with *p*-toluenesulfonic acid provided lactam **66** in excellent yield (Calcimycin-8).

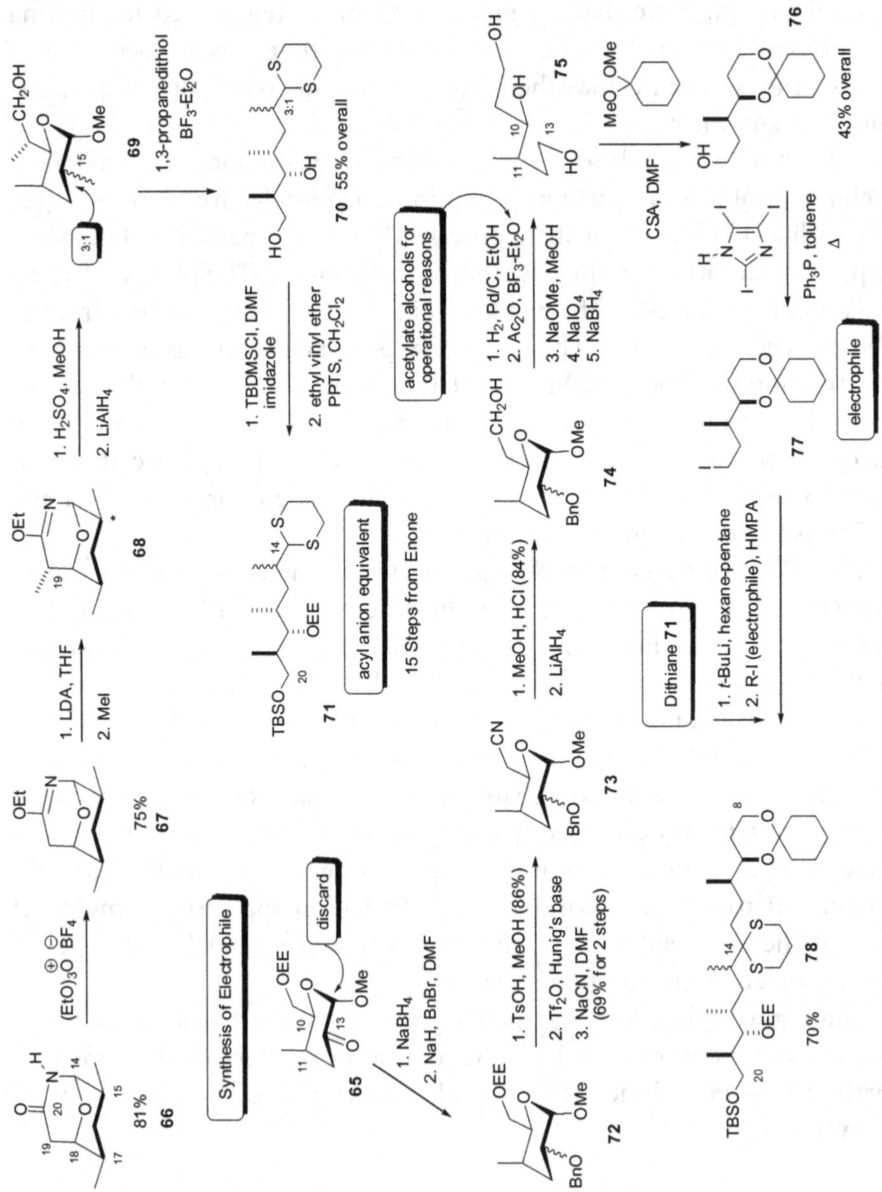

Calcimycin-8

Calcimycin-8

The rigid bicyclic framework of **66** was used to introduce the C_{19} methyl group with control of stereochemistry. Thus, conversion of **66** to imidate **67** was followed by deprotonation and alkylation to provide **68**. If you compare C_{15} of **68** with the target (**57N**) you will see that it has the incorrect configuration. This was adjusted (in part) by treating **68** with acidic methanol. This gave methyl glycoside **69**, in which partial epimerization had occurred at C_{15}. Recall that it was already known that this stereogenic center could be equilibrated to provide the needed isomer in calcimycin itself (Calcimycin-2). The reaction of **69** with propan-1,3-dithiol gave 1,3-dithiane **70**. This is a standard move in carbohydrate chemistry, used to open a pyranose or furanose to a derivative of the acylic polyhydroxyaldehyde. Sequential protection of the primary and secondary alcohols completed the synthesis of **71** (as a 3:1 mixture of diastereomers at C_{15}).

Iodide **77** was chosen to play the role of the electrophile (**57E**) in the projected coupling with dithiane **71**. The plan was to use **65** as a point of departure. The anomeric carbon was to be discarded, the ketone was to become the iodide (C_{13} of calcimycin), and C_4 and C_5 of this monosaccaride derivative were to become the two stereogenic centers in **77**. Reduction of **65** with sodium borohydride and protection of the resulting alcohols as benzyl ethers gave **72**. Homologation of **72** to alcohol **74**, via nitrile **73**, was accomplished using chemistry similar to that used in the preparation of **71**. The anomeric carbon was then excised using a five-reaction sequence. The resulting triol (**75**) was converted to acetal **76** and the free hydroxyl group was converted to iodide **77**. The coupling of **71** and **77** proceeded well. Metallation of **71** was followed by alkylation with **77** to give **78** in 70% yield.

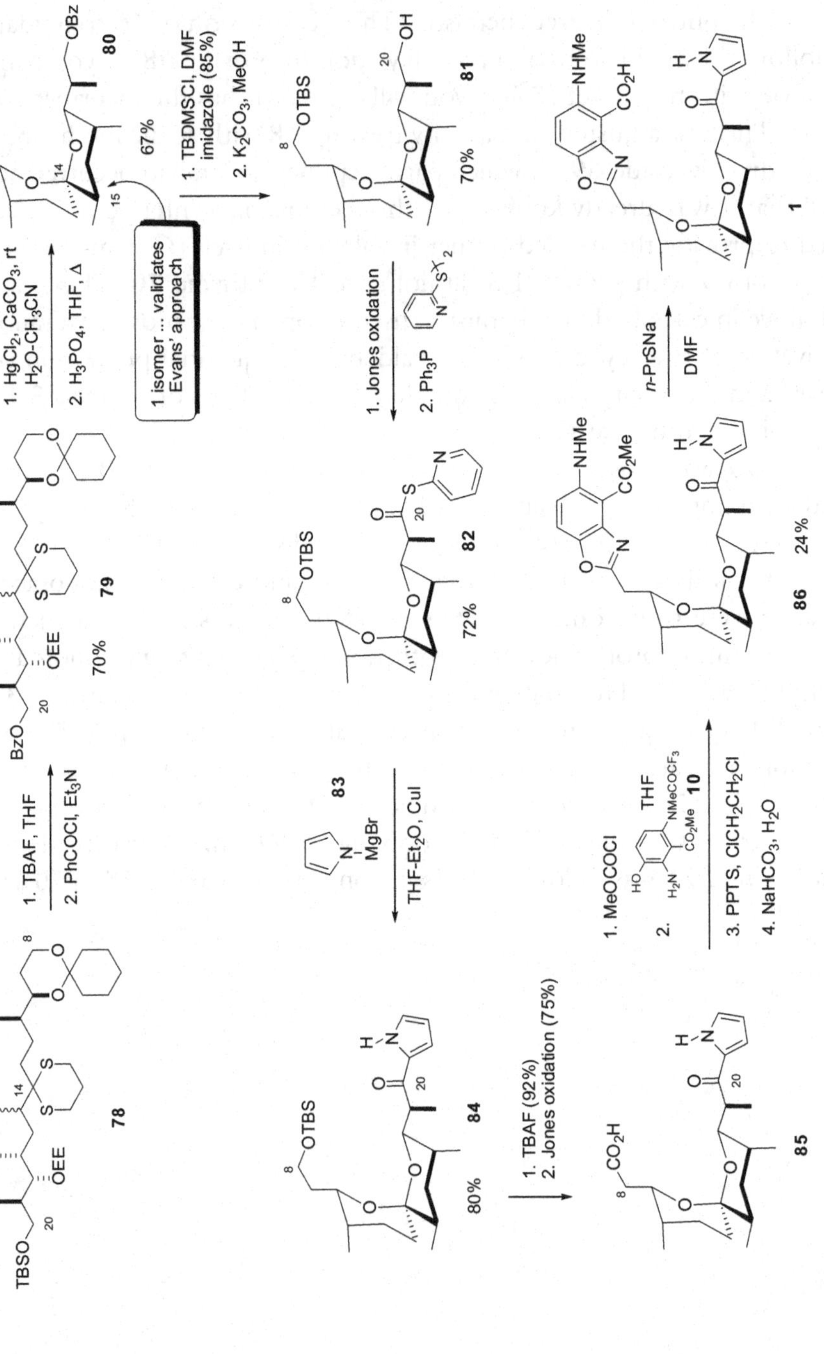

Calcimycin-9

Calcimycin-9

The C_{20} protecting group was exchanged to provide **79**. Hydrolysis of the thioketal and the acetal protecting groups provided **80** in good yield as a single stereoisomer (recall Evans' plan for thermodynamic control of C_{15} stereochemistry). The C_8 alcohol was protected as a TBS ether and the C_{20} ester was hydrolyzed to set the stage for introduction of the pyrrole. Alcohol **81** was oxidized to the corresponding acid and then converted to thioester **82**. Copper promoted coupling of **82** with pyrrole derivative **83** gave 2-acylpyrrole **84**. The C_8 protecting group was removed and a Cr^{VI} oxidation gave acid **85**. Acylation of *o*-aminophenol **10**, followed by acid promoted ring closure and hydrolysis of the trifluoroacetamide, gave **86**. The synthesis of calcimycin (**1**) was completed in the usual manner.

Looking back at the Nakahara synthesis, it relies heavily on the use of cyclic compounds to set relative stereochemistry. The absolute stereochemistry at C_{10} and C_{18} came from D-glucose. The stereochemistry at C_{11}, C_{16} and C_{19} was controlled by constraints imposed by cyclic compounds. And the stereochemistry at C_{14} and C_{15} was controlled by thermodynamics. We will now look at a synthesis that uses a different strategy for construction of the spiroacetal substructure of calcimycin.

An Alternative Approach to Formation of the Spiroketal

Negri, D. P.; Kishi, Y. "A Total Synthesis of Polyether Antibiotic (-)-A23187 (Calcimycin)" *Tetrahedron Lett.* **1987**, *28*, 1063-1067

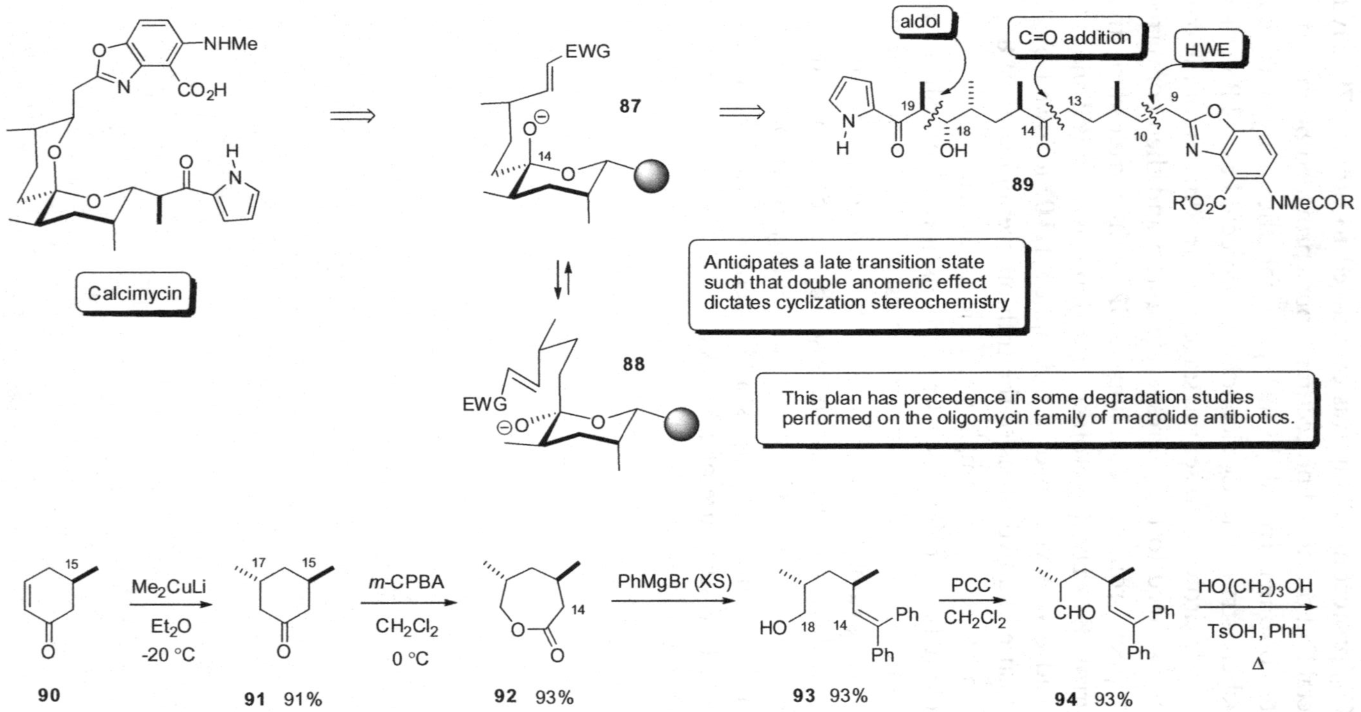

Calcimycin-10

Calcimycin-10

Kishi and Negri reported a synthesis of calcimycin that revolved around a base-initiated cyclization of **89**. The hope was that the C_{18} hydroxyl group of **89** would add to the C_{14} carbonyl group to provide an equilibrium mixture of **87** and **88**. It was anticipated that these might undergo intramolecular conjugate addition to provide spiroketals related to calcimycin. It was hoped that the rate of cyclization from **87** might exceed the cyclization rate from **88** because the double anomeric effect would be felt in that cyclization transition state (and not in the transition state derived from **88**). This zipper reaction has some precedence in degradation studies performed on the oligomycin family of macrolide antibiotics.[10] The cyclization substrate was to be prepared using an aldol condensation to set the C_{18}-C_{19} bond, a carbonyl addition reaction to prepare the C_{13}-C_{14} σ-bond, and a Horner-Wadsworth-Emmons to prepare the C_9-C_{10} double bond. Cyclohexenone **90**, which can be prepared from the monoterpene pulegone, was to supply the C_{11} and C_{15} stereogenic centers, and provide a handle for introduction of the C_{17} stereogenic center.

An early goal was preparation of ketone **100** (Calcimycin-11). The olefin was to be a handle for introduction of the benzimidazole. An aldehyde was to be generated from the acetal and used to introduce the pyrrole substructure via an aldol condensation. Ketone **100** was to be prepared from aldehyde **93** and bromide **99**, both of which were to be prepared from cyclohexenone **90**. Moving forward, a cuprate addition gave C_2-symmetric ketone **91**. The plan called for the carbonyl carbon of **91** to be oxidatively excised, leaving C_{14} and C_{18} at the aldehyde oxidation state, but differentially protected. With this in mind, Baeyer-Villiger oxidation of **91** provided lactone **92**. Reaction of **92** with excess phenylmagnesium bromide gave alcohol **93** after facile dehydration of an intermediate tertiary alcohol. Oxidation of the alcohol was followed by protection of **94** to give acetal **95**. Ozonolysis of the olefin liberated the needed C_{14} aldehyde (Calcimycin-11).

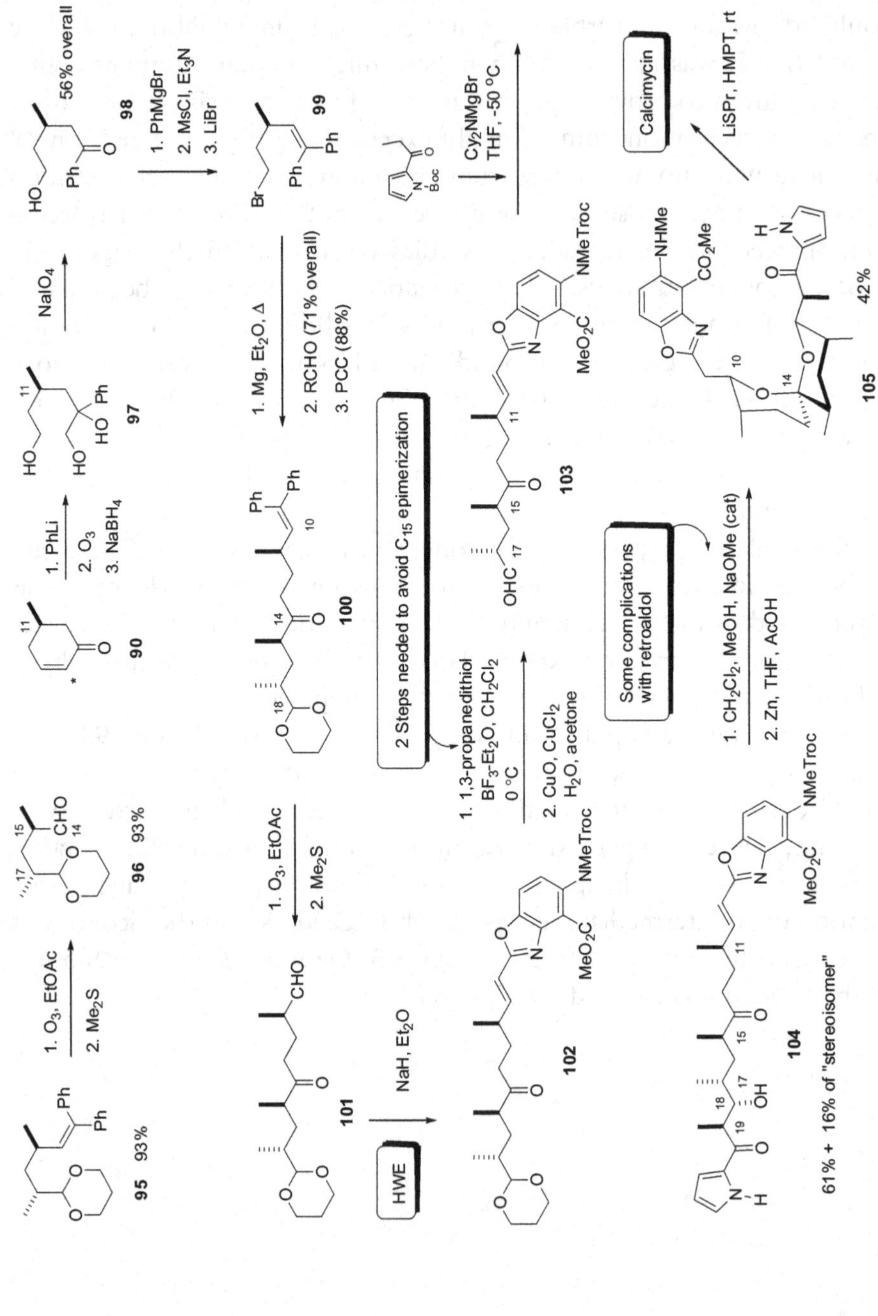

Calcimycin-11

Calcimycin-11

Conversion of **90** to **99** called for excision of the carbon marked with an asterisk. This was set up by converting **90** to triol **97** using a straightforward reaction sequence. Periodate cleavage of the vicinal diol accomplished the required one-carbon excision. The resulting ketone (**98**) was converted to **99** in a straightforward manner. Bromide **99** was converted to the corresponding Grignard reagent, reacted with aldehyde **96**, and an oxidation completed the synthesis of **100**.

Next the benzimidazole was introduced. Thus, cleavage of the olefin liberated C_{10} as an aldehyde. The HWE reaction proceeded without incident. The C_{15} ketone was protected and the C_{18} acetal hydrolyzed to give aldehyde **103**. An aldol condensation then provided the Felkin-Ahn (or Cram) product **104**, in good overall yield. The key cyclization occurred as anticipated using sodium methoxide as a catalyst to trigger the process (**104** to **105**). The Troc protecting group was removed and the methyl ester was converted to the acid to provide calcimycin (**1**).

Another Synthesis with All But C_{15} Stereochemistry Set

Boeckman, R. K. Jr.; Charette, A. B.; Asberom, T.; Johnston, B. H. "A Convergent General Synthetic Protocol for Construction of Spirocyclic Ketal Ionophore: An Application to the Total Synthesis of (-)-A-23187 (Calcimycin)" *J. Am. Chem. Soc.* **1987**, *109*, 7553-7555.

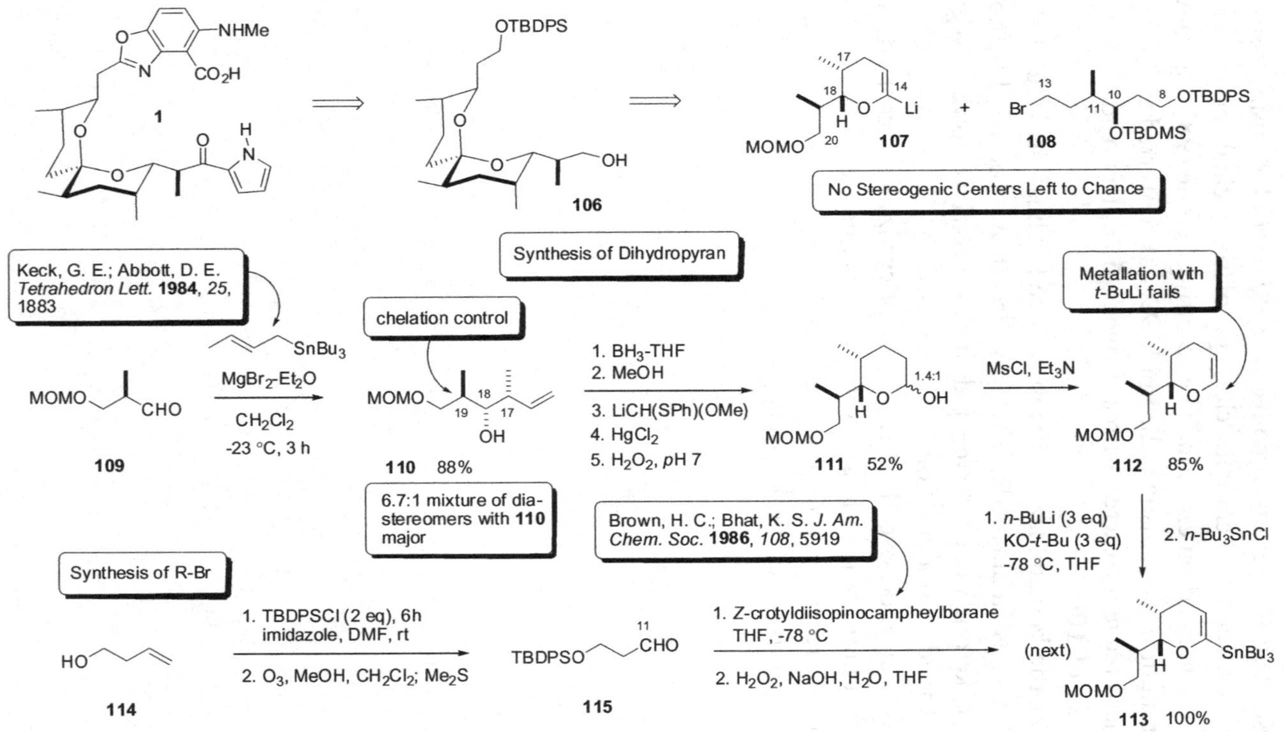

Calcimycin-12

Calcimycin-12

The last synthesis of calcimycin we will examine was reported by the Boeckman group (Rochester) at about the same time as the Kishi and Nakahara syntheses. The synthesis was (1) convergent, (2) featured metallated dihydropyran chemistry developed by the Boeckman group, (3) set all of the stereogenic centers with the exception of C_{14} and C_{15} prior to spiroacetal synthesis and (4) used crotylation-oxidative cleavage as a strategy for introduction of propionate units. The plan was to prepare calcimycin from spiroacetal **106**, which was to be assembled from metallated dihydropyran **107** and bromide **108**. The C_{17}-C_{18} and C_{10}-C_{11} bonds of these respective targets were to be prepared using the aforementioned crotylation chemistry.

Let's begin with the preparation of a **113**, the precursor of **107**. Known aldehyde **109** was treated with *trans*-crotyltributylstannane in the presence of a Lewis acid to provide **110**. The relative stereochemistry between C_{18} and C_{19} was chelation controlled. The relative stereochemistry between C_{17} and C_{18} resulted from presumed addition of the stannane to the aldehyde via a chair-like transition state with Sn coordinated to the carbonyl oxygen. An interesting procedure was used to convert **110** to the homologous aldehyde, which cyclized to a mixture of anomeric hemiacetals **111**. Dehydration of the mixture provided **112**. Metallation of **112** using Schlosser's base, followed by stannylation of the resulting vinyl anion, provided **113**.[11]

The synthesis of **108** began with homoallylic alcohol **114**. Protection of the alcohol was followed by ozonolysis to give aldehyde **115**. Reaction of the aldehyde with *Z*-crotyldiisopinocampheylborane, a chiral crotylborane, gave **116** in high yield with excellent diastereo- and enantioselectivity (see Calcimycin-13).[12]

Note that oxidative cleavage of the olefins in **110** and **116** would provide α-methyl-β-hydroxy carbonyl compounds. This is the "repeating unit" found in polypropionate natural products. This methodology has been widely used in syntheses of such natural products. The same unit can be produced by aldol methodology (for example see **19** to **20** on Lasonolide-4). This methodology has also been widely used.

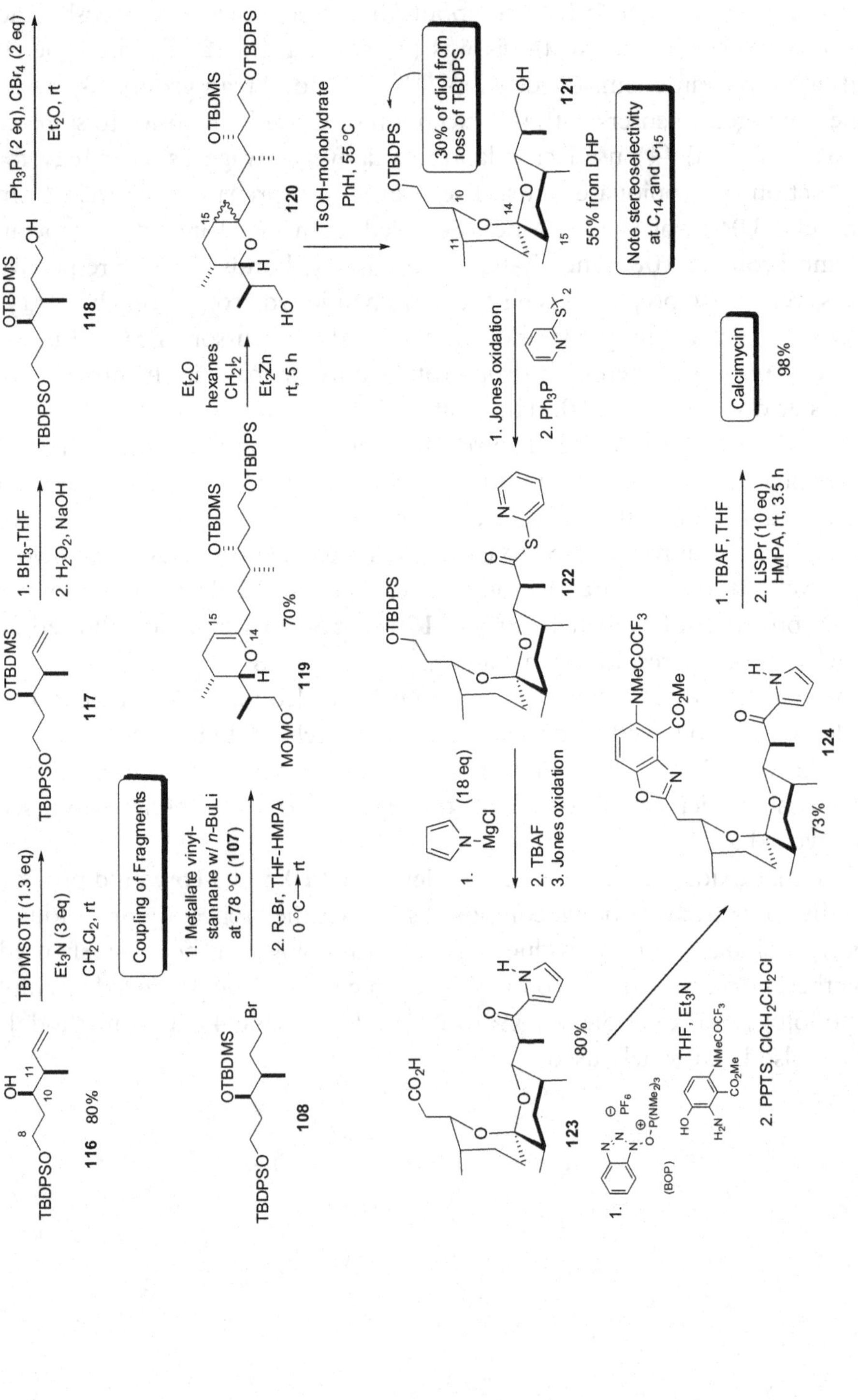

Calcimycin-13

Calcimycin-13

The conversion of **116** → **108** was straightforward. The coupling of fragments was accomplished by converting stannane **113** → **107**, and alkylating **107** with bromide **108**. This provided dihydropyran **119**. The C_{15} methyl group was installed by cyclopropanation of **119** to provide **120** as a mixture of diastereomers. Treatment of **120** with *p*-toluenesulfonic acid accomplished protonolysis of the cyclopropane, and cyclization of the resulting oxocarbenium ion to give **121** as a single diastereomer at C_{14} and C_{15}. Epimerization at C_{15} with thermodynamic control of stereochemistry was clearly at work once again. The rest of the synthesis followed chemistry we have largely seen before, and is outlined in Calcimycin-13 without comment.[13]

References

1. For two syntheses see the following articles and references cited therein: Fukuyama, T.; Akasaka, I.; Karanewsky, D. S.; Wang, C. L. J.; Schmid, G.; Kishi, Y. "Synthetic Studies on Polyether Antibiotics. 6. Total Synthesis of Monensin. 3. Stereocontrolled Total Synthesis of Monensin" *J. Am. Chem. Soc.* **1979**, *101*, 262–263. Collum, D. B.; McDonald, J. H., III; Still, W. C. "Synthesis of the Polyether Antibiotic Monensin. 3. Coupling of Precursors and Transformation to Monensin" *J. Am. Chem. Soc.* **1980**, *102*, 2120–2121.
2. Shanzer, A.; Libman, J.; "Total Synthesis of Enterobactin via an Organotin Template" *J. Chem. Soc., Chem. Commun.* **1983**, 846–847. Rastetter, W. H.; Erickson, T. J.; Venuti, M. C. "Synthesis of Iron Chelators, Enterobactin, Enantioenterobactin, and a Chiral Analog" *J. Org. Chem.* **1981**, *46*, 3579–3590. Corey, E. J.; Bhattacharyya, S. "Total Synthesis of Enterobactin, a Macrocyclic Iron Transporting Agent of Bacteria" *Tetrahedron Lett.* **1977**, *19*, 3919–3922.
3. Gelin, M.; Bahurel, Y.; Descotes, G. "Dipole Moment Studies of Substituted 2-Alkoxytetrahydropyrans" *Bull. Soc. Chim. Fr.* **1970**, 3723–3729.
4. Cram, D. J.; Greene, F. D. "Stereochemistry. XX. Steric Control of Asymmetric Induction in the Preparation of the 3-Cyclohexyl-2-butanol System" *J. Am. Chem. Soc.* **1953**, *75*, 6005–6010. Eliel, E. L. "Application of Cram's Rule: Addition of Achiral Nucleophiles to Chiral Substrates" in *Asymmetric Synthesis*, Morrison, J. D., Ed.; Academic Press, New York, New York, **1983**, *2*, 125–155.
5. Nguyen, T. A.; Eisenstein, O. "Theoretical Interpretation of 1,2-Asymmetric Induction. The Importance of Antiperiplanarity" *Nouv. J. Chim.* **1977**, *1*, 61–70. Cherest, M.; Felkin, H.; Prudent, N. "Tortional Strain Involving Partial Bonds. The Stereochemistry of the Lithium Aluminum Hydride Reduction of Some Simple Open-Chain Ketones" *Tetrahedron Lett.* **1968**, *10*, 2199–2204.
6. Gerlach, H.; Oertle, K.; Thalmann, A.; Stefano, S. "Synthesis of Nonactin" *Helv. Chim. Acta* **1975**, *58*, 2036–2043.
7. Ireland, R. E.; Anderson, R. C.; Badoud, R.; Fitzsimmons, B. J.; McGarvey, G. J.; Thaisrivongs, S.; Wilcox, C. S. "Total Synthesis of Ionophore Antibiotics. A Convergent Synthesis of Lasalocid A (X537A)" *J. Am. Chem. Soc.* **1983**, *105*, 1988–2006.
8. Ireland, R. E.; Willard, A. K. "Stereoselective Generation of Ester Enolates" *Tetrahedron Lett.* **1975**, *16*, 3975–3978. Ireland, R. E.; Mueller, R. H.; Willard, A. K. "Ester Enolate Claisen Rearrangement. Construction of the Prostanoid Skeleton" *J. Org. Chem.* **1976**, *41*, 986–996.
9. S. Hanessian "Total Synthesis of Natural Products: The Chiron Approach", Pergamon Press, **1983**.

10. Prouty, W. F.; Thompson, R. M.; Schnoes, H. K.; Strong, F. M. "Oligomycin: Degradation Products and Part Structure of Oligomycin B1" *Biochem. Biophys. Res. Commun.* **1971**, *44*, 619–627.
11. Rauchschwalbe, G.; Schlosser, M. "Selective Synthesis with Organometallics. IV. Hydroxylation of Allyl Positions" *Helv. Chim. Acta* **1975**, *58*, 1094–1099.
12. Brown, H. C.; Bhat, K. S.; Randad, R. S. "Chiral Synthesis via Organoboranes. 21. Allyl- and Crotylboration of α-Chiral Aldehydes with Diisopinocampheylboron as the Chiral Auxiliary" *J. Org. Chem.* **1989**, *54*, 1570–1576. Brown, H. C.; Bhat, K. S. "Chiral Synthesis via Organoboranes. 7. Diastereoselective and Enantioselective Synthesis of *Erythro*- and *Threo*-β-methylhomoallyl alcohols via Enantiomeric (*Z*)- and (*E*)-Crotylboranes" *J. Am. Chem. Soc.* **1986**, *108*, 5919–5923.
13. For other syntheses of calcimycin see: Ziegler, F. E.; Cain, W. T. "Formal Synthesis of (−)-Calcimycin (A-23187) via the 3-Methyl-γ-butyrolactone Approach" *J. Org. Chem.* **1989**, *54*, 3347–3353.

Problems

1. Propose a mechanism for the exchange of the hydrogens at C_{13} and C_{15} of **1** upon treatment with DCl in dioxane. (Calcimycin-2)
2. Provide the structure of intermediates and a mechanism that explains how **21** was converted to **22**. (Calcimycin-3)
3. Show your understanding of the Cram and Felkin-Ahn models for asymmetric induction in carbonyl addition reactions by predicting (or rationalizing) the stereochemical course of the following reactions. (Calcimycin-3)

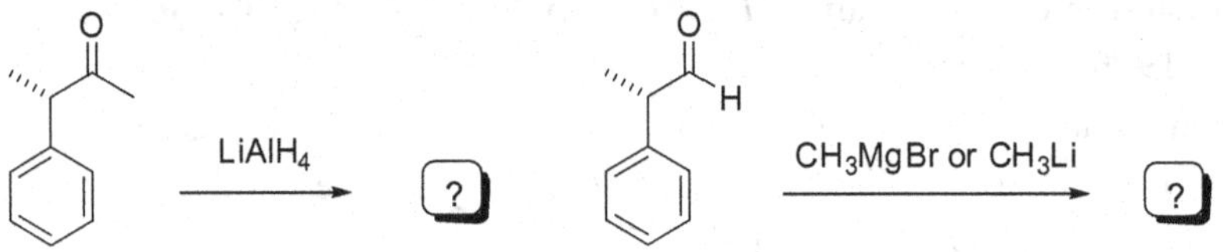

4. Examine all of the difunctional relationships from $C_8 \rightarrow C_{20}$ in compound **28**. Classify them as "odd" or "even". Go through the Evans synthesis and determine how each difunctional relationship was constructed, or if they were purchased. (Calcimycin-3)
5. Predict the stereochemical course of the following enolate alkylations. (Calcimycin-4)

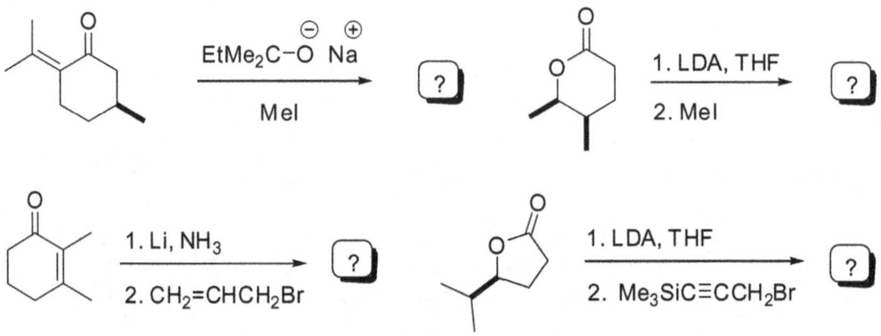

6. Provide the structure of intermediates along the route from **40** → **42**; from **41** → **42**. How would you convert **40** to the C_{11} epimer of **42**? **41** to the C_{11} epimer of **42**? (Calcimycin-5)
7. Provide a mechanistic explanation for the conversion of **44** → **45** including stereochemistry at C_{14}. (Calcimycin-5)
8. Provide the structure of the spiroacetal derived from the "undesired" C_{10} diastereomer of **51**. Suggest why this diastereomer might not have cyclized to this spiroacetal. (Calcimycin-6)

9. Outline a synthesis of **64** from D-glucose. (Calcimycin-7)
10. Provide the structure of all intermediates in the conversion of **74** → **75**. (Calcimycin-8)
11. Propose a mechanism for the conversion of **76** → **77**. (Calcimycin-8)
12. The analysis provided for the conversion of **89** to a spiroketal is abbreviated. What other chair conformations are available to **87** and **88**? What products would result from cyclization through **88** and/or these additional conformations? What would be the expected product if thermodynamics controlled the entire cyclization process? (Calcimycin-10)
13. Suggest tactics that would accomplish the following transformation. (Calcimycin-10)

14. Provide an explanation for the stereochemical course of the conversion of **90** → **91**. (Calcimycin-10)
15. Provide the structure of the HWE reagent needed to convert **101** → **102**. (Calcimycin-11)
16. Provide a mechanism for removal of the Troc [($Cl_3CCH_2OC(=O)$)] protecting group in the penultimate step of the Kishi synthesis. (Calcimycin-11)
17. Examine all of the difunctional relationships from C_8 to C_{20} in compound **104**. Classify them as "odd" or "even". Go through the Kishi synthesis and determine how each difunctional relationship was constructed. Indicate where A-functions were used (see Chapter 6). (Calcimycin-11)
18. Propose (with structures) a transition state model that explains the stereochemical course of the conversion of **109** → **110**. (Calcimycin-12)
19. Provide the structure of the product expected from the following reaction sequence. (Calcimycin-12)

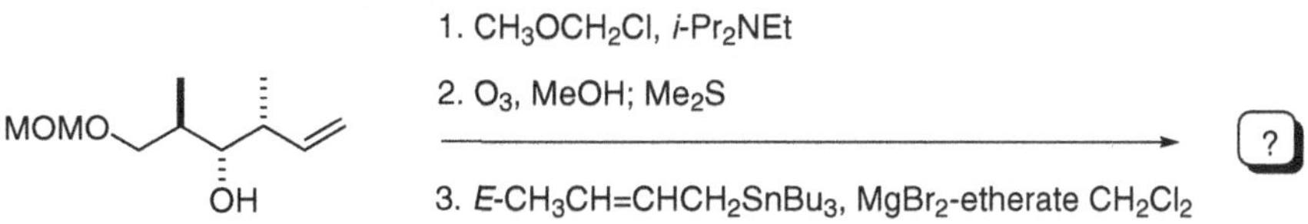

20. Propose a mechanism for the conversion of **110** → **111**. (Calcimycin-12)
21. Propose a mechanism for the conversion of **120** → **121**. (Calcimycin-13)

Outline a synthesis of [illegible] Calculate [illegible]

10. Provide the structures of all intermediates in the [illegible] (Calculate in 5.)

11. Propose a mechanism for [illegible] of [illegible] (Calculate in 6.)

12. [illegible] 5% [illegible] What [illegible] What products [illegible] we [illegible] cyclization through [illegible] additional [illegible] What [illegible] (Calculate in 10.)

13. Suggest [illegible] (Calculate in 1.1.)

[illegible]

14. Propose [illegible] (Calculate in 12.)

[illegible]

[illegible]

21. Propose [illegible] (Calculate in [illegible])

CHAPTER 14

Erythromycin A Aglycone

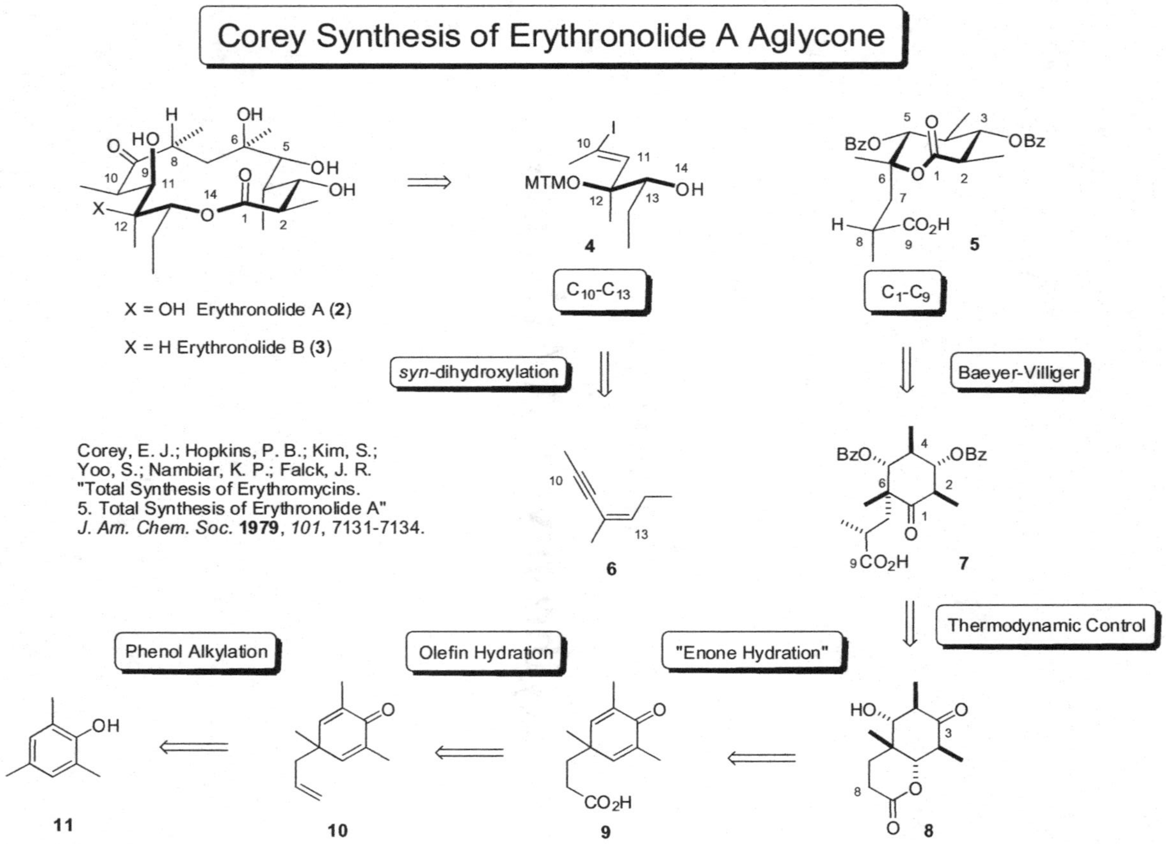

Erythronolide-1

Erythronolide-1

Erythromycin A (**1**) is a member of an important family of therapeutically useful antibiotics. The structure of **1** is shown on Erythronolide-5. This natural product consists of an "aglycone" (see **2** in Erythronolide-1) linked to two unusual monosaccharides (L-cladinose and D-desosamine) via glycosidic bonds to the C_3 and C_5 hydroxyl groups, respectively. We will conclude by examining two syntheses, one of erythronolide A (**2**) and the other of erythromycin A. Before jumping into the synthesis, I will make a few comments.

Erythronolide A (**2**) contains seven propionic acid units linked from the α-carbon of one to the carbonyl carbon of the next unit. Using polymer terminology, **2** is an oligomer derived from propionic acid (the monomer). The hydroxy acid derived from a macrolide is refered to as the *seco*-acid (see Erythronolide-5 for the *seco*-acid of **2**). Lactonization of the *seco*-acid is the most obvious way to approach any macrolide, although not the only way as we saw with lasonolide A (Chapter 12). To cyclize the *seco*-acid of **2**, most of the hydroxyl groups spattered along the C_1-C_{13} chain would have to be protected. For a long time it was felt that such lactonizations would not be feasible, but a number of macrolactonization methods have been developed over the years (see Lasonolide-9 for one method) and macrolactonization is now considered a viable step in a synthetic strategy.

Erythronolide A (**2**) has 10 stereogenic centers distributed along the 13-carbon chain. Setting these centers with the proper relative and absolute stereochemistry is the major challenge in assembling **2** (and a myriad of other polypropionate and polyacetate derived natural products). With lasonolide A and calcimycin we saw several broad approaches to this challenge: (1) set stereochemistry in a ring and open the ring (2) set stereochemistry in intermolecular reactions (3) use convergence to advantage where possible (make small pieces and couple them using reactions that either do or do not generate stereogenic centers). We will examine two historically significant syntheses of erythronolide A that employ all three of these strategies at one point or another.

The Corey group was the first to report a synthesis of **2**. The plan was to cyclize an appropriately protected *seco*-acid using macrolactonization methodology developed in house.[1] The *seco*-acid was to be assembled from C_{10}-C_{13} fragment **4** and C_1-C_9 fragment **5**. The smaller fragment (**4**) was to be prepared from enyne **6** via *syn*-dihydroxylation of the olefin and formal regioselective *syn*-addition of HI to the alkyne. The more complex fragment (**5**) was to be prepared by Baeyer-Villiger oxidation of **7** (migration of the more highly substituted carbon to control regiochemistry). Cyclohexanone **7**

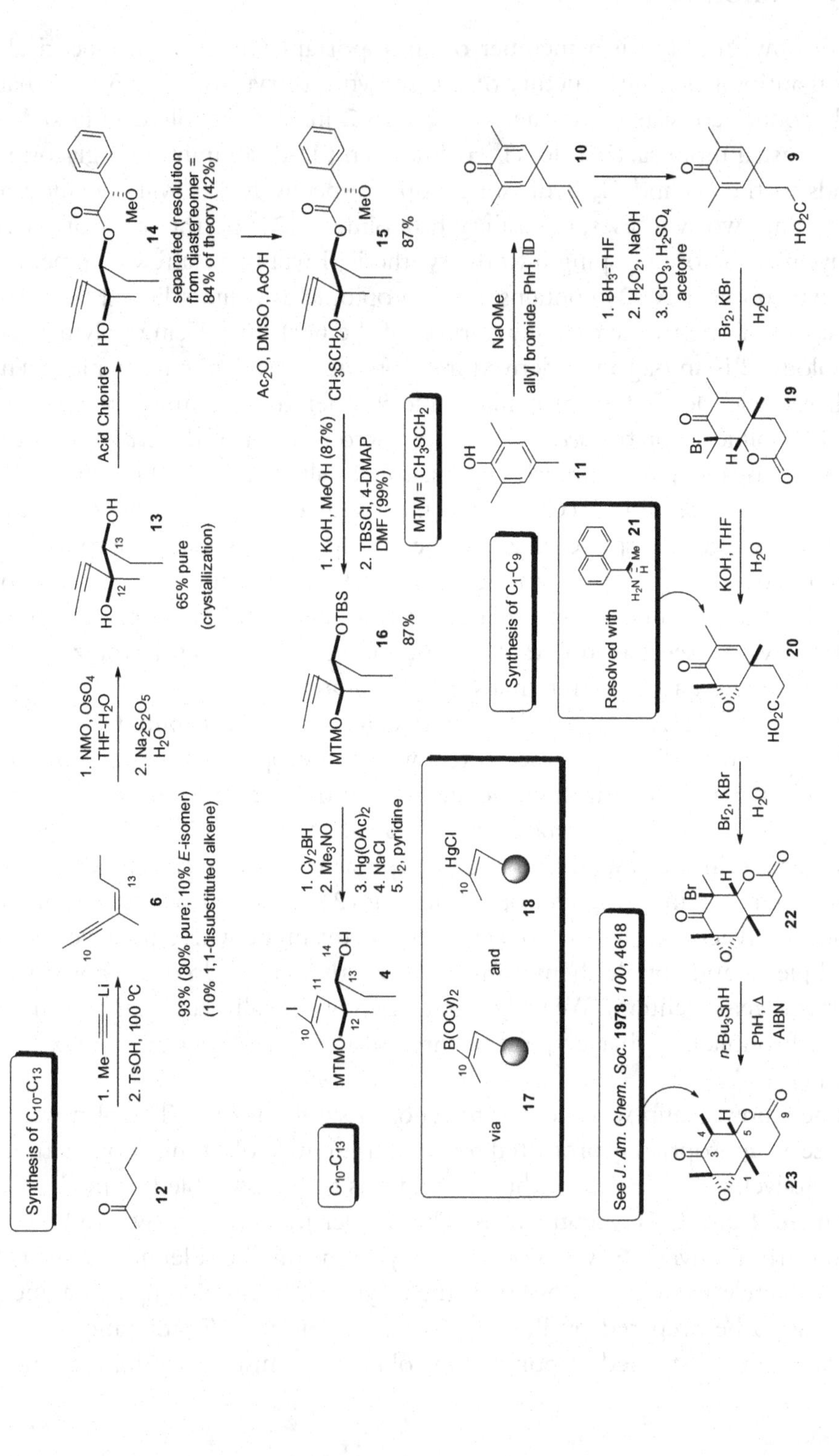

Erythronolide-2

was to come from *cis*-oxadecalin **8**, using the ring system to control introduction of stereogenic centers at C_3 and C_8. A very creative aspect of this approach was recognition that **8** was to be prepared from symmetrical cyclohexadienone **9**. The transformation of **9** to **8** requires directed *syn*-hydration of the two olefins and desymmetrization at some point via lactone formation. Dienone **9** was to come from **10** which was to be prepared using a Claisen-Cope rearrangement to dearomatize **11**. The plans for both pieces (**4** and **5**) require that either asymmetric induction be used to establish absolute stereochemistry, or classical resolutions be performed along the way. In addition, a novel feature of the plan was to install the C_{10} and C_{11} stereogenic centers after macrolactonization by "hydration" of the C_{10}-C_{11} olefin present in fragment **4**. Let's move to the synthesis of **4**.

Erythronolide-2

Addition of 1-lithiopropyne to 2-pentanone (**12**) was followed by dehydration to give **6**. Although this material was contaminated with some of the *E*-isomer and the regioisomeric dehydration product, it was possible to move forward and separate at the next step. Vicinal dihydroxylation of **6** was accomplished using a VanRheenan oxidation (developed at UpJohn in Kalamazoo) — osmium tetraoxide was used in catalytic amounts and *N*-methylmorpholine-*N*-oxide (NMO) was used as a stoichiometric oxidant. This provided racemic diol **13** which was purified by crystallization. Esterification of the least hindered alcohol with the acid chloride derived from *O*-methylmandelic acid, afforded **14** as a single diastereomer. Thus, a classical resolution was effective in handling the absolute stereochemistry issue in the synthesis of **4**. The tertiary alcohol of **14** was protected as a methylthiomethyl (MTM) ether. Ester **15** was hydrolyzed and the liberated hydroxyl group was protected as a TBS ether, setting the stage for the formal addition of HI across the alkyne. Regioselective hydroboration of **16** (controlled by steric effects) was followed by oxidation with trimethylamine oxide to give vinylboronic acid derivative **17**. Treatment of **17** with mercuric acetate and then sodium chloride gave vinyl mercurial **18**, which reacted with iodine to provide the desired vinyl iodide **4**.

The synthesis of the C_1-C_9 fragment began with phenol **11** [the oxidation product derived from 1,3,5-trimethylbenzene (mesitylene)]. Reaction of **11** with allyl bromide and base in benzene at reflux gave cyclohexadienone **10**. Hydroboration-oxidation of the least hindered olefin gave **9**, setting the stage for the aforementioned directed hydrations. Halolactonization of **9** gave **19**, and lactone hydrolysis gave epoxide **20**. The sequence moving from **9** to **20**

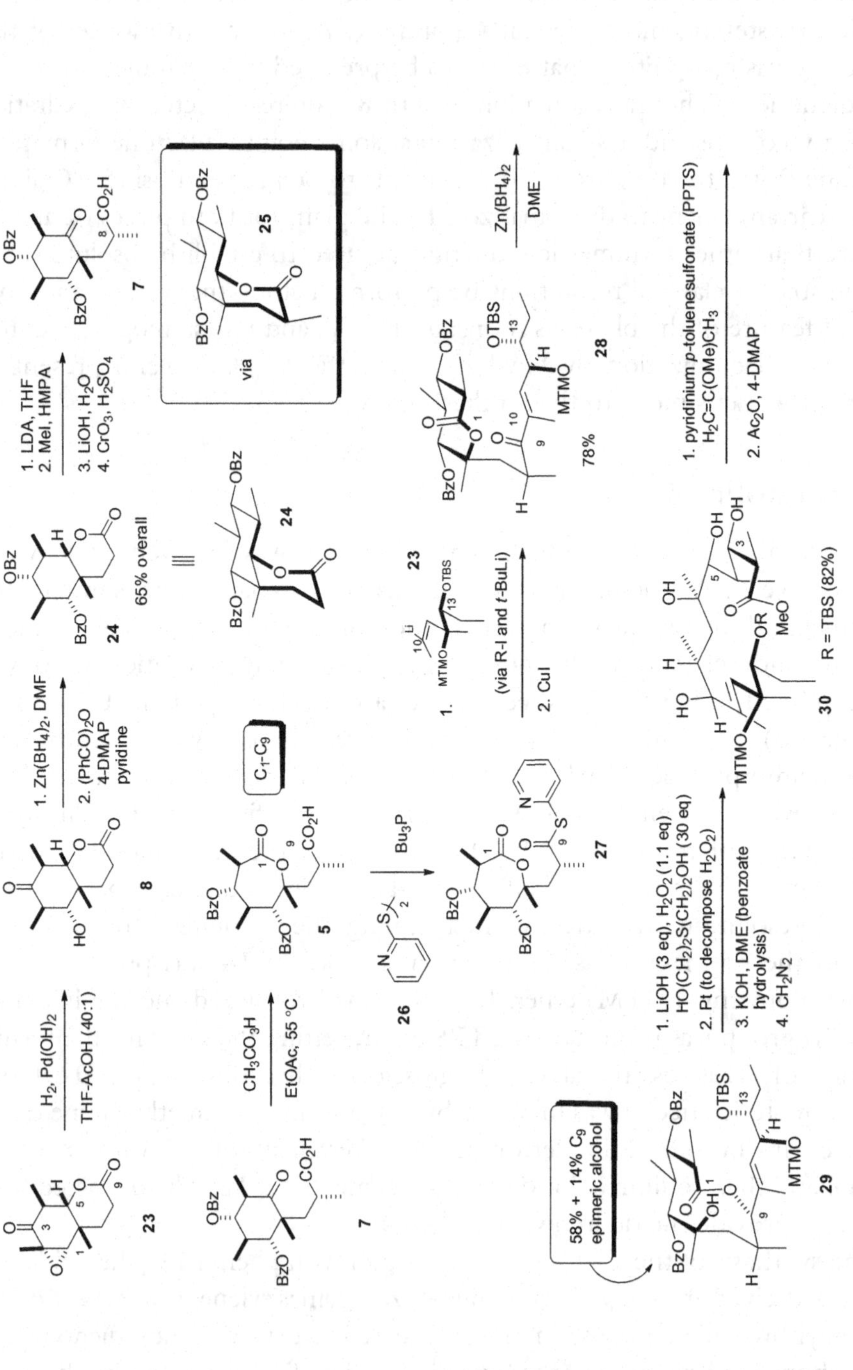

Erythronolide-3

destroys symmetry and provides racemic **20**. The acid, however, was resolved using amine **21** to provide a single enantiomer of **20**. Thus, once again a classical resolution was used to establish absolute stereochemistry. Halolactonization onto the remaining olefin provided **22**, and reduction of the halide with tri-*n*-butyltin hydride gave **23**. Whereas it is debatable whether or not the stereochemistry of the reduction of **22** was controlled by kinetics, **23** is certainly the thermodynamically most stable C_4 epimer.

Erythronolide-3

Hydrogenolysis of epoxide **23** using a Pd catalyst in the presence of acid provided **8**. Axial delivery of hydride to the C_3 ketone (also from the convex face of the *cis*-oxadecalin) and conversion of the resulting alcohol to a benzoate ester gave **24**. The C_8 methyl group was next introduced by alkylation of the enolate derived from **24**. The product was the thermodynamically more stable **25**. I presume that either the C_6 methyl group encouraged alkylation of the enolate via a boat-like transition state, or epimerization occurred under the basic reaction conditions to provide the product expected on the basis of thermodynamics. Hydrolysis of the lactone and oxidation of the alcohol gave ketone **7**. A Baeyer-Villiger oxidation completed the synthesis of the C_1-C_9 fragment (**5**).

The two fragments were coupled via a Cu mediated coupling of thioester **27** (derived from **5** and disulfide **26**) and the lithium reagent derived from vinyl iodide **4**. Reduction of the resulting ketone (**28**) provided a mixture of C_9 diastereomers from which lactone **29** was isolated in 58% yield. The lactone was hydrolyzed using basic hydrogen peroxide. Excess peroxides were destroyed using Pt wire. The benzoates were hydrolyzed and the resulting acid was converted to methyl ester **30** with diazomethane. The hydroxyl groups at C_3 and C_5 were converted to the corresponding acetonide and the remaining secondary hydroxyl group was acetylated to provide **31a**.

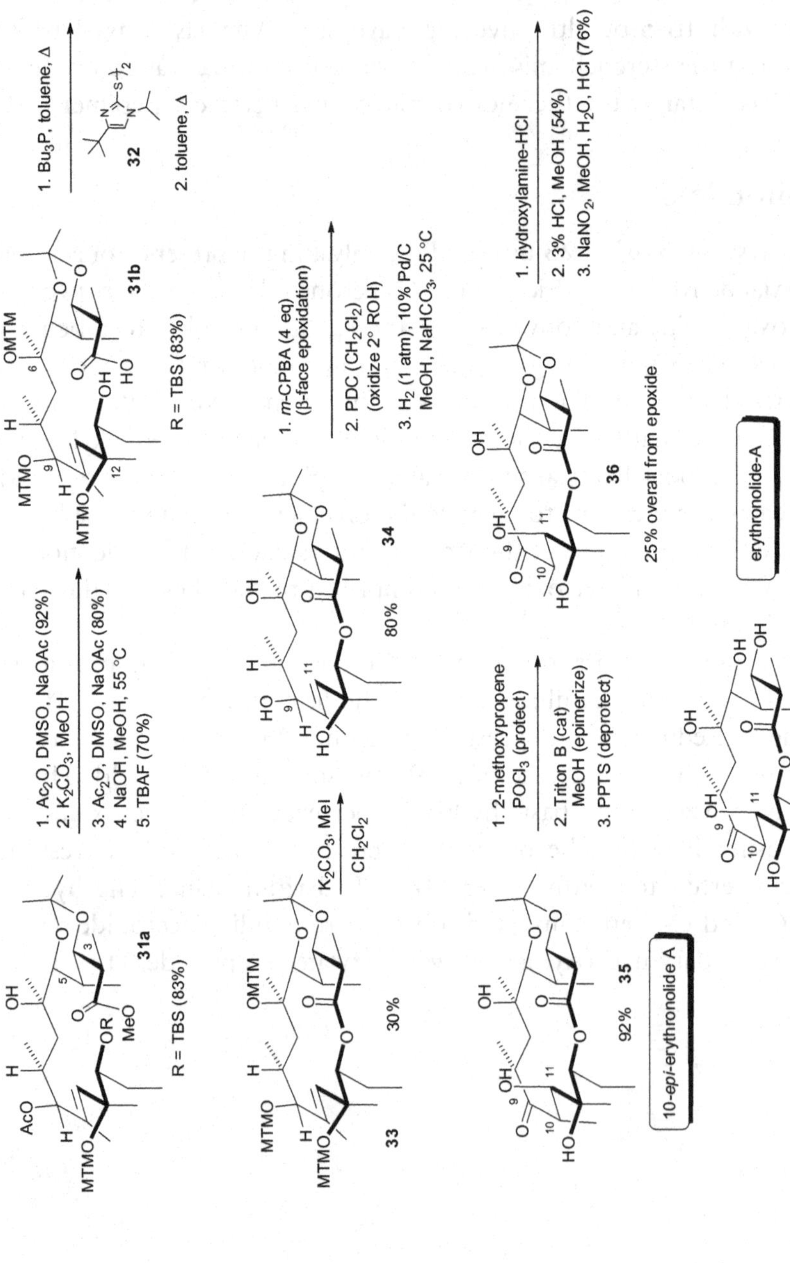

Erythronolide-4

Erythronolide-4

The next task was to modify **31a** such that the macrolactonization could be accomplished. The ultimate cyclization substrate was hydroxy acid **31b**. An MTM protecting group was first introduced at C_6. The C_9 acetate was hydrolyzed and replaced with an MTM group. The methyl ester was hydrolyzed and then the silicon protecting group at C_{13} was removed. Treatment of **31b** with disulfide **32** and tri-*n*-butylphosphine provided the hydroxy thioester, and heating in toluene under high dilution conditions gave the macrocyclic lactone **33** albeit in modest yield. We later see that this is a very demanding macrolactonization!

Lactone **33** is not conformationally mobile because many conformational changes introduce transannular and torsional strain. Thus, it was possible to functionalize the C_{10}-C_{11} olefin with predictable control over stereochemistry. In preparation for this chemistry, the MTM groups were first removed by *S*-alkylation, followed by hydrolysis using potassium carbonate in aqueous acetone. The choice of the MTM protecting group was most likely because these deprotections had to be accomplished in the presence of the lactone and acetal groups. Triol **34** was epoxidized with *m*-CPBA (hydroxyl-directed epoxidation), the secondary alcohol at C_9 was oxidized to a ketone, and the epoxide was hydrogenolyzed to give 10-*epi*-erythronolide A (**35**). The C_{10}-C_{11} vicinal diol group was protected as an acetonide, C_{10} was epimerized to the thermodynamically more stable pseudo-equatorial isomer, and the acetonide was removed to provide **36**. The keto-acetonide was unstable to acid and thus, the 9-keto group was protected as an oxime prior to acetonide hydrolysis. Finally, oxidative removal of the oxime provided erythronolide A (**2**). Overall this was a superb plan. I imagine this summary does not do justice to the trial and error and effort that must have gone into translating this plan into reality.

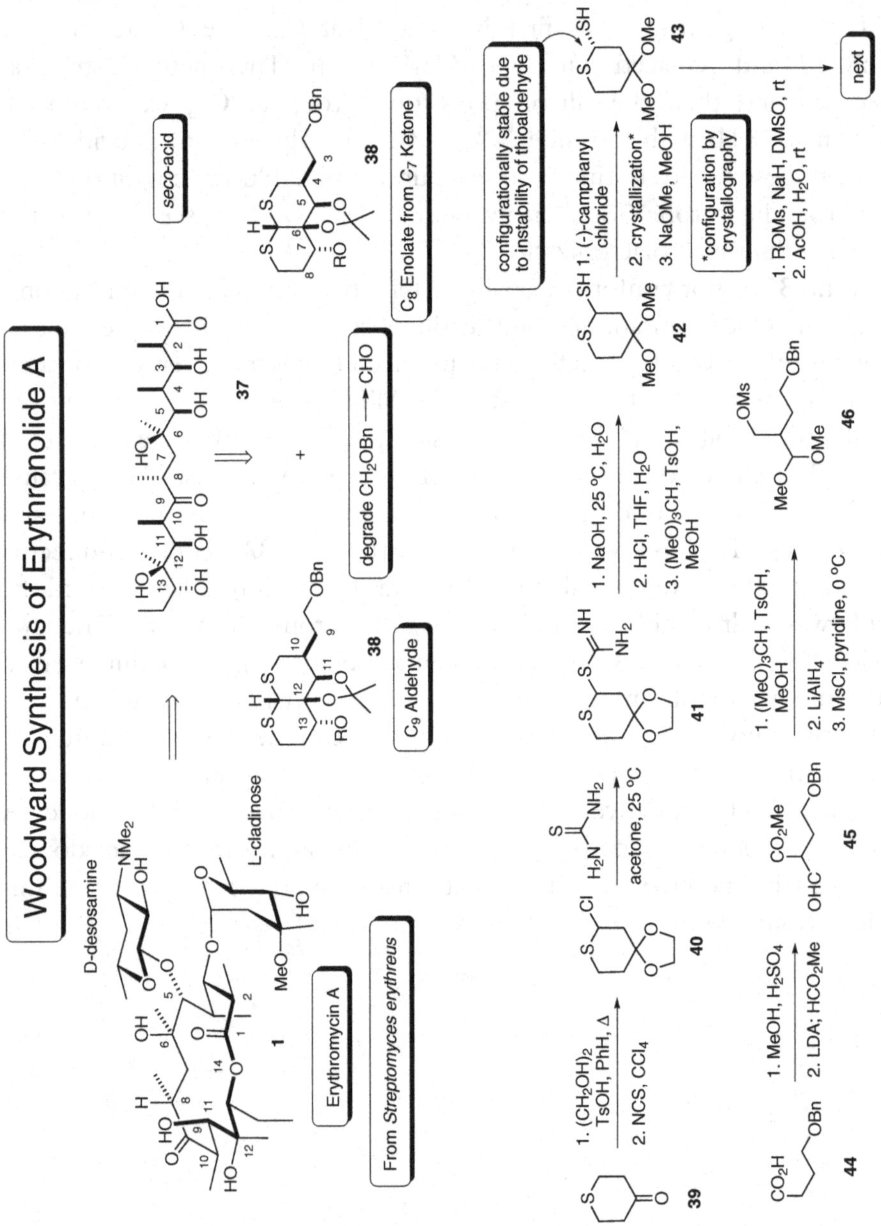

Erythronolide-5

Erythronolide-5

I will return to Woodward for the final synthesis of this book. This synthesis of Erythromycin A was published after his death in 1978. It is a synthesis in the true Woodwardian style (after all it is a Woodward synthesis), making elegant use of symmetry, tying a complex molecule into knots and then springing loose the target as an intimate part of the endgame. I remember hearing Woodward lecture on this topic at the 1977 National Organic Symposium held at the University of West Virginia in Morgantown, the only time I heard the master give a lecture. He used slides, an unusual practice for Woodward, who was famous for chalk-talks where he never had to erase anything.

The overall strategy was to cyclize an appropriate *seco*-acid derivative prior to installation of the monosaccharides. The *seco*-acid derivative was to be prepared from **38**. It was hoped that **38** would provide both the C_3-C_8 and C_9-C_{13} portions of the *seco*-acid. The marriage of **38** with itself was to involve construction of the C_8-C_9 bond via an aldol condensation (the original plan involved nitrile oxide cycloaddition chemistry to join the two pieces, but the aldol is what ultimately worked). Detection of identical fragments (C_4-C_6 and C_{10}-C_{12}) within the *seco*-acid was critical to development of this plan. It was also planned to set most of the stereogenic centers in conformationally well-defined molecules. The methyl groups at C_4, C_6, and C_8 as well as C_{10}, C_{12}, and the terminal methyl of the C_{13} ethyl group, were joined to accomplish this task. Because of the convergent nature of the synthesis, the synthesis of **38** demanded enantioselectivity.

The synthesis of **38** began with **39**. Acetal formation and chlorination adjacent to sulfur provided **40**. Thiourea was used to introduce sulfur. Hydrolysis of **41** provided the free thiol and a ketal exchange (hydrolysis-protection) gave **42**. This compound was configurationally stable at the "anomeric center" and thus, was resolved via the thioester derived from reaction with (−)-camphanyl chloride. The absolute configuration of the proper enantiomer was established by X-ray crystallography of this thioester. *S*-Alkylation of **43** with racemic mesylate **46** provided a mixture of diastereomers **47** (Erythronolide-6).

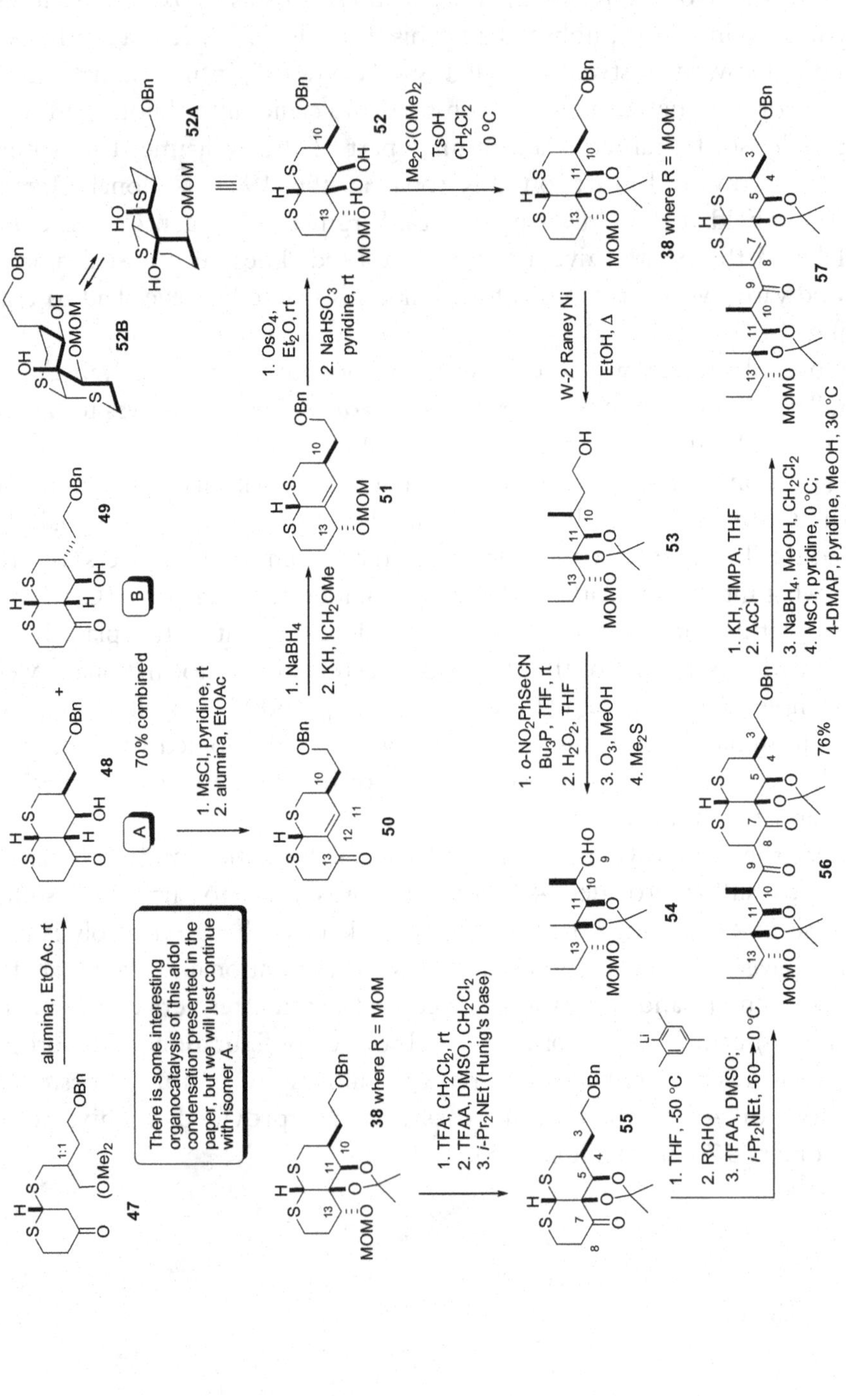

Erythronolide-6

Erythronolide-6

Treatment of **47** with alumina gave a separable 1:1 mixture of intramolecular aldol condensation products **48** and **49**. The required isomer (**48**) was converted to the mesylate and dehydration afforded **50**. Reduction of **50** from the convex face of the dithiaoctalone gave **51** after protection of the alcohol with a MOM group. Vicinal dihydroxylation of the olefin from the most accessible face, and acetonide formation, completed the synthesis of **38** (R=MOM).

The next portion of the synthesis involved conversion of **38** into derivatives that would be suitable for joining via the aforementioned aldol condensation. Thus **38** was desulfurized and debenzylated using Raney-Ni to provide alcohol **53**. The primary alcohol was subjected to the Grieco-Sharpless dehydration procedure (via the selenide and selenoxide) and resulting olefin was cleaved with ozone to provide aldehyde **54** (the C_9-C_{13} fragment).[2] The C_3-C_8 fragment was prepared from **38** by removal of the MOM group, and oxidation of the secondary alcohol to a ketone using a Moffat-type oxidation.

Ketone **55** and aldehyde **54** were then joined via a crossed aldol condensation. The resulting alcohol was oxidized to give **56**. The enolate derived from *β*-dicarbonyl **56** was *O*-acylated at C_9. The C_7 carbonyl group was reduced to the alcohol oxidation state with concomitant reductive cleavage of the enol acetate. Treatment of the resulting *β*-hydroxy ketone with mesyl chloride gave **57**.

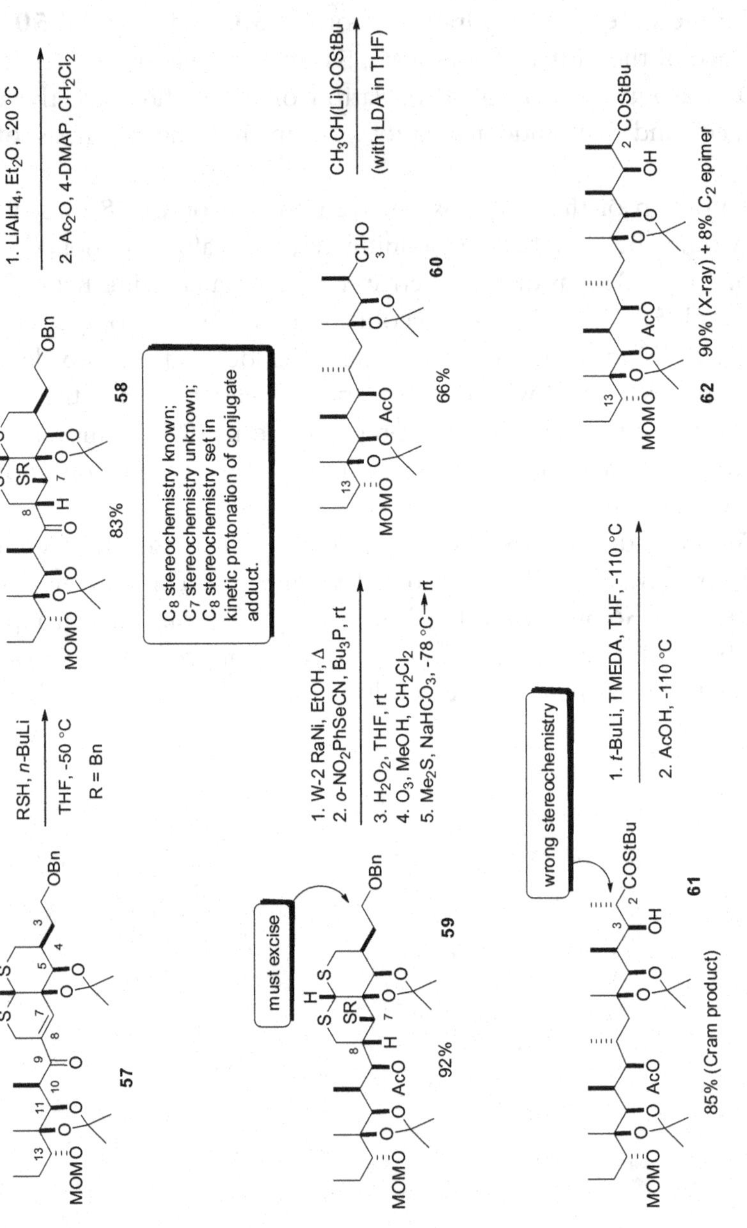

Woodward, R. B.; et al (including T. V. Rajanbabu) "Asymmetric Total Synthesis of Erythromycin. 1. Synthesis of Erythronolide A Seco Acid Derivative via Asymmetric Induction" *J. Am. Chem. Soc.* **1981**, *103*, 3210.

Erythronolide-7

Erythronolide-7

The next stage of the synthesis required reduction of the C_7-C_8 double bond with control over stereochemistry at C_8. The tactics ultimately used to accomplish this transformation involved conjugate addition of thiophenoxide to the enone to provide **58** with C_7 stereochemistry that was never established. The critical stereochemistry (C_8), however, was clean and presumably controlled by kinetic protonation of the intermediate enolate. Reduction of the C_9 ketone was followed by esterification to provide acetate **59** as a single stereoisomer (C_7 stereochemistry still not defined). Reduction of the C_7 thiol was followed by excision of the extra carbon in the usual manner to provide aldehyde **60**. The final carbons of the *seco*-acid were introduced via crossed condensation of the enolate derived from a thioester of propionic acid, with aldehyde **60**. This reaction provided the proper stereochemistry at C_3, but the undesired stereoisomer at C_2. The C_2 stereochemistry was corrected by kinetic protonation of the enolate derived from **61** with acetic acid. The structure of the resulting *seco*-acid derivative (**62**) was established by X-ray crystallography.

Woodward, R. B. et al "Asymmetric Total Synthesis of Erythromycin. 2. Synthesis of an Erythronolide A Lactone System" *J. Am. Chem. Soc.* **1981**, *103*, 3213

Erythromycin —degradation→ 17 Substrates (all thiopyridyl esters) of which only 3 could be lactonized. Two gave low yields but ...

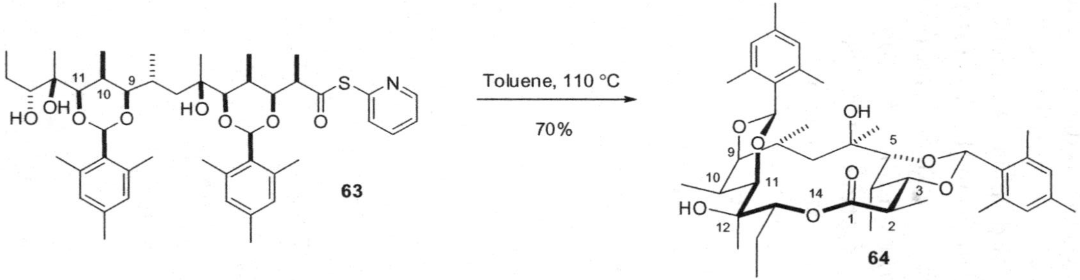

Important Features: (1) *S*-configuration at C_9 (2) $C_{3,5}$ and $C_{9,11}$ must be cyclic acetals. Eventually the Woodward group settled on the following compound which was prepared from the natural product and cyclized in 70% yield.

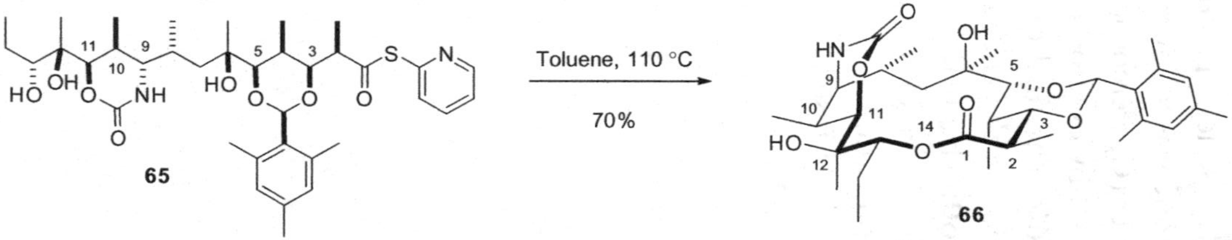

The next task was to convert the "bis-acetonide" into the cyclization substrate.

Erythronolide-8

In principle, all that remained to do to reach erythronolide (**2**) was to deprotect the C_{13} hydroxyl group of **62**, conduct the macrolactonization, remove the acetonides, and carry out an oxidation state adjustment at C_9. Of course this was easier said than done. The macrolactonization turned out to be very difficult. The Woodward group degraded Erythromycin A (**1**) to 17 thiopyridyl ester substrates, only three of which underwent lactonization under the Corey-Nicolaou conditions. Two of these derivatives gave low yields, but **63** cyclized to **64** in good yield. This effort established that (1) the *S*-configuration was essential at C_9 and (2) the C_3-C_5 and C_9-C_{11} diol units had to be protected as cyclic acetals. Based on this information, carbamate-acetal **65** was eventually prepared from the natural product and found to cyclize to **66** in 70% yield. Thus, to complete a total synthesis, it was necessary to convert *bis*-acetonide **62** to **65**, and move **66** forward to erythromycin A (**1**).

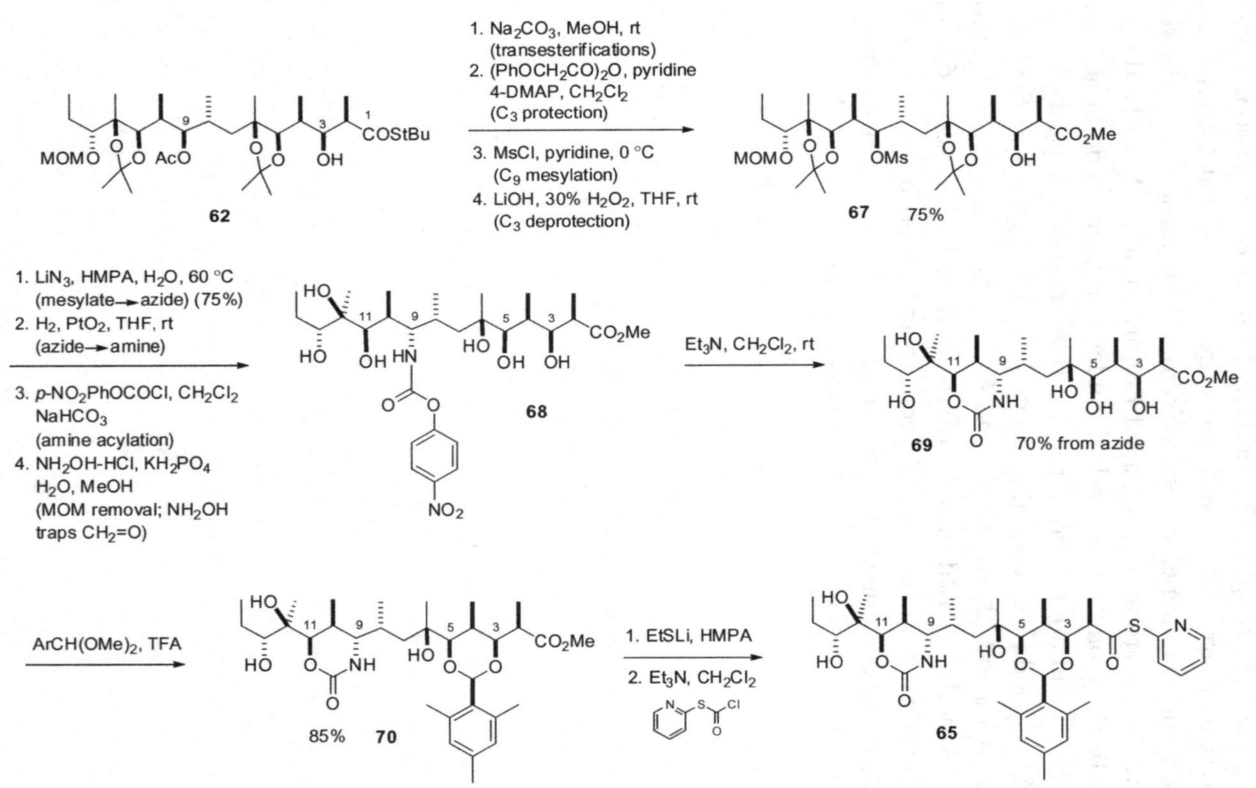

For lactonization methodology see Corey, E. J.; Nicolaou, K. C. "An Efficient and Mild Lactonization Method for the Synthesis of Macrolides" *J. Am. Chem. Soc.* **1974**, *96*, 5614.

Erythronolide-9

Erythronolide-9

The 12-step conversion of **62** to **65** began with methanolysis of the C_1 thioester and C_9 acetate. The C_3 hydroxyl group was then selectively protected as the α-phenoxyacetate. The C_9 hydroxyl group was converted to a mesylate and the C_3 ester was removed to provide **67**. The mesylate was displaced with azide, the azido group was reduced to the amine, the amine was acylated, and the acetonide and MOM protecting groups (all acetals) were removed under mild acidic conditions. The resulting polyol **68** cyclized to carbamate **69** upon treatment with triethylamine. The C_3-C_5 acetal was installed to give **70**. The methyl ester was then converted to the corresponding acid and then thioester **65**.

Woodward, R. B. et al "Asymmetric Total Synthesis of Erythromycin. 3. Total Synthesis of Erythromycin" *J. Am. Chem. Soc.* **1981**, *103*, 3215-3217.

We will not discuss the completion of the synthesis in detail. The fundamental operations, however, can be summarized as follows: (1) removal of the cyclic carbamate and acetal groups maintaining carbamate protection of the C_9 amino group (2) sequential introduction of the desosamine and cladinose units (3) oxidation of the C_9 amine to the required C_9 ketone via *N*-chloroamine and imine intermediates.

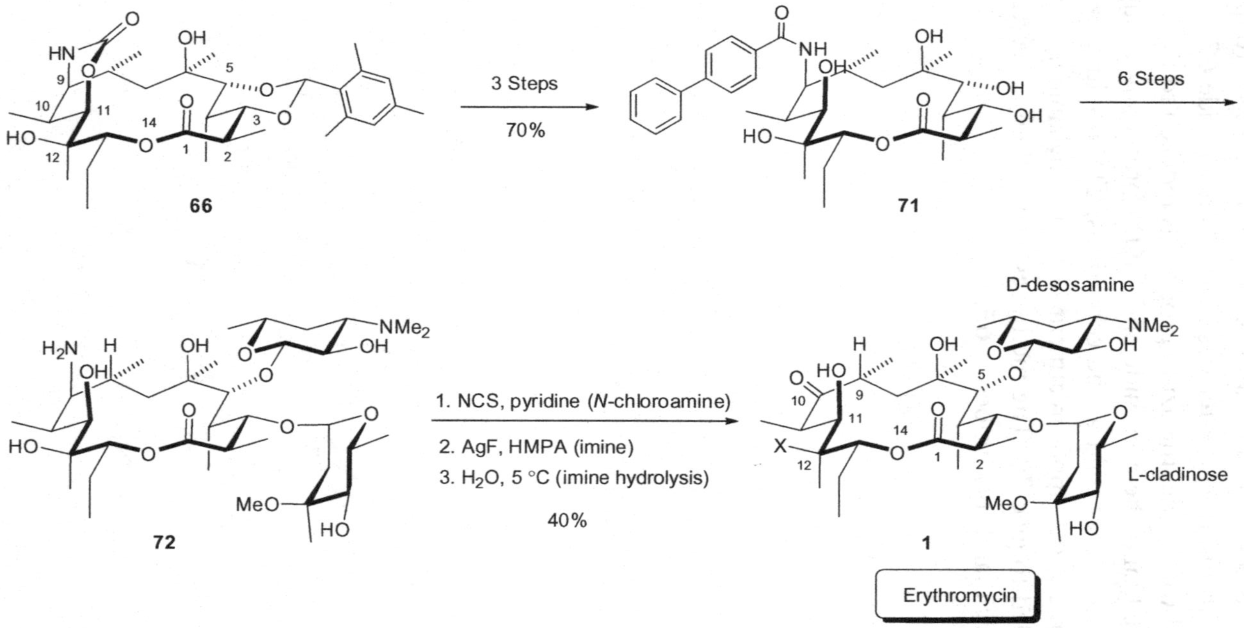

Erythronolide-10

Erythronolide-10

I will not discuss the completion of the erythromycin A synthesis in detail. The fundamental operations, however, can be summarized as follows. The cyclic carbamate and acetal protecting groups of **66** were removed while maintaining protection of the C_9 amino group. The C_5 and C_3 hydroxyl groups of **71** were sequentially glycosylated with suitable desosamine and cladinose derivatives, respectively. The "sugars" were deprotected and the C_9 amide was hydrolyzed to provide **72**. Finally the C_9 amino group was converted to the C_9 ketone via dehydrohalogenation of an intermediate chloramine and hydrolysis of the intermediate imine (recall Corey prostaglandin synthesis).

Several other syntheses of erythromycins have been reported. These all contain unique features and good lessons for students of organic synthesis. I encourage you to look some of these up and perhaps prepare your own flow sheets and commentary to add to this chapter.[3]

References

1. Corey, E. J.; Nicolaou, K. C. "An Efficient and Mild Lactonization Method for the Synthesis of Macrolides" *J. Am. Chem. Soc.* **1974**, *96*, 5614–5616.
2. Grieco, P. A.; Gilman, S.; Nishizawa, M. "Organoselenium Chemistry. A Facile One-Step Synthesis of Alkyl Aryl Selenides from Alcohols" *J. Org. Chem.* **1976**, *41*, 1485–1486. Sharpless, K. B.; Young, M. W. "Olefin Synthesis. Rate Enhancement of the Elimination of Alkyl Aryl Selenoxides by Electron-Withdrawing Substituents" *J. Org. Chem.* **1975**, *40*, 947–949.
3. Bernet, B.; Bishop, P. M.; Caron, M.; Kawamata, T.; Roy, B. L.; Ruest, L.; Sauve, G.; Soucy, P.; Deslongchamps, P. "Formal Total Synthesis of Erythromycin A. Part I. Total Synthesis of a 1,7-Dioxaspiro[5.5]undecane Derivative of Erythronolide A" *Can. J. Chem.* **1985**, *63*, 2810–2814 and following two papers in *CJC*. Kinoshita, M.; Arai, M.; Ohwawa, N.; Nakata, M. "Synthetic Studies of Erythromycins. III. Total Synthesis of Erythronolide A Through (9*S*)-Dihydroerythronolide A" *Tetrahedron Lett.* **1986**, *27*, 1815–1818 preceeding two articles in this series. Stork, G.; Rychnovsky, S. D. "Concise Total Synthesis of (+)-(9*S*)-Dihydroerythronolide A" *J. Am. Chem. Soc.* **1987**, *109*, 1565–1567. Kochetkov, N. K.; Sviridov, A. F.; Ermolenko, M. S.; Yashunsky, D. V.; Borodkin, V. S. "Stereocontrolled Synthesis of Erythronolides A and B from 1,6-Anhydro-β-D-glucopyranose (levoglucosan). Skeleton assembly in (C_9-C_{13}) + (C_7-C_8) + (C_1-C_6) Sequence" *Tetrahedron* **1989**, *45*, 5109–5136. Hikota, M.; Tone, H.; Horita, K.; Yonemitsu, O. "Stereoselective Synthesis of Erythronolide A via an Extremely Efficient Macrolactonization by the Modified Yamaguchi Method" *J. Org. Chem.* **1990**, *55*, 7–9. Hikota, M.; Tone, H.; Horita, K.; Yonemitsu, O. "Chiral Synthesis of Polyketide-Derived Natural Products. 31. Stereoselective Synthesis of Erythronolide A by Extremely Efficient Lactonization Based on Conformational Adjustment and High Activation of Seco-Acid" *Tetrahedron* **1990**, *46*, 4613–4628. Stürmer, R.; Hoffmann, R. W. "Stereoselective Synthesis of Alcohols. XLVIII. Linear Synthesis of (9*S*)-Dihydroerythronolide A" *Chem. Ber.* **1994**, *127*, 2519–2526. Muri, D.; Lohse-Fraefel, N.; Carriera, E. M. "Total Synthesis of Erythronolide A by Mg(II)-mediated Cycloadditions of Nitrile Oxides" *Angew. Chem., Int. Ed.* **2005**, *44*, 4036–4038.

Problems

1. Suggest why **6** is the major product from the dehydration of the adduct of **12** and 1-lithiopropyne. (Erythronolide-2)
2. The conversion of **14** → **15** is an example of a Pummerer reaction. Provide a mechanism for this transformation. (Erythronolide-2)
3. Provide a mechanistic explanation for the conversion of **16** → **17**. (Erythronolide-2)
4. Provide a mechanism for the conversion of **11** → **10** that begins with the expected O-alkylation. (Erythronolide-2)
5. Propose a mechanism for the removal of the MTM groups *en route* from **33** → **34**. (Erythromycin-4)
6. Propose a mechanism for the conversion of **39** → **40**. (Erythronolide-5)
7. What is the purpose of the trimethyl orthoformate in the conversions of **41** → **42** and **45** → **46**? (Erythronolide-5)
8. Provide a mechanistic explanation for the stereochemistry of the crossed condensation reactions that converted **60** → **61** → **62**. (Erythronolide-7)

Concluding Remarks

Additional Targets

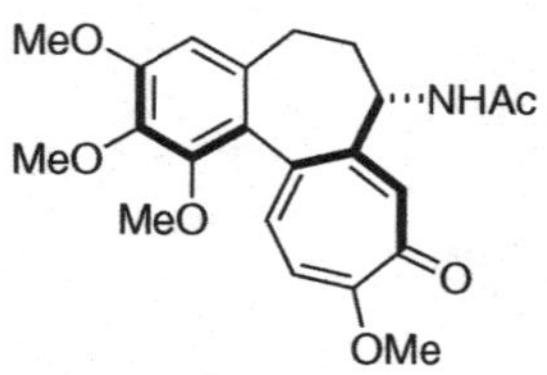

1

colchicine

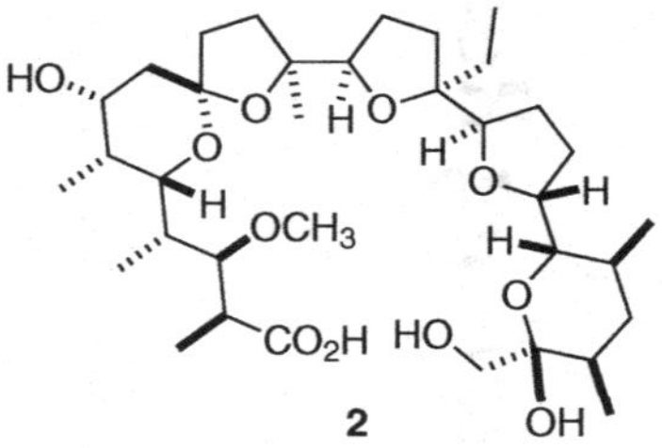

2

Monensin

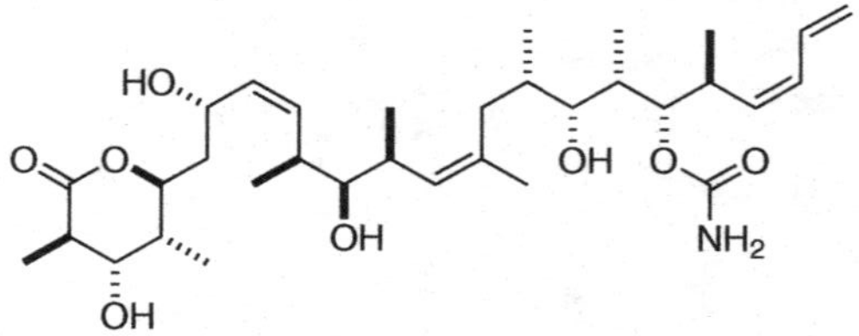

3

Discodermolide

Concluding Remarks

Concluding Remarks

This ends our look at some selected natural product syntheses. At the onset of this course I had intended to cover three additional natural products: the alkaloid colchicine (**1**), the ionophore monensin (**2**) and disocodermolide (**3**). Time did not allow me to cover these topics. It turns out that an analytical and up-to-date compilation of colchicine syntheses has been published, and monensin has also been nicely covered elsewhere.[1,2] Discodermolide is a modern example where synthesis met the demand for supply of a complex natural product with medicinal promise. I encourage students and teachers alike to write an additional chapter that discusses syntheses of discodermolide within the context of what appears in Chapters 1–14. This is only a suggestion, and I am sure that you will all eventually have your favorites that you can add to the collection of work described in this book.

For those of you beginning careers in organic synthesis, I think it is important to learn from others (see how practitioners of synthesis have solved problems), and to develop and improve your skills through practice. In this regard, I asked students to work a few problems during the course of the quarter (about 2% of what appears in this book) and also to propose a synthesis of a natural product.

For teachers, I think it is important to provide students with guidance. For this class, I provided students with one way (there are many) to identify a target for synthesis. I chose a method that was well-defined and was manageable within the context of a course (I directed students to the website of the journal *Heterocycles* for a searchable compilation of natural product structures). I provided students with help navigating the website, feedback on their target selection, and also provided feedback on their synthesis plan at an intermediate point in the course. I hope you find this information useful.

References

1. Graening, T.; Schmalz, H.-G. "Total Synthesis of Colchicine in Comparison: A Journey Through 50 Years of Synthetic Organic Chemistry" *Angew. Chem. Int. Ed.* **2004**, *43*, 3230–3256.
2. Nicolaou, K. C.; Sorensen, E. J. "Classics in Total Synthesis: Targets Strategies, Methods", VCH, 1996.

Index

The following index can be used to find topical information within the text. A comprehensive index of compounds, reagents, reactions and topics covered in graphical form on even-numbered pages is available at the following website: http://www.worldscibooks.com/chemistry/7815.html